Characterization of Cereals and Flours

FOOD SCIENCE AND TECHNOLOGY

A Series of Monographs, Textbooks, and Reference Books

Additional Volumes in Preparation

Characterization of Cereals and Flours
Properties, Analysis, and Applications

edited by
Gönül Kaletunç
The Ohio State University
Columbus, Ohio, U.S.A.

Kenneth J. Breslauer
Rutgers University
Piscataway, New Jersey, U.S.A.

CRC Press
Taylor & Francis Group
Boca Raton London New York

CRC Press is an imprint of the
Taylor & Francis Group, an **informa** business

First published 2003 by Marcel Dekker, Inc.

Published 2018 by CRC Press
Taylor & Francis Group
6000 Broken Sound Parkway NW, Suite 300
Boca Raton, FL 33487-2742

First issued in paperback 2019

No claim to original U.S. Government works

ISBN 13: 978-0-367-45453-1 (pbk)
ISBN 13: 978-0-8247-0734-7 (hbk)

Visit the Taylor & Francis Web site at
http://www.taylorandfrancis.com

and the CRC Press Web site at
http://www.crcpress.com

Library of Congress Cataloging-in-Publication Data
A catalog record for this book is available from the Library of Congress.

To my parents, Nevin and Fethi Kaletunç,
Janet and George Plum,
my husband, Eric,
and my son, Barış,
for their support and encouragement

Gönül Kaletunç

To my wife, Sherrie Schwab,
and my two sons, Danny and Jordan Breslauer,
for their patience, support, and special spirit for life

Kenneth J. Breslauer

Preface

Cereal-based foods comprise a substantial portion of the world's food supply, despite regional, economical, and habitual differences in consumption. In the human diet, cereals are considered excellent sources of fiber and nutrients (e.g., starches, proteins, vitamins, and minerals). In many developing countries, cereals provide as much as 75% of human dietary energy. In 1992, the U.S. Department of Agriculture emphasized the importance of cereal-based foods in the human diet by introducing the Food Guide Pyramid. This graphical guideline organizes foods into five groups and recommends daily consumption of 6–11 servings of bread, cereals, rice, and pasta (two to three times more than the number of servings for other food groups), thereby stressing the relative significance of the grains group. As economical and abundant raw materials, cereals have long been used for the production of a wide range of food and nonfood products, including breads, cookies, pastas, breakfast cereals, snack foods, malted cereals, pharmaceuticals, and adhesives.

The improvement and development of cereal products and processes require an understanding of the impact of processing and storage conditions on the physical properties and structure of pre- and postprocessed materials. In this book, we focus on techniques used to characterize the influence on the physical properties of cereal flours of several cereal processing technologies, including baking, pasta extrusion, and high-temperature extrusion, as well as cookie and cracker production. This text facilitates viewing the impact of various cereal processing technologies on cereal flours from three complementary perspectives: characterization of thermal, mechanical, and structural properties. Establishing quantitative relationships among the various physical observables and between the physical properties and the sensory attributes of end products should provide a rapid and objective means for assessing the quality of food materials, with the overall goal of improving this quality. To this end, a fourth perspective is also included: namely, sensory end-product attributes of significance to the consumer. In several chapters, in fact, correlations between sensory attributes and physical properties are reported.

Cereal processing consists basically of mixing cereal flours with water, followed by heating to various temperatures, cooling, and storing. Consequently, for the purpose of improving processing, it would be most useful if one could predict the physical properties of pre- and postprocessed cereal flours when subjected to varied processing and storage conditions. Part I of this book, which includes Chapters 1 through 5, focuses on discussions of thermal analysis techniques to assess the impact of various cereal processing conditions on the physical properties of cereal flours in high-temperature extrusion, cookie manufacturing, and baking.

Chapters 1, 2, and 3 describe thermally induced transitions (glass, melting, gelatinization) in cereal flours as a function of conditions relevant to cereal processing technologies. Chapter 4 addresses the influence of moisture on the processing conditions and the physical properties of the product. The final chapter of Part I (Chapter 5) focuses on the utilization of a database created from the studies described in the previous chapters to establish state diagrams that define the state of the cereal flour prior to, during, and after processing. This chapter also describes the application of such state diagrams to map the path of processes, to assess the impact of processing conditions, and, ultimately, to design processing conditions that achieve desired end-product attributes.

Part II includes Chapters 6 through 10 and focuses on the characterization of mechanical properties of cereal flours, prior to, during, and after processing. Chapter 6 reports on the assessment of the stability of cereal flours in terms of caking or loss of flowability as a result of moisture sorption or exposure to elevated temperatures during storage. Chapter 7 covers the rheological characteristics of cereal flours during processing and their relation to end-product physical properties such as expansion of extrudates. Chapters 8 and 9 describe the mechanical properties of postprocessed cereal flours as a function of processing conditions, additives, and postprocessing storage conditions in relation to pasta drying, textural attributes, and shelf life of extruded products. Mechanical properties of biopolymers change as their physical state is altered during processing or storage. Chapter 10 focuses on the application of this information in product and process development.

The third and final part includes studies exploring the microscopic determinants of macroscopic properties. These studies employ techniques such as light and electron microscopy and nuclear magnetic resonance (NMR) spectroscopy. Chapters in this part focus on the development of correlations between the microscopic structural features of pre- and postprocessed food biopolymers and their macroscopic physical properties. Chapter 11 describes how image analysis techniques can be used to evaluate macrostructures created in expanded extrudates as a function of formulation and processing conditions. The cell structure and cell size distribution in these products are responsible for the characteristic crispy texture of cereal products.

Macroscopic observables do not reveal whether an observed order–disorder transition reflects a change in the overall structure or whether the transition is specific for local structural domains. As with all food materials, compositional and microstructural heterogeneity are intrinsic characteristics of pre- and postprocessed cereal flours. Consequently, it is most useful to characterize chemical and structural composition at a microscopic level. Chapter 12 focuses on the use of microscopy as a tool to gain such information about structural organization, as well as the distribution of various domains within proteins, starches, and other components in pre- and postprocessed cereal flours. Chapters 13 and 14 focus on probing the relationships between structure, dynamics, and function using NMR and phosphorescence spectroscopy. Due to the noninvasive character of NMR and the richness of its information content, its use to study pre- and postprocessed cereal biopolymers has increased in the past decade. Such NMR studies range from structural characterization of starch granules to observations of changes in water mobility in staling bread. Phosphorescence spectroscopy is a promising emerging technique for studying the molecular dynamics of the glassy state in which the mobility is very limited. Chapter 15 is devoted to two converging lines of starch research with implications for the cereal processing industry. Chemical studies link the molecular characterization of starch granules and starch-bound proteins to the properties of starch-based products. Biochemical and genetic studies provide information on starch modification and biosynthesis with the ultimate objective being to enhance the starch yield and quality.

All of the chapters in this book are designed

1. To develop a fundamental understanding of the influence of processing on cereal flours by creating a database via systematic studies of the physical properties of pre- and postprocessed cereal flours
2. To demonstrate how this knowledge can be used as a predictive tool for evaluating the performance of cereal flour during processing, and, ultimately, for adjusting, in a rational fashion, the formulation of raw materials and processing parameters so as to achieve desired end-product attributes

This book bridges the gap between basic knowledge and application. We believe it will prove to be a comprehensive and valuable teaching text and reference book for students and practicing scientists, in both academia and industry.

Gönül Kaletunç
Kenneth J. Breslauer

Contents

Contributors

Karin Autio VTT Biotechnology, Espoo, Finland

G. V. Barbosa-Cánovas Department of Biological Systems Engineering, Washington State University, Pullman, Washington, U.S.A.

Ann H. Barrett Combat Feeding Program, U.S. Army Natick Soldier Center, Natick, Massachusetts, U.S.A.

Peter Belton School of Chemical Sciences, University of East Anglia, Norwich, U.K.

Kenneth J. Breslauer Department of Chemistry and Chemical Biology, Rutgers University, Piscataway, New Jersey, U.S.A.

Lilia S. Collado Institute of Food Science and Technology, University of the Philippines Los Baños, College, Laguna, Philippines

Paul Colonna Plant Products Processing, Institut National de la Recherche Agronomique, Nantes, France

Harold Corke   Department of Botany, The University of Hong Kong, Hong Kong, China

Guy Della Valle   Plant Products Processing, Institut National de la Recherche Agronomique, Nantes, France

Ann-Charlotte Eliasson   Department of Food Technology, Center for Chemistry and Chemical Engineering, Lund University, Lund, Sweden

Alex Grant   Institute of Food Research, Norwich, U.K.

Brian Hills   Institute of Food Research, Norwich, U.K.

Victor T. Huang   General Mills, Inc., Minneapolis, Minnesota, U.S.A.

James Ievolella*   Nabisco, Kraft Foods, East Hanover, New Jersey, U.S.A.

Kirsi Jouppila   Department of Food Technology, University of Helsinki, Helsinki, Finland

Gönül Kaletunç   Department of Food, Agricultural, and Biological Engineering, The Ohio State University, Columbus, Ohio, U.S.A.

Harry Levine   Nabisco, Kraft Foods, East Hanover, New Jersey, U.S.A.

Richard D. Ludescher   Department of Food Science, Rutgers University, New Brunswick, New Jersey, U.S.A.

Martin Okos   Department of Agricultural and Biological Engineering, Purdue University, West Lafayette, Indiana, U.S.A.

Yrjö Henrik Roos   Department of Food Science, Food Technology, and Nutrition, University College Cork, Cork, Ireland

Marjatta Salmenkallio-Marttila   VTT Biotechnology, Espoo, Finland

Louise Slade   Nabisco, Kraft Foods, East Hanover, New Jersey, U.S.A.

---

* Retired.

Andrew C. Smith Institute of Food Research, Norwich, U.K.

Bruno Vergnes Centre de Mise en Forme des Materiaux (CEMEF), Ecole des Mines de Paris, Sophia-Antipolis, France

Martha Wang Nabisco, Kraft Foods, East Hanover, New Jersey, U.S.A.

Betsy Willis School of Engineering, Southern Methodist University, Dallas, Texas, U.S.A.

H. Yan Department of Biological Systems Engineering, Washington State University, Pullman, Washington, U.S.A.

1

Calorimetry of Pre- and Postextruded Cereal Flours

Gönül Kaletunç
The Ohio State University, Columbus, Ohio, U.S.A.

Kenneth J. Breslauer
Rutgers University, Piscataway, New Jersey, U.S.A.

I. INTRODUCTION

Extrusion processing is widely utilized in the food and feed industries for the manufacture of value-added products. Extrusion processing is a versatile technology producing a wide range of products, including confectionery products, pasta, ready-to-eat (RTE) cereals, flat bread, snack products, texturized proteins, and pet foods. A broad range of operating parameters is used to manufacture products with a large variety of structures and textures, ranging from high moisture (up to 75%)–low temperature (as low as 50°C)–low shear in texturized vegetable and pasta production to low moisture (as low as 11%)–high temperature (as high as 180°C)–high shear in breakfast cereal and snack production.

High-temperature extrusion processing finds wide application in the food industry for the preparation of breakfast cereals and snack foods. Starch-and protein-based cereal flours are frequently encountered as major components of the raw material mixtures. Rice, wheat, oat, corn, and mixed grain cereal flours or meals are commonly utilized for extrusion processing. During extrusion, as a result of shear and high temperatures, usually above 140°C, cereal flours are transformed into viscoelastic melts. Upon extrusion, the melt expands and cools rapidly due to vaporization of moisture, eventually settling into an expanded solid foam. Because extrusion processing is associated with thermal manipulation (mainly heating and some cooling for unexpanded materials) of the materials, thermal characterization of cereal flours and their biopolymer components will lead to data that can be related directly to the processing protocols. Furthermore, thermal characterization of extruded products as a function of storage conditions (relative humidity–temperature) allows evaluation of the impact of such treatment.

In this chapter, we review the characterization by calorimetry of thermally induced conformational changes and phase transitions in pre- and postextruded cereal flours and the use of calorimetric data to elucidate the macromolecular modifications that these materials undergo during extrusion processing. The use of calorimetric data as a tool to evaluate the impact of formulation, processing, and storage on end-product attributes will be demonstrated.

II. CALORIMETRY

Differential scanning calorimetry (DSC) is a thermal analysis technique that detects and monitors thermally induced conformational transitions and phase transitions as a function of temperature. A pair of matching crucibles or sample pans, one containing the sample and one serving as reference, are heated in tandem. As a crucible is heated, its temperature increases, depending on the heat capacity of the contents of the crucible. At temperatures where an endothermic transition occurs, the thermal energy supplied to the crucible is consumed by that transition and the temperature of the sample cell lags behind the reference cell temperature. Conversely, the reference cell temperature lags when an exothermic transition occurs in the sample. A temperature difference between the cells results in heat flow between the cells. DSC measures the differential heat flow between the sample and reference crucibles as a function of temperature at a fixed heating rate. DSC thermograms are normalized to yield the specific heat capacity (C_p) as a function of temperature (1).

At temperatures where crystalline regions of cereal flour components undergo order–disorder transitions, peaks are observed in the heat flow vs. temperature diagrams, either as heat absorption (endotherm) or as heat release (exotherm). Endotherms are typically associated with the melting of mono-, di-, oligo-, and polysaccharides, denaturation of proteins, and gelatinization of starch. Exotherms are observed for crystallization of carbohydrates and aggregation of denatured proteins. When both crystalline and amorphous structures are present, which is typical in cereal flours, an additional transition is observed prior to the exothermic and endothermic transitions. This transition, known as a glass transition, is associated with amorphous materials or amorphous regions of partially crystalline materials. With DSC, the glass transition is observed as a sharp decrease of the heat capacity on cooling and a sudden increase in heat capacity on heating. A typical DSC thermogram, displaying glass, endothermic, and exothermic transitions, is given in Figure 1.

The glass transition temperature indicates a change in the mobility of the molecular structure of materials. Because cooperative motions in the molecular structure are frozen below the glass transition temperature, for partially crystalline materials exothermic and endothermic events are not observed until the glass transition is completed. Slade and Levine (2) discussed in detail that crystallization (exothermic event) can occur only in the rubbery state and the overall rate of crystallization (net rate of nucleation and propagation) in polymer melts is maximized at a temperature midway between the glass transition and melting temperatures. Furthermore, it has been demonstrated that for partially crystalline polymers the ratio of the melting to glass transition temperatures (T_m/T_g) varies from 0.8 to greater than 1.5. This ratio is shown to correlate with the glass-forming tendency and crystallizability of the polymers, because it predicts the relative mobilities of polymers at T_g and at $T \gg T_g$. More specifically, polymers with $T_m/T_g \gg 1.5$ readily crystallize, while the polymers with $T_m/T_g \ll 1.5$ have a high glass-forming tendency. Food biopolymers such as gelatin, native starch, and dimers or monomers such as galactose and fructose are reported to exhibit behavior similar to synthetic polymers with $T_m/T_g \ll 1.5$, which demonstrates a large free-volume requirement and thus a large temperature increase required for mobility.

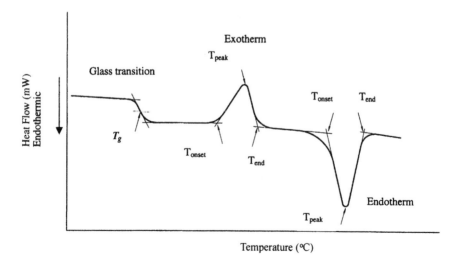

Figure 1   Typical DSC curve for partially crystalline materials.

In DSC thermograms similar to the one in Figure 1, the glass transition is detectable by a step change of the heat capacity. Although T_g can be observed experimentally by measuring physical, mechanical, or electrical properties, it is important to point out that DSC alone supplies thermodynamic information about T_g (3). The thermodynamic property of interest in DSC measurements is the change in the heat capacity, which reflects changes in molecular motions. It should be emphasized that the formation and behavior of the glassy state is a kinetic phenomenon. However, the rubbery state on the high-temperature side of the glass transition is at equilibrium and can be described by equilibrium thermodynamics. Equilibrium thermodynamics also can be applied well below the glass transition temperature because the response of internal degrees of freedom to external effects is very slow. However, during the glass transition, both intrinsic and measurement variables occur on the same time scale, the measured quantities become time dependent, and equilibrium thermodynamics cannot be applied to analyze the system. The system has a memory of its thermal history, which results in the occurrence of relaxation phenomena if the heating and cooling rates are different. It is not possible to get equilibrium values for T_g and the heat capacity change at the glass transition by extrapolating to zero scanning rate because these quantities depend on the thermal history, which includes the scanning rate, annealing temperature, and time.

A complete characterization of the glass transition can be achieved using several parameters. These parameters, as described by Höhne et al. (1), include the temperatures corresponding to vitrification ($T_{g,f}$) and devitrification ($T_{g,i}$) of the material upon cooling and heating, extrapolated onset temperature ($T_{g,e}$), heat capacity change, and the temperature corresponding to the midpoint of the heat capacity change between the extrapolated heat capacity of the glassy and rubbery states ($T_{g,1/2}$). The specific heat capacity versus temperature curve derived from a DSC thermogram of a typical glass transition and the parameters describing the glass transition are given in Figure 2.

T_g is also reported as the inflection point of the heat capacity versus temperature curve. The inflection point that corresponds to T_g is the temperature corresponding to the peak in the dC_p/dT vs. T curve. However, it should be kept in mind that the glass transition curve typically has an asymmetric shape and that the temperature corresponding to the inflection point and ($T_{g,1/2}$) are not the same. Therefore, it is a good practice to report the approach by which T_g is defined. Höhne and coauthors (1) indicate that $T_{g,e}$ and $T_{g,1/2}$ cannot describe the nonequilibrium nature of the glass transition, especially if the "enthalpy relaxation peaks" appear in the glass transition curve. However, these authors discuss at length that the glass transition, although a kinetically controlled parameter, can be unambiguously defined thermodynamically using a temperature called the *fictive* temperature. This thermodynamically defined T_g is based on the equality of enthalpy of the glassy and rubbery states at T_g. Further discussion of this subject

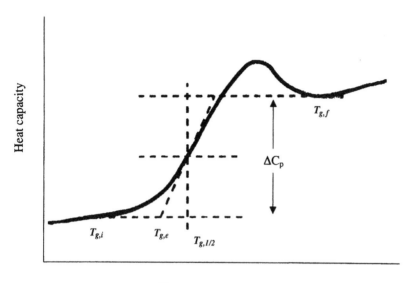

Figure 2 Glass transition.

is beyond the scope of this chapter, and the reader is referred to the book by Höhne et al. (1).

In addition to the conventional linearly increasing temperature protocol utilized in DSC, a recently introduced modulated differential scanning calorimetry (MDSC) employs a temperature protocol utilizing an oscillating sine wave of known frequency and amplitude superimposed onto the linear temperature increase applied to the sample and reference pans (4). The MDSC output signal, heat flow, can be deconvoluted to evaluate the contributions from thermodynamically reversible (reversing heat flow) and from irreversible or kinetically controlled transitions (nonreversing heat flow) within the time scale of the DSC experiment. This attribute enables the separation of overlapping complex transitions. One of the primary applications of this feature is the separation of the relaxation endotherm from the glass transition in amorphous materials (5–8). Figure 3 shows the total heat flow, reversing heat flow, and nonreversing heat flow deconvoluted from an MDSC experiment carried out with corn flour extrudate (9). It is apparent from Figure 3 that MDSC is an effective method of characterizing the glass transition of an amorphous extrudate, allowing the separate characterization of T_g and endothermic relaxation in one heating cycle.

At plasticizer levels that cause the partial or complete masking of the glass transition by an endothermic relaxation endotherm, the protocol used to deter-

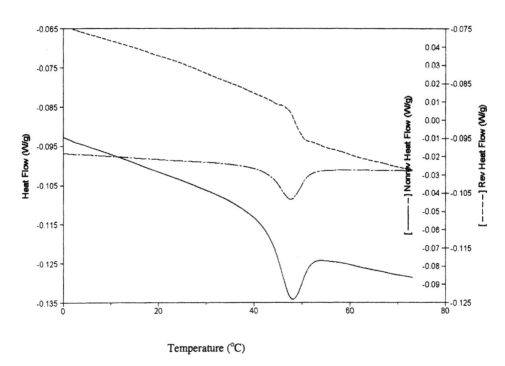

Figure 3   Glass transition of corn extrudate using MDSC.

mine the glass transition temperature using conventional DSC is to make a partial scan to just above the glass transition, followed by cooling and a second scan. The glass transition temperature is determined from the second heating scan (10). Another benefit of this technique over standard DSC is an increase of resolution and sensitivity due to the cycling instantaneous heating rates, which enables one to detect weak transitions. Although the use of MDSC expands the capabilities of DSC and allows one to measure heat capacities and characterize reversible/ nonreversible thermal transitions, one should remember that the deconvolution of the total heat flow into reversing and nonreversing components is affected by experimental parameters. In addition to linear heating rate, the operational parameter in conventional DSC, MDSC requires the choice of modulation amplitude and modulation period. The selection of the best combination of modulation parameters, modulation period, and modulation amplitude, and the underlying linear heating rate for the specific sample under investigation, is critical in the generation of reliable data and the correct analysis and interpretation of results. MDSC should be applied with caution for the analysis of melting transitions (due to the difficulty of maintaining controlled temperature modulation throughout the fusion) and for sharp transitions (due to the difficulty of achieving at least four modulations through the thermal event). For a detailed discussion of application of the MDSC technique to biopolymers and food materials, readers are referred to book chapters and review articles in the literature (11–14).

Although the DSC is a commonly used technique to monitor the glass transition, other techniques, based on the observation of various physical properties sensitive to the molecular mobility of polymers, also are reported. Studies involving various measuring techniques that are sensitive to changes in segmental mobility in biopolymers and complex food systems have been reviewed (15, 16). Some techniques are dynamic in nature and involve the measurement of a property as a function of scanning temperature, such as in dynamic mechanical analysis (DMA or DMTA) and thermodielectrical analysis (TDEA). DMA monitors the change in loss modulus and the elastic modulus to calculate tan δ, a measure of structural relaxation. And NMR spectrometry is performed at various temperatures, with equilibration of the sample at a given temperature prior to measurement. Typically, the T_2 relaxation time, which defines the rigid lattice limit temperature (T_{RLL}), rather than T_g, is measured. The moisture content that results in transition from a glassy state to rubbery state at ambient temperature in model or real food systems also can be evaluated by mechanical and optical techniques. Measurement of mechanical properties as a function of moisture content employs a three-point bend test to evaluate the change in Young's modulus, which is a macroscopic parameter sensitive to microscopic mobility. Spectroscopic techniques that use optical probes (specific molecules with well-characterized spectroscopic properties) to evaluate the emission, intensity, and decay kinetics of phosphorescence or fluoresecence as a function of the sample moisture content report on changes of mobility of the probe. Because the various techniques are sensitive to the mobility of different scales of distance and time, differences in the reported glass transition temperatures are expected. The studies comparing several techniques demonstrate that the NMR transition occurs 5–35°C below the DSC transition and, the DSC midpoint generally is observed between the tan δ peak and the drop in elastic modulus by DMA for amylopectin (17). Blanshard (15) claims that the consumer perceives significant changes in texture at T_{RLL} detected by NMR, but he does not report supporting evidence. The mechanical and spectroscopic techniques used to evaluate the glass transition are discussed in detail in other chapters in this book.

It is important to recognize that the detection of the glass transition temperature depends on several factors.

1. *Sensitivity of the observable to the mobility in the system*: In high-molecular-weight materials the sidechains will have a greater degree of mobility than the backbone. The mobility in local domains might occur at a lower temperature than the mobility in the total system.

Therefore, techniques sensitive to the mobility in local domains, such as molecular probes used in spectroscopic techniques and in NMR, will detect the change in observables at a lower temperature.

2. *Time scale*: Especially in dynamic measurement systems, for events to be detected the experimental time scale should match the time scale of the relaxations in the molecular structure.

3. *Magnitude of the observable*: The magnitude of the change in the physical observable can be different for different techniques. The decrease in the elastic modulus can be several orders of magnitude through the glass transition, which makes it easily identifiable. However, if the energy associated with the glass transition is small, the heat capacity change can be small, which leads to ambiguities in its detection.

4. *Moisture loss during experiment*: T_g is highly influenced by the moisture content of cereal systems. If the sample is not sealed well, the moisture content of the sample will change due to evaporation during the course of experiment. This may lead to overestimation of the glass transition temperature in techniques such as DMA.

Therefore, the combined use of different but complementary experimental techniques is recommended as the most powerful approach to study glass transitions in model and real food systems (18).

The temperatures for the endothermic and exothermic transitions and the heat involved in such transitions are measured in DSC experiments. The transition temperatures (T_{peak}) are points of maximum heat capacity of endotherms or minimum heat capacity of exotherms. From a DSC thermogram, heat capacity (C_p) vs. temperature curve, one can extract values for the thermal (temperature of transition) and thermodynamic changes in free energy (ΔG), enthalpy (ΔH), entropy (ΔS), and heat capacity (ΔC_p) of the various transitions, in addition to determination of the bulk heat capacity of the material.

The enthalpy change of a transition, at any temperature T, is extracted using

$$\Delta H = \Delta H_{T_o} + \int_{T_o}^{T} \Delta C_p \, dT \tag{1}$$

where T_o is the midpoint of the transition and ΔC_p is the difference in heat capacity between the pre- and posttransition material.

A similar analysis is employed to extract the entropy change (ΔS) for the transition:

$$\Delta S = \frac{\Delta H}{T_o} + \int_{T_o}^{T} \frac{\Delta C_p}{T} \, dT \tag{2}$$

The free energy change is obtained from the relation

$$\Delta G = \Delta H - T \, \Delta S \tag{3}$$

Taken together these data provide a complete thermodynamic characterization of the material. The advantages of the calorimetric approach to studying thermodynamics are that direct measurements are made of ΔH and ΔC_p, the data collection and analysis are not specific to particular materials, and the materials require neither destructive nor elaborate sample preparation before analysis.

Differential scanning calorimetry (DSC) is used to detect and define those temperatures that correspond to significant thermally induced transformations (glass, melting, and gelatinization transitions), over temperature and moisture content ranges that simulate extrusion processing. The basis for thermodynamic study of biopolymers is that the relevant initial and final states can be defined and the energetic and/or structural differences between these states can be measured using calorimetric instrumentation. Comparison of various final states achieved under different extrusion processing conditions starting from the same initial state will allow one to predict the impact of various processing conditions on the creation of new structures and textures. Furthermore, in complex systems individual biopolymers as well as the interactions among biopolymers in macromolecular assemblies can be assessed.

III. THERMALLY INDUCED TRANSITIONS IN PRE- AND POSTEXTRUDED CEREALS

A. Glass Transition

Cereal flours are partially crystalline biopolymer systems, comprising mainly starch, but also containing protein and lipid. Being a partially crystalline polymer system, cereal flours display thermally induced transitions typical of both amorphous and crystalline materials, as shown in Figure 1. Glass transitions are observed in noncrystalline regions of partially crystalline polymers such as starches and proteins (19–22). The temperature interval of the glass transition over which the "freezing in" of long-range molecular motions, including translational and rotational motions, occurs depends on the physical and chemical structure of the molecules and their interactions (1).

1. Molecular Weight Dependence

It has been demonstrated that T_g increases with increasing molecular weight (3). Levine and Slade (23) showed that T_g of samples of amorphous linear poly(vinyl acetate) approach an asymptote around an average molecular weight of 10^5. For homologous series of linear polymers, empirical equations are developed expressing T_g as a function of the molecular weight of the polymer, Eq. (4), Fox and Flory equation (24), or degree of polymerization, Eq. (5) (25):

$$T_g = T_g(\infty) - \frac{K}{M_n} \tag{4}$$

where $T_g(\infty)$ is the value of T_g when the molecular weight goes to infinity, K is a constant, and M_n is the number average molecular weight.

$$\frac{1}{T_g} = \frac{1}{T_g^0} + \frac{a}{\mathrm{DP}} \tag{5}$$

where T_g^0 is the high-molecular-weight limit for the glass transition temperature, a is a constant, and DP is the degree of polymerization.

Buera et al. (26) applied the Fox and Flory equation to predict the molecular weight dependence of T_g for amorphous poly(vinylpyrrolidone). Furthermore, studies on maltodextrins over a wide range of molecular weights demonstrated that the T_g of maltodextrins increases with increasing molecular weight (27, 28). Roos and Karel (27) determined the T_g values of glucose homopolymers with average molecular weights of 343, 504, and 991 g/mol experimentally to calculate the constants for the Fox and Flory equation. Constants for the Fox and Flory equation, $T_g(\infty)$ (243°C) and K (52,800) were predicted. Roos and Karel reported the $T_g(\infty)$ as the T_g of anhydrous starch. Orford et al. (28) measured T_g values of malto-oligomers of increasing degree of polymerization and concluded that the increases in T_g of oligomers are marginal from maltoheptose to amylose and amylopectin. These investigators, using the equation proposed by Ueberreiter and Kanig (25), estimated the T_g of amylose and amylopectin to be 227°C using Eq. (5). The concept of molecular weight dependence of T_g is important for high-temperature extrusion processing applications because during the process the molecular weight decreases due to fragmentation. Discussion of extrusion-induced fragmentation and its relation to T_g will be deferred to a later section.

2. Chemical Structure Dependence

For synthetic polymers with different physical and chemical structures, the T_g is shown to be specific to each anhydrous material (3). Lillie and Gosline (29) indicated that T_g of proteins may occur over a wide range of temperatures, depending on whether the material is dry or in the presence of small amounts of water. The average molecular weight and T_g have been reported for several dry proteins, including elastin 70,000 and above 200°C (19), wheat gluten above 160°C (21, 30), gliadin 30,000–60,000 and ~157°C, zein 30,000 and 150°C (31), dry casein 23,000 and 144°C (32). Some of these data are not for anhydrous protein but are extrapolated to zero water content using the Gordon–Taylor (33) equation or curve fitting. It should be emphasized that for proteins, in addition to molecular weight, primary (amino acid sequence) and secondary (α-helix, β-sheet, β-turn, random coil) structure are expected to influence the T_g. Lillie and Goslin (29) state that the heterogeneity in protein structure due to nonuniform amino acid distribution may create microenvironments with various thermal stabilities that manifest themselves as multiple or broad glass transitions.

Although the correlation between molecular weight and dry T_g is well established for carbohydrates, the T_g of like-molecular-weight carbohydrates with different chemical structures differ (16). Specifically, differences in T_g were reported between glucose and fructose (both 180.2 g/gmole) and among maltose, sucrose, and lactose (all 342.2 g/gmole).

Difficulty is experienced in determining glass transition temperatures of some high-molecular-weight biopolymers because the glass transition tends to be broad with a small heat capacity change (27), and for many biopolymers thermal degradation may occur before the glass transition is reached (19). These limitations were proposed to be overcome by measuring T_g for oligomers followed by extrapolation to high molecular weight (27, 28) or by studying T_g as a function of diluent content followed by extrapolating to zero percent moisture (27, 28). In the latter approach, it is important to emphasize that T_g values should be measured as close to the dry condition as possible, because the slope of the T_g vs. moisture content curve may get steeper as the dry condition is approached. If T_g is not determined at sufficiently low moisture content, the dry T_g value may be underestimated.

3. Thermal History Dependence

Both the temperature and the magnitude of the glass transition event are important. The glass transition is associated with increased energy in the system, which manifests itself as an increase in the heat capacity. Wunderlich (34) states that microscopically T_g is either the temperature of the liquid where motions with the large amplitudes stop during cooling or the temperature of the solid where motions with the large amplitudes start during heating. It is apparent that, for a glass transition to be observed, the experimental time scale should match the time scale required for the molecules to adjust to the new conditions. If the experimental time is too short for the molecules to adjust to the changes in temperature, the glass transition is observed (34). In DSC, an overly fast scanning rate can result in an overestimated T_g value during heating and an underestimated T_g value during cooling. Furthermore, because they are not at equilibrium, if the glasses are prepared at different cooling rates, each will have a different free energy state. Even glasses with identical chemical structure and molecular weight but different thermal history might display different T_g values. The T_g value data for low-molecular-weight carbohydrates compiled from the literature by Levine and Slade (16) clearly demonstrate the influence of molecular weight, chemical structure, and thermal history on T_g.

4. Effect of Crystallinity

In synthetic polymers, crystallinity is reported to affect the glass transition temperature, leading to the measurement of an apparent glass transition temperature in partially crystalline materials. In semicrystalline polymers, the measured T_g value appears to be higher, because the molecular mobility in the amorphous regions is restricted by the surrounding crystalline regions. In highly crystalline polymers, T_g may appear to be masked (3). Boyer (35) reports the presence of more than one T_g value, where one attributed to the completely amorphous state and correlates with the chemical structure and where the other has a higher apparent value and depends on the extent of crystallinity and the morphology.

During high-temperature extrusion, a viscous melt is produced. Upon exiting the die, the melt cools very rapidly and settles into a solid state. Because the time to reach the solid state is faster than the time required for crystallization, high-temperature extrusion produces extrudates in a glassy state. The amorphous characteristics of extruded products in the glassy state may be demonstrated by X-ray diffraction (10, 36, 37), polarized-light microscopy (38), and calorimetric studies (10, 37). X-ray diffraction techniques are utilized to monitor the loss of crystallinity in starches and cereal flours as a result of extrusion cooking (10, 36, 37). These investigators concluded that the crystalline structure of all of the materials studied were destroyed partially or completely, depending on the amylose–amylopectin ratio and the extrusion variables, including moisture, shear, and temperature. Specifically, Kaletunç and Breslauer (10) studied the X-ray diffraction patterns of extrudates as a function of specific mechanical energy (SME), which is used to quantify the extent of mechanical stress applied per unit of material processed. Kaletunç and Breslauer report that X-ray diffraction patterns of extrudates produced over an SME range of 452–1,386 kJ/kg were similar to that of grease, which consists of smooth curves rather than oscillating or periodic sawtooth fine structure. Charbonniere et al. (36) report that while reduced crystallinity is observed at an extrusion temperature of 70°C, for extrusion temperatures above 100°C the crystalline structure of starch is completely destroyed, leading to an X-ray diffraction pattern typical of an amorphous state. McPherson et al. (37) studied the effect of extrusion on the loss of crystallinity in unmodified, hydroxypropylated, and cross-linked hydroxypropylated starches. X-ray diffraction patterns showed a partial crystallinity loss for all of the extrudates produced at a 60°C extrusion temperature, while the crystalline peaks were absent in products extruded at 80 and 100°C. When lipids are present in the system, extrudates display a crystal structure (V-type) typical of amylose–lipid complexes (37, 39).

Absence of ordered structure is also reported as a loss of birefringence in wheat flour extrudates using polarized-light microscopy (38). The amorphous characteristics of corn and wheat flour extrudates are further verified by calorimetric results, in which the glass transition is the only thermally induced transition detectable in the extrudates (10, 40).

5. Moisture Plasticization

Water acts as a plasticizer for cereal-based foods. Moisture plasticization causes the loss of a characteristic crispy texture as well as the shrinkage of extruded cereal flours, which in turn affects the quality and shelf life stability of such products (40). The glass transition temperature has been used to evaluate thermal stability, with the purpose of assessing the quality and shelf stability of extruded foods, especially those that have low to intermediate moisture content. Below T_g the material is in the glassy state, whereas above T_g it is in the rubbery state. The extreme sensitivity of T_g to water plasticization (water induces a lowering of T_g), especially in the 0–10% water content range, underscores the importance of determining the moisture sorption characteristics of food materials. The plasticizing effect of increasing moisture content at constant temperature is identical to the effect of increasing temperature at constant moisture content, as reported by Slade and Levine (2). This fact makes the selection of storage conditions extremely critical, especially for materials that display low T_g values even under dry conditions.

Moisture sorption may occur in extruded products due to exposure to various relative humidity conditions during postprocessing storage. It is well established that the thermal stability of the glassy state (T_g) of the extrudates is highly sensitive to moisture plasticization (moisture induces a lowering of T_g), especially in the 0–10% water content range (2, 10, 20, 40, 41). Specifically, Kaletunç and Breslauer (10) reported a rapid drop of T_g by about 8°C/(% of moisture added) for an extrudate of high-amylopectin corn flour (Fig. 4).

Slade and Levine (2) have discussed the influence of water as a plasticizer on water-compatible, amorphous, and partially crystalline polymers, an effect that has been observed by a number of investigators as a depression in T_g. Typically, at low moisture content ($\sim{<}10\%$ water), a 5–10°C/(% of moisture) reduction of T_g is reported. Biliaderis (42) reported water plasticization [$\sim$7.3°C/(% of moisture added), up to 20%] for rice starch, a partially crystalline biopolymer system. Kalichevsky and co-workers published glass transition curves for amorphous amylopectin samples with moisture contents in the 10–25% range (17) and for amorphous wheat gluten samples with moisture contents up to 16% (30). Both studies demonstrated that the T_g-depressing effect of water continues beyond the 10% moisture content range [7°C/(% of moisture), over 10–25% moisture, for amylopectin (17) and 10°C/(% of moisture), over 0–16% moisture, for wheat

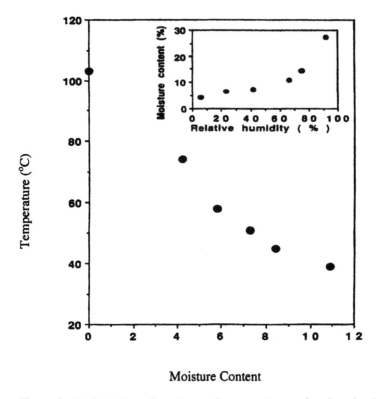

Moisture Content

Figure 4 T_g of a high-amylopectin corn flour extrudate as a function of moisture content. (From Ref. 10.)

gluten (30). A similar trend was also reported by Kaletunç and Breslauer (40) for wheat flour.

Empirical and theoretical equations were proposed to predict the T_g of miscible polymer blends (33, 43). These equations are also utilized to predict the glass transition temperature of polymer–water mixtures (44). The empirical equation of Gordon and Taylor (33) relates T_g to the composition of miscible polymer blends:

$$T_g = \frac{w_1 T_{g1} + kw_2 T_{g2}}{w_1 + kw_2}$$ (6)

where w and T_g are, respectively, the weight fraction and glass transition temperatures of each component and k is an empirical constant. The Gordon–Taylor equation has been applied by many investigators to predict the plasticizing effect of moisture on amorphous materials (26, 27, 45). These investigators fitted the experimental T_g data obtained as a function of moisture content to determine the empirical constant, k. For a binary system of amorphous solid (1) and water (2), the k values reported in the literature are 2.66–2.82 for poly(vinylpyrrolidone) (26), 6 for maltose and 6–7.7 for maltodextrins (27), 2.3 for casein, 1.3 for sodium caseinate (32), 5 for gluten (30), and 4.5 for amylopectin (45). It is apparent that the k value is related to the effectiveness of the diluent to plasticize the amorphous solid, with a higher k value indicating easier plasticization of the amorphous material or a greater decrease in T_g as a function of moisture content. The Gordon–Taylor equation is demonstrated to describe successfully the glass transition temperature of food biopolymers as a function of water content. A significant limitation of this equation remains its applicability to only binary systems.

The Couchman–Karasz equation (43) predicts the glass transition temperature of a mixture of compatible polymers from the properties of the pure components based on a thermodynamic theory of the glass transition in which the entropy of mixing is assumed to be continuous through the glass transition:

$$T_g = \frac{w_1 \Delta C_{p1} T_{g1} + w_2 \Delta C_{p2} T_{g2}}{w_1 \Delta C_{p1} + w_2 \Delta C_{p2}} \tag{7}$$

where w, T_g, and ΔC_p are, respectively, weight fraction, glass transition temperature, and heat capacity change at the glass transition of each pure component. The Couchman–Karasz equation is also applied to predict the T_g of moisture-plasticized amorphous materials from the properties of the solid and water. This equation requires that the glass transition temperatures and the heat capacity changes at the glass transition temperature of the solid and water are known.

6. Glass Transition Temperature of Complex Systems

There are numerous studies in the literature designed to characterize the thermally induced transitions (glass, melting, gelatinization, crystallization, denaturation) in starch, protein, and lipids, the biopolymer components of cereal flours, as a function of moisture content, with the ultimate goal of developing an understanding of structure–function relationships (16). Kaletunç and Breslauer (40) discussed the potential pitfalls of using thermal data on the individual components to interpret complex systems, such as cereal flours. This approach may overlook interactions between the components. Furthermore, thermal properties of individual biopolymers may be altered, depending on the severity of conditions applied during the isolation processes used to prepare the individual components. Kaletunç and Breslauer (40) suggested that the flour itself should provide the appropriate thermodynamic reference state for process-induced alterations in the thermal properties of the processed material and that experiments can be designed to elucidate the differential thermal properties of pre- and postextruded flours (a real food system) as a function of central factors that may greatly influence the quality attributes of the extruded products. These factors include moisture content of pre- and postextruded cereals and extrusion processing parameters.

Although cereal flours are a mixture of biopolymers, usually only a single apparent T_g is reported (40). This does not necessarily mean that there is only one glass transition. If the heat capacity change associated with a glass transition is too small, that transition may not be detectable by DSC. Multiple independent glass transitions may occur within a narrow temperature range; the resultant superimposed changes in the observable may appear to be a single transition.

Proteins are present as a small fraction of cereal flours, ranging from 5 to 6% in rice, 9 to 10% in corn, and up to 15% in wheat flour. Proteins in flour go through conformational transitions during extrusion processing and contribute to both structural and functional properties of extruded products. Kaletunç (46), using light microscopy and FTIR microspectroscopy, showed that proteins form fiberlike structures aligned in the extrusion direction, while starch acts as a filler. Antila et al. (47) proposed an empirical model incorporating protein content in addition to moisture content to predict radial expansion of extrudates.

At high extrusion temperatures (150–200°C), proteins denature, aggregate, and cross-link through covalent or noncovalent bonding. Upon exiting the die and cooling, they assume the glass state. Extruded products typically exhibit broad glass transitions. This may indicate a series of closely spaced transitions that coalesce and appear to be a single, broad transition. The glass transition of proteins may not be observed if it overlaps with the transition due to the more abundant search. Even if the glass transition occurs at a different temperature, the associated change in the heat capacity may be too small to detect due to the low concentration of protein in the extruded product.

B. Melting and Gelatinization of Starches

Melting and gelatinization are phase transitions that are associated with the transformation of the crystalline regions of partially crystalline starch to amorphous liquid. These endothermic transitions can be detected and monitored using DSC. It is well established that the thermal stability of the crystalline phase in starch decreases by increasing the amount of water present in the system. Furthermore, while, at low (<20%) and high (>65%) moisture contents only one endothermic transition is detectable by DSC in lipid-free starch, at moderate water content more than one endothermic peak is reported for starches from various origins, including rice (48), potato (49), and wheat (2, 40). Figure 5, extracted from the wheat flour state diagram given by Kaletunç and Breslauer (40), shows the peak temperatures of starch melting endotherms observed in wheat flour as a function of moisture content.

It is apparent from Figure 5 that at low moisture contents (less than 20%), a single endotherm, of which the transition temperature is very sensitive to the

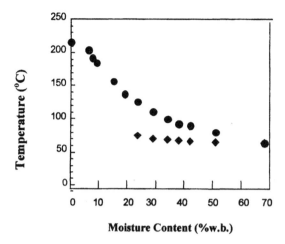

Figure 5 Thermal stability of melting endotherms in wheat flour as a function of moisture content. (From Ref. 40.)

moisture content of wheat flour, appears. A second endotherm, with a comparatively lower transition temperature and less sensitive to moisture increase, becomes visible above 23% moisture. The two endotherms coalesce above 67% moisture, and the transition temperature of the resultant endotherm is independent of further increase in moisture content. The presence of two endotherms is sometimes (40, 42, 50) but not always (51) reported for the starch–water system.

The influence of water on the thermal stability of the starch crystalline phase has been studied by many investigators. Several models have been proposed to predict the influence of water on melting transitions in starch. In earlier studies, the Flory–Huggins equation (52), which is used to describe the melting-temperature-depressing effect of diluents in synthetic polymers, was applied to predict the melting temperature of starch as a function of water content (53–55). Several investigators reported that melting temperatures predicted using the Flory–Huggins equation were in agreement with the experimentally determined values for volume fractions of water between 0.1 and 0.7 (42). However, a significant deviation between theoretical and experimental melting temperatures was reported for rice starch above a volume fraction of water of 0.7. Also, the melting temperature for dry starch was underestimated (54). Furthermore, the presence of more than one endotherm in DSC thermograms cannot be predicted or explained using the Flory–Huggins equation. After the pioneering work of Slade and Levine (50), it was recognized that starch melting is a *nonequilibrium* phenomenon and, therefore, it cannot be analyzed by Flory–Huggins theory based on equilibrium thermodynamics. Slade and Levine's study (50) on wheat and waxy corn starch with 55% (wt) water demonstrated that the melting process is irreversible, kinetically controlled, and mediated indirectly by water plasticization, as it affects the stability of glassy regions. The proposed "fringe micelle" network model for starch structure supports the observation of a glass transition preceding the crystalline melting, as the melting of interconnected microcrystallites depends on the mobility of the continuous glassy regions on which water exerts a plasticizing effect. Observation of more than one melting endotherm at limited moisture content (~25–60%) can be attributed to nonuniform moisture distribution or the presence of microcrystalline domains with different thermal stabilities.

Gelatinization enthalpies must be interpreted with caution when used as indices of starch crystallinity, because they represent net thermodynamic quantities of different events: granule swelling and crystallite melting (endothermic) and hydration and recrystallization (exothermic). Furthermore, significant contributions to the ΔH value from amorphous regions have been also suggested (42). Gelatinization enthalpies are used to quantify the starch modification as a result of extrusion processing. The degree of starch conversion (DC) is described by the reduction in the area of the gelatinization endotherm before and after extrusion (56, 57):

$$DC(\%) = \frac{\Delta H_0 - \Delta H_i}{\Delta H_0} \times 100 \tag{8}$$

where ΔH_0 is the enthalpy of gelatinization of the native starch and ΔH_i is the enthalpy of gelatinization of the extruded starch.

Starch conversion is sometimes used synonymously with degree of gelatinization (58). It should be remembered that the application of mechanical energy during extrusion results in structural changes to starch, including granule disruption and fragmentation of starch molecules into smaller-size polymers. Although some gelatinization is expected, the reduction in the gelatinization peak area cannot be attributed only to the gelatinization of native starch during extrusion.

C. Amylose–Lipid Complex

Lipids are present naturally in cereal flours in varying amounts (1.2% in wheat flour, 1.4% in degermed corn flour, 5–7% in oat flour). Furthermore, lipids are added during extrusion processing, mostly in the form of emulsifiers. The addition of emulsifiers is reported to modify the product characteristics, including expansion, cell size and distribution, and texture (59–61). Emulsifiers are reported to form complexes with amylose during extrusion processing (62). Extrusion operating conditions influence the creation of a product, with its structure and physical properties both depending on the severity of treatment and on the provision of suitable conditions for the ingredients to react with each other. Colonna et al. (63) claim that molecular modification is reduced because lipids may act as lubricants. Each type of lipid has a distinct effect on the material properties during extrusion processing that requires modification of the extrusion operating parameters.

Most emulsifiers are compounds that have both hydrophobic and hydrophilic ends on the same molecule. Glycerol monostearate (GMS) and sodium steroyl lactylate (SSL) are two small-molecular-weight emulsifiers commonly used in food applications. The hydrophobic ends of emulsifiers are believed to form a complex with the amylose fraction of starch during cooking, retarding starch gelatinization and decreasing swelling. Starch granule swelling and solubility decline with an increase in complex formation (64). Galloway et al. (62) reported that the formation of amylose–GMS complex resulted in a decreased degree of gelatinization, water solubility, and expansion of wheat flour extrudates. Scanning electron microscopy studies of wheat flour extrudate microstructure showed that addition of GMS and SSL to wheat flour extrudate increases the size and uniformity of the cells (60).

Amylose forms complexes with iodine (65), alcohols (66), and fatty acids. Amylose–lipid complexes are formed by mixing the complexing agent with hot (60–90°C) dilute aqueous amylose solution. Several studies confirm that amylose–lipid complexes also form during high-temperature extrusion processing at low water content (39, 62, 67–71). Amylose–lipid complexes form during the extrusion process and crystallize during the cooling process, displaying a V-type X-ray diffraction pattern, characterized by three main peaks, the major one being located at 9°54′ (Θ). Mercier et al. (67) suggest that amylose–lipid complexes have a helical structure, with six glucose residues per cycle in a hexagonal network. Both Mercier et al. (39) and Meuser et al. (72) report an upper concentration limit for lipid complexation of about 3% with saturated and unsaturated fatty acids (C_2 to $C_{18:2}$), GMS, and SSL involved in the complexation. Schweizer et al. (69) report that triglycerides do not contribute significantly to the complex formation.

DSC is used to study the formation of amylose–lipid complexes as well as to characterize complexes formed in solution and during high-temperature extrusion processing (62, 69, 73). The presence of amylose–lipid complexes manifests itself in DSC thermograms as a reversible endothermic transition. The thermal stability, shape, and energy associated with the transition are highly influenced by the water content of the extrudate. Melting of amylose–lipid complexes at high moisture content (above 70%) is highly cooperative, yielding a single transition (48). As the moisture content decreases, two endothermic events separated by an exothermic event are observed. The thermal stability of the complex decreases with increasing moisture content for wheat flour–GMS extrudates at 50 and 90% hydration. The observation of lower thermal stability and apparent enthalpy of transition upon rescanning of such samples indicates nonequilibrium melting due to the formation of various metastable states depending on water content and cooling and heating rates during scanning.

Amylose–lipid complex formation affects the physical properties, quality attributes, and nutritional characteristics of extruded products. Amylose–lipid complexes display very low susceptibility to amylase hydrolysis in vitro. Hydrolysis of the complex with increased enzyme concentration and incubation time indicates a slow rate of digestion. Although an inverse relationship between the amount of amylose–lipid complex in extruded wheat flour (measured by X-ray diffraction and iodine-binding capacity) and in vitro hydrolysis by pancreatic amylase is reported, higher glycemic responses are observed with extruded products in comparison with boiled and baked products (74). These results indicate that the disintegration of the granular structure and the fragmentation leading to increased starch solubilization dominate starch digestion rather than the formation of amylose–lipid complexes.

IV. EVALUATION BY DSC OF THE IMPACT OF EXTRUSION PROCESSING AND STORAGE ON EXTRUDED CEREALS

A. Effect of Formulation

Sugars are the second major component, after flours, in the formulation of presweetened cereals. In directly expanded products, the total concentration of sugar added during extrusion ranges from 0 to 28% (75). Sugars are added to RTE cereals principally for flavor. However, sugars also contribute to the color, structure, and texture of extruded products (76–79). At the high temperatures of extrusion, reducing sugars, such as glucose, fructose, maltose, and lactose, take part in Maillard browning reactions with the amino groups of proteins and peptides. Depending on the balance of reactants and the conditions of processing, Maillard reactions contribute to the desirable color and flavor of products such as breakfast cereals.

Sucrose is the most commonly used and investigated sugar. Replacement of a fraction of the starch-containing material by sucrose changes the extrusion process parameters, such as SME and die pressure, which in turn leads to changes in the structure of the extrudates (76, 79). Specifically, extrudate bulk density increases with a concomitant reduction in expansion as sucrose concentration increases (76–78, 80, 81). The change in extrudate physical structure strongly influences the fracturability characteristics and sensory texture (82, 83).

Maltodextrins are used in extrusion processing formulations at concentrations as high as 15% by weight (79). Maltodextrins have average chain lengths of 10–100 glucose units, while amylose has chain lengths of 70–350 glucose units (84, 85). Maltodextrins do not contribute sweetness to the end product, and they are fairly inert in Maillard reactions because of their limited number of reducing groups. In extrusion processes, maltodextrins cause a decrease in the melt viscosity, because they replace starch polymers, as does sucrose.

Plasticization of starch systems by small-molecular-weight constituents, in addition to water, is a widely reported phenomenon (16). Barrett et al. (76) showed that the T_g of corn flour extrudates, as measured by differential scanning calorimetry and by dynamic mechanical spectrometry, decreases rapidly with increasing sucrose content. Addition of sugar would be expected to reduce SME, leading to less fragmentation, higher T_g, and a crispier product. However, the observed reduction in T_g indicates that plasticization by sucrose overwhelms the effect of reduced fragmentation on the thermal stability of the glassy state.

In addition, sugars, such as sucrose, fructose, glucose, and xylose, reduce the T_g of nonextruded amylopectin (32, 86, 87) observed a progressive reduction in the stress–strain functions of extruded wheat starch during bending tests with increasing levels of added glucose.

B. Effect of Extrusion Processing Conditions

Cereal flours are the structure-forming materials in extruded products, comprising a dispersed protein phase, usually in a fibrous form aligned with the extrusion direction, in a continuous starch phase. The extent of fragmentation during extrusion processing influences the structure formation of the extruded products. In the extruder, cereal flours are subjected to thermomechanical stress that may lead to the depolymerization of biopolymer components, depending on the severity of the extrusion conditions. The review by Porter and Casale (88) on the stress-induced degradation of synthetic polymers emphasizes that as the molecular weight of polymers increases, mechanical energy is stored in the molecule rather than dissipated as heat. The concentration of mechanical energy into a smaller number of bonds results in bond rupture (89). Several investigators report that extruded starch has lower average molecular weights and significantly different molecular weight distribution in comparison to unextruded starch (37, 63, 90–99). Fragmentation is more significant at high temperatures, high screw speeds, and low extrusion moisture. Colonna et al. (63) compared the weight-average molecular weights ($\overline{M}w$) of pre- and postprocessed starch exposed to extrusion processing and drum drying and reported a decrease in the molecular weight of extruded samples. However, because the temperature during extrusion processing can be as high as 180°C, depolymerization of starch during extrusion processing may occur due to thermal degradation as well as mechanical degradation. Tomasik et al. (100) noted that depolymerization occurs in dry starch below 300°C,

resulting in dextrin formation. However, the time required for thermal dextriniza-
tion of starch is several order of magnitues longer than the residence time of
starch in the extruder during processing.

Macromolecular degradation is also reported for cereal flours during extrusion processing by several investigators (10, 40, 101–104). Kaletunç and Breslauer (10, 40) monitored the extrusion processing–induced fragmentation by
reductions in the extrudate glass transition temperature relative to the glass transition temperature of preextruded flour. These investigators designed their experiments so that extrusion-induced fragmentation would be the primary cause of T_g
reduction. The general trend observed was a reduction of T_g as a function of SME
for high-amylose and high-amylopectin corn flours and wheat flour (Fig. 6).

Figure 6 shows the range of SME values (200–1400 kJ/kg) achieved in a
pilot-plant-size twin-screw extruder and the corresponding T_g values for freeze-
dried extrudates of corn and wheat flour (10, 40). Kaletunç and Breslauer (40)
also demonstrated that for wheat flour extrudates the weight-average molecular
weight decreases in a trend similar to the decrease in T_g value as a function
of SME. Figure 7 shows the dependence of the $\overline{M}w$ of wheat flour extrudates
(determined by gel permeation chromatography) (38) as well as the calorimetrically determined T_g on SME (40).

A similar observation of molecular weight reduction with increasing SME
for extruded wheat starch was reported by Meuser et al. (96). Using their data,

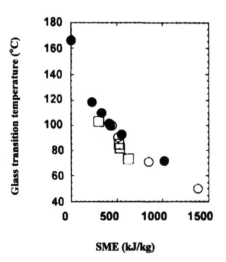

Figure 6   SME dependence of glass transition temperature for ● wheat flour, □ high-
amylose corn flour, ○ high-amylopectin corn flour. (Combined from Refs. 10 and 40.)

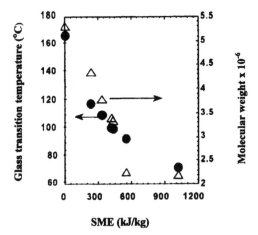

Figure 7 SME dependence of glass transition temperature and molecular weight for wheat flour. (From Ref. 40.)

these investigators developed a mathematical model predicting the molecular breakdown of wheat starch, quantified as mean molecular weight, as a function of SME. Kaletunç and Breslauer (40) observed that T_g values of extrudates with similar SME values (416 and 432 kJ/kg) but different extrusion processing conditions were indistinguishable. $\overline{M}w$ for these two extrudates were virtually identical, 3.34×10^6 for 432 kJ/kg and 3.41×10^6 for 416 kJ/kg (38). These investigators concluded that for a given flour, the extent of extrusion-induced fragmentation is related to the mechanical energy generated in the extruder as quantified by SME. A similar conclusion was reached by Meuser et al. (96); that is, the energy history, primarily SME, causes a significant change in the molecular structure of a given material, in turn defining the characteristic attributes of the final product.

Figure 6 also shows the T_g and molecular weight values of freeze-dried, unextruded cereal flours corresponding to a zero SME value. Note that T_g of the extrudates is less than that of the corresponding flour, even for the lowest SME value of 236 kJ/kg, indicating fragmentation in the extruder. It was shown in a previous study on molten maize starch (94) that macromolecular degradation first appears when the mechanical energy reaches about 10^7 J/m^3 (~10 kJ/kg). The lowest SME of 236 kJ/kg was reported by Kaletunç and Breslauer, corresponding to ~10^8 J/m^3, which is above the minimum mechanical energy required to induce fragmentation. At higher SME values, the difference between the T_g values of wheat flour and its extrudate becomes larger. Input of mechanical energy also causes damage to starch granules. Significant granule damage in starch is reported when SME is greater than 500–600 kJ/kg, while even below 250–360 kJ/kg, a large fraction of flattened and sheared granules were reported (105, 106).

Starch degradation is investigated using viscometry, gel permeation chromatography (GPC), and light-scattering techniques. Although the fragmentation of starch molecules during extrusion processing is well established, the extent of molecular fragmentation and the involvement of amylose and amylopectin in fragmentation are not well characterized. While Davidson et al. (92) proposed that molecular degradation in starch is due to debranching of amylopectin molecules, Colonna et al. (107), using iodine binding and hydrolysis by β-amylase, determined that the percentage of $\alpha(1–6)$ linkages did not change as a result of extrusion. Therefore, Colonna et al. (107) concluded that depolymerization of starch molecules is by random chain scission and that amylose and amylopectin have the same susceptibility to degradation. Later, Sagar and Merrill (98) used GPC coupled with light scattering to study the molecular fragmentation of starch. Their results indicated that amylose also loses branches during extrusion processing. Several investigators report that depolymerization did not produce sugar monomers, but the end products were composed of shorter-length macromolecules (63, 67, 103). Viscometry and light scattering are more sensitive to the higher molecular fractions, so contributions from small-molecular-weight fractions and from sugar monomers, if any exist, will not be reflected in the average molecular weight. Because the molecular weight resolution by GPC increases at high elution volumes, this technique is not sensitive to very large or very small molecules. Sagar and Merrill (98) discussed extensively the limitations of both GPC and viscometric methods in determining the molecular weight distributions of complex polymer systems, such as starch.

Kaletunç and Breslauer (10, 40) reported that the reduction of T_g with increasing SME, as assessed by $\Delta T_g/\Delta SME$, differed significantly for various types of flours. They reported $\Delta T_g/\Delta SME$ values of 0.05°C/(kJ/kg) for high-amylopectin corn flour, 0.09°C/(kJ/kg) for high-amylose corn flour, and 0.06°C/(kJ/kg) for wheat flour. The amylose–amylopectin ratios of these flours were 0, 2.3, and 0.35 respectively. These investigators noted the increased sensitivity of T_g to SME with increasing amylose–amylopectin ratio. They proposed that this observation may indicate a greater stress-induced reduction in $\overline{M}w$ values due to greater fragmentation in high amylose flours than in high-amylopectin flours. They further hypothesized that the breakage of linear amylose chains would produce relatively large fragments, which may decrease drastically the $\overline{M}w$ of the starch. The breakage in amylopectin may occur at branch points, and the breakage of small-molecular-weight branches may not affect the overall molecular weight of amylopectin. This hypothesis is in agreement with the findings of Porter and Casale (88), who reported that the extent of stress-induced reaction depends on the imposed strain, chain topology, cross-link density, and chemical composition and also that the junctions of long branches of branched polymers are most susceptible to stress reactions.

As discussed by Kaletunç and Breslauer (10, 40), SME can be utilized to quantify the extent of applied mechanical stress during extrusion processing. SME is a processing parameter that combines the effects of equipment (screw speed), process (mass flow rate), and material (viscosity) properties. The viscosity of the flour–water mixture in the extruder is affected by the applied shear, temperature, and moisture content. Thus, SME reflects the collective influences of various operating conditions (e.g., temperature, screw speed, mass flow rate, moisture content, and additives) on the extent of fragmentation. In direct expanded cereals, mechanical energy conversion is important in creating the product structure, because the mean residence time is on the order of seconds. The influence of extrusion operating variables on the extent of fragmentation is indirect. Each operating variable affects SME (76), and their combined effect, in terms of SME, defines the extent of fragmentation.

C. Effect of Postextrusion Storage Conditions

To define the optimal storage conditions that maintain or improve the quality of a product, one first must characterize its properties as a function of conditions that simulate the environment to which the product will be exposed during storage. Development of a general correlation describing the moisture sorption isotherms of food materials at different temperatures is of practical value in drying and storing of these materials. Among the existing isotherm equations, reviewed by van den Berg and Bruin (108), it is generally concluded that at present the Guggenheim–Anderson–de Boer (GAB) isotherm model provides the best description of food isotherms up to 0.9 water activity (109). The GAB isotherm equation has the form

$$\text{EMC} = \frac{M_0 C_1 Ka_w}{(1 - Ka_w)(1 - Ka_w + C_1 Ka_w)} \tag{9}$$

where EMC is the equilibrium moisture content, a_w is the water activity, M_0 is the monolayer value, and C_1 and K are constants.

Figure 8 shows the experimental equilibrium moisture content of wheat flour extrudates as a function of water activity and the fitted curve described by the GAB isotherm equation at 35°C (110). The three parameters of the GAB model, M_0, C_1, and K, evaluated by fitting are given in the inset to the figure. Knowledge of the isotherm describing the moisture sorption characteristics of extrudates provides a rapid means of assessing the moisture content of an extrudate if the water activity is known, or vice versa.

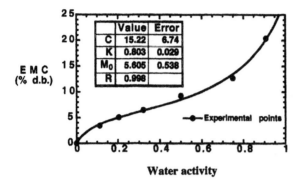

Figure 8 Fitting GAB isotherm equation to experimental moisture sorption data of wheat flour extrudate.

D. Use of T_g as a Tool to Design Processing and Storage Conditions

Extrudate moisture sorption isotherms can be used in conjunction with glass curves of extrudates to define storage conditions (temperature, relative humidity) that enhance the stability, and therefore the shelf life, of extruded products. Roos (111) proposed the combined use of the Gordon–Taylor equation and the GAB sorption isotherm as a function of water activity to predict water plasticization of various food components and foods exposed to various relative humidity environments. Roos indicated that such a diagram is useful in locating critical values of water activity and moisture content at which the glass transition temperature decreases below ambient temperature. He proposed to use the state diagram defining the glass and rubbery states for amorphous foods, together with moisture sorption isotherm, as a "map" for the selection of storage conditions for low- and intermediate-moisture foods. The utilization of glass curves in conjunction with moisture sorption isotherms as a tool to predict the critical water content and water activity values required to achieve stability was applied to wheat flour extrudates by Kaletunç and Breslauer (112). The relationship between the product's T_g and the temperature is critical to storage. If the storage temperature is greater than T_g at a given relative humidity, molecular mobility in the product increases and the time scale for deteriorative changes decreases, resulting in an unstable product in that storage environment. Knowledge of T_g as a function of moisture content can be used to design appropriate packaging and to define the temperature and relative humidity environments required to maintain a stable product.

V. RELATIONSHIP BETWEEN SENSORY ATTRIBUTES AND THE GLASS TRANSITION TEMPERATURE

One of the major criteria by which a consumer judges the quality of low-moisture cereal products, such as breakfast cereal and puffed snack products, is the sensory-textural attribute of crispness. Because the crispness of the extruded product is highly influenced by the moisture of the environment, the determination of crispness is essential to assess the quality loss and the shelf life of these products. The crispness of extruded products has been evaluated by sensory and instrumental methods (113, 114). Instrumental measurements of crispness are based on mechanical (115–117) and acoustic (118, 119) experiments. Several studies are available on the sensory evaluation of crispness in extruded cereals and crackers (116, 120–122). The crispness is a sensory attribute associated with brittle foods. Brittle foods are in the glassy state and are highly influenced by increases of temperature or moisture. Plasticization by temperature or moisture increase causes a loss of crispy texture. Roos et al. (123) report that although there is a critical water activity specific to each product at which crispness is lost, a change in crispness is generally reported over a water activity range between 0.35 and 0.50.

A theory based on the glass transition has been used to describe textural changes in brittle foods as a function of moisture content (122, 124). Peleg (125) proposed an empirical model to describe the sigmoid relationships between the sensory crispness and water activity for breakfast cereals. Levine and Slade (23) and Noel et al. (18) noted that T_g is relevant to food processing operations and

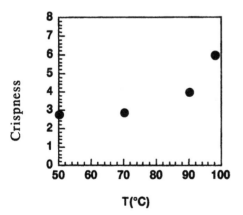

Figure 9 Relationship between the sensory attribute of crispness and T_g for corn flour extrudates. (From Ref. 10.)

correlates well with quality attributes. Kaletunç and Breslauer (10) reported the existence of a relationship between T_g and end-product attributes for corn flour extrudates. More specifically, the T_g values of corn flour extrudates exhibited trends related to sensory attributes (crispness and denseness) evaluated using quantitative descriptive analysis. Specifically, an increase in T_g (increase in thermal stability of the glassy state) is related to an increase in the sensory-textural attribute of crispness (Fig. 9). Kaletunç and Breslauer indicate that a threshold T_g value may be reached before crispness increases with T_g. By contrast, the sensory attribute denseness decreases as T_g increases (10).

The relationship between T_g and crispness may provide a criterion for adjusting extruder operating conditions to produce extrudates with optimal crispness. Thus, knowledge of T_g values corresponding to desirable levels of product crispness can be of value to a manufacturer.

VI. CONCLUSION

Calorimetry can be successfully applied to characterize the impact of processing and storage conditions on the stability and shelf life of extruded products. The observation that processing-induced fragmentation is reflected in both the apparent molecular weight distribution and the temperature of the glass transition emphasizes the utility of T_g as a means of assessing the extent of fragmentation in extruded cereal flours. Moisture sorption isotherms can be used in conjunction with the glass curves of extruded products to define storage conditions that maintain the stability and enhance the shelf life of the product. Determination of the thermal properties of products, specifically T_g, can be useful for quality control in the manufacture of extruded products. Taken together these applications of calorimetric data make differential scanning calorimetry a useful tool for the evaluation and rational design of formulation, processing conditions, storage conditions, and packaging for extruded products.

REFERENCES

1. GWH Höhne, W Hemminger, HJ Flammersheim. Differential Scanning Calorimetry: An Introduction for Practitioners. New York: Springer-Verlag, 1996, p 21.
2. L Slade, H Levine. Structural stability of intermediate moisture foods—a new understanding. In: JMV Blanshard, JR Mitchell, eds. Food Structure: Its Creation and Evaluation. London: Butterworths, 1988, pp 115–147.
3. LH Sperling. Introduction to Physical Polymer Science. New York: Wiley, 1992, pp 303–381.
4. M Reading. Modulated differential scanning calorimetry—a new way forward in materials characterization. Trends Polym Sci 1:248–253, 1993.

5. LN Bell, DE Touma. Glass transition temperatures determined using a temperature-cycling differential scanning calorimeter. J Food Sci 61:807–810, 828, 1996.
6. VL Hill, DQM Craig, LC Feely. Characterization of spray-dried lactose using modulated differential scanning calorimetry. Int J Pharm 161:95–107, 1998.
7. PG Royall, DQM Craig, C Doherty. Characterization of the glass transition of an amorphous drug using modulated DSC. Pharm Res 15:1117–1121, 1998.
8. L Di Gioia, B Cuq, S Guilbert. Thermal properties of corn gluten meal and its proteic components. Int J Biol Macromol 24:341–350, 1999.
9. G Kaletunç. Unpublished data, 1999.
10. G Kaletunç, KJ Breslauer. Glass transitions of extrudates: relationship with processing-induced fragmentation and end-product attributes. Cereal Chem 70:548–552, 1993.
11. A Boller, C Schick, B Wunderlich. Modulated differential scanning calorimetry in the glass transition region. Therm Acta 266:97–111, 1995.
12. B Wunderlich, A Boller, I Okazaki, S Kreitmer. Modulated differential scanning calorimetry in the glass transition region. Part II. The mathematical treatment of the kinetics of the glass transition. J Therm Anal 47:1013–1026, 1996.
13. T Hatakeyama, FX Quinn. Thermal analysis: fundamentals and applications to polymer science. Chichester, New York: Wiley, 1999, pp 5–25.
14. NJ Coleman, DQM Craig. Modulated temperature differential scanning calorimetry: a novel approach to pharmaceutical thermal analysis. Int J Pharm 135:13–29, 1996.
15. JMV Blanshard. The glass transition, its nature and significance in food processing. In: ST Beckett, ed. Physico-chemical Aspects of Food Processing. New York: Blackie Academic & Professional, 1995, pp 17–49.
16. L Slade, H Levine. Glass transitions and water–food structure interactions. In: JE Kinsella, SL Taylor, eds. Advances in Food Nutrition and Research, 1995, pp 103–269.
17. MT Kalichevsky, EM Jaroszkiewicz, S Ablett, JMV Blanshard, PJ Lillford. The glass transition of amylopectin measured by DSC, DMTA, and NMR. Carbohydr Polym 18:77–88, 1992.
18. TR Noel, SG Ring, MA Whittam. Glass transition in low moisture foods. Trends Food Sci Technol 1:62–67, 1990.
19. SR Kakivaya, CA Hoeve. The glass point of elastin. Proc Nat Acad Sci USA 72:3505–3507, 1975.
20. A Eisenberg. The glassy state and the glass transition. In: JE Mark, A Eisenberg, WW Gaessley, L Mandelkern, JL Koenig, eds. Physical Properties of Polymers. Washington, DC: Am Chem Soc, 1984, pp 55–95.
21. RC Hoseney, K Zeleznak, CS Lai. Wheat gluten: a glassy polymer. Cereal Chem 63:285–286, 1986.
22. KJ Zeleznak, RC Hoseney. The glass transition in starch. Cereal Chem 64:121–124, 1987.
23. H Levine, L Slade. Collapse phenomena a unifying concept for interpreting the behavior of low-moisture foods. In: JMV Blanshard, JR Mitchell, eds. Food Structure: Its Creation and Evaluation. London: Butterworths, 1988, pp 149–180.
24. TG Fox, PJ Flory. Second-order transition temperatures and related properties of polystyrene. I. Influence of molecular weight. J Appl Phys 21:581–591, 1950.

25. K Ueberreiter, G Kanig. Self-plasticization of polymers. J Coll Sci 7:569–583, 1952.
26. MP Buera, G Levi, M Karel. Glass transition in poly(vinylpyrrolidone): effect of molecular weight and diluents. Biotechnol Prog 8:144–148, 1992.
27. Y Roos, M Karel. Water and molecular weight effects on glass transitions in amorphous carbohydrates and carbohydrate solutions. J Food Sci 56:1676–1681, 1991.
28. PD Orford, R Parker, SG Ring, AC Smith. Effect of water as a diluent on the glass transition behaviour of malto-oligosaccharides, amylose, and amylopectin. Int J Biol Macrom 11:91–96, 1989.
29. MA Lillie, JM Gosline. The effects of swelling solvents on the glass transition in elastin and other proteins. In: JMV Blanshard, PJ Lillford, eds. Glass State in Foods. Nottingham, UK: Nottingham University Press, 1993, pp 281–301.
30. MT Kalichevsky, EM Jaroszkiewicz, JMV Blanshard. The glass transition of gluten. 1. Gluten and gluten–sugar mixtures. Int J Biol Macromol 14:257–266, 1992.
31. G Kaletunç. Unpublished data, 1998.
32. MT Kalichevsky, JMV Blanshard, PF Tokarczuk. Effect of water content and sugars on the glass transition of casein and sodium caseinate. Int J Food Sci Technol 28:139–151.
33. M Gordon, JS Taylor. Ideal copolymers and the second-order transitions of synthetic rubbers. I. Noncrystalline polymers. J Appl Chem 2:493–500, 1952.
34. B Wunderlich. Thermal Analysis. San Diego, CA: Academic Press, 1990, pp 101–104.
35. R Boyer. Transitions and relaxations in amorphous and semicrystalline organic polymers and copolymers. In: Encyclopedia of Polymer Science and Technology, Suppl. Vol II. New York: Wiley, 1977, pp 745–839.
36. R Charbonniere, P Duprat, A Guilbot. Changes in various starches by cooking-extrusion processing. II. Physical structure of extruded products. Cereal Sci Today 18:286, 1973.
37. AE McPherson, TB Bailey, J Jane. Extrusion of cross-linked hydroxypropylated corn starches. I. Pasting properties. Cereal Chem 77:320–325, 2000.
38. ML Politz, JD Timpa, AR White, BP Wasserman. Non-aqueous gel permeation chromatography of wheat starch in dimethylacetamide (DMAC) and LiCl: Extrusion-induced fragmentation. Carbohydr Polym 20:91–99, 1994.
39. C Mercier, R Charbonniere, J Grebaut, JF De La Gueriviere. Formation of amylose–lipid complexes by twin-screw extrusion cooking of manioc starch. Cereal Chem 57:4–9, 1980.
40. G Kaletunç, KJ Breslauer. Construction of a wheat-flour state diagram: application to extrusion processing. J Thermal Analysis 47:1267–1288, 1996.
41. PD Orford, R Parker, SG Ring. Aspects of the glass transition behavior of mixtures of carbohydrates of low molecular weight. Carboh Res 196:11–18, 1990.
42. CG Biliaderis. Thermal analysis of food carbohydrates. In: VR Harwalkar, C-Y Ma, eds. Thermal Analysis of Foods. London: Elsevier Applied Science, 1990, pp 168–220.
43. PR Couchman, FE Karasz. A classical termodynamic discussion of the effect of composition on glass-transition temperatures. Macromolecules 11:117–119, 1978.

44. G ten Brinke, FE Karasz, TS Ellis. Depression of glass transition temperatures of polymer networks by diluents. Macromolecules 16:244–249, 1983.
45. MT Kalichevsky, EM Jaroszkiewicz, JMV Blanshard. A study of the glass transition of amylopectin–sugar mixtures. Polymer 34:346–358, 1993.
46. G Kaletunç. FTIR microspectroscopy of wheat extrudate compositional microstructure. IFT Annual Meeting Chicago, abs 60-10, 1999.
47. J Antila, K Seiler, P Linko. Production of flat bread by extrusion cooking using different wheat/rye ratios, protein enrichment and grain with poor baking ability. J Food Eng 2:189–210, 1983.
48. CG Biliaderis, CM Page, TJ Maurice, BO Juliano. Thermal characterization of rice starches: a polymeric approach to phase transitions of granular starch. J Agric Food Chem 34:6–14, 1986.
49. JW Donovan, K Lorenz, K Kulp. Differential scanning calorimetry of heat-moisture treated wheat and potato starches. Cereal Chem 60:381–387, 1979.
50. L Slade, H Levine. Non-equilibrium melting of native granular starch. I. Temperature location of the glass transition associated with gelatinization of A-type cereal starches. Carbohydr Polym 8:183–208, 1988.
51. RF Tester, SJJ Debon. Annealing of starch—a review. Int J Biol Macromol 27: 1–12, 2000.
52. PJ Flory. Principles of Polymer Chemistry. Ithaca, NY: Cornell University Press, 1953, pp 563–571.
53. J Lelievre. Theory of gelatinization in a starch–water–solute system. Polymer 17: 854–858, 1976.
54. JW Donovan. Phase transitions of the starch–water system. Biopolymers 18:263–275, 1979.
55. CG Biliaderis, TJ Maurice, JR Vose. Starch gelatinization studied by differential scanning calorimetry. J Food Sci 45:1969–1974, 1980.
56. SS Wang, WC Chiang, AI Yeh, BL Zhao, IH Kim. Kinetics of phase transition of waxy corn starch at extrusion temperatures and moisture contents. J Food Sci 54: 1298–1301, 1326, 1989.
57. R Wulansari, JR Mitchell, JMV Blanshard. Starch conversion during extrusion as affected by gelatin. J Food Sci 64:1055–1058, 1999.
58. S Bhatnagar, MA Hanna. Starch–stearic acid complex development within single- and twin-screw extruders. J Food Sci 61:778–782, 1996.
59. MJ Harper. Food extruders and their applications. In: C Mercier, P Linko, JM Harper, eds. Extrusion Cooking. St Paul, MN: Am Assoc Cereal Chem, 1989, pp 1–15.
60. GH Ryu, CE Walker. Cell structure of wheat flour extrudates produced with various emulsifiers. Lebensm-Wiss Technol 27:432–441, 1994.
61. G Moore. Snack food extrusion. In: ND Frame, ed. The Technology of Extrusion Cooking. London: Blackie Academic and Professional, 1994, pp 110–143.
62. GI Galloway, CG Biliaderis, DW Stanley. Properties and structure of amylose–glyceryl monostearate complexes formed in solution or on extrusion of wheat flour. J Food Sci 54:950–957, 1989.
63. P Colonna, J Tayeb, C Mercier. Extrusion cooking of starch and starchy products.

In: C Mercier, P Linko, JM Harper, eds. Extrusion Cooking. St Paul, MN: Am Assoc Cereal Chem, 1989, pp 247–319.

64. GH Ryu, PE Neumann, CE Walker. Effect of emulsifiers on physical properties of wheat flour extrudates with and without sucrose and shortening. Lebensm Wiss Technol 27:425–431, 1994.

65. RE Rundle, JF Foster, RR Baldwin. On the nature of the starch–iodine complex. J Am Chem Soc 66:2116, 1944.

66. RE Rundle, FC Edwards. The configuration of starch in the starch-iodine complex. IV. An x-ray diffraction investigation of butanol precipitated amylose. J Am Chem Soc 65:2200, 1943.

67. C Mercier, R Charbonniere, D Gallant, A Guilbot. Structural modifications of various starches by extrusion cooking with a twin-screw French extruder. In: JMV Blanshard, JR Mitchell, eds. Polysaccharides in Food. London: Butterworths, 1979, pp 153–170.

68. P Colonna, C Mercier. Macromolecular modifications of manioc starch components by extrusion cooking with and without lipids. Carbohydr Polym 3:87–108, 1983.

69. TF Schweizer, S Reimann, J Solms, AC Eliasson, NG Asp. Influence of drum-drying and twin-screw extrusion cooking on wheat carbohydrates. II. Effects of lipids on physical properties, degradation and complex formation of starch in wheat flour. J Cereal Sci 4:249–260, 1986.

70. LB Guzman, TC Lee, CO Chichester. Lipid binding during extrusion cooking. In: JL Kokini, CT Ho, MV Karwe, eds. Food Extrusion Science and Technology. New York: Marcel Dekker, 1992, pp 693–709.

71. S Bhatnagar, MA Hanna. Extrusion processing conditions for amylose–lipid complexing. Cereal Chem 71:587–593, 1994.

72. F Meuser, W Pfaller, B Van Lengerich. Technological aspects regarding specific changes to the characteristic properties of extrudates by HTST extrusion cooking. In: C O'Connor, ed. Extrusion Technology for the Food Industry. London: Elsevier Applied Science, 1987, pp 35–53.

73. R Stute, G Koneiczny-Janda. DSC investigation of starches. Part II. Investigations on amylose–lipid complexes. Staerke 35:340–347, 1983.

74. I Bjorck, NG Asp, D Birkhed, I Lundquist. Effect of processing on starch availability in vitro and in vivo: extrusion cooking of wheat flour and starch. J Cereal Sci 2:91–103, 1984.

75. ND Frame. Operational characteristics of the co-rotating twin-screw extruder. In: ND Frame, ed. The Technology of Extrusion Cooking. London: Blackie Academic and Professional, 1994, pp 1–50.

76. A Barrett, G Kaletunç, S Rosenburg, KJ Breslauer. Effect of sucrose on the structure, mechanical strength and thermal properties of corn extrudates. Carbohydr Polymers 26:261–269, 1995.

77. PA Sopade, GA Le Grays. Effect of added sucrose on extrusion cooking of maize starch. Food Control 2:103–109, 1991.

78. D Moore, A Sanei, E Van Hecke, JM Bouvier. Effect of ingredients on physical/structural properties of extrudates. J Food Sci 55:1383–1387, 1990.

79. RCE Guy. Raw materials for extrusion cooking process. In: ND Frame, ed. The

Technology of Extrusion Cooking. London: Blackie Academic and Professional, 1994, pp 53–72.

80. GH Ryu, PE Neumann, CE Walker. Effects of some baking ingredients on physical and structural properties of wheat flour extrudates. Cereal Chem 70:291–297, 1993.

81. Z Jin, F Hsieh, HE Huff. Extrusion cooking of corn meal with soy fiber, salt and sugar. Cereal Chem 71:227–234, 1994.

82. AH Barrett, M Peleg. Extrudate cell-structure–texture relationships. J Food Sci 57: 1253–1257, 1992.

83. AH Barrett, AV Cardello, LL Lesher, IA Taub. Cellularity, mechanical failure, and textural perception of corn meal extrudates. J Texture Studies 25:77–95, 1994.

84. TE Luellen. Structure, characteristics, and uses of some typical carbohydrate food ingredients. Cereal Foods World 33:924–927, 1988.

85. KJ Valentas, L Levine, JP Clark. Food Processing Operations and Scale-Up. New York: Marcel Dekker, 1991.

86. MT Kalichevsky, JMV Blanshard. The effect of fructose and water on the glass transition of amylopectin. Carbohydr Polym 20:107–113, 1993.

87. L Ollett, R Parker, AC Smith. Deformation and fracture behavior of wheat starch plasticized with glucose and water. J Mater Sci 26:1351–1356, 1991.

88. RS Porter, A Casale. Recent studies of polymer reactions caused by stress. Polym Eng Sci 25:129–156, 1985.

89. AM Basedow, KH Ebert, H Hunger. Effect of mechanical stress on the reactivity of polymers: shear degradation of polyacrylamide and dextran. Makromol Chem 180:411–427, 1979.

90. MH Gomez, JM Aquilera. Changes in the starch fraction during extrusion-cooking of corn. J Food Sci 48:378–381, 1983.

91. MH Gomez, JM Aquilera. A physicochemical model for extrusion of corn starch. J Food Sci 49:40–43, 1984.

92. VJ Davidson, D Paton, LL Diosady, LJ Rubin. A model for mechanical degradation of wheat starch in a single-screw extruder. J Food Sci 49:1154–1157, 1984.

93. K Kim, MK Hamdy. Depolymerization of starch by high-pressure extrusion. J Food Sci 52:1387–1390, 1987.

94. B Vergnes, JP Villemaire. Rheological behavior of low-moisture molten maize starch. Rheol Acta 26:570–576, 1987.

95. R Chinnaswamy, MA Hanna. Macromolecular and functional properties of native and extrusion-cooked corn starch. Cereal Chem 67:490–499, 1990.

96. F Meuser, N Gimmler, B van Lengerich. A systems analytical approach to extrusion. In: JL Kokini, C Ho, MV Karwe, eds. Food Extrusion Science and Technology. New York: Marcel Dekker, 1992, pp 618–630.

97. G Della Valle, Y Boché, P Colonna, B Vergnes. The extrusion behavior of potato starch. Carbohyd Polym 28:255–264, 1995.

98. AD Sagar, EW Merrill. Starch fragmentation during extrusion processing. Polymer 36:1883–1886, 1995.

99. O Myllymaki, T Eerikainen, T Stuortti, P Forssell, P Linko, K Poutenen. Depolymerization of barley starch during extrusion in water glycerol mixtures. Lebensm Wiss Technol 30:351–358, 1997.

100. P Tomasik, W Stanislaw, P Mieczyslaw. The thermal decomposition of carbohy-

drates. Part II. The decomposition of starch. Adv Carbohydr Chem 47:279–343, 1989.

101. C Mercier. Comparative modifications of starch and starchy products by extrusion cooking and drum drying. In: C Mercier, C Cantarelli, eds. Pasta and Extrusion-Cooked Foods. London: Elsevier Applied Science, 1986, pp 120–130.

102. TF Schweizer, S Reiman. Influence of drum drying and twin-screw extrusion cooking on wheat carbohydrates. I. A comparison between wheat starch and flours of different extraction. J Cereal Sci 4:193–203, 1986.

103. P Rodis, LF Wen, BP Wasserman. Assessment of extrusion-induced starch fragmentation by gel-permeation chromatography and methylation analysis. Cereal Chem 70:152–157, 1993.

104. L Wen, P Rodis, BP Wasserman. Starch fragmentation and protein insolubilization during twin-screw extrusion of corn meal. Cereal Chem 67:268–275, 1990.

105. RCE Guy, AW Horne. Extrusion and co-extrusion of cereals. In: JMV Blanshard, JR Mitchell, eds. Food Structure: Its Creation and Evaluation. London: Butterworths, 1988, pp 331–349.

106. G Della Valle, A Kozlowski, P Colonna, J Tayeb. Starch transformation estimated by the energy balance on a twin-screw extruder. Lebensm-Wiss Technol 22:279–286, 1989.

107. P Colonna, JL Doublier, JP Melcion, F De Monredon, C Mercier. Extrusion cooking and drum drying of wheat starch. I. Physical and macromolecular modifications. Cereal Chem 61:538–543, 1984.

108. C van den Berg, S Bruin. Water activity and its estimation in food systems: theoretical aspects. In: LB Rockland, GF Stewart, eds. Water Activity: Influences on Food Quality. New York: Academic Press, 1981, pp 1–61.

109. LN Bell, TP Labuza. Moisture sorption: Practical aspects of isotherm measurement and use. St. Paul, MN: Am Assoc Cereal Chem, 2000, p 51.

110. G Kaletunç, KJ Breslauer. Influence of storage conditions (temperature, relative humidity) on the moisture sorption characteristics of wheat flour extrudates. IFT Annual Meeting, June 14–18, 1997, Orlando, FL, abs 18-3.

111. YH Roos. Water activity and physical state effects on amorphous food stability. J Food Process Preserv 16:433–447, 1993.

112. G Kaletunç, KJ Breslauer. Use of glass transition (T_g) as a tool to evaluate the effect of processing-induced fragmentation on the storage stability of wheat flour extrudates. IFT Annual Meeting, June 22–26, 1996, New Orleans, abs. 86-10.

113. BV Pamies, G Roudaut, C Dacremont, M Le Meste, JR Mitchell. Understanding the texture of low-moisture cereal products: mechanical and sensory measurements of crispness. J Sci Food Agric 80:1679–1685, 2000.

114. JFV Vincent. The quantification of crispness. J Sci Food Agric 78:162–168, 1998.

115. EE Katz, TP Labuza. Effect of water activity on the sensory crispness and mechanical deformation of snack food products. J Food Sci 46:403–409, 1981.

116. GE Attenburrow, AP Davies, RM Goodband, SJ Ingman. The fracture behavior of starch and gluten in the glassy state. J Cereal Sci 16:1–12, 1992.

117. A Barrett, MD Normand, M Peleg, E Ross. Characterization of the jagged stress-strain relationships of puffed extrudates using fast Fourier transform and fractal analysis. J Food Sci 57:227–232, 235, 1992.

118. ZM Vickers. Pleasantness of sounds. J Food Sci 48:783–786, 1983.
119. ZM Vickers. Sensory, acoustical, and force-deformation measurements of potato chip crispness. J Food Sci 52:138–140, 1987.
120. F Hsieh, L Hu, HE Huff, IC Peng. Effects of water activity on textural characteristics of puffed rice cake. Lebens Wiss Technol 23:471–473, 1990.
121. F Sauvageot, G Blond. Effect of water activity on crispness of breakfast cereals. J Textur Studies 22:423–442, 1991.
122. MK Karki, YH Roos, H Tuorila. Water plasticization of crispy snack foods. IFT Annual meeting, Atlanta, GA, June 25–29, 1994, Paper 76-10.
123. YH Roos, M Karel, JL Kokini. Glass transitions in low-moisture and frozen foods: effects on shelf life and quality. Food Technol 50:95–108, 1996.
124. KA Nelson, TP Labuza. Glass transition theory and the texture of cereal foods. In: JMV Blanshard, PJ Lillford, eds. Glass State in Foods. Nottingham, UK: Nottingham University Press, 1993, pp 513–517.
125. M Peleg. A mathematical model of crunchiness/crispness loss in breakfast cereals. J Texture Stud 25:403–410, 1994.

2

Application of Thermal Analysis to Cookie, Cracker, and Pretzel Manufacturing

James Ievolella,* Martha Wang, Louise Slade, and Harry Levine
Nabisco, Kraft Foods, East Hanover, New Jersey, U.S.A.

I. INTRODUCTION

Commercial low-moisture baked goods, such as cookies, crackers, and hard pretzels, derive their characteristic textural crispness from the glassy solid state of their crumb structures (1–4). In polymer science terms, products of this class can be described as rigid foams consisting of air cells surrounded by a glassy matrix with "fillers," or discontinuous inclusions. The matrix of a crisp cookie is a continuous sucrose-water glass with embedded ungelatinized starch granules, undeveloped gluten, and fat. Similarly, the matrix of a cracker or pretzel is a continuous glassy network of (partially) gelatinized starch, (partially) developed gluten, amorphous sucrose, and included fat. To achieve the kinetically metastable state of a glass, such products are deliberately formulated and processed so that the possibility for glass-to-rubber transitions within the matrix is minimized; i.e., the glass transition temperature (T_g) is brought well above ambient temperature. Thus, variations in the crumb matrix—and perceived crispness—of these products are critically dependent on formulation, processing, and storage. For example, commercial cookies formulated with bread flour in place of soft wheat flour (the standard cookie formula ingredient) are typically hard or tough rather than crisp, a sensory difference traceable to the effect on T_g of high levels of damaged starch and pentosans typically found in bread flours (2, 3, 5). In regard to processing, as will be discussed in detail in this chapter, cracker and pretzel matrices can vary considerably, depending on the extent of starch gelatinization occurring

---

* Retired

in the preparation and baking of doughs. Storage conditions play a profound role in baked-goods crispness, because, as many have documented, the matrix is highly sensitive to absorbed moisture (see, e.g., Ref. 6). Since crispness is perhaps the most important sensory attribute of these products (7), the technologist must pay special attention to textural properties, in efforts to create new varieties or to improve the quality of existing varieties.

Biopolymer systems such as these readily lend themselves to study and evaluation by thermal and thermomechanical techniques. Hence, differential scanning calorimetry (DSC) has proved to be a particularly valuable analytical tool for a broad range of baked products. Much past work has focused on high-moisture products such as bread (e.g., see Refs. 1, 5, 8–11 and references therein) and cakes (see, e.g., Ref. 12), whereas DSC studies of low-moisture baked-goods systems have appeared infrequently in the literature until recent years (13). The next section briefly summarizes reported applications of DSC analysis to cookie and cracker systems.

II. BACKGROUND

Abboud and Hoseney (14, 15) were among the first to apply DSC to cookie doughs—in effect, using the instrument as a "microbaking" oven to record the occurrence of thermal transitions. Typical sugar-snap cookie doughs (60% sucrose, flour weight basis [f.b.]) produced three endothermic peaks, identified as fat-melting (30°C), sugar-dissolving (70°C), and starch gelatinization (120°C). Although evidence was later presented showing that the third, high-temperature peak actually resulted from water vaporization (Ref. 2 and infra), it was noteworthy that no starch gelatinization endotherm was present below the maximum internal temperature reached by doughs during baking (about 100°C); this finding explained the virtual absence of gelatinized starch in such cookie products.

Kulp et al. (16) reported similar results in a DSC study of wire-cut cookies (45% sucrose, f.b.), namely, that starch in the cookie dough remained unchanged during the baking process. The DSC curves "showed no apparent difference" in peak areas for both gelatinization and amylose-lipid melting, between starch samples isolated from cookie flour and from the corresponding baked products. In confirmation of the DSC results, no significant differences were found between the two isolated starch samples when examined by scanning electron microscopy (SEM) and when tested for such physicochemical properties as viscosity (by Visco-Amylograph), swelling power and solubility, and iodine affinity.

Wada et al. (17) evaluated cracker-type products prepared from model doughs containing gluten/corn starch mixtures and no added sugar. The

DSC curves of the doughs differed according to the type of corn starch used: In doughs with high-amylose corn starch (HACS), no thermal events were visible up to about 120°C, other than a fat-melting peak at 36°C; with waxy corn starch (WCS), however, a broad, shallow gelatinization endotherm starting at about 88°C and peaking at about 114°C also appeared. This difference was reflected both in the degree of gelatinization (by enzymatic assay) of the baked products obtained from HACS and WCS doughs (i.e., 0.8% and 2.0%, respectively) and in the gelatinization temperatures of the two native starches (lower for WCS).

Doescher et al. (18) used DSC to evaluate the effects of various cookie dough ingredients on the T_g of commercial hard wheat gluten. Formula water alone (20%, gluten basis) produced a significant depression in the gluten T_g, from 82°C originally down to 33°C; sodium bicarbonate (around 1%) shifted the T_g nearly back up to its original temperature; glucose (60%) in water also shifted the T_g upward, but only to 68°C; the combination of water (20%), glucose (60%), soda (1%), salt (1%), nonfat dry milk (3%), and fat (30%) produced nearly the same shift in T_g (to 66°C), as did the water–glucose pair. Results of their study led these workers to propose a mechanism for the "setting" of cookie dough during baking: i.e., gluten goes through a glass (to rubber) transition and expands into a continuous web, causing an increase in viscosity sufficient to halt further flow of the dough piece (19). This hypothesis has met with some disagreement, based in part on data that indicate the actual occurrence of a reduction in cookie dough viscosity during baking (5, 13).

Piazza and Schiraldi (20) investigated the effect of dough resting time on the T_g of semisweet biscuit doughs (19% sucrose, f.b.). They determined by DSC analysis that the T_g of doughs allowed to rest for 120 minutes after mixing increased by about 2°C over the T_g of freshly mixed doughs. The increase in T_g conformed to rheological data showing comparable changes with resting time in elongational viscosity, tensile elastic modulus, and tensile stress at break, changes that implied that structural ripening and the growth of protein networks (gluten polymerization) had occurred in the doughs during resting. Perhaps unexpectedly, in three-point bending tests on the baked biscuit products, values for elastic modulus decreased with dough resting times—i.e., the biscuits made from rested doughs were more leathery and lower in breaking strength. Since the moisture content of the baked products also increased with dough resting times (from about 0.2% for 1-minute doughs to 2.3% for 120-minute doughs), plasticization of the biscuit matrix by residual water, which reduces the T_g (1), had evidently occurred to a greater extent in products obtained from the longer-resting doughs.

In a study of moisture effects on the textural attributes of model flour-water crackers, Given (21) used DSC to determine the unfreezable water content (UWC) of samples prepared with variations in dough moisture, oven temperature, and mixing time. The maximum UWC was essentially constant (ca. 26%) and well above the baked sample moistures, which ranged from 3 to 20%; hence, all of the water present in the test crackers was unfreezable. Moisture content, i.e., UWC, correlated with sensory texture scores for perception of moistness ($R^2 = 0.96$), fracturability (-0.90), and hardness (-0.73). It was suggested that unfreezable water, acting as a plasticizer, effected the incremental reductions in T_g of the cracker system, which were responsible for the macroscopic changes in texture. Aubuchon (22) demonstrated by modulated DSC (MDSC) that the T_g of a commercial cookie sample decreased from 60°C when fresh from the package to below room temperature (about 10°C) after overnight exposure to ambient conditions, which resulted in moisture uptake and plasticization. And MDSC was reported to be more effective than standard DSC in resolving the glass transition of this product.

Thermal analysis methods other than DSC have also been used for the product types discussed in this chapter. Nikolaidis and Labuza (23), for example, have studied the glass transitions of commercial crackers and their doughs by means of dynamic mechanical thermal analysis (DMTA). The T_g of crackers was reported to decrease with increasing moisture content, from ca. 155° to 12°C, a drop of 15°C per 1% moisture. T_g-vs.-moisture curves were very similar for the baked cracker and its dough, an indication that oven temperatures did not significantly modify ingredient functional behavior and that gluten or its glutenin fraction (rather than gelatinized starch) was primarily responsible for the mechanical behavior of the dough.

Miller et al. (19) determined the apparent T_g of model cookie dough systems by use of a thermomechanical analyzer (TMA), which also was reported to be more sensitive than DSC for lower-intensity thermal transitions, such as the glass transition. Doughs (25% moisture) made with hard wheat flour had a lower apparent T_g, compared to doughs made with soft wheat flour (78° vs. 71°C). The difference between doughs could not be attributed to inherent differences in flour properties, however, since T_g values for the two flours were identical (ca. 30°C, at 13% moisture). Decreasing the level of sucrose in the doughs from 60% (control) down to 50%, 40%, or 30% (f.b.) brought about corresponding decreases in T_g as well as in cookie "set time" and diameter.

In the following sections, we review an application of DSC as a diagnostic assay, or analytical "fingerprinting" method, that has been used to characterize the thermal properties of wheat starch in low-moisture, wheat-flour-based baked products, including cookies, crackers, and hard pretzels (13). This use of DSC has enabled us to relate starch thermal properties, on the one hand, to starch structure, and on the other hand, to starch functionality, in terms of baking performance and finished-product quality. Such DSC "fingerprinting" has been used as an aid to successful product development efforts (e.g., Ref. 25), by identifying matches between appropriate ingredient functionality/baking performance and superior finished-product quality.

III. EXPERIMENTAL METHODS

A Perkin-Elmer (PE) model DSC 7 differential scanning calorimeter, equipped with PE model TAC 7 Instrument Controller, PE model 7700 Professional Computer with PE TAS 7 Thermal Analysis Software, and Intracooler II (FTS Systems) subambient temperature controller, was used for all DSC measurements described here. Indium, benzophenone, and a series of National Bureau of Standards melting-point standards (52–164°C) were used to calibrate temperature and enthalpy of melting.

Dough samples for DSC analysis were filled into PE large-volume stainless steel DSC pans, which were then hermetically sealed and weighed (to 0.001 mg) on a PE AD-6 Autobalance. Sample weights were typically 35–45 mg.

Baked products for DSC analysis were prepared by grinding samples (e.g., cookies, crackers, or pretzels with low "as is" moisture contents) to a powder in a Krups coffee grinder; adding to the powdered product (or to samples of wheat flour of the type used to make the product) an equal weight of water (distilled, deionized) (10); and mixing powder and water together by hand with a spatula to the consistency of a homogeneously hydrated slurry. Slurry samples (35–45 mg) were immediately filled into the stainless steel DSC pans, which were then sealed and weighed as described earlier.

Duplicate sample pans were analyzed by DSC (against an empty stainless steel reference pan) within 1 hour of sample preparation, in order to ensure reproducibility of experimental results (9). After loading and temperature equilibration of pans in the DSC 7, samples were heated from 15° to 180°C (doughs) or from 15° to 130°C (baked products) at 10°C/min. In some experiments, after this initial scan, samples were immediately cooled (at 320°C/min, nominal instrumental rate) to 15°C and rescanned to 130°C, at 10°C/min.

Materials analyzed by DSC included the following samples. Flours, of the types used to make the cookie, cracker, and pretzel products, were typical, commercial, soft-wheat-based, cookie/cracker flours with "as is" moisture contents around 13%. The rotary-molded, high-sugar cookie was a typical commercial product [with a formula of a type similar to that of the standard AACC "sugar-snap" cookie (24)], with a moisture content less than about 3%. The fat-free, fermented ("sponge-and-dough" type) saltine cracker was a patented commercial product (25). The pretzel and full-fat, fermented ("sponge-and-dough" type) saltine cracker were typical commercial products with moisture contents below about 5%. Doughs used in the preparation of these baked products were obtained directly from production-scale runs and frozen until ready for analysis. For such doughs and baked goods, general aspects of formulation and processing are familiar to those skilled in the art. Proprietary aspects of the products and flour samples analyzed are not germane to the subject of this chapter.

Actual "as is" moisture contents of all baked products and flours analyzed by DSC were determined by a standard method of vacuum oven drying at 70°C for 18–48 hr. Throughout this chapter, compositions expressed in percent or by ratio represent weight percentages or ratios, unless otherwise indicated.

IV. DIFFERENTIAL SCANNING CALORIMETRY AS A DIAGNOSTIC TOOL FOR COOKIE DOUGHS

In order to develop a diagnostic DSC assay and facilitate the interpretation of complex DSC profiles that exhibit multiple thermal events, model systems of increasing complexity and variable water content were explored, as illustrated by the DSC heating profiles shown in Figure 1A (2). For a model batch-mix cookie dough containing 100:54:27:24 parts by weight flour:sucrose:fat:water, the schematic DSC profile (top curve) in Figure 1A shows three broad endothermic transitions (labeled endo-1, endo-2, and endo-3, with peak temperatures centered around 55°, 90°, and 135°C, respectively), followed by an exothermic transition with a peak around 160°C (labeled exo). An actual DSC profile (bottom curve) for a model continuous-mix dough of the same composition is also shown in Figure 1A.

To identify the thermal events underlying the four transitions manifested in the DSC profiles in Figure 1A, various two- or three-component model mixtures were examined. For example, DSC analysis of the melting/dissolution behavior of sucrose is illustrated in Figure 1B. The top curve shows the DSC profile for the melting of pure sucrose (USP grade), with a crystalline T_m peak at 192°C. In contrast, the bottom curve shows the DSC profile for a sample of 71.4 wt% sucrose in water (i.e., above the saturation concentration at room temperature), which revealed a very broad, shallow endotherm centered around 90°C, followed by an exothermic peak at about 170°C. From a knowledge of the locations of the solidus and vaporous curves in the state diagram for sucrose-water (2), we may conclude that the endotherm corresponds to a convolution of (a) the melting/dissolution of that small portion of the total sucrose that was undissolved at room temperature, and (b) the subsequent vaporization of water at $T > 100°C$, while the exotherm corresponds to sucrose recrystallization from the resulting supersaturated solution. Thus, these three sequential thermal events, sucrose melting/dissolution, water vaporization, and sucrose recrystallization, appear to correspond to the series of peaks labeled endo-2, endo-3, and exo in Figure 1A. This conclusion is also consistent with the description of the baking process for high-

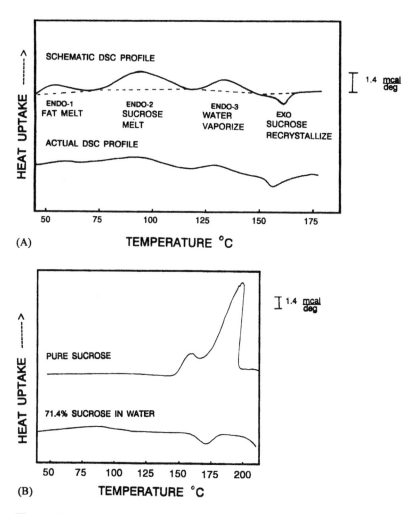

Figure 1 Typical DSC curves for representative samples of: (A) a model high-sugar cookie dough—top curve is a schematic representation for a batch-mix dough, bottom curve is an actual profile for a continuous-mix dough; (B) top curve—the melting of pure sucrose, bottom curve—the behavior of a sample of 71.4 wt% sucrose in water. (From Ref. 2.)

sugar cookie doughs positioned on the sugar-water state diagram (see Ref. 2). The peak labeled endo-1 in Figure 1A was routinely confirmed to be due to fat melting.

It should be noted, as mentioned earlier, that Abboud and Hoseney (15) had previously published a DSC thermogram, for a high-sugar cookie dough, similar to the schematic profile in Figure 1A. However, while they had also attributed endo-1 to fat melting and endo-2 to sucrose dissolving, they had identified the third endotherm (the one at $T > 100°C$) as resulting from starch gelatinization rather than from water vaporization. We suggested (2) that their identification of endo-3 cannot be correct, for the following reason. Under the high-sugar/low-moisture conditions that exist in such cookie doughs, the "antiplasticizing" effect of the sucrose-water cosolvent on wheat starch gelatinization would be quite pronounced, as described in detail elsewhere (8). Thus, starch gelatinization would be severely retarded, and the glass transition and subsequent crystalline melting transition that represent the sequential thermal events of gelatinization would not occur at temperatures below 175°C. In contrast,

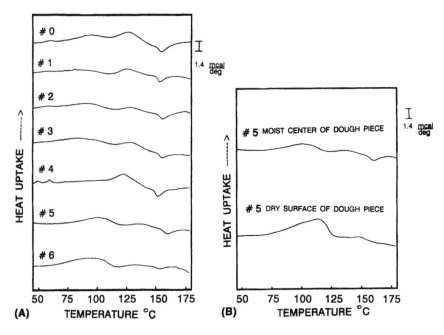

Figure 2   Typical DSC curves for representative samples of: (A) a series of continuous-mix cookie doughs produced with varying water contents and mixing parameters; (B) for one of the doughs in A, samples from the moist center and the dry surface of a dough piece. (From Ref. 2.)

moisture loss and consequent increased concentration of sucrose sufficient to return to a supersaturated condition must have preceded the recrystallization of sucrose between 150°C and 175°C, evidenced by the DSC profiles in Figure 1A.

As illustrated in Figure 2 (2), the diagnostic DSC assay was used to demonstrate that variations in water level and mixing parameters cause wide variations in the functional behavior of fat, sucrose, and water in continuous-mix doughs (Fig. 2A). These variations were exaggerated by surface drying of the doughs (Fig. 2B). As illustrated in Figure 3 (2), by varying mixing parameters and water level and by predissolving some of the sucrose before addition to the continuous mixer, it was possible to produce a dough that matched the batch-mix control. Again, various trial doughs showed significant variations in the functional behavior of the fat, sucrose, and water (Fig. 3A). However, the DSC assay showed that only the successful continuous-mix dough (#4) exhibited the same functional profile as the batch-mixed control (Fig. 3B).

The diagnostic DSC assay has proven useful because it provides a profile

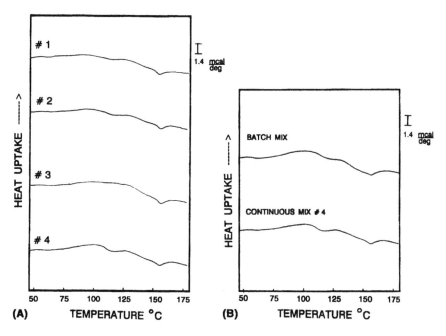

Figure 3 Typical DSC curves for representative samples of: (A) a series of continuous-mix cookie doughs produced with varying water contents, mixing parameters, and mode of addition of sucrose and water; (B) a control batch-mix dough and its matching continuous-mix dough from A. (From Ref. 2.)

of the state of crystallinity of the fat and sucrose, the temperatures of melting of the fat and of melting/dissolution of the sucrose, and the resulting effect on the vaporizability of the water. The amount of sucrose dissolved during mixing affects the (a) rate and extent of hydration of flour components during mixing and machining, (b) subsequent temperature and extent of further melting/dissolution of sucrose during baking, (c) temperature and extent of moisture evaporation during baking, and (d) finished-product texture and shelf life associated with moisture management and sugar recrystallization. The DSC assay was shown to be capable of assessing the effects of order of addition of sugar and water on the functional differences between continuous- and batch-mix doughs.

V. DIFFERENTIAL SCANNING CALORIMETRY OF WHEAT FLOUR, COOKIES, AND CRACKERS

Figure 4 (13) shows typical DSC thermograms for representative samples (1:1 mixtures with water) of a cookie/cracker flour, a commercial, rotary-molded, high-sugar cookie, and a commercial saltine cracker. Wheat flours of the type used to make such baked goods typically contain about 13% moisture (wet basis) and approximately 85% starch (dry basis), the latter in the native form of partially crystalline, partially amorphous granules containing the two starch polymers amylopectin (Ap) and amylose (Am) (14). Thus, the flour sample represented in Figure 4A comprises a 40% starch-in-water slurry, and the appearance of the DSC thermogram is that widely recognized to be typical of native granular wheat starch in mixtures of approximately 1:1 with water (8–10, 14 and references therein). The characteristic biphasic endotherm, with onset temperature (T) of 53°C, completion T of 88°C, peak T of 64°C, and shoulder at about 75°C, is generally acknowledged to represent a combination of glass transition, of water-plasticized amorphous regions, followed by nonequilibrium melting, of microcrystallites of the partially crystalline glassy Ap in the "fringed-micelle" structure of the granules of native wheat starch, in a process known as starch gelatinization (1, 5, 8–11, 26–31 and references therein). The appearance of this gelatinization endotherm in the thermogram in Figure 4A signifies, as expected, that the starch in the flour was native, until it was gelatinized as a consequence of heating to 90°C during the DSC measurement. The characteristic higher-T endotherm, with onset T of 89°C, completion T of 117°C, and peak T of 104°C, is recognized to represent the melting of amylose-lipid (Am L) crystalline inclusion complex [1, 5, 8–11, 32 and references therein].

In Figure 4B, the DSC thermogram for the high-sugar cookie sample shows two principal endothermic events, aside from a pair of small fat-melting endotherms below 40°C (representing the fat ingredient in the cookie formula, which

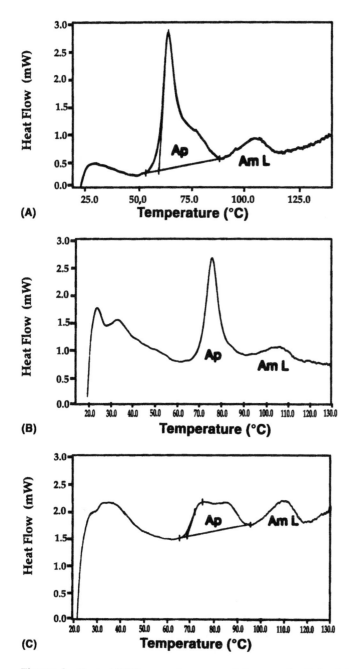

Figure 4 Typical DSC curves for representative samples (1:1 mixtures with water) of: (A) a cookie/cracker flour; (B) a commercial rotary-molded high-sugar cookie; and (C) a commercial saltine cracker. See text for explanation of peak labels. (From Ref. 13.)

is not of interest in this discussion). The smaller of the two endotherms of interest, occurring above 100°C, had previously been observed in the thermogram of wheat starch isolated from a wire-cut cookie, as reported by Kulp et al. (see Fig. 6 in Ref. 16). As described for Figure 4A, this endotherm represents the melting of crystalline Am L complex. The larger endotherm, with onset T of 66°C, completion T of 88°C, and peak T of 77°C, resembles the gelatinization endotherm in Figure 4A (same completion T), although it is obviously narrower and shifted to higher peak T. In fact, the appearance of this endotherm is characteristic of that for the gelatinization of native granular wheat starch in the presence of sucrose-water solution rather than water alone (8). Based on the known amounts of flour and sucrose in the cookie formula, and thus the corresponding known amounts of flour, sucrose, and water contained in the DSC sample pan, the appearance of this endotherm can be explained, to begin with, by recognizing that the sample pan contained, in essence, a 1:4 mixture of starch and 20% sucrose solution.

It is well known that the presence of sucrose causes the gelatinization temperature of starch [taken as the peak T of the gelatinization endotherm in Figure 4A (8–11 and references therein)] to be elevated (29–31). This effect of sucrose has been explained by a concept of "antiplasticization" (by sugar-water relative to water alone) (8), which has received wide support in recent years (28, 33 and references therein). According to this "antiplasticization" mechanism (8), sugar, in the presence of native starch and excess water, behaves as a plasticizing cosolvent with water, such that the sugar-water coplasticizer, of higher average molecular weight (MW) [and lower free volume, so higher glass transition temperature (T_g) (34)] than water alone, plasticizes [i.e., depresses the temperature of the glass transition of the amorphous regions, which immediately precedes the gelatinization of native, partially crystalline starch (8, 9, 26)] starch less than does water alone. Thus, the gelatinization temperature (as well as the T_g that precedes it) in the presence of sugar is elevated (hence, "anti") relative to the gelatinization temperature of starch in water alone. Moreover, with increasing concentration of sugar in the three-component mixture [thus, a sugar-water coplasticizer with increasing average molecular weight (MW), decreasing free volume, and increasing T_g, relative to water alone], the magnitude of the antiplasticizing effect increases, and so do T_g and the gelatinization temperature (8).

Thus, based on previously reported DSC results for native granular wheat starch in mixtures with sucrose solutions of varying concentration (8), the major endotherm in Figure 4B for the cookie sample is interpreted as representing the gelatinization of starch in 20% sucrose solution. This interpretation was confirmed by DSC analysis of a sample of flour prepared so as to represent a 1:4 mixture of starch in 20% sucrose (thermogram not shown). Once normalized for sample weight, that thermogram for flour (containing fully native starch) in 20%

sucrose was found to be essentially identical in appearance to the one in Figure 4B (except for the fat-melting peaks below 40°C) for the 1:1 mixture of high-sugar cookie in water, in terms of both areas and temperature ranges for both of the endothermic peaks (labeled Ap and Am L) in Figure 4B. It should be noted that we prefer to express peak area in terms of delta Q (for the change in total heat uptake), rather than the conventional delta H (change in enthalpy) terminology used by the PE DSC 7 software (and unavoidably listed as such in the printouts from the instrument), when a peak is known to comprise multiple thermal events, such as the glass and crystalline melting transitions represented within the gelatinization endotherm of starch (8, 9, 26). Further comparison of the peak areas for the Ap and Am L endotherms [first normalized for total sample weight, and then further normalized for flour (and therefore starch) weight] in the thermogram for flour: 20% sucrose [or in the equivalent (after normalization) thermogram in Figure 4B] with the corresponding peak areas in the thermogram for flour: water in Figure 4A demonstrated that both the Ap and Am L peak areas in Figure 4B represent 100% of those for the native wheat starch represented in Figure 4A. This result indicates that the starch in the baked cookie was completely native prior to DSC analysis and was first gelatinized during DSC heating, thus demonstrating that the starch was not gelatinized at all during baking of this high-sugar cookie. This finding is in agreement with the conclusion reached previously in other studies (2, 3, 16, 29–31). To anticipate a question about why, if the native starch represented in Figure 4B could be completely gelatinized by heating to 88°C in the DSC, was it not gelatinized during baking of the cookie, wherein the internal temperature reached approximately 100°C, we point out that the gelatinization temperature of 77°C in Figure 4B was measured for starch in a 1: 4 mixture with 20% sucrose solution. Thus, for the 20% starch slurry, plasticization by the sucrose solution (present in fourfold excess) depressed the gelatinization temperature down to 77°C. In contrast, during baking of the cookie dough (with a dry-basis composition of 87:54:35 flour: sucrose: water), the native starch in the flour was in an aqueous environment composed of a small excess of total solvent [= sum of sucrose + water (4)] comprising a 61% sucrose solution. Under these conditions (i.e., less effective plasticization of the starch, by a lesser amount of a more concentrated sucrose solution), the gelatinization temperature would be expected, based on previously reported DSC results for starch: sucrose: water model systems, to be elevated to well above 100°C (8). Therefore, the native starch of the flour in the cookie dough would be unaffected (i.e., not gelatinized at all) by the maximum temperature reached by the cookie during baking.

In Figure 4C, the DSC thermogram for the saltine cracker sample shows two principal endothermic events, aside from a broad fat-melting endotherm, centered about 35°C (representing the fat ingredient in the cracker formula, which again is not of interest in this discussion). The smaller of the two endotherms of interest, with onset T of 96°C, completion T of 118°C, and peak T of 109°C, is again identified (as in Figs. 4A and 4B) as that representing the melting of crystalline Am L complex. The larger endotherm, with onset T of 65°C, completion T of 95°C, and peak T of 75°C, again resembles the gelatinization endotherm in Figure 4A, although it is obviously smaller in peak area and shifted to a higher temperature range. We can begin to explain these differences by first noting that a saltine cracker is typically formulated with no added sugars (the presence of which would elevate the starch gelatinization temperature) and with enough water [i.e., at least about 27 parts water to 73 parts dry wheat starch (8)] in the dough (at least, early in baking) to allow starch in the flour to gelatinize during baking of the cracker, wherein the internal temperature reaches at least about 100°C. If we assume that the peak labeled Ap in Figure 4C represents what remained of the full gelatinization endotherm in Figure 4A, after some but not all of the starch in the cracker was gelatinized during baking, we can calculate the percentage remaining native Ap structure, by comparing the Ap peak areas in Figures 4C (1.49 J/g) and 4A (3.92 J/g) and then normalizing first for total sample weight and second for flour (and thus starch) weight. We obtain a value of 40%, indicating that the extent of starch gelatinization during baking of the cracker was 60%. Apparently, it could not reach 100%, because, as the content of plasticizing water in the dough decreased as baking progressed, the gelatinization temperature would have increased, evidently to well above 100°C by the end of baking. Since not all of the starch was gelatinized during baking, the portion remaining native was evidently subject to annealing (at temperatures within the range from T_g to T_m at the end of Ap crystallite melting), due to the heat-moisture treatment constituted by baking (1, 8, 9, 26). The expected effect of this annealing treatment (35) is manifested by the Ap peak in Figure 4C, which is narrower by about 5°C and up-shifted by about 10°C, relative to the corresponding Ap peak in Figure 4A. As with the Ap peak areas, we can compare the Am L peak areas in Figures 4A (0.63 J/g) and 4C (0.71 J/g), and, after the same normalizations as before, we obtain a value of 121% native Am L structure in the cracker sample represented in Figure 4C. Thus, evidently as a consequence of gelatinization of some of the granular starch during cracker baking, some previously uncomplexed amylose was made available for forming additional Am L complex (32) in the cracker. The Am L peak in Figure 4C is narrower by about 6°C and up-shifted by about 5°C, relative to the corresponding Am L peak in Figure 4A, apparently as a consequence of the same annealing treatment during baking, which similarly influenced the Ap peak. As a final remark about the thermogram in Figure 4C, we point out what we view as the salient DSC features of this cracker sample (which is taken to represent an excellent eating-quality, commercial product with optimum properties): 40% remaining native Ap structure and 121% native Am L structure. As will be developed further in the discussion of Figure 5 that follows, these features illustrate the basis of our application of DSC analysis as a diagnostic "fingerprinting" method that has allowed us to relate starch structure and thermal properties to starch function in, and associated finished-product quality of, low-moisture baked goods (13).

VI. DIFFERENTIAL SCANNING CALORIMETRY OF CRACKERS

Figure 5 (13) shows typical DSC thermograms for representative samples (1:1 mixtures with water) of a cracker flour (of the same type as described earlier with regard to Fig. 4A), a commercial full-fat saltine cracker of excellent eating quality (of the same type as described earlier with regard to Fig. 4C), a prototype

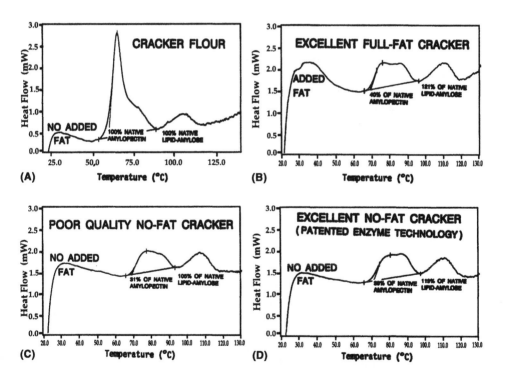

Figure 5   Typical DSC curves for representative samples (1:1 mixtures with water) of: (A) a cracker flour; (B) a commercial full-fat saltine cracker of excellent eating quality; (C) a prototype no-fat saltine cracker of poor eating quality; and (D) a patented (25) commercial no-fat saltine cracker of excellent eating quality. (From Ref. 13.)

no-fat saltine cracker of poor eating quality, and a patented (25), commercial no-fat saltine cracker of excellent eating quality. The critical aspect of Figure 5 lies in the comparison among the "fingerprint" thermograms in parts B, C, and D. When one tries to produce a no-fat version of the saltine cracker represented in Figure 5B by simply omitting from the formula the added fat (the presence of which would normally allow the cracker dough to be soft enough to be ma-chined on commercial equipment), one must ordinarily add extra water, to obtain a dough of softness and machinability equal to that of its full-fat analog (25). The result of adding extra water to the dough is revealed in the thermogram in Figure 5C. Evidently, more of the starch in the flour is gelatinized during baking of the dough, because sufficient plasticizing water was present for a longer por-tion of the baking time, thus keeping the gelatinization temperature depressed, for a longer time, below the maximum temperature reached during baking. As a result, only 31% native Ap structure remained (to be detected by its DSC fingerprint) in the no-fat cracker after baking, rather than 40%, as in the full-fat cracker. Furthermore, less additional Am L complex was able to form in the no-fat dough during baking (resulting in 105% native Am L structure in the cracker) than that formed in the full-fat dough (resulting in 121% in the cracker), possibly because of time/temperature/moisture conditions during baking that were less favorable to the sequestering of solubilized Am in Am L crystalline complex (1, 5, 8). The consequence of these differences in the resulting starch structure—more Ap gelatinized and less Am sequestered in discrete crystallites of Am L complex, rather than free, within the gelatinized starch network, to contribute toughness to the matrix—was a deleterious effect on product texture, as assessed by sensory analysis. The no-fat cracker represented in Figure 5C was less tender, more brittle, and tougher than the target full-fat cracker represented in Figure 5B (25).

In contrast, when Nabisco's patented pentosanase-enzyme technology (36, 37) was applied in the commercial production of the no-fat saltine cracker (25), the DSC fingerprint of the resulting cracker (Fig. 5D: 39% native Ap structure, 119% native Am L structure) was found to be a virtual match for the fingerprint of the corresponding full-fat cracker (Fig. 5B: 40% native Ap structure, 121% native Am L structure). In the case of the cracker represented in Figure 5D, even though the fat normally included in the formula was omitted, so that extra water was needed to produce a machinable dough, this extra water in the dough did not result in a significantly increased extent of Ap gelatinization or in a reduced extent of Am L complex formation, during baking, because of the beneficial effect of the pentosanase enzyme in the dough (25, 36, 37). By hydrolyzing the highly water-holding pentosans (nonstarch polysaccharides) in the flour during dough mixing, and thereby markedly reducing the water-holding capacity of the resulting dough, this enzyme caused a facilitation of moisture loss from the dough during baking. Thus, more of the plasticizing water was removed, more rapidly, from the dough during baking, so that it was not available to make possible excessive Ap gelatinization or reduced Am L complex formation and the deleterious effect on product texture, which would otherwise have resulted. In the absence of excessive starch gelatinization, the resulting no-fat cracker had a tender, nonbrittle, nontough texture, comparable to that of the full-fat target product (25).

As a final remark about Figure 5, in the context of DSC as a "fingerprinting" method, it is worth noting that, even if one did not know the identity or cause of the two principal endotherms in Figure 5B, one would logically assume (and, at least in this case, turn out to be correct) that the DSC sample represented in Figure 5D was a much closer match (in terms of its structure–thermal property relationships and thus, presumably, in its functional characteristics) to the target represented in Figure 5B than was the DSC sample represented in Figure 5C.

VII. DIFFERENTIAL SCANNING CALORIMETRY OF PRETZELS

Figures 6–8 (13) illustrate the results of diagnostic DSC analysis of commercial pretzel doughs and products. The pretzel samples represented in Figures 6A and 6B were produced from a cookie/cracker flour of the type represented in Figure 4A and were formulated with added fat (i.e., not fat-free). Thus, the three endothermic peaks evident in the thermograms in both Figure 6A and Figure 6B can be assumed to correspond to the three similar endotherms in Figure 5B for the full-fat cracker sample. The lowest-T, broad endotherm, centered around 35°C in Figures 6A and 6B, is again assigned to fat-melting. This peak is seen to reappear, with a slightly lower peak T and narrower width, in the immediate rescan in Figure 6C. Such thermal behavior is well known to be characteristic of the kinds of polymorphic crystalline fats typically used in such baked goods. Since the same type and amount of fat were used in both pretzel formulas (in fact, the formulas were identical in all aspects), we must assume that fat played no direct role in distinguishing the good product (Fig. 6B) from the bad one (Fig. 6A). Therefore, we again turn our attention away from fat and focus it on the two peaks—the one below 100°C, assigned earlier to Ap, and the one above 100°C, assigned earlier to Am L—arising from starch in the wheat flour. We note in passing that the Am L peak reappears in the immediate rescan in Figure 6C, while the Ap peak does not. Such thermal behavior is quite familiar and well established for wheat starch:water mixtures and has been explained in detail elsewhere (1, 3–5, 8–11 and refs. therein). With the Ap peak completely absent in the rescan in Figure 6C, the appearance of the curved baseline in the 70–100°C range is revealed, thus demonstrating that the Ap peaks in Figures 6A and 6B, while small in height and area, are unquestionably real. It is also worth

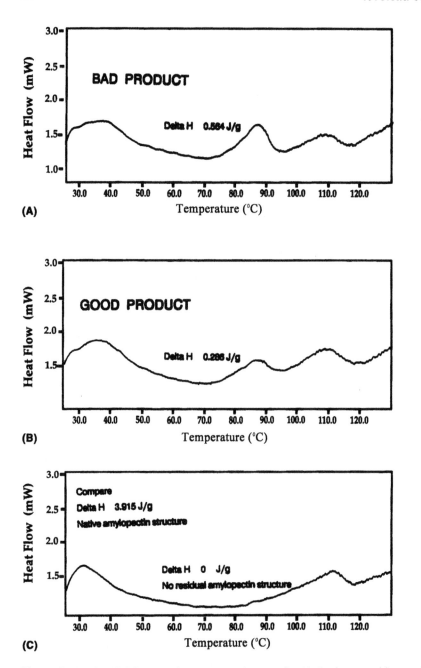

Figure 6 Typical DSC curves for representative samples (1:1 mixtures with water) of: (A) a prototype pretzel product of poor eating quality; (B) a commercial pretzel of good eating quality; and (C) an immediate rescan of the sample in (B). (From Ref. 13.)

mentioning that these Ap peaks are much smaller in area and somewhat narrower than the corresponding Ap peaks for the cracker samples represented in Figures 5B–D, thus indicating greater extent of starch gelatinization and annealing during baking of the pretzels. It is tempting to suggest that such differences in starch structure and thermal properties must correlate with functional differences and must therefore be related to the obvious textural differences between pretzels and saltine crackers, both of which are produced from doughs formulated with flour and water (but little added fat and virtually no added sugars) as the predominant elements. However, as discussed further later, the Ap peaks may not tell the whole story.

The critical functional distinction between the pretzel samples represented in Figures 6A and 6B concerned eating quality. The product whose thermogram is shown in Figure 6A had poor eating quality, as assessed by sensory panel evaluation; its texture was described as too hard, and it had a dry, mealy, pasty mouthfeel. In contrast, the product represented in Figure 6B had good texture (crisp but not too hard) and eating quality. Once again, if we examine the DSC thermograms in Figures 6A and 6B as fingerprints, whose differences might correlate with, and account for, the differences in texture and eating quality of the two products, we see immediately that the only obvious difference is that the Ap peak areas (normalized for sample weight) differ by a factor of about 2. Further rough comparison of these Ap peak areas (0.56 J/g in Fig. 6A, 0.29 J/g in Fig. 6B) with that for the flour in Figure 4A (3.92 J/g $\equiv$ 100% native Ap structure) showed that the poor product had 14.4% remaining native Ap structure, while the good one had only 7.3% remaining native Ap structure after baking. The logical implication of these results is that the extent of starch gelatinization during production of the bad pretzel was insufficient (i.e., only 86%, rather than 93%) for optimum product texture and eating quality.

However, more careful examination of the thermograms in Figures 6A and 6B revealed that less obvious differences involving the Am L peaks may also have been instrumental in differentiating the good and bad products. It can be seen, even without consulting the delta Q's, that the Am L peaks are quite similar in area, but the one in Figure 6B looks slightly larger. (This was verified by Figure 7C, as discussed later.) Probably more important, and certainly more evident, is the fact that, in Figure 6A, the Ap peak is much larger, in both height and area, than the Am L peak, while in Figure 6B, in contrast, this is not the case. The Ap and Am L peaks are much more similar in size; in fact, the Am L peak is slightly larger in area. If we recall the earlier discussion about the crackers represented in Figure 5, wherein a benefit to texture accrued from enhanced Am L complex formation during baking, such that the resulting Am L peak was both larger in area and closer in size to that of the optimum Ap peak, we may infer that the good pretzel represented in Figure 6B, like the good crackers represented in Figures 5B and 5D, enjoyed a similar textural benefit arising from increased sequestering of available Am in Am L crystalline complex. We will return to this point again later. For now, it is important to also recall that the good and bad pretzels were formulated identically. Thus, differences in their finished-product quality were assumed to have arisen because of differences in their manufacturing process.

Typical pretzel production involves running a pretzel dough through a bath of hot caustic solution [lye (NaOH)] prior to baking. The caustic-bath treatment is responsible for the glossy brown surface appearance of a typical pretzel (hard, low-moisture type), by a process involving gelatinization of starch on the surface of the pretzel dough, via sufficient contact with the hot caustic solution (14). Thus, in a typical pretzel dough, starch in the wheat flour can be gelatinized during caustic-bath treatment and also, of course, during baking. The thermograms in Figure 7 (measured for DSC samples of virtually the same weight) reveal the progress, in two discrete but evidently additive processing steps, of increasing extents of starch gelatinization and annealing during manufacture of a different sample of the same type of commercial pretzel product (with good texture and eating quality) as described before with regard to Figure 6B. In Figure 7A, the thermogram for the pretzel dough, prior to caustic-bath treatment, can be seen to resemble quite closely the one for flour in Figure 4A. The reason for this was alluded to earlier: It is simply that a pretzel dough is, in essence (i.e., aside from a bit of fat and a few other minor ingredients that need not concern us here), a flour-water dough. Thus, the major biphasic endotherm with a peak T of 69°C can be assigned without question to Ap (as verified, in part, by the expected appearance of the immediate rescan), and its peak area, corresponding to a delta Q of 4.22 J/g, can be taken to represent 100% native Ap structure. Applying the "fingerprint" analysis to the DSC thermograms in Figure 7, we see in Figure 7B that, as a consequence of the caustic-bath treatment, the Ap peak area (2.26 J/g) was reduced to 53.6% of its original value for native Ap structure for the in-process sample of dough taken prior to baking. For the finished product, after both caustic-bath treatment and baking, the area of the Ap peak (0.33 J/g) in Figure 7C demonstrated that the remaining native Ap structure had reached a final value of 7.8%, in reasonable agreement with the value of 7.3% from Figure 6B for a different sample of the pretzel with good eating quality.

Figure 7   Typical DSC curves for representative samples (1:1 mixtures with water) of: (A) a commercial pretzel dough before caustic-bath treatment—scan and immediate rescan; (B) the same pretzel dough after caustic-bath treatment but before baking; and (C) the same pretzel dough after caustic-bath treatment and baking, representing a finished product of good eating quality. (From Ref. 13.)

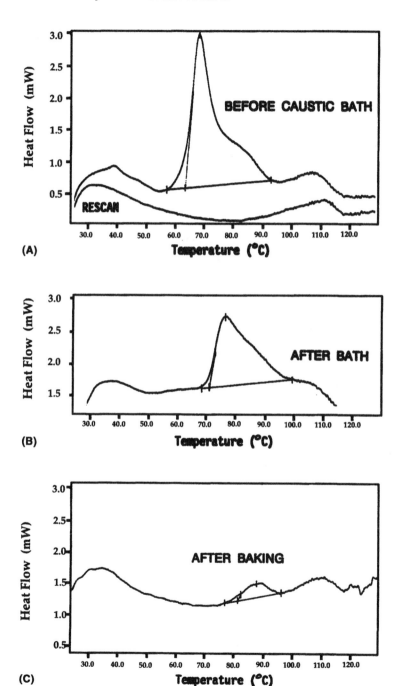

(A)

(B)

(C)

We also note that the Am L peak in Figure 7C (in its characteristic location above 100°C) is slightly larger in area than the Ap peak (this time, the delta Qs are given in the figure, showing the Am L peak to be 8% larger), as was also the case for the good-textured product represented in Figure 6B.

The DSC thermograms in Figure 8 reveal how the prototype pretzel, with poor texture and eating quality, had been subjected to a presumably less-than-optimal process of starch conversion (i.e., combination of gelatinization and annealing), via a processing path that contrasted significantly with the one followed by the pretzels (Figs. 6B and 7C) with good eating quality. Again, by applying the "fingerprint" approach to the DSC results in Figure 8, in order to compare them to the corresponding results in Figure 7, we see that, as a consequence of the progression from untreated dough (Fig. 8A) to dough after caustic-bath treatment but before baking (Fig. 8B) to finished product after caustic treatment and baking (Fig. 8C), the percentage remaining native Ap structure, as reflected by the Ap peak area, decreased from 100% (4.30 J/g) to 61.8% (2.66 J/g) to 15.8% (0.68 J/g), with the last value of 15.8% again being in reasonable agreement with the earlier value of 14.4% obtained from Figure 6A for a different sample of these same poor-eating-quality pretzels. The values of 61.8% (Fig. 8B) and 15.8% (Fig. 8C) can be said to contrast significantly with the respective values of 53.6% and 7.8% obtained from Figures 7B and 7C. [That is, if we can infer a measure of significance from the reproducibility (actually, of one sample of commercial product to another) of experimental results compared earlier— 7.3 vs. 7.8% and 14.4 vs. 15.8%.] Thus, we are led to surmise that the inferior pretzel (represented in Fig. 8) had been subjected to too little starch conversion in the caustic bath, possibly resulting from (a) too short a residence time, (b) too cool a bath, and/or (c) too low an NaOH concentration. Was this the direct and sole cause of its poor texture and eating quality? It seems clear from the DSC "fingerprints" that the caustic-bath treatment was certainly a critical processing step that distinguished the good and bad products, in terms of their starch structure–thermal property relationships, on the one hand, and their starch functional characteristics and concomitant finished-product quality attributes, on the other hand.

But what about what happened during baking? Interestingly, the DSC "fingerprint" results revealed that the extent of starch conversion caused by bak-

---

Figure 8 Typical DSC curves for representative samples (1:1 mixtures with water) of: (A) a prototype pretzel dough before caustic-bath treatment; (B) the same pretzel dough after caustic-bath treatment but before baking; and (C) the same pretzel dough after caustic-bath treatment and baking, representing a finished product of poor eating quality. (From Ref. 13.)

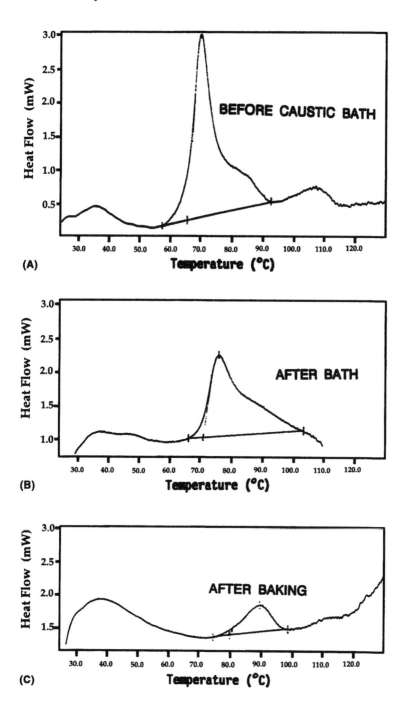

ing was essentially the same for the good and bad products; i.e., the change in percentage native Ap due to baking was 45.8% for the good product (53.6–7.8%—Figs. 7B and 7C) vs. 46.0% for the bad product (61.8–15.8%—Figs. 8B and 8C). However, as noted earlier with regard to the thermogram in Figure 6A representing the other sample of bad product, the Am L peak in Figure 8C is smaller (this time, obviously much smaller) than the Ap peak, which again contrasts with the situation for the good product, as illustrated in Figure 7C. We conclude that the baking process for the bad product must also (like the caustic-bath treatment) have been deficient, at least in the sense that Am L complex formation [which would be expected to be favored by the higher product temperature reached in the oven (8) rather than in the caustic bath] was evidently not enhanced. But what if the inferior product had been more optimally baked? For example, what if it had been baked more intensively (assuming this were possible, without burning it or causing its moisture content to be too low, even though its texture was already too hard), to a final 7.5% remaining native Ap structure (presumably, the target value), in an attempt to compensate for insufficient starch conversion in the caustic bath? Could its texture and eating quality have been salvaged in this way? Fortunately (and in contrast to what usually happens in the "real world" of commercial baked-goods production), we were able to obtain the appropriate product samples required to answer these questions. And the answer was no; after insufficient starch conversion in the caustic bath, even baking of product down to about 7.5% remaining native Ap structure appeared (thermogram not shown) to be insufficient to produce a significantly enlarged Am L peak in the thermogram, and the quality of this product was still inferior. The unavoidable conclusion from this part of the study was that both the caustic-bath treatment and baking must be optimal, with respect to both starch conversion and Am L complex formation, in order to produce a pretzel with optimal texture and eating quality.

VIII. CONCLUSIONS

In this chapter, we have demonstrated that: (a) DSC can be applied as an analytical "fingerprinting" method to characterize the thermal properties of wheat starch in cookies, crackers, and pretzels; (b) these thermal properties can be related to both starch structure and function in such low-moisture baked goods; and (c) such DSC "fingerprinting" can be used as a valuable time- and labor-saving research aid to successful product development efforts, by screening prototype products by DSC rather than by more traditional trial-and-error methods (referred to in the food industry as "cook-and-look"), in order to identify promising matches between appropriate ingredient functionality/baking performance and superior finished-product quality, which can be used to help guide subsequent developmental work.

Nevertheless, as is always the case when a new or different research approach is advocated, the interpretations and conclusions expressed in this chapter are still open to further research and to new challenges to the claimed utility of this DSC "fingerprinting" method.

REFERENCES

1. L Slade, H Levine. Beyond water activity: recent advances based on an alternative approach to the assessment of food quality and safety. Crit Rev Food Sci Nutri 30: 115–360, 1991.
2. L Slade, H Levine. Structure–function relationships of cookie and cracker ingredients. In: H Faridi, ed. The Science of Cookie and Cracker Production. New York: Chapman and Hall, 1994, pp 23–141.
3. L Slade, H Levine. Water and the glass transition—dependence of the glass transition on composition and chemical structure: special implications for flour functionality in cookie baking. J Food Eng 24:431–509, 1995.
4. L Slade, H Levine, J Ievolella, M Wang. The glassy state phenomenon in applications for the food industry. Application of the food polymer science approach to structure–function relationships of sucrose in cookie and cracker systems. J Sci Food Agric 63:133–176, 1993.
5. H Levine, L Slade. Influences of the glassy and rubbery states on the thermal, mechanical, and structural properties of doughs and baked products. In: H Faridi, JM Faubion, eds. Dough Rheology and Baked Product Texture. New York: Van Nostrand Reinhold, 1989, pp 157–330.
6. ME Zabik, SG Fierke, DK Bristol. Humidity effects on textural characteristics of sugar-snap cookies. Cereal Chem 56:29–33, 1979.
7. L Piazza, P Masi. Development of crispness in cookies during baking in an industrial oven. Cereal Chem 74:135–140, 1997.
8. L Slade, H Levine. Recent advances in starch retrogradation. In: SS Stiva, V Crescenzi, ICM Dea, eds. Industrial Polysaccharides. New York: Gordon and Breach Science, 1987, pp 387–430.
9. L Slade, H Levine. Non-equilibrium melting of native granular starch: Part I. Temperature location of the glass transition associated with gelatinization of A-type cereal starches. Carbohydr Polym 8:183–208, 1988.
10. L Slade, H Levine. Thermal analysis of starch. CRA Scientific Proceedings, Corn Refiners Association, Washington, DC, 1988, pp 169–244.
11. L Slade, H Levine. A food polymer science approach to selected aspects of starch gelatinization and retrogradation. In: RP Millane, JN BeMiller, R Chandrasekaran, eds. Frontiers in Carbohydrate Research—1: Food Applications. London: Elsevier Applied Science, 1989, pp 215–270.
12. JW Donovan. A study of the baking process by differential scanning calorimetry. J Sci Food Agric 28:571–578, 1977.

13. L Slade, H Levine, M Wang, J Ievolella. DSC analysis of starch thermal properties related to functionality in low-moisture baked goods. In: H Levine, L Slade, eds. Recent Advances in Applications of Thermal Analysis to Foods, special issue of J Thermal Anal 47:1299–1314, 1996.

14. RC Hoseney. Principles of Cereal Science and Technology. St. Paul, MN: American Association of Cereal Chemists, 1986, p 258.

15. AM Abboud, RC Hoseney. Differential scanning calorimetry of sugar cookies and cookie doughs. Cereal Chem 61:34–37, 1984.

16. K Kulp, M Olewnik, K Lorenz. Starch functionality in cookie systems. Starke 43: 53–57, 1991.

17. Y Wada, T Kuragano, H Kimura. Effect of starch characteristics on the physical properties of cookies. J Home Econ Jpn 42:711–717, 1991.

18. LC Doescher, RC Hoseney, GA Milliken. A mechanism for cookie dough setting. Cereal Chem 64:158–163, 1987.

19. RA Miller, R Mathew, RC Hoseney. Use of a thermomechanical analyzer: study of an apparent glass transition in cookie dough. In: H Levine, L Slade, eds. Recent Advances in Applications of Thermal Analysis to Foods, special issue of J Thermal Anal 47:1329–1338, 1996.

20. L Piazza, A Schiraldi. Correlation between fracture of semi-sweet hard biscuits and dough viscoelastic properties. J Texture Studies 28:523–541, 1997.

21. PS Given. Molecular behavior of water in a flour–water model system. In: H Levine, L Slade, eds. Water Relationships in Foods: Advances in the 1980s and Trends for the 1990s. New York: Plenum Press, 1991, pp 465–483.

22. S Aubuchon. Modulated DSC of foods. TA Instruments 1:2–3, 1997.

23. A Nikolaidis, TP Labuza. Use of dynamic mechanical thermal analysis (DMTA): glass transitions of a cracker and its dough. In: H Levine, L Slade, eds. Recent Advances in Applications of Thermal Analysis to Foods, special edition of J Thermal Anal 47:1315–1328, 1996.

24. AACC Method 10–52. Approved Methods of the American Association of Cereal Chemists. 8th ed. St Paul, MN: AACC, 1983.

25. SAS Craig, PR Mathewson, MS Otterburn, L Slade, H Levine, RT Deihl, LR Beehler, P Verduin, AM Magliacano. US Patent 5,108,764: Production of crackers with reduced or no added fat, 1992.

26. TJ Maurice, L Slade, C Page, R Sirett. Polysaccharide–water interactions—thermal behavior of rice starch. In: D Simatos, JL Multon, eds. Properties of Water in Foods. Dordrecht, The Netherlands: Martinus Nijhoff, 1985, pp 211–227.

27. L Slade, H Levine. Water relationships in starch transitions. Carbohydr Polym 21: 105–131, 1993.

28. L Slade, H Levine. Glass transitions and water–food structure interactions. In: SL Taylor, JE Kinsella, eds. Advances in Food and Nutrition Research, vol 38. San Diego: Academic Press, 1995, pp 103–269.

29. JMV Blanshard. The significance of the structure and function of the starch granule in baked products. In: JMV Blanshard, PJ Frazier, T Gaillard, eds. Chemistry and Physics of Baking. London: Royal Society of Chemistry, 1986, pp 1–13.

30. JMV Blanshard. Starch granule structure and function: physicochemical approach. In: T Gaillard, ed. Starch: Properties and Potential. New York: Wiley, 1987, pp 16–54.

31. JMV Blanshard. Elements of cereal product structure. In: JMV Blanshard, JR Mitchell, eds. Food Structure—Its Creation and Evaluation. London: Butterworths, 1988, pp 313–330.
32. CG Biliaderis, CM Page, L Slade, RR Sirett. Thermal behavior of amylose–lipid complexes. Carbohydr Polym 5:367–389, 1985.
33. L Slade, H Levine. Mono- and disaccharides: selected physicochemical and functional aspects. In: AC Eliasson, ed. Carbohydrates in Food. New York: Marcel Dekker, 1996, pp 41–157.
34. JD Ferry. Viscoelastic Properties of Polymers. 3rd ed. New York: John Wiley & Sons, 1980.
35. B Wunderlich. Macromolecular Physics, vol 2—Crystal Nucleation, Growth, Annealing. New York: Academic Press, 1976.
36. L Slade, H Levine, S Craig, H Arciszewski, S Saunders. US Patent 5,200,215: Enzyme-treated low moisture content comestible products. 1993.
37. L Slade, H Levine, S Craig, H Arciszewski. US Patent 5,362,502: Reducing checking in crackers with pentosanase. 1994.

3

Utilization of Thermal Properties for Understanding Baking and Staling Processes

Ann-Charlotte Eliasson
Lund University, Lund, Sweden

I. INTRODUCTION

The baking of bread starts with mixing, a process in which wheat flour (and/
or some other cereal flour), water, yeast and other ingredients are transformed
into a dough during mechanical treatment. The mixing results in a foam, i.e.,
a gas phase dispersed into an aqueous matrix, and the volume of this foam
increases during fermentation. The final step is the application of heat in the
oven; when the final volume is reached, the crumb structure is fixed and the
crust gets its color. For other baked products the recipe, the mixing, and fermentation differ, but there is always a step involving the application of heat. This
can be in order to dry the product, as in pasta making, or it can be to bring about
the transformation from dough to bread, as in baking. For an understanding
of cereal processing it is thus necessary to understand the thermal transitions
that cereal components can undergo as a result of a change in temperature. The
present chapter deals with transitions like glass transitions, starch gelatinization and retrogradation, protein denaturation, and crystallization. Chemical reactions are also of importance for bread making—just think of the chemical
reactions resulting in the production of carbon dioxide and ethanol, or the aroma
components in the crust. However, the structure of the baked product and
sensorial attributes related to "texture" or "consistency" are to a great extent
related to the physical changes and phase transitions occurring in the cereal
components.

That the transformation of a wheat flour dough into a bread does not occur without the application of heat is self-evident. The result of the baking process thus depends on how the different components in the dough behave in relation to the increase in temperature. The temperatures reached during the baking process differ for crumb and crust. In the crumb the temperature will not exceed 100°C as long as water is still present, whereas the temperature in the crust can be considerably higher (1). The difference in temperature between crumb and crust means that completely different transitions can occur in the different parts of the loaf. In the crust a range of chemical reactions will occur that results in the color and aroma of the crust. In the crumb, where the temperature is lower, the physicochemical transitions will dominate. The thermal transitions related to the baking process can be measured using thermal methods at conditions that are as close as possible to the real baking situation. However, when staling is studied, a thermal transition is developed in order to indicate what has happened during storage. The glass transition is used in the same way: It is not supposed to occur during processing but can be used to predict what can happen during storage.

Because the baking process requires the application of heat, it is not surprising that thermal methods have been used for studying the thermal behavior of cereal components, doughs, and breads. Of the different methods available, differential scanning calorimetry (DSC) has become the most frequently used. The first thermal method to be used in the studies of problems related to baking and staling was probably differential thermal analysis (DTA). In this method the sample is heated at a fixed scanning rate, and the difference in temperature between sample and reference due to a thermal transition in the sample is registered as a function of temperature. This technique was applied in some early studies of retrogradation of concentrated starch gels, and knowledge was gained concerning the influence of storage temperature on the retrogradation (2, 3). The DSC technique has the advantage that the transition enthalpy can be measured in a simple way from the area of the transition endotherm (or exotherm). Since the first studies using DSC, this technique has come to dominate the studies of starch-related transitions (4, 5). Other transitions, like protein denaturation and order–disorder transitions in nonstarch polysaccharides, are also studied using DSC (6, 7). The development of the modulated DSC (MDSC) (8) would in principle give new possibilities for studying the thermal behavior of cereal components, but so far the method is still being developed for such applications, both in our laboratory and in others. In MDSC a cycled temperature is developed on top of the ordinary temperature increase in the DSC. In the evaluation of the thermogram it should be possible to differentiate between reversible and nonreversible events. In both DTA and DSC, the thermal behavior is studied during heating (or cooling). For studying the thermal behavior at constant conditions, microcalorimetry has recently been used in studies of starch gelatinization (9), retrogradation (10), and the formation of amylose–lipid complexes (11). Thermogravimetry analysis (TGA), where the weight loss from a sample is registered during heating, seems not to have been used much for studies of baking and staling processes (12). Mechanical properties can also be measured in order to detect thermal transitions, by using methods like dynamic mechanical thermal analysis (DMTA) and thermomechanical analysis (TMA) (13, 14). These methods have mostly been used with the aim of studying glass transition temperatures (T_g) in starch and other cereal systems.

The most frequently used cereal in baking is wheat, and therefore this chapter will deal mainly with thermal transitions of wheat components and with wheat flour doughs and baked products. However, examples will also be included of other cereals, where appropriate. Still, the knowledge we have about thermal transitions of cereal components in relation to baking and staling is to an overwhelming extent obtained for wheat components. This chapter starts with a description of the individual flour components—starch, proteins, lipids, arabinoxylans—and the thermal transitions they can undergo when heated in the presence of water. Then these simple model systems will be made more complicated by adding other components. To be more precise, the addition of wheat flour components to starch will be described, since starch dominates the thermal transitions occurring in doughs and baked products. Also, a few examples from real doughs and products will be given.

II. THERMAL BEHAVIOR OF CEREAL COMPONENTS

Although wheat starch dominates the composition of a wheat flour dough, there are other components present that can be expected to exhibit thermal transitions during heating. The composition of wheat flours is given in Table 1, and it can

Table 1 Composition of Wheat Flours

Component	Mean value (weight %)	Range (weight %)
Starch[a]	80.5	73.9–85.8
Protein[a]	12.5	9.4–18.4
Ash[a]	0.61	0.51–0.71
Lipids[b]		1.1–1.2
Nonstarch polysaccharides[a]	3.3	2.5–4.5

[a] Data collected from 49 wheat flour samples. (*Source*: Ref. 184.)
[b] *Source*: Data compiled from Ref. 185.

thus be expected that during heating starch gelatinizes, proteins denature, and fats and lipids undergo phase transitions. Helix-coil transitions might be expected for the nonstarch polysaccharides, in accordance with observations for certain gums (15). However, as will be shown in this chapter, the main transitions registered on a DSC thermogram for a suspension of a wheat flour in water are all related to starch (16). Of course, if a high amount of solid fat is added, the melting of fat will be observed; but for most ingredients used in baking, their effect on the DSC thermogram will be their effect on starch-related transitions. As described later, several transitions related to starch are possible to analyze using thermal methods, whereas for the other cereal components thermal methods give much less information due to the lack of thermal transitions during conditions corresponding to baking. In this section, examples will be given on results obtained when thermal methods are used to study the individual cereal components and their importance for the baking process and baked products.

A. Starch

Starch is present in wheat flour as particles, starch granules, with diameters in the range 1–40 μm (17, 18). In fact, two populations of granules are present: large, lenticular A-granules (mean diameter around 14 μm), and small B-granules (mean diameter around 4 μm) (19). It has been observed that the ratio between A- and B-granules is important for baking performance, with an optimum proportion of B-granules at about 25–35% by weight (20).

In wheat starch the proportion of amylose is about 29%, the remainder being amylopectin (19, 21). Amylose is an essentially linear and unbranched glucose polymer, whereas amylopectin is highly branched (22). The degree of polymerization (DP) for wheat amylose has been determined to be in the range 830–1570 anhydro glucose units, with 4.8–5.5 branch linkages per molecule (22). However, the slight branching of amylose has no influence on its physicochemical properties; it will still behave as a linear polymer. Amylopectin is a much larger molecule, with DP in the range of 10^4–10^5. The average chain length (CL) in wheat amylopectin is 20–21. Debranching of amylopectin results in a distribution of chain lengths, and the maximum in the distribution profile for wheat amylopectin was found at CL values of 12, 18, 47, and 86 (21). Recently, a new wheat variety has been developed in which the starch lacks amylose (23, 24). This is thus a waxy wheat starch, and it can be expected to exhibit somewhat different thermal behavior from the wheat starch of normal amylose content.

Cereal starches also contain other components, such as polar lipids and proteins (25). These can influence the functional properties of the wheat starch; the polar lipids especially will have a great influence on the thermal transitions that can be detected using DSC (5). Polar lipids form an inclusion complex with amylose (26), and this complex gives rise to thermal transitions at temperatures around 100°C in excess water (i.e., >70% water) (27).

Starch is partially crystalline, and for native wheat starch the A-pattern is observed in X-ray diffraction studies (28–30). The crystalline order is a result of the organization of double helices formed by amylopectin branches (26, 31). The amylose, as well as amylose–lipid complexes, is present in amorphous parts of the starch granule. The amylopectin branches are also located in amorphous regions, although one branching point fits into the crystalline domain (32). In most applications of starch, including baking, the crystalline structure is lost in the process known as *gelatinization*. During storage of starch, recrystallization might occur, a process known as *retrogradation*. This recrystallization leads to the formation of the B-pattern, observed in X-ray diffraction studies. For gelatinized wheat starch, the B-pattern is observed together with the V-pattern of the amylose–lipid complex (23, 24). Contrary to the crystallinity in the native starch granule, crystallization during retrogradation can involve both amylopectin and amylose.

1. Gelatinization

Differential scanning calorimetry has become perhaps the most widely used method for studying starch gelatinization. Before the DSC epoch, starch gelatinization was usually studied by measuring the change in viscosity during the application of heat, and several instruments have been designed for such measurements. The heating rate and stirring rate are usually fixed, thus resulting in shear rates that cannot be varied (33). Moreover, there are limitations in how high the starch concentration can be, and the rather diluted starch suspensions with a starch concentration around 5–15% (w/w) are far from the dough with a water content of around 40%. Viscosity measurements have been used for checking flour quality, for example, for studying the influence of enzymes, mainly α-amylase (34). However, although the change in rheological properties is measured, and this is of considerable importance for the baking behavior (35), the measurements have to be performed at conditions far from the real baking situation. When DSC was first used for the study of starch, the measurements were carried out at excess water conditions, and an endotherm assigned to starch gelatinization was observed (36). In fact, when potato starch and water are mixed, an exothermic process occurs at temperatures below 55°C, whereas the mixing is endothermic at temperatures above 60°C (9). Gelatinization involves a range of events with somewhat different onset temperatures, and in the DSC only part of all the aspects of gelatinization is measured (25). The changes in rheological properties occur mostly at temperatures higher than the range corresponding to the gelatinization temperature range measured in DSC (37). However, there is a relation between the temperatures of the DSC endotherm and the development of rheological properties for concentrated starch gels, because the deformability of the starch granule will change with the melting of the crystalline domains in the starch granule, and the latter is measured in DSC (38, 39).

When measurements on potato and wheat starches were performed over a range of water contents, the usefulness of the DSC method became evident (4, 5). For potato starch it was found that when the water content was reduced, a shoulder emerged at the high-temperature side of the endotherm that hitherto had been assigned to gelatinization (4). At even lower water content the gelatinization endotherm was more or less split into two endotherms. The low-temperature endotherm was designated the G endotherm, and the high-temperature endotherm the M1 endotherm. The result for wheat starch was similar, although transition enthalpy values and temperatures differed from potato starch. But, moreover, a third endotherm was observed at temperatures at or above 100°C, depending on the water content (Fig. 1) (5). This third endotherm (the M2 endotherm) is thus always located at temperatures above the gelatinization temperature range. This endotherm could later be assigned to a transition of the amylose–lipid complex (40).

The DSC scan is not performed during equilibrium conditions, and therefore the heating rate influences the DSC results for starch (41, 42). The gelatinization process is reported to be independent of the scanning rate when the starch–

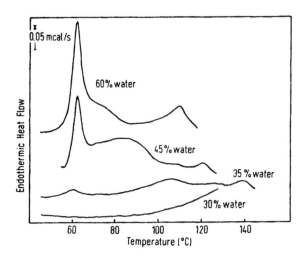

Figure 1   DSC thermograms of wheat starch at different starch–water ratios. (Reproduced with permission from Ref. 5.)

water suspension is heated at a scan rate lower than 0.5 K/min (9), whereas in many DSC experiments the heating rate is 10°C/min (4, 5, 21, 43–45). An increase in heating rate will cause an increase in the gelatinization temperatures T_o, T_m, and T_c (for explanations, see Fig. 2). Microcalorimetry (isothermal calorimetry) was used to evaluate the heat of gelatinization for potato starch, and it was found to be 17.6 ± 1.7 J/g at 61°C, in good agreement with the DSC value (9). When using DSC for analyzing starch gelatinization, the conditions could be made to simulate baking conditions not only in the composition of the sample, but also in the heating rate and the temperature range of heating. The heating rate during the most rapid increase in temperature in the crumb during baking is similar to the heating rates used in DSC experiments (46). The DSC scan can easily be interrupted at 100°C, a temperature that will not be exceeded in baking as long as there is water left in the bread. The DSC scan can of course also be interrupted at a lower temperature, and the heating continued at a fixed temperature, should such treatments be of interest for simulating a real baking situation. The starch gelatinization is irreversible, and when the sample is cooled only an exothermic transition of the amylose–lipid complex will be observed (i.e., if the

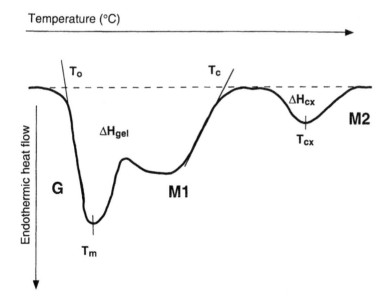

Figure 2 The definition of transition temperatures in the DSC thermogram obtained for the heating of wheat starch. The transition enthalpies, ΔH_{gel} and ΔH_{cx}, are calculated from the area below the transition endotherm.

temperature for the M2 transition was reached during the heating scan). An immediate reheating will then again show only the (endothermic) transition of the
amylose–lipid complex.

The presence of the double endotherm (the G and M1 endotherms, Fig. 2)
has been discussed ever since these two endotherms were first detected, and different explanations for their origin have been suggested. Donovan (4) interpreted
the endotherm as showing the plasticizing effect of water: The water content
influenced the melting temperature of starch crystallites. This approach could be
criticized because the DSC scan is not performed during equilibrium conditions
(13, 47). The nonequilibrium conditions as such have been used as an explanation; the double endotherm would then arise from an overlapping exothermic
process resulting from transitions between different polymorphic forms (43).
However, there is no change in the X-ray diffraction pattern, except for the decrease in intensity with increasing temperature, in the temperature interval corresponding to the DSC gelatinization temperature range (30). The double endotherm has also been described as resulting from a redistribution of water between
crystalline domains and gelatinized starch (48). The first crystallites to melt will
melt at excess water conditions, thus at a low temperature. However, the melted
starch will absorb water that will not be available for the remaining crystallites,
and therefore these will melt at a higher temperature. There are also indications
of a glass transition, occurring at the low-temperature side of the G endotherm
(30, 42, 47).

The DSC characteristics of the gelatinization endotherms differ between
starches of different botanical origin, in the gelatinization temperature range, the
enthalpy, and the appearance of the endotherm; some examples are given in Table
2. The different temperatures obtained from the DSC thermogram are defined in
Figure 2. The transition enthalpy, ΔH_{gel}, is calculated from the area below the
transition endotherm and is expressed in J/g starch or J/g amylopectin. To some
extent the differences observed are related to the chemical composition of the
starch; for example, the gelatinization temperature was found to be negatively
correlated with the amylose content (21). It can be concluded from Table 2 that
differences are observed not only between starches from different species, but
also between varieties of the same species. It is even possible to detect differences
in gelatinization DSC characteristics between the small and large granules in
wheat (18) and barley (49). However, the differences are rather small. It should
also be pointed out that it is difficult to compare different studies, for the conditions during the DSC analyses differ.

When other ingredients that are used in baking are present, they will all
affect the DSC characteristics of starch gelatinization, as illustrated with a few
examples in Table 3. When the added ingredient has a thermal transition of its
own there might be overlap between endotherms, and thus the evaluation is difficult or even impossible (50). It is evident that most added ingredients increase

Table 2 DSC Characteristics of the Gelatinization Endotherm of Some Cereal Starches

Starch	Water content (weight %)/ scanning rate (°C/min)	T_o (°C)	T_m (°C)	T_c (°C)	ΔH_{gel} (J/g starch)	Ref.
Wheat	50/10	51	56	83	17.5[a]	21
Wheat ($n = 4$)[b]	80/5	56.2–59.0	62.2–64.3	73.7–78.1	7.7–9.5	24
Waxy wheat	80/5	59.8	65.6	83.6	9.1	24
Rye	50/10	51	56	88	15.8[a]	21
Rye ($n = 9$)	60/10		54.8–60.3		10.6–12.0	71
Barley ($n = 2$)	60/10		58.3–58.6		11.9–12.1	186
Waxy barley ($n = 2$)	60/10		59.8–61.3		14.0–14.3	186
High-amylose barley	50/10	55	63	91	16.1[a]	21
Oats ($n = 6$)	80/10	44.7–47.3	56.2–59.5	68.7–73.7	11.4–13.1[a]	187

[a] Calculated as J/g amylopectin.
[b] n = Number of varieties.

Table 3 Influence of Different Baking Ingredients on the Gelatinization Behavior of Wheat Starch

Ingredient	Effect on T_o	Effect on ΔT ($= T_c - T_o$)	Effect on ΔH_{gel}	Ref.
Sucrose	Increase	Decrease	n.r.	112
Lactose	Increase	n.r.	Decrease at low levels, increase at high levels of addition	117
Salt (NaCl)	Increase at low level, decrease at high levels of addition	n.r.	Same as on T_o	48
Fat	Unchanged	n.r.	n.r.	188
Oils	Unchanged	Unchanged	Unchanged	180
Milk proteins	Increase	n.r.	Unchanged	117

n.r. = Not reported.

T_o and T_m, whereas ΔH_{gel} is usually unaffected. In the case of salt, gelatinization might even occur at room temperature; however, this would be at salt concentrations or with types of salts that are not feasible in food products (48).

2. Retrogradation

Starch gelatinization is an irreversible process, and if a wheat starch–water suspension is gelatinized, cooled, and reheated immediately in the DSC, only the transition due to the amylose–lipid complex will be registered during the second heating (51). However, if the sample is stored for at least a day before reheating, an endotherm will be observed in about the same temperature interval as the gelatinization endotherm (Fig. 3). This endotherm is attributed to the melting of recrystallized amylopectin (52, 53). For detecting recrystallized amylose, the heating must continue to well above 100°C (Fig. 4) (54–56). Both the recrystallization of amylopectin and amylose contribute to the process known as *retrogradation* (57), whereas the recrystallization of amylopectin is claimed to be one of the most important reasons, if not the only one, for the staling of bread (58). Thermal methods, and especially DSC, thus offer very convenient tools for measuring the staling of bread. As in the case of gelatinization, DSC will not register all aspects of the retrogradation, but it can still contribute to a basic understanding of the process.

The retrogradation of starch is essentially a process that involves the reorganization of molecules that eventually will lead to the development of crystallinity

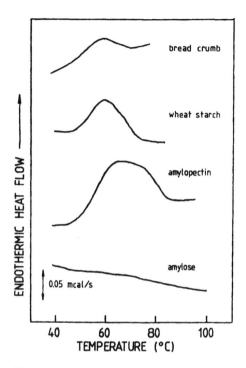

Figure 3 DSC thermogram for melting of recrystallized amylopectin in different starch products. (Reproduced with permission from Ref. 52.)

(58). During gelatinization, the X-ray diffraction pattern of the native starch disappears; during storage, an X-ray diffraction pattern, the B-pattern, emerges. The B-pattern is observed independent of the X-ray pattern of the native starch. For cereal starches that contain lipids, a V-pattern due to the amylose–lipid complex will also be detected (59).

   When DSC is used to study retrogradation of starch, the retrogradation is followed by melting the crystallized amylopectin during heating in the DSC, and the retrogradation is evaluated from the melting enthalpy, ΔH_c (21, 52, 53). The temperature range for melting recrystallized amylopectin is rather similar to the gelatinization temperature range (compare Fig. 3 and Table 2), and heating to a temperature above 80°C will melt all recrystallized amylopectin for most starches (60). Depending on the storage conditions, the quality of crystallites can be affected, and the temperature of the endotherm will then be influenced, as illustrated in Table 4. It is evident that when the storage temperature is cycled between the low temperature (6°C), facilitating the nucleation, and a higher temperature (30 or 40°C), facilitating the growth of crystallites, the quality of the crystallites is

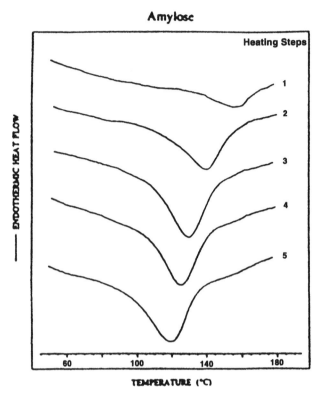

Figure 4 DSC thermogram for melting recrystallized amylose. (Reproduced with permission from Ref. 55.)

improved, and they are melting at a higher temperature. They are also becoming more homogeneous, because $\Delta T\ (= T_o - T_c)$ decreases.

For studying the retrogradation of amylose, the heating in the DSC has to be continued to temperatures above 100°C. A transition peak temperature of 153.6°C and a transition enthalpy of 5.7 J/g was reported for amylose of a DP of 5500, and for defatted amylomaize the corresponding values were 148.7°C and 3.3 J/g (56, 61). Both recrystallized amylose and amylopectin give the B-pattern in X-ray diffraction. Thus, an analysis using the X-ray diffraction technique also has to be performed at a temperature above the melting temperature of recrystallized amylopectin and below the melting temperature of recrystallized amylose in order to differentiate between the contributions to retrogradation from the individual components. Using DSC, the individual contributions from amylo-

Table 4 Temperature Range of the Amylopectin-Retrogradation-
Related Endotherm Measured at a Heating Rate of 10°C/min for
Wheat Starch Gels Stored at Different Conditions

Storage conditions[a]	T_o (°C)	T_c (°C)	ΔT $(= T_c - T_o)$ (°C)
6/6	40.4	63.4	23
6/6/6/6	37	63	26
6/30	48.5	64.7	16
6/30/6/30	49.3	65.4	16
6/40	58.0	68.5	11
6/40/6/40	58.2	69.0	11

Source: Data from Refs. 21 and 60.
[a] Each number indicates the temperature for one day of storage; 6/30 thus
means one day at 6°C, followed by one day at 30°C.

pectin and amylose can easily be separated due to the differences in melting
temperature.

The DSC method shows the history of the sample: when the DSC scan is
run, the retrogradation has already taken place, and it is not possible to tell from
the thermogram whether most of the retrogradation occurred in the beginning or
toward the end of the storage time, or whether the retrogradation rate was the
same during the storage period. To obtain such information, samples are stored
and reheated after different storage times, in order to construct a plot of ΔH_c
as a function of time. When using microcalorimetry, the recrystallization is followed in real time (Fig. 5). For maize starches differing in amylose content studied in microcalorimetry after gelatinization, amylomaize (80.4% amylose) was
observed to give a large exothermic effect during the first 4–5 hours of storage,
whereas the waxymaize (0% amylose) gave a very low exothermic effect, and
the normal maize (26.2% amylose) gave an intermediate response (10). After
5 h the largest exothermic effect was observed for the normal maize. This result was interpreted as showing that amylose dominated the initial recrystallization, whereas amylopectin dominated the recrystallization during longer storage
times.

Many factors are known to influence the amylopectin retrogradation, both
factors related to the storing conditions, such as temperature and time, and factors
related to the sample composition, such as water content and other ingredients.
Certain additives are used to increase the softness of bread during storage, for
example, emulsifiers and enzymes, and these additives will also influence the

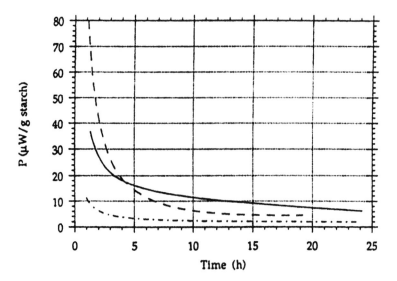

Figure 5 *P–t* traces from isothermal microcalorimetric analysis of maize (——), amylo-maize (-----) and waxy maize (-·-·-·-) starches. (Reproduced with permission from Ref. 10. Copyright John Wiley & Sons Limited.)

amylopectin retrogradation, which will be discussed in detail in Sec. V of this chapter (Staling of Bread).

Although the amylopectin retrogradation rate is greatest during the first days of storage, crystallization will continue for many days (53, 62, 63). The storage temperature influences the extent of retrogradation, as seen in Table 4. It turns out that the maximum amylopectin retrogradation occurs in the temperature range 0–10°C (52, 64, 65). The water content of the starch gel influences the extent of retrogradation, and maximum retrogradation is observed at intermediate water contents, i.e., at water contents in the range 40–50% (66, 67). It is the water content during retrogradation that is important, not the water content during gelatinization (67). This maximum in amylopectin retrogradation at certain concentrations has been observed for different storage temperatures, including −4°C (68).

The retrogradation behavior differs between starches from different botanical sources, and potato starch is known to show a greater extent of amylopectin retrogradation, compared to many other starches (60, 69). Among cereal starches, rye and oat starch retrograde to a lesser extent than wheat starch (21, 70, 71). In the case of oat starch, at least part of the low retrogradation could be explained by the lipid content of this starch (70). Oat starch of high lipid content retrograded to a lesser extent than an oat starch of lower lipid content; when the starches were defatted, the extent of retrogradation increased for both. However, the difference between oat starches of high and low lipid content also remained after defatting. In case of rye and wheat starches, the lipid content cannot explain the lower retrogradation of rye starch, for the lipid content of rye starch is lower than that of wheat starch, as observed both from determination of starch-bound lipids and from the transition of the amylose–lipid complex (21, 71). The retrogradation tendency has also been found to differ to a large extent between rye varieties (71).

The differences in retrogradation behavior between starches have been related to the chemical composition of the amylopectin, and especially to the average chain length, where longer chains seem to facilitate retrogradation (21, 72–74). It has been found that the melting enthalpy of retrograded amylopectin correlates to certain unit-chain-length populations. Amylopectin unit-chain lengths with DP of 6 and DP around 18–19 were positively correlated to ΔH_c, whereas unit-chain lengths with DP of 8–11 were negatively correlated (60). Another factor of importance seems to be the amylose content, and it has been suggested that at certain conditions amylose could facilitate the retrogradation of amylopectin (63, 75, 76). It has been suggested that the amylose network could act as seed nuclei for the amylopectin aggregation and crystallization. Such an effect could perhaps explain the considerable retrogradation observed for high-amylose barley, despite its low average amylopectin chain length (21).

3. Glass Transition

A glass transition expresses the transition between a glassy, brittle state and a rubbery state. As will be discussed later, the glass transition temperature (T_g) is important for storage stability and retaining of crispness. The T_g can be evaluated from the diffuse steplike transition in heat capacity observed in DSC, and then either the initial, midpoint or endpoint temperature of the transition can be used to define T_g (77, 78). Another suitable method is DMTA, where T_g can be defined and measured in different ways from the elastic modulus (E'), loss modulus (E''), or tan δ $(= E''/E')$ measured as a function of temperature. The cereal components that have been of most interest in this connection are starch and gluten, but doughs and baked products have also been investigated.

Because the starch granule is partially crystalline and partially amorphous, it is to be expected that a glass transition related to the amorphous parts of the starch granule would be detected (79). The T_g of starch is very sensitive to the water content, and it decreases when the water content increases (see Fig. 6). The T_g range broadens when the water content decreases, making an exact deter-

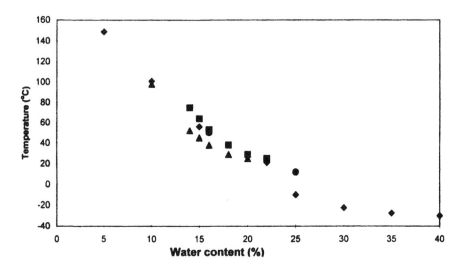

Figure 6 Glass transition temperature (T_g) as a function of water content, determined in different studies for native and gelatinized starch samples. ◆: Native wheat starch, TMA (from Ref. 14); ■: Native wheat starch, DSC (from Ref. 189); ▲: Gelatinized wheat starch, DSC (from Ref. 189); ●: Waxymaize amylopectin (from Ref. 80).

mination of the temperature very difficult (77). Moreover, not all starches show a well-defined T_g, probably due to the heterogeneous nature of the amorphous parts of the starch granule (68). It has been suggested that the presence of crystallinity makes the measurement of T_g even more difficult and that it should be easier to determine T_g for the native starch granule than for retrograded starch, because in the native starch the heterogeneous starch is still heterogeneous in an organized way, whereas this is not the case in the retrograded starch (80). Some examples of results obtained using DSC and TMA are presented in Figure 6. For amorphous amylopectin it was found that T_g measured in DSC was between the values obtained from tan δ and E' measurements in DMTA (80). Biliaderis et al. (42), using DSC, detected a glass transition just before the gelatinization endotherm for waxy rice starch. Heating of wheat starch suspensions [55% (weight %) water] to different temperatures within the DSC gelatinization temperature range, followed by cooling and reheating, revealed that the position of the glass transition was in the range 54–63°C (47).

The position of T_g is of interest because it puts the lower temperature limit where changes in the starch crystallinity can occur. The T_g has to be passed before melting (or, in other words, gelatinization) can occur. However, processes like retrogradation and annealing can also occur only in the temperature interval between the T_g and the onset of gelatinization/melting of crystallites (47, 81).

4. Amylose–Lipid Complexes

At temperatures above the gelatinization temperature range, i.e., at 90–100°C at excess-water conditions, an endothermic transition is detected in the DSC thermogram for wheat starch (Fig. 1) (5). This transition is related to a transition of the amylose–lipid complex (40) and will be present in lipid-containing starches (21), and for lipid free starches when lipids such as emulsifiers are added (50). The peak temperature (T_{cx}) and enthalpy (ΔH_{cx}) of the transition depend on the water content, and at certain conditions multiple endotherms are observed (5, 82). The nature of the polar lipid in the complex will also influence the DSC characteristics of the transition (83, 84). For monoglycerides, the T_{cx} increases with increasing chain length of the monoglyceride and decreases with increasing degree of unsaturation (83). The polar head will also influence the transition temperature, and for an ionic group the influence will depend on the chain length, so that for short chains (below 12 carbons) the complex with a charged group will have a lower transition temperature than the complex with an uncharged group (84). The DSC characteristics of the transition related to the amylose–lipid complex thus differ between starches, and some examples are given in Table 5. The ΔH_{cx} value is related to the amylose content, and, in fact, it has been used to estimate the amylose content of starch (85, 86). In the first study the starch was saturated with lysolecithin, and ΔH_{cx} for the starch–lysolecithin transition was compared with the value obtained for amylose–lysolecithin at saturation (lysolecithin/amylose >0.2) (85). It was found that a second heating was required to obtain the maximum enthalpy. The amylose content determined in this way agreed well with other methods for certain starches, but gave unexpected high values for wheat starch (37%!). However, with modifications of the original method, including higher heating temperatures in the DSC and using the exothermic transitions during cooling for calculations, the method has been improved, and it also seems to be reliable for the determination of amylose content for wheat starch (86, 87).

Table 5 DSC Characteristics of the Transition Related to the Amylose–Lipid Complex for Some Cereal Starches

Starch	Conditions	T_{cx} (°C)	ΔH_{cx} (J/g)
Wheat	50% water	110.1	1.4
Rye	50% water	107.8	0.8
Barley	50% water	110.3	1.8
High-amylose barley	50% water	110.8	2.8
Waxy barley	50% water	ND[a]	ND[a]

Source: Data from Ref. 21.
[a] Not detected.

Defatting of wheat starch, as expected, removed the transition due to the amylose–lipid complex (56). The defatting procedure was also found to affect the gelatinization endotherm. Of the different solvents used, 80% methanol was found to have the smallest effect, T_o increased from 53.2 to 54.9°C, T_m increased from 59.0 to 59.3°C, and ΔH_{gel} decreased from 11.1 to 10.1 J/g. The largest effect was observed for 70% propanol, where T_o increased to 62.2°C and T_m to 67.7°C, and ΔH_{gel} decreased to 1.3 J/g. These results could be interpreted as showing that partial gelatinization had occurred during defatting, and the degree of gelatinization was larger in the case of 70% propanol. Such a result seems to support the generally held belief that for effectively defatting starch it has to be gelatinized (88). However, also with 80% methanol, which affected the starch gelatinization very slightly, it was possible to defat the starch so that no transition due to the amylose–lipid complex was detected.

Microcalorimetry has been used to study the complex formation between amylose and certain ligands (11). It was found that for the surfactant sodium dodecylsulfate (SDS), complex formation occurred with amylose of a DP as low as 9. A reaction between amylopectin and SDS was demonstrated by microcalorimetry. The ΔH value of the reaction was in the order corresponding to complexation of an amylose molecule with a DP of 13.

The transition of the amylose–lipid complex occurs at temperatures that will never be reached during baking, at least not in the crumb. The transition will thus probably not directly affect the baking process. However, more complexes seem to be formed during heating (89), and it could therefore be speculated that polar lipids that otherwise would contribute to the baking performance might not be available because of the complexation (25). The extra complexation has been suggested to occur at a temperature interval corresponding to the gelatinization temperature range, because the addition of complex forming agents usually causes a decrease in ΔH_{gel} (27, 90, 91).

The possibility to study the amylose–lipid complex is of special interest because of the use of emulsifiers to decrease retrogradation (92). When an emulsifier is added to starch, the retrogradation during storage decreases, whereas the transition enthalpy of the complex is not changed during storage (52, 59, 62). However, the explanation for the reduction in ΔH_c is not as simple as the presence of an amylose–lipid complex. The addition of a complex-forming agent to a waxy starch will also decrease ΔH_c of such a starch, compared to a control without addition, although no amylose–lipid complexes are formed (93). When preprepared amylose–lipid complexes were added to a starch–water suspension before or after gelatinization, the results presented in Table 6 were obtained (94). It thus seems that the complex as such does not influence the retrogradation. Another explanation that has been put forward is that by complex formation the amylose, which at certain amylose/amylopectin ratios facilitates the retrogradation of amylopectin (63, 95), would be made unavailable for affecting amylopec-

Table 6 Influence of an Added Amylose–Surfactant Inclusion Complex on the Retrogradation of Waxy Maize Starch

Treatment	Effect on ΔH_c	T_{cx} (°C)	ΔH_{cx} (J/g starch)
1. Complex added to dry starch, water added, heating to 100°C before storage.	Decrease	96.8 ± 0.8	0.7 ± 0.15
2. Same as 1, except that heating was to 85°C.	A small decrease	105.1 ± 2.0	0.7 ± 0.3
3. Starch and water were mixed and heated to 100°C and cooled to room temperature before addition of complex.	Increase	96.3 ± 0.7	0.6 ± 0.2
4. Same as 3, except that the samples were reheated to 100°C after the addition of the complex.	Decrease	100.5 ± 1.5	0.6 ± 0.3

Source: Data collected from Ref. 94.

tin retrogradation. However, such a mechanism does not explain the results obtained for waxy starches, and it has thus been suggested that amylopectin–lipid interactions could be an explanation (96).

B. Protein

When gluten, isolated from a wheat flour, is heated in the DSC at the water content present after isolation, only very small endotherms are observed (Fig. 7) (97, 98). Moreover, some of the endotherms that are possible to register can be assigned to starch. A consequence of this is that the DSC thermogram of a mixture of wheat flour and water will look very similar to the thermogram for a starch suspension at the corresponding water content (16).

One explanation for the lack of evident denaturation endotherms for gluten proteins could be that endothermic and exothermic events occur simultaneously, thus canceling each other (98). However, changing the polarity of the solvent (from water to a mixture of glycerol:water), which was thought to affect endothermic and exothermic transitions to different degrees, did not cause any peak to be observed. It could be argued that the proteins lack long-range order or a three-dimensional structure that can be denatured. However, circular dichroism spectroscopy measurements on purified gliadins indicate considerable content of α-helix, at least 30–35% in α-, β-, and γ-gliadins and 10% β-sheet in α-gliadins

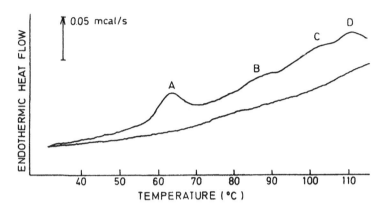

Figure 7 DSC thermogram for the heating of gluten at 10°C/min. (Reproduced with permission from Ref. 97.)

(99). These structures could then be expected to be very stable and not be denatured by heat, at least not at temperatures below 100°C. But heating of these proteins resulted in partial loss of the α-helical content as observed by circular dichroism spectroscopy. Electrophoretic methods indicate that a large number of proteins are present in gluten (25). This could result in a broad range of denaturation temperatures, which are thus not possible to detect in the DSC. That denaturation peaks are not observed in the DSC thermogram does not mean that gluten proteins are not affected by heat treatment. When gluten was heat-treated before use in baking experiments, a decrease in loaf volume was observed when the treatment temperature was 55°C or above (100). It was suggested that the observed effects on functionality were due to formation of aggregates involving sulphydryl/disulphide interchange.

For albumins and globulins from oats, typical denaturation thermograms have been registered (101). The albumin fraction gave rise to a broad endotherm with a peak maximum at 87°C, and the globulin fraction gave rise to a sharp endotherm at about 110°C. For the globulin transition, ΔH was calculated to about 23 J/g.

The glass transition temperature has been measured for gluten as well as gluten fractions, and some results are summarized in Figure 8. T_g as a function of moisture content has been determined for gluten (102, 103) and for gliadin and glutenin (104, 105). Determination of T_g for gluten separated from a soft and a hard wheat flour revealed no significant differences (106). However, when T_g was determined for the protein in the wheat flour–water mixture, i.e., without the fractionation, it was found that at equal moisture contents T_g was lower for the hard wheat flour than for the soft flour. T_g of isolated gluten was about 30°C lower than for gluten in flour. It was speculated that these differences are related to the mechanism for the setting of the cookie dough.

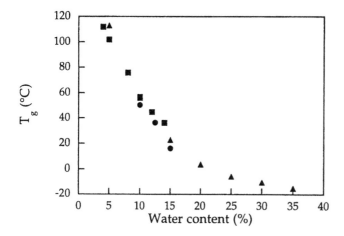

Figure 8 T_g for gluten collected from ▲: TMA (Ref. 14); ●: DSC (Refs. 102 and 190);
■ DMTA (Ref. 191).

When different glutens, gliadins, and glutenins were analyzed (102–105) it was found that T_g for gliadin was lower than for glutenin or gluten, which was attributed to differences in molecular weight. For gliadin and glutenin, transitions from the rubbery zone to an entangled state was observed, and then to a reaction zone where the protein system became cross-linked into a network. For gliadin the reaction zone was found at temperatures above 70°C at moisture contents above 25% (104). Fructose was found to plasticize gluten (102), whereas lipids and emulsifiers were found to have little or no effect on T_g, except for the low-molecular-weight compounds like caproic and hydroxycaproic acids, which had a plasticizing effect (103).

C. Lipids

Polar lipids can give rise to DSC endotherms due to melting of crystalline lipids or to transitions between liquid crystalline phases. When a fat is present there is also the possibility of an exothermic transition between different polymorphic forms. The melting enthalpy, or enthalpy of fusion (ΔH_f), is rather high for these substances. The melting of β-crystals of monoglycerides (monopalmitin, mono-stearin, and monoelaidin), resulting in the formation of the lamellar liquid crys-talline phase, involves enthalpy values in the range 155–195 J/g lipid (83). For solid fats (triglycerides) ΔH_f is in the range 80–170 kJ/mol (107). The transition enthalpies involved in a transition between two liquid crystalline phases are much smaller, and for monolinolein a value of 0.4 J/g lipid was reported for the transition cubic → hexagonal II, and 1.0 J/g lipid for the transition hexagonal II → L2 (83).

Lipids extracted from a wheat flour were reported to give rise to an endo-thermic transition just below −20°C, and a small double endotherm at about 15°C, when heated in the DSC at a scanning rate of 20°C (108). The fatty acid composition of wheat lipids is dominated by linoleic acid (about 60%), palmitic acid (about 20%), and oleic acid (about 15%) (109). The thermal behavior of the extracted lipids seems thus to be dominated by the melting of lipids containing linoleic and oleic acid, respectively. Because the lipid content is low (see Table 1), it is not to be expected that these transitions would be detected when a wheat flour suspension or a wheat flour dough is heated. If a transition should occur between different liquid crystalline phases, for example, between a lamellar liquid crystalline phase and a hexagonal liquid crystalline phase (110), the transition enthalpy will presumably be too small to be detected.

D. Nonstarch Polysaccharides

The most abundant nonstarch polysaccharides (NSPs) in wheat and rye are pento-sans (arabinoxylans and arabinogalactans), whereas in barley and oats, β-glucans are also found (25). As seen in Table 1, the levels of these components are low. They have not been reported to give rise to transitions that are detected by thermal methods. However, their influence on starch gelatinization and retrogradation has

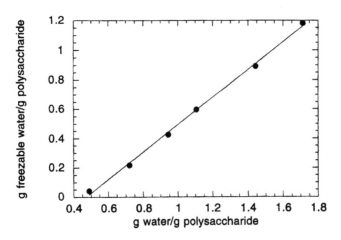

Figure 9 The amount of unfreezeable water is calculated for a pentosan fraction from rye (unpublished data). DSC is used to determine the amount of freezeable water for a range of total amounts of water. Linear regression gave the equation $y = -0.441 + 0.935 * x$ ($r = 0.9989$), and with the amount of freezeable water set to 0 the amount of unfreezeable water was calculated to be 0.47 g/g polysaccharide.

been studied. The hypothesis is that polysaccharides could be used to decrease the retrogradation of starch because they have a better water-holding capacity (WHC) than starch. This is certainly true if the WHC is measured as non-freezeable water, using, for example, DSC (Fig. 9). For water-soluble nonstarch polysaccharides from wheat, barley, oats, rye, and triticale, the amount of non-freezeable water was determined to be 0.41–0.47 g/g dry matter (111), as compared with 0.30–0.38 g/g dry matter for wheat starch (52). When NSP is added to starch, it is not evident what the effect should be, for amylopectin retrogradation has a maximum at a certain water content (66). If NSP holds water, the outcome could then be a decrease as well as an increase in ΔH_c (25).

III. THERMAL ANALYSIS OF MODEL DOUGHS

When two or more flour components are mixed together, thermal transitions resulting from interactions between the components might be observed as well as the transitions arising from the individual components. In this section, measurements on mixtures of different flour components will be described. However, strictly speaking, as indicated by the title of this section, there are not very many studies that involve the measurements on real wheat flour doughs, including yeast.

A. Influence of Wheat Flour Components on the Thermal Transitions of Starch

The DSC thermogram of a wheat flour–water mixture will show endotherms related to the transitions of starch. There is usually a difference in the gelatinization temperature range between the starch and flour, in that T_o, T_m, and T_c are shifted toward higher temperatures in the wheat flour suspension compared to the starch–water suspension (16). The difference in T_m can be around 5°C. This means that components present in the flour will affect starch gelatinization. At low water content (water-to-starch ratio 0.70–0.72) there is also a difference in the appearance of the endotherm, in that the second peak (M1, see Fig. 2) becomes much more pronounced in the wheat flour suspension (112). If the whole wheat kernel is heated, the DSC thermogram will show the endotherm expected for starch gelatinization at the actual water content (113). For a gelatinization endotherm to emerge it can thus be necessary to boil or steam the grain in order to increase the water content inside the kernel.

When a complete dough is studied it can be expected that other ingredients added to the dough will also exert an influence. This means that, for example, an increasing protein content can shift T_c of the starch to temperatures above 100°C; i.e., the starch will not be completely gelatinized during the baking pro-

cess. If this is the case, this will have an effect not only on the rheological properties of the baked product, but also on the retrogradation behavior of the starch
(52, 114). Ungelatinized starch can also be of interest from a nutritional point
of view, for nongelatinized starch has a lower bioavailability than gelatinized
starch (115). Certain ingredients are added with the aim of decreasing staling,
and they should thus be expected to influence the retrogradation-related endotherm. The addition of emulsifiers could be expected to show up in a transition
related to the amylose–lipid complex, and the addition of enzymes could also
be expected to result in changes in the thermogram. So far, no systematic study
has been presented on the influence of mixing and other processing conditions
on the DSC thermogram, although it has been reported that the amount of
freezeable water was higher in a fully developed dough than in unmixed flours
at the same moisture level (108).

1. Gluten

The main component, besides starch, in flour is of course protein (Table 1). As
discussed earlier, the gluten proteins do not give rise to any detectable endotherms
during the DSC scan, but they will shift T_o and T_m for the thermal transitions
related to starch gelatinization (116). It was found that for the two peaks obtained
at water conditions where the gelatinization endotherm has the shoulder (G and
M1 in Fig. 2), gluten causes a linear increase in the peak temperatures when
added to wheat starch, at least at a level of addition up to 0.4 g gluten/g starch.
For the starch and gluten used in that investigation, the following relation was
obtained for the G endotherm (116):

$$T_m(°C) = 60.48 + 7.54 * GS \qquad (1)$$

where GS is the amount of gluten expressed as g gluten/g starch. Gluten from
a flour of poor baking performance affected the temperatures in the same way
as the gluten from the flour of good baking performance used for Eq. (1). It was
also found that the enthalpy (ΔH) for the G endotherm decreased with increasing
gluten level:

$$\Delta H = -1.76 * GS \qquad (2)$$

When starch and gluten were mixed in a 1:1 ratio and analyzed at excess water
conditions in the DSC, a slight increase in T_o and T_m was observed (117). However, when the water-soluble fraction (obtained when the wheat flour was hand-
washed into prime starch, tailings starch, gluten, and water solubles) was mixed
with starch, the effects were much greater: T_o increased from 57.8 to 65.9°C, and
T_m from 64.1 to 71.8°C. In the case of the starch:gluten mixture, T_{cx} was found
to decrease, whereas it increased for the mixture of starch and the water-soluble

fraction. The effect of the water-soluble fraction could be understood from its content of solutes, including sugars.

When the retrogradation was studied for wheat starch–gluten mixtures it was found that ΔH_c decreased somewhat in the presence of gluten, especially after long times (7 days) and high levels of gluten (0.4 g gluten/g starch) (118). The source of gluten did not seem to influence the retrogradation significantly.

Determination of T_g for 1:1 mixtures of gluten and amylopectin was used to study their miscibility (80). At most conditions, two well-separated transitions occurred, with the T_g at the highest temperature attributed to amylopectin or an amylopectin-rich phase, and T_g at the lowest temperature attributed to the gluten-rich phase. When the water content of the mixtures increased, the two T_g values approached each other and eventually overlapped. It was concluded that gluten and amylopectin could then be miscible under certain conditions.

2. Polar Lipids

When a wheat flour–water suspension is heated in the DSC, an endotherm due to the transition of the amylose–lipid complex is detected. This endotherm usually increases in size after a second heating (27), and the presence of uncomplexed lipids in the wheat flour explains an increase in ΔH_{cx} for flour, compared with starch (16, 89). Part of the complexation is thus believed to occur during the heating in the DSC. When a flour is heated, more lipids are present that can form the complex than when starch is heated. In wheat starch the main polar lipid is lysolecithin, whereas in the wheat flour other polar lipids are also present (109).

The ΔH_{cx} of the complex is affected in cereal processing. Drum drying has been found to increase ΔH_{cx} (89), whereas extrusion cooking seems to decrease ΔH_{cx} (119). In this connection the presence of different polymorphic forms of the complex is of interest (120). Depending on the processing conditions, complexes of different crystallinity could be expected, and it could be speculated that such complexes will differ in digestibility (120).

3. Nonstarch Polysaccharides

When 1 or 2% of a soluble arabinoxylan fraction was added to wheat starch at water contents corresponding to the situation in dough (50.8–53.1% water), T_m increased whereas ΔH_{gel} was unaffected (121). The increase in T_m was 2°C when 2% of the soluble arabinoxylan fraction was added. When the water content was much higher (around 75%) there was no observable effect of the added arabinoxylan. For 40% (w/w) aqueous wheat starch slurries, the addition of 1% arabinoxylan or 1% β-glucan did not significantly change the T_m, whereas the gelatinization temperature range and ΔH_{gel} increased (122). The transition of the amylose–lipid complex was not affected by the addition of nonstarch polysaccharides.

When the retrogradation behavior of wheat starch–arabinoxylan mixtures was studied, it was found that an increase in retrogradation could be obtained at certain conditions (121). The interpretation of the results is complicated because the addition of an arabinoxylan fraction to starch will also change the dry matter content, and this will affect starch retrogradation, as discussed earlier.

B. Influence of Added Ingredients on the Thermal Transitions of Starch

A broad range of ingredients is added to a wheat flour dough, and they can all be expected to exert an influence on the gelatinization behavior of the starch and thus have an influence on the DSC endotherm. Moreover, some added ingredients can certainly give rise to a transition of its own in the DSC thermogram, thus complicating the picture even more.

1. Sugar

When sucrose or any other mono- or disaccharide is added to starch, an increase in both T_o and T_m occurs (48, 123, 124). How large the shift will be depends on the type of sugar and on the concentration. When the DSC scan is performed at water conditions giving rise to the double endotherm for starch gelatinization (for example, at a starch:water ratio of 1:1), the double endotherm is shifted into a single endotherm at a starch:water:sugar ratio of 1:1:1 (124, 125). At sugar levels high enough (1.7 g sucrose/g starch in the case of potato starch), almost the whole gelatinization temperature range is shifted to temperatures above 100°C. For products with a high sugar content, part of the starch might thus be ungelatinized even when the temperature has reached 100°C during baking.

The amount of unfreezeable water was determined for starch suspensions in the presence of sugars, and it was found that the onset temperature of gelatinization (determined using microscopy) increased when the amount of unfreezeable water increased (125).

The addition of glucose or glucose syrup will shift T_g of gluten to higher temperatures, compared to the hydrated gluten alone (106).

2. Salt

The effect of added salt on the thermal transitions is rather complicated. For certain salts and concentrations, a decrease in T_o and T_m is observed, whereas for other salts or other concentrations an increase is observed (48, 126). However, in the case of NaCl, effects are observed at concentrations that are not possible for use in bread.

3. Lipids

The presence of endotherms due to the melting of fats or monoglycerides can be a problem when the effect of these components on gelatinization or retrogradation of starch is studied. One way to overcome this problem is of course to use lipids containing unsaturated fatty acids. For applications involving triglycerides, the presence of fat crystals must then not be a prerequisite for the effect (127). In case of monoglycerides, the complexing ability with amylose could be important, and an unsaturated monoglyceride could be expected to perform more poorly than the saturated monoglyceride. However, the poorer complexing ability is a result of the phase behavior of the monoglyceride and not necessarily of its configuration (128). Therefore, unsaturated monoglycerides can be used as models for the saturated ones as long as the conditions are such that the complexing is made possible. Monoglycerides can be transferred into liposomes by dispersing them in sodium cholate solutions in order to facilitate the complex formation (128).

The change in DSC gelatinization characteristics observed for some lipid additives are given in Table 7. The addition of an emulsifier like sodium stearoyl lactylate (SSL) results in an increase in T_o and T_m, whereas lysolecithin decreases these temperatures. The effect of saturated monoglycerides (SMGs) is somewhat difficult to interpret because of the overlapping between the endotherm due to the melting of SMG and the gelatinization endotherm. The decrease in ΔH_{gel} has been interpreted as a result of an exothermic complex formation occurring in the

Table 7 Effect of Added Lipids/Emulsifiers on the Gelatinization Characteristics of Wheat Starch Measured by DSC at Different Water Contents

Additive[a]	Change in		
	T_o (°C)	T_p (°C)	ΔH_{gel} (J/g)
Starch:water = 1:3			
SSL[b]	1.4	1.0	−2.4
SMG[c]	−0.3	−0.8	0.1
Lysolecithin	−1.3	−0.7	−5.2
Lecithin	−0.3	−0.5	−0.5
Starch:water = 1:1			
SSL[b]	−0.5	3.3	−4.0
Lecithin	−3.5	−0.3	−1.2

[a] Addition at the 5% level, calculated on starch.
[b] Sodium stearoyl lactylate.
[c] Saturated monoglycerides.
Source: Data compiled from Refs. 16 and 50.

same temperature range as the gelatinization (27). That lecithin also causes a decrease in ΔH_{gel} indicates that even a diacyl lipid might take part in complex formation, although not to the same extent as the monoacyl lipids (50).

When wheat starch is heated in corn oil, no transitions at all are detected in the DSC thermogram (129). The addition of low levels of shortening (6% calculated on flour) does not influence the gelatinization of starch (112). Even an increase in the amount of triglyceride to 50% will not influence the gelatinization behavior of starch. It was found that at a wheat flour–water ratio of 1:1 and at a level of addition of oil of 50% calculated on wheat flour, no effects on T_o, T_m, or ΔH_{gel} were obtained (127). However, if the fat was added dispersed as an L2 phase, the gelatinization endotherm decreased in size and shifted into a higher temperature range (130). At high levels of added triglycerides (coconut oil), an endotherm due to the melting of the fat is observed (131). Depending on the mixing temperature and hydration, the appearance of the thermogram will differ.

4.   Enzymes

The effect of pretreatment of barley starch with α-amylase was investigated by DSC. Starch was kept for 3 h at 50, 55, or 60°C, either with enzyme added or without (132). There was a slight increase in T_m for the enzyme-treated starches, whereas the ΔH_{gel} values did not differ significantly.

When wheat starch was annealed, i.e., kept at a temperature below T_o, an increase in T_o and T_m was observed, whereas ΔT decreased and ΔH_{gel} remained unchanged (133). Pancreatin hydrolysis of the starches caused a slight increase in T_o and T_m of native starch, whereas the DSC thermograms of annealed and hydrolyzed starch were essentially unaffected. If the starch was hydrolyzed before annealing, the annealing effect on T_o and T_m increased.

5.   Hydrocolloids

The influence of hydrocolloids on the gelatinization behavior of starch has been investigated for some cereal starches, including corn (68). It was found that T_o, T_m, and T_c increased with increasing levels of hydrocolloid, resulting in a broadening of ΔT. ΔH_{gel} was found to increase, but because the water content changed at the same time as the addition of the hydrocolloid, the interpretation was difficult. However, when the samples were compared at the same volume fraction of water it was found that ΔH_{gel} was independent of the addition of a hydrocolloid. The change in gelatinization temperature was interpreted as showing that water was less available for the starch in the presence of the hydrocolloid. The additions of certain hydrocolloids (guar gum and xanthan gum) at the 1% level did not affect the gelatinization peak temperature, whereas the gelatinization temperature range and ΔH_{gel} increased (122). The transition of the amylose–lipid complex was not affected by the presence of hydrocolloids.

IV. RELATIONS BETWEEN THERMAL BEHAVIOR OF STARCH AND PRODUCT QUALITY

The thermal behavior of cereal components, or more or less complicated doughs, is of course studied in order to learn more about the baking process and the baked product. In this section a few examples will be given on how thermal behavior has been used to understand, and even predict, the properties and quality of the end product. The staling of bread, which is routinely assayed using DSC, will be described in the next section.

A. Changes in Starch Properties Due to Processing

1. Damaged Starch

During milling, part of the starch can be mechanically damaged, which causes these granules to absorb water and swell even at room temperature (134). Such starch granules are more easily digested by enzymes, and methods for the determination of damaged starch include either the enzymatic availability or the extraction of water-soluble components (135, 136). However, the presence of damaged starch can also be detected in the DSC endotherm (16, 137). Starch damage results in a loss of crystalline order and a loss in double helix content (137). This is possible to detect in the DSC thermogram as a decrease in ΔH_{gel} and a decrease in T_o as well as in T_m and T_c. The appearance of the DSC endotherm changes: The G endotherm decreases and eventually disappears with increasing levels of starch damage (25). After severe damage, neither the G endotherm nor the M1 endotherm is detected (137).

2. Annealing and Heat-Moisture Treatments

The DSC characteristics of the gelatinization of starch can be affected by annealing and heat-moisture treatments (133, 138). Annealing involves keeping a starch–water suspension at temperatures above T_g but below T_o. Annealing of wheat starch at 50°C and 75% water resulted in an increase in T_o, T_m, and T_c, a decrease in ΔT, and an increase in ΔH_{gel} (45, 81). An effect on the temperatures was observed after only 0.5 h, but an effect on ΔH_{gel} was not observed until after 48 h. When the moisture content during annealing is low enough to allow the appearance of the M1 endotherm for the native starch, it has been found that the appearance of the endotherm will change so that it looks more like the endotherm obtained at excess-water conditions, albeit shifted into higher temperature (139). An increase in ΔH_{gel} can sometimes be observed, indicating an increase in crystallinity (140), whereas an unchanged ΔH_{gel} is interpreted as indicating that no gelatinization has taken place (139). When the treatment is performed at a lower water content and a higher temperature than corresponding to the annealing conditions, the treatment is described in the literature as heat-moisture treatment (133, 138). However, in principle there is no difference between the treatments; when the water content decreases, both T_g and T_o increase (Figs. 1 and 6), and thus the treatment temperature has to be increased. For wheat starch, heat-moisture treated at 100°C for 16 h at moisture contents of 10, 20, and 30%, T_o, T_m, and T_c were found to increase with increasing moisture level, whereas ΔT and ΔH_{gel} were more or less unchanged (141). The duration of the treatment did not influence the DSC characteristics after the first few hours.

Annealing and heat-moisture conditions could in principle be used to modify the properties of starch and thus produce physicochemically modified starch. However, for baking purposes, such starch seems not to offer any advantages, except perhaps in the case of staling. Steeping of wheat starch at temperatures in the range of 25–50°C for up to 72 h did not improve the baking quality of the starch; on the contrary, prolonged heating times at the highest temperature resulted in deteriorated baking performance (142). However, the crumb was softer and did not firm as fast as with the bread baked from nontreated wheat.

Annealing or heat-moisture treatment conditions can be applied unintentionally to starch. Annealing conditions can be present during drying of grains with hot air, which results in an increase in T_m and in ΔH_{gel} (143). Depending on the climate during the growth period, the starch gelatinization properties might change, and an increase in the environment temperature results in an increase in T_m (144). ΔH_{gel} was not found to be affected by the growing temperature, but it differed between seasons.

B. Gelatinization Temperature and Fixation of Crumb Structure

The different DSC characteristics need to be correlated with baking performance in order to be translated into relevant information for the baking process. One interesting parameter is the onset of gelatinization, T_o. This temperature indicates the temperature when changes in the starch component begin. Because the temperature range of the DSC endotherm coincides with the temperature range of the disappearance of the X-ray diffraction pattern and the loss of molecular order (30, 31), T_o could be said to indicate the start of the primary events during gelatinization. Then, at higher temperatures, swelling, leaking, and, consequently, changes in rheological properties will occur (25). Rheological measurements indicate that during heating of a wheat flour dough there will first be a decrease in viscosity, followed by an increase (145). The increase in viscosity was attributed to the starch gelatinization and was observed to start at a temperature of 50–65°C, depending on the heating rate. At a heating rate of 9°C/min (corresponding to many DSC studies), the minimum in apparent viscosity was observed at about 64°C. From this it can be concluded that the fixation of the crumb structure will not occur until T_o is passed. It thus follows that if T_o is high there will be a longer period of time for volume expansion to occur, for the starch gelatinization and crumb fixation stop the volume expansion. A relation was indeed found for the gelatinization peak temperature (T_m) and loaf volume ($r = +0.964$, $P < 0.05$) for six starches used in reconstitution baking experiments (20). ΔH_{gel} did not correlate to loaf volume. The variation in both T_m and ΔH_{gel} was rather small; T_m was in the range 58.9–64.4°C, whereas ΔH_{gel} was in the range 9.4–11.5 J/g starch. For these experiments, the starches were extracted and then used together with a common source of gluten and water solubles. In another study, T_m and ΔH_{gel} were determined for starches extracted from 10 different wheat varieties and correlated with the baking result (146). In that study, no correlation was found, neither for T_m nor for ΔH_{gel}. The range in T_m was even smaller, only 58.4–61.1°C. The ranges in T_m covered in these two investigations are thus small, 5.5°C and 2.7°C. However, one important aspect of these two studies is that in both cases the DSC characteristics have been determined for extracted starch, whereas in baking all other components, from the flour and other added ingredients, are present. As discussed earlier, the main difference between the DSC thermogram for starch and flour at the same water content is that in the flour the transitions are shifted to a higher temperature (16). When the baking experiments were carried out as reconstitution experiments, all starches were used in the same flour environment; and if other components shifted T_o and T_m, the shift would presumably be the same for all starches in the investigation.

In a study of 69 different wheat flours, where the DSC characteristics were determined for wheat flour–water suspension, the correlation between T_o and baking quality was not significant, whereas for T_m and baking quality a significant correlation (at the 0.05 level) was found, with $r = 0.44$ (147). A significant correlation (at the 0.01 level) was also found for baking quality and ΔH_{gel}, with $r = -0.54$. When the material was divided into smaller groups and a group of durum wheat was included, it was found that the highest T_m was obtained for durum flours, followed by good baking wheats. The lowest values were observed for winter wheats aimed for feed and biscuit. The range in temperatures was again small, as in the studies cited earlier. For durum, T_m was in the range 60.6–65.6°C, with an average of 63.4°C; for winter wheats of good baking quality the range was 60.3–65.5°C, with an average of 62.1°C; for feeding and biscuit wheats the range was 59.6–63.6°C, with an average of 60.8°C. Although the differences are small, the results seem to indicate that high T_o and T_m are positive for good baking results, and perhaps also a low ΔH_{gel}. Such a result has several implications. Additives, such as emulsifiers, usually increase T_o and T_m (see Table 7). The effect is small. But because the difference in temperatures between wheat varieties very well can be of this magnitude, perhaps such an increase is enough for the emulsifiers to have an effect on baking volume, also due to the influence on starch gelatinization temperature.

If our conclusion, that high T_o and T_m together with low ΔH_{gel} improves baking performance, is valid, then the effect of climate on the baking quality of wheat flour (148) might also be due to starch gelatinization properties.

C. Gelatinization Enthalpy (ΔH_{gel}) and Degree of Gelatinization

The gelatinization enthalpy (ΔH_{gel}) is a measure of the amount of starch crystallites and of their quality. However, the presence of a glass transition complicates the interpretation of the endotherm, and it is not possible to assign the total enthalpy value to a certain event. It has even been suggested that ΔQ, signifying a total thermal effect, should be used instead, because a sum of thermal effects is measured (149). However, it is evident that crystallinity remains until the complete gelatinization temperature range has been passed (30). It should then be possible to calculate a degree of gelatinization (%DG) from a partial enthalpy (ΔH_{part}), either remaining at a certain temperature or after a process, compared to the total enthalpy (ΔH_{gel}) according to Eq. (3):

$$\%DG = \left(1 - \frac{\Delta H_{part}}{\Delta H_{gel}} \right) * 100 \tag{3}$$

Figure 10 shows %DG plotted against temperature for a wheat starch–water suspension heated in the DSC.

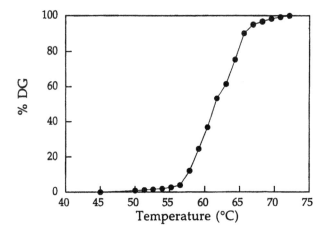

Figure 10   %DG as a function of temperature during the gelatinization of wheat starch at a starch:water ratio of 1:3. (Reprinted from Ref. 192 with permission from Elsevier Science.)

The degree of gelatinization, calculated according to Eq. (3), has been compared with %DG determined using enzymatic methods for wheat starch samples at excess-water conditions (115). For starch suspensions heated to 47°C, the DSC method and the enzymatic method gave similar results (%DG was 14.2 and 14.1%, respectively). However, when the heating was performed to higher temperatures, the enzymatic method gave higher values of %DG, up to 59°C. Then the samples were almost completely gelatinized. In other studies it has been observed that X-ray diffraction is not as sensitive as enzymatic methods for detecting the initial swelling of starch granules (150). In the case of α-amylase degradation, there seems to be a good correlation between the gelatinization enthalpy determined by DSC and the amount of starch degraded with α-amylase (151). This correlation, which was negative, was found for starches from different raw materials, as well as for wheat starch in different products.

D. Some Real Doughs and Products

1. Bread Doughs

A bread dough is of course much more complicated than the model samples discussed so far, but most investigations have been carried out on samples that are not complete bread doughs. As shown in previous sections, most added ingredients will influence the thermal transitions of starch. However, although many examples are found concerning the influence of an individual component, much less is known about combinations of several components. For a wheat flour–water dough it was observed that at the addition of 2% NaCl, T_o increased from 56 to 62°C, with 6% sucrose to 61°C, with both NaCl and sucrose to 66°C, and with also 3% shortening to 67°C (112).

Differential scanning calorimetry has been used to determine the amount of unfreezeable water for flour and dough (108). It was found that more freezeable water was present in the fully developed dough, compared with the wheat flour suspension. The amount of unfreezeable water has been determined to be 0.3 g/g dry matter for several wheat flours (152). There was no difference between flours of different mixing strength.

It might have been expected that thermal methods, including the construction of state diagrams, would gain much interest in the studies of frozen-dough stability. However, this seems not to have been the case, although the frozen-dough market is expanding and is still facing many problems. The possibility of studying the behavior of a frozen dough from a food polymer approach was discussed by Kulp in 1995 (153). It was concluded that interactions between components are likely to complicate the description of the complete dough. The main concern regarding frozen-dough stability seems to be the survival of the yeast cells (154, 155), although more recently attention has been paid to structural considerations (156, 157). The water distribution in doughs has been analyzed, and an attempt to measure T_g for frozen doughs showed T_g values around $-30°C$ (158). There was a great variation in T_g, both between methods and between different flours, stressing the difficulties in determining T_g for such a complicated material as a wheat flour dough.

2. Cake Batters

A cake batter differs from a bread dough in its high fat and sugar contents. During the baking of cakes, the temperature inside the batter will reach 100°C at different times, depending on the position, but the crumb temperature will never exceed 100°C (159). The baking of angle food cake was simulated in the DSC, and it was observed that protein denaturation (egg white proteins) and starch gelatinization was shifted to higher temperatures in the presence of sucrose at concentrations corresponding to a cake recipe (160). This allowed the volume expansion to continue until the maximum temperature during baking was obtained, before the setting of the structure took place.

An unemulsified cake batter gave a DSC thermogram very similar to that of wheat starch in the corresponding sucrose solution (129). The T_c of the endotherms was just below 100°C. When an emulsifier (composed of 85–90% glycerol monostearate and 10–15% glycerol monopalmitate) was added, an endotherm just below 60°C was detected, whereas the addition of an unsaturated monoglyceride did not give rise to any new peaks.

V. STALING OF BREAD

The staling of bread is defined as "almost any change, short of microbiological spoiling, that occurs in bread during storage which makes it less acceptable to a consumer" (58). The staling of bread has been recognized as a problem during centuries, and the typical changes occurring in crumb and crust during storage of bread are well known. It seems that the changes in the crust can be explained from a change in T_g because of water uptake from the crumb or from the surroundings (161). As the moisture content increases in the crust, this could be expected to influence T_g so that it decreases and approaches room temperature (162), thus causing the change from crisp to rubbery texture. In the center of the crumb, the moisture content will decrease during storage (161), but T_g will still be well below room temperature at the moisture content of bread (162). The increase in crystallinity of starch, occurring in parallel with other changes during staling, points to the involvement of starch retrogradation in the staling of bread. However, several mechanisms, besides the involvement of starch retrogradation, have been proposed, including water redistribution (163), gluten changes (164), and protein–starch interactions (165). In the present chapter only those changes that have been analyzed using thermal methods (mainly DSC) are dealt with.

When bread crumb is heated in the DSC, an endotherm in the temperature range 50–80°C is obtained (52), and this endotherm has been termed the *staling endotherm* (53). An important question is whether the staling endotherm and the crumb firmness develop independent of each other. As will be discussed later, there need not be a direct relation between the amount of starch retrogradation (measured as ΔH_c) and crumb firmness. The size of crystallites and their ability to form three-dimensional networks are also of importance.

A. The Staling Endotherm

The staling endotherm of bread crumb is present at the same conditions as the endotherm due to the melting of retrograded amylopectin (see Fig. 3). The development of this endotherm in bread is affected by the same conditions as in a starch gel and will thus be influenced by, for example, water content (161) and the addition of emulsifiers (53, 62). The staling endotherm has been concluded to be a reliable measure of the retrogradation of amylopectin. However, an exothermic transition, at about 20°C, has also been detected for stale bread (12). This exotherm, which was partially reversible, was interpreted to be related to the presence of two binding states of water, which were detected using TGA. The mechanism for many antistaling additives was suggested to be related to their influence on redistribution of water.

The development of the staling endotherm and the development of firmness has been shown to occur in parallel, at least for certain types of breads (53, 62), although the increase in firmness seems to continue for a longer period of time than the starch retrogradation (166). Keetels and co-workers (167–169) used gluten-free bread as a model for studying starch crystallization and its influence on the mechanical properties of bread crumb. The mechanical properties of recrystallized amylopectin can easily be studied in a starch gel, but the results are not easily translated from the gel to the foam structure of bread, and therefore gluten-free breads were baked. A potato starch bread was found to be much stiffer than a wheat starch bread, and its stiffness increased during storage. The melting enthalpy of recrystallized amylopectin, ΔH_c, was higher for the potato starch bread than for the wheat starch bread. The Young modulus was also found to increase when ΔH_c increased, but there was not a linear relationship. Moreover, a certain ΔH_c value for wheat starch corresponded to a higher increase in stiffness, compared with potato starch (although the same ΔH_c was reached after a much shorter time for the potato starch bread). As pointed out in the introduction to this chapter, the ability of the amylopectin crystallites to form three-dimensional networks should be taken into account. This means that firmness could be related to the formation of a cross-linked network and not necessarily to the amount of amylopectin crystals. Such a mechanism has also been suggested to explain the influence of enzymes (170) and emulsifiers (25) on bread staling.

At the same time as the staling endotherm is measured it is possible to analyze the amylose–lipid complex (53, 62, 118). There seems to be no change in ΔH_{cx} with time (53, 60), but it has been shown that ΔH_{cx} is higher near the crust, compared to the center of the bread loaf (161).

B. Water Relations

Transport of water between crust and crumb takes place during the storage of bread, and drying of the crumb will enhance all other negative changes occurring during storage (161, 171). However, a transport of water, or redistribution of water, between starch and gluten has also been suggested to be involved in the staling of bread (163). As a means to investigate the state of water during staling, measurements of unfreezeable water have been performed using DSC. The amount of "bound" or unfreezeable water can be determined for starch (and other macromolecules) using DSC (172). The experimental conditions involve keeping the sample for a certain period of time at a temperature well below the freezing point of water, and then from the ice melting endotherm during heating calculate the amount of water frozen during the treatment (152). With knowledge about the total amount of water present, the percentage of unfreezeable water can be calculated.

The amount of unfreezeable water was found to increase during the storage of concentrated starch gels (172). In bread samples the freezeable water was found to decrease, whereas the unfreezeable water was more or less unchanged. Such an effect could be explained by an uneven distribution of water between the components in bread and a redistribution during storage. Freezeable water for standard white bread was determined using DSC, and the amount of freezeable water was found to follow the equation

$$y = 1.51 * x - 50.1 \tag{4}$$

where x is moisture content in % at a moisture content above 33% (173). Bread of different water contents was studied by dynamic mechanical analysis (DMA). When freezeable water (according to DSC) was present, the transitions in DMA were sharp; there was an evident peak in tan δ and a sharp drop in E' with increasing temperature. The transition was located around 0°C and was attributed mainly to the melting of ice. At moisture contents when no freezeable water was present, the curves of tan δ and E' against temperature became flat, and it was difficult to define any transition temperature. This behavior was discussed in relation to bread being a complex system that is phase separated. During storage the same changes in the DMA curves were obtained only when the breads were allowed to dry during storage. TGA studies have indicated the presence of two binding states of water (12).

C. Influence of Baking and Storage Conditions

Starch retrogradation is influenced by conditions during the baking process. In simple starch gels (wheat starch–water mixtures), the retrogradation (measured as ΔH_c) will depend on the final temperature reached during gelatinization; the higher the degree of gelatinization, the more retrogradation (52). This is of course self-evident—if the crystallites didn't melt in the first place, there can be no recrystallization. However, for bread crumb a relation between baking temperature and retrogradation has also been found: The higher the baking temperature, the more retrogradation (166). When wheat starch gels of differing water content were stored it was found that the maximum enthalpy developed at 50–70% water when stored at 4°C, and at 50% water when stored at 26 or 32°C (174). The maximum temperature in the retrogradation endotherm was at a much higher temperature when the gels were stored at 26 and 32°C, compared to 4°C. There was also an influence of water content, and usually the higher T_m was obtained at the lower moisture content. It was concluded that the enthalpy was influenced by the nucleation rate, and the peak temperature by the propagation rate. When bread crumb was analyzed after storage at 4, 25, and 40°C, respectively, it seemed that with time the ΔH-values all converged to the same value (65). The breads were first stored in plastic bags for the appropriate storage time, then frozen and lyophilized. Before DSC analysis, the dried crumb was mixed with water (1 part crumb : 2 parts water). When the bread storage temperature increased, the endothermic transition of the melting of recrystallized amylopectin was shifted toward a higher temperature and it became narrower. This was attributed to annealing when the samples were stored at temperatures close to the melting temperature. The pronounced narrowing of the endotherm was taken to show that more perfect crystals had formed.

The relevance of storage temperature is easily understood from the fact that starch retrogradation is a crystallization process. The nucleation benefits from supercooling, i.e., a temperature as far below the crystallite melting temperature as possible, whereas the growth of crystallites benefits from a temperature as close to the melting temperature (T_m) as possible (64). It is observed in Figure 3 that the onset of melting of amylopectin crystallites stored at room temperature is about 50°C. This is thus the upper limit for retrogradation to occur. Storage at a temperature T above this will cause immediate melting. The lower limit of storage temperature is fixed by the T_g of the gelatinized starch. Below this temperature the crystallization will be extremely slow, if it occurs at all (175). The crystallization rate as well as the extent of crystallization will thus depend on $T - T_g$. At low $T - T_g$ conditions, the extent of crystallization was found to be low, because the crystal growth was restricted, although the nucleation was fast; at high $T - T_g$ conditions, the crystal growth was fast, whereas nucleation was slow (175).

D. Influence of Ingredients

Many of the ingredients used in baking will affect the rate and extent of staling, and these ingredients could also then be expected to influence starch retrogradation. However, although DSC has been used to follow starch retrogradation, there are very few systematic studies on the influence of ingredients, except for emulsifiers. In the following, some examples will be given on results obtained using mainly DSC.

1. Sugars

The influence of sugar on starch retrogradation has been studied, but no simple correlation between the presence of a sugar and its effect on retrogradation has been found. When sucrose, glucose, or fructose was added to isolated amylopectin from wheat or corn, no effect on the ΔH_c values for storage during 4 weeks was observed (74). An increase in T_o for the melting of retrograded amylopectin was observed, however.

For wheat starch gels, ribose was found to completely inhibit the development of the B-pattern, when added at the level starch:sugar:water 1:1:1 (176). Ribose also decreased the firmness of the gels considerably. Glucose and sucrose also decreased the development of crystallinity and gel firmness, sucrose more so than glucose. Both ribose and xylose seem to suppress the rate of amylopectin crystallization (177). For fructose, results were obtained that indicated an increase in crystallization of the components in starch. Low-resolution nuclear magnetic resonance has shown that the addition of sucrose to wheat starch gels decreases the development of the signal from solid phase (178).

2. Salt

The value of ΔH_c decreased with increasing salt concentration in the range investigated (0–4.43%) (179). NaCl [4% (w/w)] was found to have no influence on the retrogradation of amylopectin from corn, but decreased the ΔH_c for wheat amylopectin (74).

3. Lipids

The addition of triglycerides (rapeseed oil, sunflower oil, and soya oil) at the 50% level, calculated on wheat flour, to a wheat flour–water mixture reduced the amylopectin retrogradation (180). The effect was largest for soya oil.

Emulsifiers and surfactants decrease ΔH_c (62, 93, 118), although at a level of 0.5% SSL had no effect on the retrogradation of corn or wheat amylopectin (74). Different surfactants, including distilled monoglycerides, SSL, and sucrose esters (the latter varying in HLB number from 7–16), were found to decrease the retrogradation of amylopectin when added to bread (181). Diacetyl tartaric acid ester of monoglycerides (DATEM) and monoglycerides both depress the development of firmness of white pan bread (62). They both also reduce the retrogradation of amylopectin, measured as ΔH_c using DSC. Monoglycerides were more effective than DATEM in reducing ΔH_c. Monoglycerides also gave higher ΔH_{cx} values than DATEM.

4. Nonstarch Polysaccharides

The NSPs are often described as decreasing the staling of bread. This is attributed to their water-holding capacity. However, regarding their effect on starch retrogradation, the observations made do not agree with this assumption. ΔH_c was unaffected after day 1 at 5°C for wheat starch gels with the addition of β-glucan, arabinoxylan, guar gum, and xanthan gum, respectively, all added at the level of 1% (122). After 2 days, xanthan gum showed decreased ΔH_c values, whereas the other additives showed increased ΔH_c values compared with the control without added gums. After 7 days of storage at 5°C, xanthan gum still gave lower values, whereas all other samples showed similar enthalpies. The retrogradation of sweet potato starch was studied in the presence of cellulose derivatives (starch:cellulose = 9:1 to correspond to the proportion between starch and fiber in sweet potato) (182). It was found that the only cellulose derivative that decreased retrogradation (measured as a decrease in ΔH_c and G') was the soluble methylcellulose. The addition of xanthan gum was found to decrease the syneresis and rheological changes of gelatinized corn starch suspension stored at temperatures below 0°C (183). The addition of xanthan gum had no effect on amylopectin retrogradation and no effect on ice formation. It was therefore concluded that xanthan–amylose interactions replaced amylose–amylose interactions, and in this way the retrogradation of amylose was inhibited.

5. pH and Acids

For wheat starch gels the largest extent of retrogradation has been found to occur at pH 5.6, and it was much lower at pH 9.4. For pH 4.4 and 7.8 the values were in between (95). Citric acid did not influence ΔH_c of corn amylopectin but increased the ΔH_c value of wheat amylopectin (74). HCl had no effect on the wheat amylopectin.

VI. CONCLUSIONS

The thermal transitions that cereal components undergo when the temperature is changed have been exemplified by mostly wheat components, and among them by wheat starch. However, other components (and cereals) also are being analyzed, and this can be expected to increase in the future. The analysis of more complicated model systems and real products can also be expected to increase in the future. That thermal properties can be analyzed in real products has been shown for the successful measuring of starch retrogradation in bread.

The dominant technique for analyzing thermal transitions of cereal components has so far been DSC. However, other techniques, such as MDSC, microcalorimetry, and TGA, have been introduced and could be expected to be used to a greater extent in the future. Microcalorimetry seems to be very suitable for studying interactions among cereal components. Combinations of different techniques should also be used more. The DSC technique does not really mimic baking, for during baking there will be a loss of moisture, whereas in the DSC this loss of moisture results in an increase in pressure. Rheological measurements could be expected to give more information when performed at different temperatures. For example, rheological measurements for accessing wheat flour quality might be improved if the measurements are also carried out at temperatures relevant for baking and not just for mixing and fermentation.

REFERENCES

1. Rask, C. The heat transfer in a convection oven—influence on some product characteristics. In: Cereal Science and Technology in Sweden, edited by N.-G. Asp. Lund, Sweden: Lund University, 1988, pp 148–157.
2. McIver, R. G., D. W. E. Axford, K. H. Colwell, and G. A. H. Elton. Kinetic study of the retrogradation of gelatinised starch. J. Sci. Food Agric. 19:560–563, 1968.
3. Colwell, K. H., D. W. E. Axford, N. Chamberlain, and G. A. H. Elton. Effect of storage temperature on the ageing of concentrated wheat starch gels. J. Sci. Fd Agric. 20:550, 1969.
4. Donovan, J. W. Phase transitions of the starch–water system. Biopolymers 18:263–275, 1979.
5. Eliasson, A.-C. Effect of water content on the gelatinization of wheat starch. Starch/Stärke 32:270–272, 1980.
6. Myers, C. D. Study of thermodynamics and kinetics of protein stability by thermal analysis. In: Thermal Analysis of Foods, edited by V. R. Harwalkar and C.-Y. Ma. New York: Elsevier Applied Science, 1990, pp 16–50.
7. Biliaderis, C. G. Thermal analysis of food carbohydrates. In: Thermal Analysis of Foods, edited by V. R. Harwalkar and C.-Y. Ma. New York: Elsevier Applied Science, 1990, pp 168–220.

8. Bell, L. N., and D. E. Touma. Glass transition temperatures determined using a temperature-cycling differential scanning calorimeter. J. Food Sci. 61:807–810, 828, 1996.

9. Shiotsubo, T., and K. Takahashi. Changes in enthalpy and heat capacity associated with the gelatinization of potato starch, as evaluated from isothermal calorimetry. Carbohydr. Res. 158:1–6, 1986.

10. Silverio, J., E. Svensson, A.-C. Eliasson, and G. Olofsson. Isothermal microcalorimetric studies on starch retrogradation. J. Thermal Analysis 47:1179–1200, 1996.

11. Kubik, S., and G. Wulff. Characterization and chemical modification of amylose complexes. Starch/Stärke 45:220–225, 1993.

12. Schiraldi, A., L. Piazza, and M. Riva. Bread staling: a calorimetric approach. Cereal Chem. 73:32–39, 1996.

13. Biliaderis, C. G., C. M. Page, T. J. Maurice, and B. O. Juliano. Thermal characterization of rice starches: a polymeric approach to phase transitions of granular starch. J. Agric. Food Chem. 34:6–14, 1986.

14. Huang, V. T., L. Haynes, H. Levine, and L. Slade. Glass transitions in starch, gluten and bread as measured. Dielectric spectroscopy and TMA methods. J. Thermal Anal. 47:1289–1298, 1996.

15. Morris, V. J. Bacterial polysaccharides. In: Bacterial polysaccharides, edited by A. M. Stephen. New York: Marcel Dekker, 1995, pp 341–375.

16. Eliasson, A.-C. Some physico-chemical properties of wheat starch. In: Wheat End-Use Properties. Wheat and Flour Characterization for Specific End-Use, edited by H. Salovaara. Helsinki: University of Helsinki, 1989, pp 355–364.

17. Morrison, W. R., and D. C. Scott. Measurement of the dimensions of wheat starch granule populations using a Coulter Counter with 100-channel analyzer. J. Cereal Sci. 4:13–17, 1986.

18. Eliasson, A.-C., and R. Karlsson. Gelatinization properties of different size classes of wheat starch granules measured with differential scanning calorimetry. Starch/ Stärke 35:130–133, 1983.

19. Soulaka, A. B., and W. R. Morrison. The amylose and lipid contents, dimensions, and gelatinization characteristics of some wheat starches and their A- and B-granule fraction. J. Sci. Food Agric. 36:709–718, 1985.

20. Soulaka, A. B., and W. R. Morrison. The bread baking quality of six wheat starches differing in composition and physical properties. J. Sci. Food Agric. 36:719–727, 1985.

21. Fredriksson, H., J. Silverio, R. Andersson, A.-C. Eliasson, and P. Åman. The influence of amylose and amylopectin characteristics on gelatinization and retrogradation properties of different starches. Carbohydr. Polym. 35:119–134, 1998.

22. Hizukuri, S. Starch: analytical aspects. In: Carbohydrates in Food, edited by A.-C. Eliasson. New York: Marcel Dekker, 1996, pp 347–429.

23. Nakamura, T., M. Yamamori, H. Hirano, S. Hidaka, and T. Nagamine. Production of waxy (amylose-free) wheats. Mol. Gen. Genet. 248:253–259, 1995.

24. Hayakawa, K., K. Tanaka, T. Nakamura, S. Endo, and T. Hoshino. Quality characteristics of waxy hexaploid wheat (*Triticum aestivum* L.): properties of starch gelatinization and retrogradation. Cereal Chem. 74:576–580, 1997.

25. Eliasson, A.-C., and K. Larsson. Cereals in Breadmaking: A Molecular/Colloidal Approach. New York: Marcel Dekker, 1993.
26. French, D. Organization of starch granules. In: Starch Chemistry and Technology (2nd ed.), edited by R. L. Whistler, J. N. BeMiller, and E. F. Paschall. Orlando: Academic Press, 1984, pp 183–247.
27. Eliasson, A.-C. Interactions between starch and lipids studied by DSC. Thermochim. Acta 246:343–356, 1994.
28. Zobel, H. F. Molecules to granules: a comprehensive starch review. Starch/Stärke 40:44–50, 1988.
29. Sarko, A., and H.-C. H. Wu. The crystal structures of A-, B- and C-polymorphs of amylose and starch. Starch/Stärke 30:73–78, 1978.
30. Svensson, E. Crystalline properties of starch. Lund, Sweden: Lund University, 1996.
31. Cooke, D., and M. J. Gidley. Loss of crystalline and molecular order during starch gelatinization: origin of the enthalpic transition. Carbohydr. Res. 227:103–112, 1992.
32. Eliasson, A.-C., K. Larsson, S. Andersson, S. T. Hyde, R. Nesper, and H.-G. von Schnering. On the structure of native starch—an analogue to the quartz structure. Starch/Stärke 39:147–152, 1987.
33. de Willigen, A. H. A. The rheology of starch. In: Examination and Analysis of Starch and Starch Products, edited by J. A. Radley. London: Applied Science Publishers, 1976, pp 61–90.
34. AACC22-10. Approved Methods of the American Association of Cereal Chemists. 8, 1983.
35. Bloksma, A. H. Rheology of the breadmaking process. Cereal Foods World 35: 228–236, 1990.
36. Stevens, D. J., and G. A. H. Elton. Thermal properties of the starch/water system. Part I. Measurement of heat of gelatinization by differential scanning calorimetry. Starch/Stärke 23:8–11, 1971.
37. Eliasson, A.-C. Viscoelastic behavior during the gelatinization of starch. I. Comparison of wheat, maize, potato and waxy-barley starches. J. Text. Stud. 17:253–265, 1986.
38. Eliasson, A.-C., and L. Bohlin. Rheological properties of concentrated wheat starch gels. Starch/Stärke 34:267–271, 1982.
39. Rolee, A., and M. Le Meste. Thermomechanical behavior of concentrated starch–water preparations. Cereal Chem. 74:581–588, 1997.
40. Kugimiya, M., J. W. Donovan, and R. Y. Wong. Phase transitions of amylose–lipid complexes in starches: a calorimetric study. Starch/Stärke 32:265–270, 1980.
41. Shiotsubo, T., and K. Takahashi. Differential thermal analysis of potato starch gelatinization. Agric. Biol. Chem. 48:9–17, 1984.
42. Biliaderis, C. G., C. M. Page, and T. J. Maurice. On the multiple melting transitions of starch/monoglyceride systems. Food Chemistry 22:279–295, 1986.
43. Biliaderis, C. G. The structure and interactions of starch with food constituents. Can. J. Physiol. Pharmacol. 69:60–78, 1991.
44. Seow, C. C., and C. H. Teo. Annealing of granular rice starches—interpretation

of the effect on phase transitions associated with gelatinization. Starch/Stärke 45: 345–351, 1993.

45. Hoover, R., and T. Vasanthan. The effect of annealing on the physicochemical properties of wheat, oat, potato and lentil starches. J. Food Biochem. 17:303–325, 1994.

46. Rask, C. Thermal properties of dough and bakery products: a review of published data. J. Food Engineering 9:167–193, 1989.

47. Slade, L., and H. Levine. Non-equilibrium melting of native granular starch: Part I. Temperature location of the glass transition associated with gelatinization of A-type cereal starches. Carbohydr. Polym. 8:183–208, 1988.

48. Evans, I. D., and D. R. Haisman. The effects of solutes on the gelatinization temperature of potato starch. Starch/Stärke 34:224–231, 1982.

49. Vasanthan, T., and R. S. Bhatty. Physicochemical properties of small- and large-granule starches of waxy, regular, and high-amylose barleys. Cereal Chem. 73: 199–207, 1996.

50. Eliasson, A.-C. On the effects of surface active agents on the gelatinization of starch—a calorimetric investigation. Carbohydr. Polym. 6:463–476, 1986.

51. Eliasson, A.-C., H. Finstad, and G. Ljunger. A study of starch–lipid interactions for some native and modified maize starches. Starch/Stärke 40:95–100, 1988.

52. Eliasson, A.-C. Retrogradation of starch as measured by differential scanning calorimetry. In: New Approaches to Research on Cereal Carbohydrates, edited by R. D. Hill and L. Munck. Amsterdam: Elsevier Science, 1985, pp 93–98.

53. Russell, P. L. A kinetic study of bread staling by differential scanning calorimetry and compressibility measurements. The effect of added monoglycerides. J. Cereal Sci. 1:297–303, 1983.

54. Sievert, D., and Y. Pomeranz. Enzyme-resistant starch. II. Differential scanning calorimetry studies on heat-treated starches and enzyme-resistant starch residues. Cereal Chem. 67:217–221, 1990.

55. Sievert, D., and P. Würsch. Thermal behavior of potato amylose and enzyme-resistant starch from maize. Cereal Chem. 70:333–338, 1993.

56. Eerlingen, R. C., G. Cillen, and J. A. Delcour. Enzyme-resistant starch. IV. Effect of endogenous lipids and added sodium dodecyl sulfate on formation of resistant starch. Cereal Chem. 71:170–177, 1994.

57. Atwell, W. A., L. F. Hood, D. R. Lineback, E. Varriano-Marston, and H. F. Zobel. The terminology and methodology associated with basic starch phenomena. Cereal Foods World 33:306–311, 1988.

58. Zobel, H. F., and K. Kulp. The staling mechanism. In: Baked Goods Freshness: Technology, Evaluation, and Inhibition of Staling, edited by R. E. Hebeda and H. F. Zobel. New York: Marcel Dekker, 1996, pp 1–64.

59. Knightly, W. H. Surfactants. In: Baked Goods Freshness. Technology, Evaluation and Inhibition of Staling, edited by R. E. Hebeda and H. F. Zobel. New York: Marcel Dekker, 1996, pp 65–103.

60. Silverio, J., H. Fredriksson, R. Andersson, A.-C. Eliasson, and P. Åman. The effect of temperature cycling on the amylopectin retrogradation of starches with different amylopectin unit-chain length distribution. Carbohydr. Polym. 42:175–184, 2000.

61. Sievert, D., and P. Würsch. Amylose chain association based on differential scanning calorimetry. J. Food Sci. 58:1332–1334, 1345, 1993.
62. Krog, N., S. K. Olesen, H. Toernaes, and T. Joensson. Retrogradation of the starch fraction in wheat bread. Cereal Foods World 34:281–285, 1989.
63. Gudmundsson, M., and A.-C. Eliasson. Retrogradation of amylopectin and the effects of amylose and added surfactants/emulsifiers. Carbohydr. Polym. 13:295–315, 1990.
64. Slade, L., and H. Levine. Recent advances in starch retrogradation. In: Industrial Polysaccharides: The Impact of Biotechnology and Advanced Methodologies, edited by S. S. Stivala, V. Crescenzi, and I. C. M. Dea. New York: Gordon and Breach, 1987, pp 387–430.
65. Zeleznak, K. J., and R. C. Hoseney. Characterization of starch from bread aged at different temperature. Starch/Stärke 39:231–233, 1987.
66. Longton, J., and G. A. LeGrys. Differential scanning calorimetry studies on the crystallization of aging wheat starch gels. Starch/Stärke 33:410–414, 1981.
67. Zeleznak, K. J., and R. C. Hoseney. The role of water in the retrogradation of wheat starch gels and bread crumb. Cereal Cehm. 63:407–411, 1986.
68. Ferrero, C., M. N. Martino, and N. E. Zaritzky. Effect of hydrocolloids on starch thermal transitions, as measured by DSC. J. Thermal Anal. 47:1247–1266, 1996.
69. Roulet, P., W. M. MacInnes, D. Gumy, and P. Würsch. Retrogradation kinetics of eight starches. Starch/Stärke 42:99–101, 1990.
70. Gudmundsson, M., and A.-C. Eliasson. Some physicochemical properties of oat starches extracted from varieties with different oil content. Acta. Agric. Scand. 39:101–111, 1989.
71. Gudmundsson, M., and A.-C. Eliasson. Thermal and viscous properties of rye starch extracted from different varieties. Cereal Chem. 68:172–177, 1991.
72. Kalichevsky, M. T., P. D. Orford, and S. G. Ring. The retrogradation and gelation of amylopectin from various botanical sources. Carbohydr. Res. 198:49–55, 1990.
73. Shi, Y.-C., and P. A. Seib. The structure of four waxy starches related to gelatinization and retrogradation. Carbohydr. Res. 227:131–145, 1992.
74. Ward, K. E. J., R. C. Hoseney, and P. Seib. Retrogradation of amylopectin from maize and wheat starches. Cereal Chem. 71:150–155, 1994.
75. Russell, P. L. Gelatinization of starches of different amylose/amylopectin content. A study by differential scanning calorimetry. J. Cereal Sci. 6:133–145, 1987.
76. van Soest, J. J. G., D. de Wit, H. Tournois, and J. F. G. Vliegenthart. Retrogradation of potato starch as studied by Fourier transform infrared spectroscopy. Starch/Stärke 46:453–457, 1994.
77. Roos, Y. H. Phase Transitions in Foods. San Diego: Academic Press, 1995.
78. Nikolaidis, A., and T. P. Labuza. Use of dynamic mechanical thermal analysis (DMTA). Glass transitions of a cracker and its dough. J. Thermal Anal. 47:1315–1328, 1996.
79. Slade, L. Starch properties of processed foods: Staling of starch-based products. 69th Annual Meeting of AACC, Minneapolis, Minn., 1984.
80. Kalichevsky, M. T., E. M. Jaroszkiewicz, S. Ablett, J. M. V. Blanshard, and P. J. Lillford. The glass transition of amylopectin measured by DSC, DMTA and NMR. Carbohydr. Polymers 18:77–88, 1992.

81. Jacobs, H., R. C. Eerlingen, W. Clauwaert, and J. A. Delcour. Influence of annealing on the pasting properties of starches from varying botanical sources. Cereal Chem. 72:480–487, 1995.
82. Biliaderis, C. G. Non-equilibrium phase transitions of aqueous starch systems. In: Water Relationships in Food, edited by H. Levine and L. Slade. New York: Plenum Press, 1991, pp 251–273.
83. Eliasson, A.-C., and N. Krog. Physical properties of amylose–monoglyceride complexes. J. Cereal Sci. 3:239–248, 1985.
84. Kowblansky, M. Calorimetric investigation of inclusion complexes of amylose with long-chain aliphatic compounds containing different functional groups. Macromolecules 18:1776–1779, 1985.
85. Kugimiya, M., and J. W. Donovan. Calorimetric determination of the amylose content of starches based on formation and melting of the amylose–lysolecithin complex. J. Food Sci. 46:765–770, 777, 1981.
86. Sievert, D., and J. Holm. Determination of amylose by differential scanning calorimetry. Starch/Stärke 45:136–139, 1993.
87. Mestres, C., F. Matencio, B. Pons, M. Yajid, and G. Fliedel. A rapid method for the determination of amylose content using differential scanning calorimetry. Starch/Stärke 48:2–6, 1996.
88. Morrison, W. R., and A. M. Coventry. Extraction of lipids from cereal starches with hot aqueous alcohols. Starch/Stärke 37:83–87, 1985.
89. Björck, I., N.-G. Asp, D. Birkhed, A.-C. Eliasson, L.-B. Sjöberg, and I. Lundquist. Effects of processing on starch availability in vitro and in vivo. II. Drum-drying of wheat flour. J. Cereal Sci. 2:165–178, 1984.
90. Villwock, V. K., A.-C. Eliasson, J. Silverio, and J. N. BeMiller. Starch–lipid interactions in common, waxy, ae du, and ae su2 maize starches examined by differential scanning calorimetry. Cereal Chem. 76:292–298, 1999.
91. Evans, I. D. An investigation of starch/surfactant interactions using viscometry and differential scanning calorimetry. Starch/Stärke 38:227–235, 1986.
92. Krog, N. Amylose complexing effect of food grade emulsifiers. Starch/Stärke 23:206, 1971.
93. Eliasson, A.-C., and G. Ljunger. Interactions between amylopectin and lipid additives during retrogradation in a model system. J. Sci. Food Agric. 44:353–361, 1988.
94. Gudmundsson, M. Effects of an added inclusion–amylose complex on the retrogradation of some starches and amylopectin. Carbohydr. Polym. 17:299–304, 1992.
95. Russell, P. L. The aging of gels from starches of different amylose/amylopectin content studied by differential scanning calorimetry. J. Cereal Sci. 6:147–158, 1987.
96. Eliasson, A.-C. Lipid–carbohydrate interactions. In: Interactions: The Keys to Cereal Quality, edited by R. J. Hamer and R. C. Hoseney. St. Paul: American Association of Cereal Chemists, 1998, pp 47–79.
97. Eliasson, A.-C., and P.-O. Hegg. Thermal stability of gluten. Cereal Chem. 57:436–437, 1981.
98. Hoseney, R. C., K. Zeleznak, and C. S. Lai. Wheat gluten: a glassy polymer. Cereal Chem. 63:285–286, 1986.

99. Tatham, A. S., and P. R. Shewry. The conformation of wheat gluten proteins. The secondary structures and thermal stabilities of α-, β-, γ-gliadins. J. Cereal Sci. 3: 103–113, 1985.

100. Schofield, J. D., R. C. Bottomley, M. F. Timms, and M. R. Booth. The effect of heat on wheat gluten and the involvement of sulphydryl–disulphide interchange reactions. J. Cereal Sci. 1:241–253, 1983.

101. Ma, C.-Y., and V. R. Harwalkar. Chemical characterization and functionality assessment of oat protein fractions. J. Agric. Food Chem. 32:144–149, 1984.

102. Kalichevsky, M. T., E. M. Jaroszkiewicz, and J. M. V. Blanshard. Glass transition of gluten. 1: Gluten and gluten–sugar mixtures. Int. J. Biol. Macromol. 14:257–266, 1992.

103. Kalichevsky, M. T., E. M. Jaroszkiewicz, and J. M. V. Blanshard. Glass transition of gluten. 2: The effect of lipids and emulsifiers. Int. J. Biol. Macromol. 14:267–273, 1992.

104. Kokini, J. L., A. M. Cocero, H. Madeka, and E. de Graaf. The development of state diagrams for cereal proteins. Trends Food Sci. Technol. 5:281–288, 1994.

105. Noel, T. R., R. Parker, S. G. Ring, and A. S. Tatham. The glass-transition behavior of wheat gluten proteins. Int. J. Biol. Macromol. 17:81–85, 1995.

106. Doescher, L. C., R. C. Hoseney, and G. A. Milliken. A mechanism for cookie dough setting. Cereal Chem. 64:158–163, 1987.

107. Larsson, K., and P. J. Quinn. Physical properties: structural and physical characteristics. In: The Lipid Handbook (2nd ed.), edited by F. D. Gunstone, J. L. Harwood, and F. B. Padley. London: Chapman & Hall, 1994, pp 401–485.

108. Daniels, N. W. R. Some effects of water in wheat flour doughs. In: Water Relations of Foods, edited by R. B. Duckworth. London: Academic Press, 1975, pp 573–586.

109. Morrison, W. R., D. L. Mann, W. Soon, and A. M. Coventry. Selective extraction and quantitative analysis of non-starch and starch lipids from wheat flour. J. Sci. Food Agric. 26:507–521, 1975.

110. Carlson, T., K. Larsson, and Y. Miezis. Phase equilibria and structures in the aqueous system of wheat lipids. Cereal Chem. 55:168–179, 1978.

111. Girhammar, U., and B. M. Nair. Certain physical properties of water soluble non-starch polysaccharides from wheat, rye, triticale, barley and oats. Food Hydrocolloids 6:329–343, 1992.

112. Ghiasi, K., R. C. Hoseney, and E. Varriano-Marston. Effect of flour components and dough ingredients on starch gelatinization. Cereal Chem. 60:58–61, 1983.

113. Stapley, A. G., L. F. Gladden, and P. J. Fryer. A differential scanning calorimetry study of wheat grain cooking. Int. J. Food Sci. Tech. 32:473–486, 1997.

114. Fisher, D. K., and D. B. Thompson. Retrogradation of maize starch after thermal treatment within and above the gelatinization temperature range. Cereal Chem. 74: 344–351, 1997.

115. Holm, J., I. Lundquist, I. Björck, A.-C. Eliasson, and N.-G. Asop. Degree of starch gelatinization, digestion rate of starch in vitro, and metabolic response in rats. Am. J. Clin. Nutr. 47:1010–1016, 1988.

116. Eliasson, A.-C. Differential scanning calorimetry studies on wheat starch–gluten mixtures. I. Effect of gluten on the gelatinization of wheat starch. J. Cereal Sci. 1: 199–205, 1983.

117. Erdogdu, N., Z. Czuchajowska, and Y. Pomeranz. Wheat flour and defatted milk fractions characterized by differential scanning calorimetry. I. DSC of flour and milk fractions. Cereal Chem. 72:70–75, 1995.
118. Eliasson, A.-C. Differential scanning calorimetry studies on wheat starch–gluten mixtures. II. Effect of gluten and sodium stearoyl lactylate on starch crystallization during aging of wheat starch gels. J. Cereal Sci. 1:207–213, 1983.
119. Schweizer, T. F., S. Reimann, J. Solms, A.-C. Eliasson, and N.-G. Asp. Influence of drum-drying and twin-screw extrusion cooking on wheat carbohydrates. II. Effect of lipids on physical properties, degradation and complex formation of starch in wheat flour. J. Cereal Sci. 4:249–260, 1986.
120. Biliaderis, C. G., and H. D. Seneviratne. On the supermolecular structure and metastability of glycerol monostearate–amylose complex. Carbohydr. Polym. 13:185–206, 1990.
121. Gudmundsson, M., A.-C. Eliasson, S. Bengtsson, and P. Åman. The effects of water soluble arabinoxylan on gelatinization and retrogradation of starch. Starch/Stärke 43:5–10, 1991.
122. Biliaderis, C. G., I. Arvanitoyannis, M. S. Izydorczyk, and D. J. Prokopowich. Effect of hydrocolloids on gelatinization and structure formation in concentrated waxy maize and wheat starch gels. Starch/Stärke 49:278–283, 1997.
123. Buck, J. S., and C. E. Walker. Sugar and sucrose ester effects on maize and wheat starch gelatinization patterns by differential scanning calorimeter. Starch/Stärke 40: 353–356, 1988.
124. Eliasson, A.-C. A calorimetric investigation of the influence of sucrose on the gelatinization of starch. Carbohydr. Polym. 18:131–138, 1992.
125. Spies, R. D., and R. C. Hoseney. Effect of sugars on starch gelatinization. Cereal Chem. 59:128–131, 1982.
126. Wootton, M., and A. Bamunuarachchi. Application of differential scanning calorimetry to starch gelatinization. III. Effect of sucrose and sodium chloride. Starch/Stärke 32:126–129, 1980.
127. Silverio, J. Retrogradation Properties of Starch. Lund, Sweden: Lund University, 1997.
128. Riisom, T., N. Krog, and J. Eriksen. Amylose complexing capacities of *cis*- and trans-unsaturated monoglycerides in relation to their functionality in bread. J. Cereal Sci. 2:105–118, 1984.
129. Cloke, J. D., J. Gordon, and E. A. Davis. Enthalpy changes in model cake systems containing emulsifiers. Cereal Chem. 60:143–146, 1983.
130. da Cruz Francisco, J., J. Silverio, A.-C. Eliasson, and K. Larsson. A comparative study of gelatinization of cassava and potato starch in an aqueous lipid phase (L2) compared to water. Food Hydrocolloids 10:317–322, 1996.
131. Le Roux, C., D. Marion, H. Bizot, and D. J. Gallant. Thermotropic behavior of coconut oil during wheat dough mixing: evidence for a solid–liquid phase separation according to mixing temperature. Food Structure 9:123–131, 1990.
132. Lauro, M., T. Suortti, K. Autio, P. Linko, and K. Poutanen. Accessibility of barley starch granules to α-amylase during different phases of gelatinization. J. Cereal Sci. 17:125–136, 1993.
133. Jacobs, H. Impact of Annealing on Physico-Chemical Properties of Starch. Leuven, Belgium: Katholieke Universiteit Leuven, 1998.

134. Evers, A. D., G. J. Baker, and D. J. Stevens. Production and measurement of starch damage in flour. Part I. Damage due to rollermilling of semolina. Starch/Stärke 36:309–312, 1984.
135. AACC76-30A. Approved Methods of the American Association of Cereal Chemists. St. Paul: American Association of Cereal Chemists, 1983.
136. AACC76-31. Approved Methods of the American Association of Cereal Chemists. St Paul: American Association of Cereal Chemists, 1983.
137. Morrison, W. R., R. F. Tester, and M. J. Gidley. Properties of damaged starch granules. II. Crystallinity, molecular order and gelatinization of ball-milled starch. J. Cereal Sci. 19:209–217, 1994.
138. Eliasson, A.-C., and M. Gudmundsson. Starch: physicochemical and functional aspects. In: Carbohydrates in Food, edited by A.-C. Eliasson. New York: Marcel Dekker, 1995, p 431.
139. Larsson, I., and A.-C. Eliasson. Annealing of starch at an intermediate water content. Starch/Stärke 43:227–231, 1991.
140. Knutson, C. A. Annealing of maize starches at elevated temperatures. Cereal Chem. 67:376–384, 1990.
141. Hoover, R., and T. Vasanthan. Effect of heat-moisture treatment on the structure and physicochemical properties of cereal, legume, and tuber starches. Carbohydr. Res. 252:33–53, 1994.
142. Lorenz, K., and K. Kulp. Steeping of starch at various temperatures—effects on functional properties. Starch/Stärke 32:181–186, 1980.
143. Zamponi, R. A., S. A. Giner, C. E. Lupano, and M. C. Anon. Effect of heat on thermal and functional properties of wheat. J. Cereal Sci. 12:279–287, 1990.
144. Tester, R. F., W. R. Morrison, R. H. Ellis, J. R. Piggott, G. R. Batts, T. R. Wheeler, J. I. L. Morrison, P. Hadley, and D. A. Ledward. Effect of elevated growth temperature and carbon dioxide levels on some physicochemical properties of wheat starch. J. Cereal Sci. 22:63–71, 1995.
145. Bloksma, A. H. Effect of heating rate on viscosity of wheat flour doughs. J. Texture Stud. 10:261–269, 1980.
146. Matsoukas, N. P., and W. Morrison. Breadmaking quality of ten Greek breadwheats. II. Relationships of protein, lipid and starch components to baking quality. J. Sci. Food Agric. 55:87–101, 1991.
147. Eliasson, A.-C., M. Gudmundsson, and G. Svensson. Thermal behavior of wheat starch in flour—relation to flour quality. Lebensm.-Wiss. u.-Technol. 28:227–235, 1995.
148. Johansson, E., and G. Svensson. Influences of yearly weather variation and fertilizer rate on bread-making quality in Swedish wheats containing HMW glutenin subunits 2 + 12 or 5+10 cultivated during the period 1990–96. J. Agric. Sci., 132:13–22, 1999.
149. Slade, L., H. Levine, M. Wang, and J. Ievolella. DSC analysis of starch thermal properties related to functionality in low-moisture baked goods. J. Thermal Anal. 47:1299–1314, 1996.
150. Varriano-Marston, E., V. Ke, G. Huang, and J. Ponte, Jr. Comparison of methods to determine starch gelatinization in bakery food. Cereal Chem. 57:242–248, 1980.
151. Wolters, M. G. E., and J. W. Cone. Prediction of degradability of starch by gelatini-

zation enthalpy as measured by differential scanning calorimetry. Starch/Stärke 44: 14–18, 1992.

152. Bushuk, W., and V. K. Mehrotra. Studies of water binding by differential thermal analysis. II. Dough studies using melting mode. Cereal Chem. 54:320–325, 1977.

153. Kulp, K. Biochemical and biophysical principles of freezing. In: Frozen and Refrigerated Doughs and Batters, edited by K. Kulp, K. Lorenz, and J. Brümmer. St. Paul: American Association of Cereal Chemists, 1995, pp 63–89.

154. Kline, L., and T. F. Sugihara. Factors affecting the stability of frozen bread doughs. I. Prepared by the straight dough method. Bakers Dig. 42(5):44–50, 1968.

155. Hsu, K. H., R. C. Hoseney, and P. A. Seib. Frozen dough. I. Factors affecting stability of yeasted doughs. Cereal Chem. 56:419–424, 1979.

156. Räsänen, J., H. Härkönen, and K. Autio. Freeze-thaw stability of prefermented frozen lean wheat doughs: the effect of flour quality and fermentation time. Cereal Chem. 72:637–642, 1995.

157. Räsänen, J., T. Laurikainen, and A. Autio. Fermentation stability and pore size distribution of frozen prefermented lean wheat doughs. Cereal Chem. 74:56–62, 1997.

158. Räsänen, J., J. M. V. Blanshard, J. R. Mitchell, W. Derbyshire, and K. Autio. Properties of frozen wheat doughs at subzero temperatures. J. Cereal Sci. 28:7–74, 1998.

159. Mizukoshi, M., T. Kawada, and N. Matsui. Model studies of cake baking. I. Continuous observations of starch gelatinization and protein coagulation during baking. Cereal Chem. 56:305–309, 1979.

160. Donovan, J. W. A study of the baking process by differential scanning calorimetry. J. Sci. Food Agric. 28:571–578, 1977.

161. Czuchajowska, Z., and Y. Pomeranz. Differential scanning calorimetry, water activity, and moisture contents in crumb center and near-crust zones of bread during storage. Cereal Chem. 66:305–309, 1989.

162. LeMeste, M., V. T. Huang, J. Panama, G. Anderson, and R. Lentz. Glass transition of bread. Cereal Foods World 37:264–267, 1992.

163. Willhoft, E. M. A. Bread staling. I. Experimental study. J. Sci. Food Agric. 22: 176–180, 1971.

164. Mita, T. Effect of aging on the rheological properties of gluten gel. Agric. Biol. Chem. 54:927–935, 1990.

165. Martin, M. L., K. J. Zeleznak, and R. C. Hoseney. A mechanism of bread firming. I. Role of starch swelling. Cereal Chem. 68:498–503, 1991.

166. Giovanelli, G., C. Peri, and V. Borri. Effects of baking temperature on crumb-staling kinetics. Cereal Chem. 74:710–714, 1997.

167. Keetels, C. J. A. M., K. A. Visser, T. van Vliet, A. Jurgens, and P. Walstra. Structure and mechanics of starch bread. J. Cereal Sci. 24:15–25, 1996.

168. Keetels, C. J. A. M., T. van Vliet, and P. Walstra. Relationship between the sponge structure of starch bread and its mechanical properties. J. Cereal Sci. 24:27–31, 1996.

169. Keetels, C. J. A. M., T. van Vliet, A. Jurgens, and P. Walstra. Effects of lipid surfactants on the structure and mechanics of concentrated starch gels and starch bread. J. Cereal Sci. 24:33–45, 1996.

170. Dragsdorf, R. D., and E. Varriano-Marston. Bread staling: x-ray diffraction studies

on bread supplemented with alfa-amylases from different sources. Cereal Chem. 57:310–314, 1980.

171. Czuchajowska, Z., Y. Pomeranz, and H. C. Jeffers. Water activity and moisture content of dough and bread. Cereal Chem. 66:128–132, 1989.

172. Wynne-Jones, S., and J. M. V. Blanshard. Hydration studies of wheat starch, amylopectin, amylose gels and bread by proton magnetic resonance. Carbohydr. Polym. 6:289–306, 1986.

173. Vodovotz, Y., L. Hallberg, and P. Chinachoti. Effect of aging and drying on thermomechanical properties of white bread as characterized by dynamic mechanical analysis (DMA) and differential scanning calorimetry (DSC). Cereal Chem. 73:264–270, 1996.

174. Jang, J. K., and Y. R. Pyun. Effect of moisture level on the crystallinity of wheat starch aged at different temperatures. Starch/Stärke 49:272–277, 1997.

175. Jouppila, K., and Y. H. Roos. The physical state of amorphous corn starch and its impact on crystallization. Carbohydr. Polym. 32:95–104, 1997.

176. I'Anson, K. J., M. J. Miles, V. J. Morris, L. S. Besford, D. A. Jarvis, and R. A. Marsh. The effects of added sugars on the retrogradation of wheat starch gels. J. Cereal Sci. 11:243–248, 1990.

177. Cairns, P., M. J. Miles, and V. J. Morris. Studies on the effect of the sugars ribose, xylose and fructose on the retrogradation of wheat starch gels by x-ray diffraction. Carbohydr. Polym. 16:355–365, 1991.

178. Le Botlan, D., and P. Desbois. Starch retrogradation study in presence of sucrose by low-resolution nuclear magnetic resonance. Cereal Chem. 72:191–193, 1995.

179. Russell, P. L., and G. Oliver. The effect of pH and NaCl content on starch gel aging. A study by differential scanning calorimetry and rheology. J. Cereal Sci. 10:123–138, 1989.

180. Silverio, J., and A.-C. Eliasson. The effect of oils on the gelatinization and retrogradation of wheat starch. Unpublished results.

181. Rao, P. A., A. Nussinovitch, and P. Chinachoti. Effects of selected surfactants on amylopectin recrystallization and on recoverability of bread crumb during storage. Cereal Chem. 69:613–618, 1992.

182. Kohyama, K., and K. Nishinari. Cellulose derivatives effects on gelatinization and retrogradation of sweet potato starch. J. Food Sci. 57:128–131, 137, 1992.

183. Ferrero, C., M. N. Martino, and N. E. Zaritzky. Corn starch–xanthan gum interaction and its effect on the stability during storage of frozen gelatinized suspensions. Starch/Stärke 46:300–308, 1994.

184. Andersson, R., E. Westerlund, A.-C. Tilly, and P. Åman. Natural variations in the chemical composition of white flour. J. Cereal Sci. 17:183–189, 1992.

185. Ziegler, E., and E. N. Greer. Principles of milling. In: Wheat Chemistry and Technology, edited by Y. Pomeranz. St Paul: American Association of Cereal Chemists, 1971, pp 115–199.

186. Gudmundsson, M., and A.-C. Eliasson. Some physical properties of barley starches from cultivars differing in amylose content. J. Cereal Sci. 16:95–105, 1992.

187. Tester, R. F., and J. Karkalas. Swelling and gelatinization of oat starches. Cereal Chem. 73:271–277, 1996.

188. Davis, E. A., J. Grider, and J. Gordon. Microstructural evaluation of model starch systems containing different types of oil. Cereal Chem. 63:427–430, 1986.
189. Zeleznak, K. J., and R. C. Hoseney. The glass transition in starch. Cereal Chem. 64:121–124, 1987.
190. Kalichevsky, M. T., and J. M. V. Blanshard. A study of the effect of water on the glass transition of 1:1 mixtures of amylopectin, casein and gluten using DSC and DMTA. Carbohydr. Polym. 19:271–278, 1992.
191. Nicholls, R. J., I. A. M. Appelqvist, A. P. Davies, S. J. Ingman, and P. J. Lillford. Glass transitions and the fracture behavior of gluten and starches within the glassy state. J. Cereal Sci. 21:25–36, 1995.
192. Eliasson, A.-C. Starch–lipid interactions studied by differential scanning calorimetry. Thermochim. Acta 95:369–374, 1985.

4

Plasticization Effect of Water on Carbohydrates in Relation to Crystallization

Yrjö Henrik Roos
University College Cork, Cork, Ireland

Kirsi Jouppila
University of Helsinki, Helsinki, Finland

I. INTRODUCTION

Amorphous materials exist in noncrystalline, metastable states in solid or liquid forms with a random molecular order in their structure. The solid amorphous materials exist in a glassy state with an extremely high viscosity and low, mainly rotational, mobility of molecules (1, 2). Amorphous, noncrystalline materials can be plasticized or softened by temperature and plasticizers. As a result of plasticization to above the glass transition, i.e., in supercooled liquid state, the viscosity decreases dramatically, allowing an increase in molecular mobility and translational motions (1, 2).

Each amorphous compound has a characteristic temperature range for the glass transition. The glass transition is often defined as the transition observed from a step change in heat capacity. The change in heat capacity occurs over a temperature range where small changes in temperature may result in substantial changes in viscosity, molecular mobility, and mechanical properties (2, 3). These changes in material properties occurring over the glass transition temperature range allow the transformation between solid, brittle, and glassy and rubbery, leathery, and sometimes even viscous, syruplike systems. Plasticizers increase molecular mobility and decrease temperatures of the glass transition range. Temperatures of the glass transition ranges of biomaterials and foods are particularly dependent on the presence and amount of water, which is the main plasticizer of carbohydrates, proteins, and other water-soluble or water-miscible compounds.

The glass transition is one of the most important changes in the physical state of concentrated biological and food materials. Various time-dependent changes, such as structural transformations, changes in food texture, and crystallization, are known to be dependent on molecular mobility and controlled, at least to some extent, by the glass transition (2, 3). Although an exact temperature for the glass transition cannot be defined or measured (4), a glass transition temperature, T_g, is often taken as the onset or midpoint temperature of the glass transition temperature range measured using a differential scanning calorimeter (DSC).

A number of low-moisture and freeze-concentrated nonlipid food solids exist either in the glassy or supercooled liquid state (2, 3, 5–8). Amorphous solids and supercooled liquids have a driving force to the stable, crystalline, equilibrium state. The relaxation times of molecular rearrangements can be related to temperature or extent of plasticization. Therefore, the T_g or the amount of plasticizer is often taken as a convenient parameter in modeling measured time-dependent changes in amorphous systems. In both low-moisture and freeze-concentrated materials, amorphous compounds tend to crystallize when the molecules have sufficient translational mobility to allow molecular rearrangements, nucleation, and crystal growth. Because translational mobility has been observed to increase rapidly at temperatures above T_g, crystallization may also exhibit increasing rates at and above the glass transition temperature range (2, 3). Most studies have reported crystallization only at temperatures above the calorimetric onset T_g. However, Yoshioka et al. (9) reported some evidence of indomethacin crystallization below the T_g.

Our studies have shown that the rate of crystallization of carbohydrate systems is dependent on temperature, the temperature difference between material temperature and the glass transition temperature, $T - T_g$, and water content (10–15). We will discuss water plasticization occurring in food and related systems and, in particular, in amorphous sugars and starch, emphasizing the relationships between glass transition, water plasticization, and time-dependent changes in crystallinity.

II. WATER PLASTICIZATION AND GLASS TRANSITION

A. Determination of Glass Transition

Water-sensitive amorphous systems may exhibit several changes in their physico-chemical properties, depending on water content and temperature. It is often agreed that the viscosity of solidlike systems with glass properties is greater than 10^{12} Pa-s, restricting the molecular motions to vibrations and sidechain rotations

(1, 2). Such molecular motions can be observed from frequency-dependent relaxations appearing in dielectric and mechanical spectra. Relaxations below the glass transition are known as β and γ relaxations, while the relaxations related to glass transition are referred to as α relaxations (1). Obviously, other techniques, such as NMR techniques and ESR as well as FT-IR and Raman spectroscopies, may also be used to follow changes in the molecular mobility in glassy systems (1, 16, 17). Increases in the translational diffusion are observed to occur over the glass transition, as can be expected from the exponentially decreasing viscosity. The rapidly increasing translational diffusivity has been measured using ESR and FRAP techniques (18–20). However, DSC has remained the predominant technique used to measure glass transition temperatures related to glass transition-related material properties, although the determination of the transition from mechanical spectra, determined using dynamic mechanical analysis (DMA), is often considered a more sensitive method (21).

1. Low-Molecular-Weight Sugars

Although the glass transition behavior of glucose has been studied fairly thoroughly in a number of early studies (e.g., Ref. 6), the importance of glass transition in controlling food material properties was mostly ignored until the pioneering work of White and Cakebread (7), Luyet and Rasmussen (22), and Slade and Levine (2). The glass transition temperatures have been reported for most common mono- and disaccharides (8, 10, 23–25). Obviously, one of the most studied sugars is sucrose. Glass transition temperatures reported for "dry" sucrose have varied from 52 to 75°C. The large variation can be explained by differences in sample preparation and analysis procedures, instrumentation, and interpretation of DSC traces. It should be noticed that extremely small variations in water content may affect the transition temperature by several degrees, and also whether the reported T_g is the onset or midpoint temperature of the glass transition may result in a 5–10°C difference in the reported T_g. Generally, the glass transition temperatures of sugars are related to molecular weight, although the T_g of sugars with the same molecular weight may be significantly different (Fig. 1). The variation in the glass transition temperatures of sugars is one of the factors explaining differences in the characteristics of sugar-containing food products.

The T_g values of anhydrous sugars are important parameters in the evaluation of their effects on physicochemical properties and transition temperatures when used as food components. Amorphous sugars are plasticized by water, which depresses the T_g. The significant effects of water on the physical state of sugars and other carbohydrates can be explained by state diagrams (2, 27, 28). State diagrams relate the phase and state transition temperatures to water content, explaining the dramatic effect of water as a plasticizer on material properties (2, 28). The state diagram of sucrose with T_g data from a number of studies is

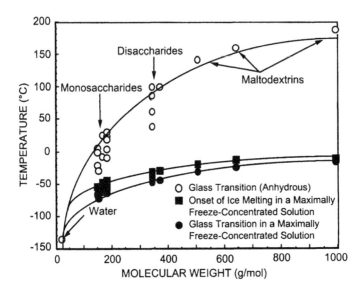

Figure 1   Glass transition temperatures, T_g, glass transition temperatures in the maximally freeze-concentrated state, T'_g, and onset temperatures of ice melting in the maximally freeze-concentrated state, T'_m, for various carbohydrates. (Adapted from Ref. 26.)

shown in Figure 2. The state diagram shows the glass transition curve following experimental T_g values from 62°C for anhydrous sucrose (29) to −135°C for pure amorphous water (33, 34). The Gordon–Taylor equation, given in Eq. (1), has been fitted to the experimental data and used to predict the T_g over the whole concentration range. In addition, the transitions, T'_g (glass transition temperature of the maximally freeze-concentrated solute phase), and T'_m (onset temperature of ice melting in the maximally freeze-concentrated state), occurring in maximally freeze-concentrated systems, as explained by Roos and Karel (29), have been shown with the equilibrium ice melting temperature, T_m, curve. Another equation often fitted to the T_g data of carbohydrates is the Couchman–Karasz Eq. (2), which includes the changes in heat capacity (ΔC_p) occurring over the glass transition of the components and uses the ΔC_p of 1.94 J/g°C for water, as reported by Sugisaki et al. (33):

$$T_g = \frac{w_1 T_{g1} + k w_2 T_{g2}}{w_1 + k w_2} \tag{1}$$

where T_g is the glass transition temperature of the mixture, k is a constant, and w_1 and T_{g1} and w_2 and T_{g2} are the weight fractions and corresponding T_g values of components 1 and 2, respectively.

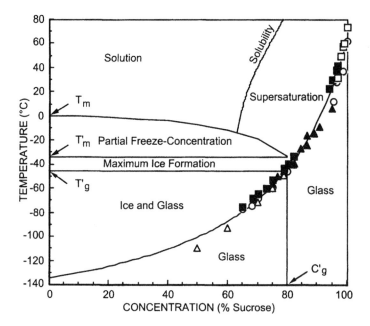

Figure 2 State diagram of sucrose. The glass transition temperatures are from Refs. 22 ($\triangle$), 29 ($\bigcirc$), 30 ($\blacktriangle$), 31 ($\blacksquare$), and 32 ($\square$). Sucrose concentration in the maximally freeze-concentrated solution with glass transition at T'_g is given by C'_g.

$$\ln T_g = \frac{w_1 \, \Delta C_{p1} \, \ln T_{g1} + w_2 \, \Delta C_{p2} \, \ln T_{g2}}{w_1 \, \Delta C_{p1} + w_2 \, \Delta C_{p2}} \tag{2}$$

where ΔC_{p1} and ΔC_{p2} are the changes in the heat capacities occurring over the glass transition of the component compounds. Note that the equation is often simplified by using only the T_g values and not the $\ln T_g$ values. Then the relationship becomes equal to Eq. (1) with $k = \Delta C_{p2}/\Delta C_{p1}$.

Water has an analogous plasticization effect on other sugars, as shown for lactose, glucose, and fructose in Figure 3. It is known empirically that lactose, trehalose, and sucrose can be fairly stable in bakery products. They can also be used to make relatively stable amorphous powders, but, for example, such materials as fruit juices containing high amounts of monosaccharides, fructose, and glucose are difficult to dehydrate and to produce into a stable powder form. Indeed, the disaccharides have T_g values above room temperature, even in the presence of some water, while the T_g values of anhydrous monosaccharides may be below room temperature, which explains many of the problems impeding processing and storage of materials containing monosaccharides.

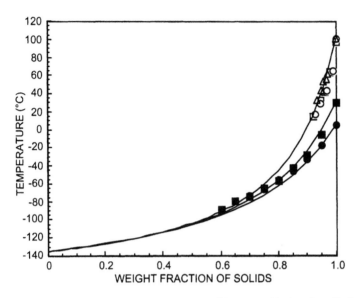

Figure 3 Glass transition temperatures of lactose, with data from Refs. 10 (○), 35 (□), and 36 (△), and glucose (■) and fructose (▲), with data from Ref. 25.

2. Oligosaccharides and Carbohydrate Polymers

It is well known that the T_g of synthetic homopolymers is dependent on molecular weight, increasing with increasing molecular weight. Fox and Flory (37) showed that the T_g of polystyrene fractions was related to the number average molecular weight, M_n, according to Eq. (3). Another equation relating T_g to molecular weight is that of Ueberreiter and Kanig (38), given in Eq. (4). Equation (4) has been reported to be more accurate over a wider molecular weight range than Eq. (3)(23).

$$T_g = T_g(\infty) - \frac{K_g}{M_n} \tag{3}$$

$$\frac{1}{T_g} = \frac{1}{T_g(\infty)} - \frac{K_g'}{M_n} \tag{4}$$

where $T_g(\infty)$ is the glass transition temperature for the material of infinite molecular weight and K_g and K_g' are constants.

In general, the T_g of sugars increases with increasing molecular weight in the order monosaccharides < disaccharides < oligosaccharides < carbohydrate polymers, although the T_g values of individual carbohydrates with the same molecular weights may differ greatly. The increase in T_g with increasing molecular weight of glucose polymers was shown by Orford et al. (23) to apply to malto-oligosaccharides. It was found that the increase followed Eq. (4) from glucose to maltohexaose. An extrapolation of the data estimated the T_g for amylose and amylopectin to be $227 \pm 10°C$. Roos and Karel (39) reported T_g values for malto-dextrins with various dextrose equivalents. The T_g of the maltodextrins increased with increasing effective molecular weight according to Eq. (3) and predicted the T_g of anhydrous starch to be $243°C$. Unfortunately, the T_g values of the high-molecular-weight carbohydrates cannot be experimentally determined due to thermal decomposition at sub-T_g temperatures.

Oligosaccharides and carbohydrate polymers, like sugars, are sensitive to water. Small amounts of water may result in significant plasticization. It has been shown that the Gordon–Taylor and Couchman–Karasz equations can also be used to predict the effects of water on the T_g of high-molecular-weight carbohy-drates (23, 27, 29) and to establish state diagrams (27, 29) (Fig. 2). High-molecu-lar-weight carbohydrates may also become plasticized by lower-molecular-weight carbohydrates. In fact, maltodextrins, as starch hydrolysis products, are mixtures of glucose, maltose, and glucose polymers of various degrees of poly-merization (39). In such materials, the overall T_g is a result of the effects of all miscible fractions contributing to the T_g of the material. Roos and Karel (39) showed that the T_g of maltodextrins with various dextrose equivalents decreased with the addition of sucrose to an extent dependent on the sucrose content in the mixture. The decrease could be modeled using the Gordon–Taylor equation. The effects of sugars are important determinants of the T_g of real amorphous foods, where small changes in composition may significantly change the processability or storage stability of products.

B. Time-Dependent Changes

The ascending translational mobility of molecules over and above the glass transi-tion temperature range is likely to affect material flow and rates of changes oc-curring in foods. Obviously, the decreasing viscosity and enhanced flow may affect food structure and texture, the rates of component crystallization, rates of other diffusion-controlled changes, and reaction kinetics (2, 3, 7). The relaxation times of structural transformations and rates of other changes are often related to the $T - T_g$ (e.g., Refs. 2, 3, 40).

1. Relaxation Times and Viscosity

The viscosity of glass-viscosity-approaching, concentrated fructose-sucrose sys-tems was successfully modeled with the well-known WLF model by Soesanto and Williams (41). The applicability of the WLF Eq. (5) to model changes in relaxation times occurring in amorphous systems over the temperature range T_g

to $T_g + 100°C$ has been emphasized by Slade and Levine (2). Indeed, viscosity and changes in relaxation times often follow the WLF relationship above the T_g. However, the WLF model has been proved to fail in predicting changes occurring over the glass transition (42). The main shortcoming of the WLF model is that it assumes a steady-state condition in the glassy state and a rapid decrease in ln a_T above the T_g.

$$\ln a_T = \frac{C_1(T - T_S)}{C_2(T - T_S)} \tag{5}$$

where a_T is the ratio of relaxation times at temperature T and a reference temperature T_S and C_1 and C_2 are constants. The T_g is often used as the reference temperature, T_S.

The WLF model suggests that the temperature dependence of viscosity (or a structural relaxation time) above glass transition differs from the Arrhenius-type temperature dependence. According to Angell (43, 44), plots of the logarithm of viscosity or a structural relaxation time against T_g/T of various glass formers may show linearity and follow the Arrhenius relationship (strong liquids) while other extremes of glass formers show nonlinearity with a strong non-Arrhenius type of temperature dependence of viscosity (fragile liquids). The classification of glass formers to strong and fragile liquids gives a measure for their sensitivity to temperature changes in relation to the glass transition. The glass transition affects very few structural changes occuring in strong liquids, but the fragile liquids tend to collapse rapidly above the T_g. Angell (43, 44) has also pointed out that strong liquids exhibit small ΔC_p over the glass transition, whereas the ΔC_p of fragile liquids is large.

The classification of glass formers to "strong" and "fragile" liquids is based on a modification of the Vogel–Tamman–Fulcher Eq. (6), which is also used to establish relationships between viscosity, η, and glass transition. The constant B in Eq. (6) can be replaced by the term DT_0, which gives Eq. (7). The parameter D controls how closely the Arrhenius relationship is followed (44).

$$\eta = \eta_0 e^{B/T-T_0} \tag{6}$$

$$\eta = \eta_0 e^{DT_0/T-T_0} \tag{7}$$

Equation (7) assumes that the temperature of viscosity divergence, T_0, is below the T_g, as determined by the parameter D. Therefore, strong liquids have a large D and the T_0 is located well below the T_g. Fragile liquids have a low D, and the divergence from infinite viscosity occurs close to the T_g. It is likely that sugars exhibiting a fragile behavior tend to crystallize above the T_g, whereas carbohydrates with strong liquid properties may form stable glasses. It may become important in the characterization of changes in food structure and texture occurring above T_g, e.g., susceptibility to stickiness, collapse, and crystallization, in terms of their fragility.

It has been shown using ESR and NMR techniques that the rotational motions of spin labels sense the calorimetric glass transition and that small molecules, such as water, remain relatively mobile in glassy carbohydrates (17, 20). Changes in molecular mobility observed with spectroscopic methods as well as changes in moduli, determined with DMA, and dielectric properties, observed using DEA, can be correlated with the calorimetric glass transition. However, it should be remembered that the changes in material properties and transitions in glassy materials are highly dependent on frequency and the T_g cannot be well defined (4, 42). The DMA and DEA have been shown to be more sensitive in detecting the glass transition than the DSC (21, 45), but it may often be difficult to define an exact frequency corresponding to the calorimetric T_g. Obviously, changes in amorphous materials occur below the T_g, e.g., physical aging, but the changes occurring in food materials during practical time frames can be related to the calorimetric T_g. The changes in stiffness-related properties, e.g., stickiness, caking, collapse, and loss of crispness, are not likely to follow the WLF relationship in the vicinity of the glass transition. Peleg (42) has shown that these changes occur over a transition temperature, water activity, or water content range following the Fermi model given in Eq. (8):

$$\frac{Y}{Y_s} = \frac{1}{1 + \exp\left[\dfrac{X - X_c}{a(X)}\right]} \tag{8}$$

where Y is the stiffness parameter, Y_s is the stiffness parameter at a reference state, e.g., in the glassy state, X is temperature or some other plasticizing parameter (plasticizer content, temperature, water activity, or water content), X_c gives the X-value resulting in a 50% change in stiffness, and $a(X)$ is a measure of the steepness of the change in stiffness as a function of X.

An important change in the texture of a number of cereal foods is the loss of crispness, which may be related to water activity (46, 47) and glass transition (2, 48, 49). The decrease in stiffness occurring over the glass transition can be accounted for by the decreasing viscosity and ascending liquid flow of the material. Figure 4 shows relationships between water activity, water content, water sorption, and the decrease in viscosity over and above the glass transition. The Fermi model predicts a decreasing viscosity even below the water activity corresponding to the T_g at 25°C. The WLF model assumes a steady viscosity in the glassy state and the dramatic decrease in viscosity above T_g. The stickiness, caking, and collapse in a practical time scale occur above the T_g (24), where the viscosity seems to follow both the Fermi and WLF models. It should be noticed that the viscosity of glassy materials has not been measured and the experimental

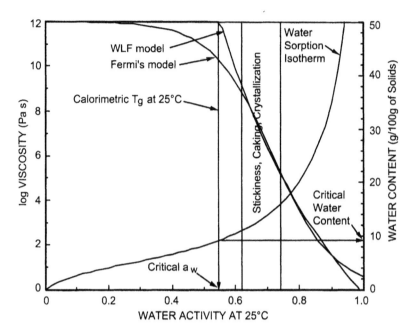

Figure 4 Water sorption and plasticization by a maltodextrin (M200) with viscosity and critical water activity and water content depressing the glass transition temperature, T_g, to 25°C. The main difference between the Fermi and WLF model predictions occurs over the glass transition. (Data from Ref. 39.)

viscosity data available for carbohydrates cover only values up to 10^9 Pa-s (41, 50, 51).

III. CRYSTALLIZATION AND GLASS TRANSITION

The driving force for crystallization is determined by the extent of supercooling below the equilibrium melting temperature or the extent of supersaturation. Amorphous food systems can be considered as supercooled and supersaturated materials, which have a temperature- and concentration-dependent driving force for crystallization. Some of the main factors controlling the rates of crystallization are probably diffusion of molecules to nucleation sites and their ability to reorient themselves into the crystal lattice structure. In concentrated systems, the diffusion of molecules may become restricted by high viscosity, and, thereby, the rates of crystallization are likely to decrease. In polymer systems, the rates of crystallization approach zero at temperatures close to the T_g (1).

A.   Crystallization of Amorphous Sugars

Water sorption studies of dairy powders have often shown that during storage at above 40% relative humidity (RH) and room temperature, large amounts of water may be sorbed initially, but during further storage the sorbed water content is reduced with storage time. For example, as early as in 1926, Supplee (5) found a gradual decrease in sorbed water content of milk powder when the material was stored at 50% RH at room temperature. Such decrease in sorbed water content has often been suggested to be a result of time-dependent crystallization of amorphous lactose (e.g., 13, 35, 52–59). Similar crystallization behavior from the amorphous state has been found to apply to other sugars, including glucose and sucrose (7, 60), and the development of crystallinity can be followed from the loss of sorbed water (32, 60, 61).

Sugar crystallization occurs in both pure amorphous sugar preparations and sugar-containing foods. Development of crystallinity of amorphous lactose and other sugars, as a result of thermal and water plasticization, has been detected using such methods as differential scanning calorimetry (11, 32, 62, 63), isothermal microcalorimetry (58, 64, 65), and X-ray diffraction techniques (13, 14). However, there has been variation between the extent of crystallinity determined with different techniques, because lactose may crystallize in several crystal forms, depending on temperature and water content (13, 14, 66). Moreover, different techniques measure either the amount of one or more of the crystal forms or the overall crystallinity. We have also shown that the crystallization of amorphous sugars can be followed using Raman spectriscopy and Raman microscopy (16).

1.   Effect of Physical State on Sugar Crystallization

Crystallization of amorphous compounds is known to be related to the glass transition (1, 2). Theoretically, crystallization ceases below the glass transition, as the molecules freeze in the solid, glassy state. Above the T_g, nucleation occurs rapidly, but the growth of crystals occurs slowly due to the high viscosity and slow diffusion (2). At temperatures below, but close to, the equilibium melting temperature, T_m, nucleation occurs slowly but crystal growth is fast, as the driving force for nucleation decreases but the mobility of the molecules increases (2, 3, 32). Therefore, the maximum rate of crystallization appears between the T_g and T_m (2, 3).

Relationships between crystallization behavior of sugars and their glass transition have been studied by several authors (e.g., 2, 7, 10, 11, 13, 14, 32, 34). The assumption in such studies has been that crystallization occurs when the temperature has been increased to above the T_g or the water content has been increased sufficiently to depress the T_g to below ambient temperature. The conditions allowing crystallization have then been obtained from the state diagrams, or a critical water activity or water content corresponding to that at which the

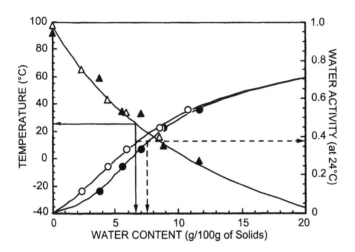

Figure 5 Water sorption and plasticization of amorphous lactose and skim milk powder. The glass transition of lactose (△) and skim milk powder (▲) decrease in a similar manner with increasing water content. The arrows indicate the critical water contents and water activity at 24°C for lactose (○) and skim milk powder (●). (Data from Ref. 35.)

T_g is at ambient temperature has been defined (67). As shown in Figure 5, we have obtained the critical water content for amorphous lactose using the Gordon–Taylor equation with T_{g1} of 97°C and k of 6.7 (35) to be 6.7 g/100 g of solids. The corresponding critical RH for amorphous lactose and lactose in milk powders at 24°C was 37%. However, the critical water content of skim milk was higher, being 7.6 g/100 g of solids, probably because of additional contribution to water sorption properties by the component proteins. Because the critical values for water activity and water content are related to the anhydrous T_g (Table 1), low-molecular-weight sugars tolerate very low amounts of water or no water at all in their structure. The critical values of disaccharides increase with increasing anhydrous T_g, and the critical values for polysaccharides are fairly high, approaching water activities, allowing the growth of microorganisms (67, 68).

2. Kinetics of Sugar Crystallization

Rates of sugar crystallization at isothermal conditions have been reported to increase with increasing storage RH (e.g., 13, 14, 35, 55–58, 60, 62, 64, 65, 69). Some studies have also found that rates of amorphous sugar crystallization increase with increasing storage temperature at constant water contents (70). These findings suggest that the crystallization of amorphous sugars above some critical water content or temperature may occur with increasing rates, as a result of both increasing temperature and increasing water content.

Table 1　Anhydrous Glass Transition Temperatures for Amorphous Carbohydrates

Compound	T_g (°C)	k	Critical m (%)	Critical m (g/100 g of solids)
Pentoses				
Arabinose	−2	3.55	—	—
Ribose	−20	3.02	—	—
Xylose	6	3.78	—	—
Hexoses				
Fructose	5	3.76	—	—
Frucose	26	4.37	0.1	0.1
Galactose	30	4.49	0.7	0.7
Glucose	31	4.52	0.8	0.8
Mannose	25	4.34	—	—
Rhamnose	−7	3.40	—	—
Sorbose	19	4.17	—	—
Disaccharides				
Lactose	101	6.56	6.8	7.2
Lactulose	79	5.92	5.4	5.7
Maltose	87	6.15	5.9	6.3
Melibiose	85	6.10	5.8	6.2
Sucrose	62	5.42	4.1	4.3
Trehalose	100	6.54	6.7	7.2
Maltodextrins				
Maltrin M365 (DE 36)	100	6.00	7.3	7.8
Maltrin M200 (DE 20)	141	6.50	10.0	11.2
Maltrin M100 (DE 10)	160	7.00	10.8	12.1
Maltrin M040 (DE 4)	188	7.70	11.7	13.2
Starch	243	5.20	20.8	26.2
Sugar alcohols				
Maltitol	39	4.75	1.8	1.8
Sorbitol	−9	3.35	—	—
Xylitol	−29	2.76	—	—

T_g, DSC onset or an estimated value; constant k of the Gordon-Taylor equation, estimated value; and critical water content, m, water content at which the T_g is 25°C.
Source: From Refs. 25 and 27.

Following the suggestions of White and Cakebread (7) and Levine and Slade (71), the first studies relating time-dependent crystallization of amorphous sugars to $T - T_g$ were those of Roos and Karel (10, 24). They used DSC to determine the T_g and isothermal DSC to measure the time to crystallization of lactose and sucrose at various temperatures above the T_g. Crystallization of the amorphous sugars was found to occur instantly at about 50°C above the T_g. The instant crystallization temperatures decreased with increasing water content following the decrease in the T_g. The time to crystallization at various $T - T_g$ decreased with increasing $T - T_g$, and the time to crystallization followed the WLF relationship. This was further confirmed to apply to amorphous lactose at relatively low $T - T_g$ values (11).

Levine and Slade (71) suggested a parabolic relationship between the rate of crystallization and $T - T_g$, emphasizing the theory that crystallization occurs slowly at both low and high $T - T_g$ conditions, and the maximum rate of crystallization is found at an intermediate $T - T_g$. Jouppila et al. (13) studied lactose crystallization in freeze-dried skim milk and found that the rate of crystallization increased, with increasing $T - T_g$, from 28 to 118°C. In a subsequent study (14), an increase in the rate of crystallization in freeze-dried lactose was found to occur at $T - T_g$ increased from 11 to 63°C. The rate of crystallization was found to be almost equal at $T - T_g$ of 63 and 105°C, suggesting that the rate of lactose crystallization probably decreases at high $T - T_g$ conditions. This was confirmed by Kedward et al. (63), who reported a parabolic relationship between the rate of crystallization and $T - T_g$ for freeze-dried lactose and sucrose.

Crystallization of amorphous materials is a complicated process requiring a sufficient time for nucleation, crystal growth, maturation of the crystals formed, sometimes transformations between different crystalline forms, and recrystallization (e.g., Ref. 72). In studies of the kinetics of crystallization, the basic assumption may be that crystallization occurs from a completely amorphous state of the material to a completely crystalline state. However, many amorphous sugars are produced by dehydration techniques, which may allow some initial nucleation, although X-ray diffraction studies have shown properly freeze-dried materials to be completely amorphous (145). The rate of crystallization is often assumed to follow the well-known Avrami equation (1, 3, 73). The Avrami Eq. (9) defines a sigmoid relationship between crystallinity, α, and the time of crystallization, t. The Avrami equation can be fitted to experimental crystallization data using nonlinear regression or the linearized form of the equation [Eq. (10)], which shows that a plot of $\ln[-\ln(1 - \alpha)]$ against $\ln t$ gives a straight line (3).

$$\alpha = 1 - e^{-kt^n} \tag{9}$$

$$\ln[-\ln(1 - \alpha)] = \ln k + n \ln t \tag{10}$$

where k is a rate constant and n is the Avrami exponent.

Roos and Karel (11) followed crystallization of amorphous lactose during storage at various $T - T_g$ conditions adjusted by temperature and water content. The extent of crystallization at intervals was followed by the determination of the latent heats of melting of the crystals formed using DSC. The rate of crystallization increased with increasing $T - T_g$, but the Avrami equation could not be fitted to the crystallization data. However, it was confirmed that the crystallization at a constant water content proceeded with a rate controlled by the $T - T_g$. But in closed containers, where water was released by the crystallizing material and subsequently sorbed by the remaining amorphous material, crystallization occurred extremely rapidly. This was also reported by Kim et al. (74), who found that the release of water as a result of crystallization caused a significant increase in the rate of nonenzymatic browning of whey powders stored in sealed containers. Obviously, crystallization of sugars in foods stored in sealed containers may cause a rapid deterioration of the materials at abusing storage conditions.

The Avrami equation was found to fit the crystallization data of amorphous lactose and lactose in skim milk, both produced by freeze-drying, when the increase in crystallinity during storage above the T_g was followed using powder X-ray diffraction detection of crystallinity (13, 14). The half-time of crystallinity, as defined by Eq. (11), was found to be a convenient parameter for the evaluation of the storage stability of lactose-containing materials. The data for half-time of crystallization reported by Jouppila et al. (13, 14) and Kedward et al. (63) suggest that the maximum rate of lactose crystallization may occur at a $T - T_g$ of 60–100°C (Fig. 6). The rate may also be affected by water content and material composition; i.e., the rates may be different at equal $T - T_g$ at various water contents and significantly dependent on other components. It seems that lactose crystallization in skim milk occurs at a substantially lower rate, even at a very high $T - T_g$, when compared with pure amorphous lactose.

$$t_{1/2} = n \sqrt{\frac{-\ln 0.5}{k}} \tag{11}$$

A number of food materials contain several amorphous carbohydrates mixed with other food components. It has been reported that the presence of other compounds with amorphous sugars may delay sugar crystallization. This is even true when anomers of the same sugar are present in the material, because they may have different solubilities and crystals formed may often contain only one of the anomers. According to Iglesias and Chirife (61), the addition of polysaccharides delayed the crystallization of amorphous sucrose. Saleki-Gerhardt and Zografi (32) found that pure amorphous sucrose crystallized more rapidly than in a mixture with lactose, trehalose, or raffinose at 32.4% RH and 30°C. Induction time for sucrose crystallization increased with increasing amount of lactose, raffinose, and trehalose. When the amount of lactose, raffinose, or trehalose exceeded 10%, no crystallization was observed during two weeks of storage. The delayed crystallization of amorphous sucrose in the presence of other sugars or corn syrup solids may be a result of effects on the T_g, effects of molecular charges, and effects of steric hindrance (75). It should also be noticed that lactose hydrolysis in dairy powders may result in delayed crystallization; e.g., we have

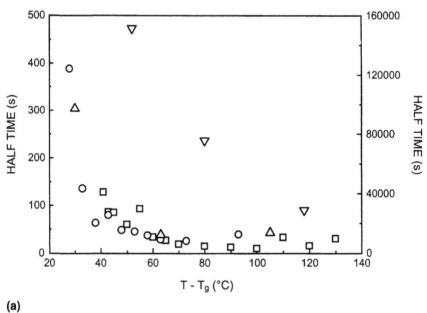

(a)

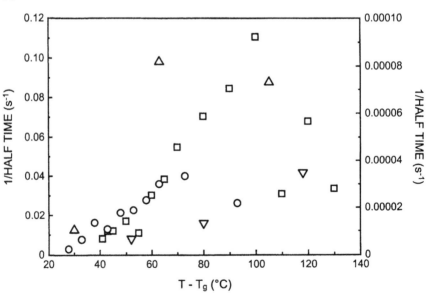

(b)

Figure 6   Half-time (a) and reciprocal half-time (b) for crystallization of amorphous lactose and sucrose, as a function of temperature difference between storage temperature and glass transition temperature, $T - T_g$. Left axis: lactose, square (Ref. 63), and sucrose, circle (Ref. 63). Right axis: lactose, up-triangle (Ref. 14), and lactose in freeze-dried skim milk, down-triangle (Ref. 13).

found no crystallization in freeze-dried skim milk containing hydrolyzed lactose (57). Therefore, sugar crystallization in food products may be controlled by addition of other sugars in addition to control by temperature and water content.

3. Extent of Sugar Crystallization

Anhydrous amorphous sugars are likely to crystallize to a full extent when exposed to conditions favoring crystallization. However, in the presence of water, the extent of crystallization may become limited by solubility. In the presence of other compounds, crystallization may be limited by intermolecular interactions between various compounds.

It has been found that the extent of isothermal sugar crystallization is dependent on the storage relative humidity (13, 14, 64, 65). Sebhatu et al. (65) found that the heat of lactose crystallization at 25°C was slightly lower for amorphous lactose stored at 57% and 100% RH than for amorphous lactose stored at 75% and 84% RH, indicating that less lactose crystallized at 57% and 100% RH than at 75% and 84% RH. According to Briggner et al. (64), the heat of crystallization of amorphous lactose at 25°C when stored at 75% and 85% RH was slightly lower than that of amorphous lactose stored at 53% and 65% RH. The highest heat of crystallization was measured for lactose stored at 65% RH (64). The results of these studies suggest that not only was the extent of lactose crystallization dependent on the storage relative humidity, but the extent of crystallization increased with storage relative humidity to a maximum value and then decreased when lactose was stored at a higher relative humidity. We have reported (14) fairly low extents of crystallization of amorphous lactose when stored at 24°C and 44.4% RH. At higher relative humidity conditions, the extent of crystallization increased, but only small differences in the extents of crystallization of amorphous lactose were found after storage at relative humidity ranging from 53.8 to 76.4%. A parabolic relationship between the extent of lactose crystallization and storage relative humidity with a maximum at 70% RH at 24°C was found when amorphous lactose crystallized in freeze-dried skim milk (13). The corresponding water content was 17%, and the $T - T_g$ value for amorphous lactose was 61°C (13). We assumed that the low extent of crystallization at the lowest relative humidity resulted from low molecular mobility and restricted diffusion of lactose molecules within the nonfat milk solids, since the crystallization occurred at a temperature relatively close to the T_g.

In addition to the overall extent of crystallization, the extent of crystallization into various crystal forms may also be dependent on $T - T_g$ and relative humidity at isothermal crystallization conditions (14). The crystal form produced as a result of amorphous sugar crystallization may also affect storage stability, in particular, when the material may crystallize as a hydrate. For example, lactose crystallization as anhydrous β-lactose crystals releases a significant amount of water compared to crystallization as α-lactose monohydrate, which contains about 5% water.

4. Crystal Forms of Sugars

Sugars may exist in anomeric forms, e.g., α- and β-glucose, and they may crystallize in anomeric crystals, crystals with mixed anomers, and hydrates with different amounts of water. For example, lactose may crystallize into several crystal forms, e.g., α-lactose monohydrate (76), anhydrous β-lactose (77), stable and unstable anhydrous α-lactose (77), and anhydrous crystals with α- and β-lactose in a molar ratio of 5:3 or 4:1 (78). Another example is glucose, which may crystallize into three crystal forms: α-glucose monohydrate, anhydrous α-glucose, and anhydrous β-glucose (79). Crystallization of sucrose usually produces anhydrous sucrose crystals. However, sucrose has been reported to crystallize also as sucrose hemipentahydrate and hemiheptahydrate at low temperatures (79, 80).

The crystal forms produced as a result of amorphous lactose crystallization have probably been paid most attention. Crystal forms produced from pure amorphous lactose or lactose in milk powders after storage at various relative humidity conditions have often been determined using X-ray diffraction techniques (e.g., 13, 14, 53, 59, 65, 66, 81). The crystal forms found have been reported to be dependent on the relative humidity and temperature during storage. In several studies, spray- and freeze-dried lactose has been reported to have crystallized as a mixture of α-lactose monohydrate and anhydrous β-lactose at 53–85% RH at room temperature (53, 59, 64, 65). Our studies have shown (14) that freeze-dried lactose crystallizes mainly as a mixture of α-lactose monohydrate and anhydrous crystals with α- and β-lactose in a molar ratio of 5:3 at 44.4–76.4% RH and at 24°C, but traces of other crystal forms were also present (Table 2).

Bushill et al. (53) studied the crystal forms resulting from amorphous lactose crystallization. They found that lactose in freeze-dried skim milk when stored at 25°C and 55% RH crystallized as anhydrous crystals consisting of α- and β-lactose in a molar ratio of 5:3. Our results (13) agreed that lactose in freeze-dried skim milk crystallized mainly into the anhydrous 5:3 α:β crystal form when freeze-dried skim milk was stored at 24°C and relative humidity of 53.8–85.8%. However, as a result of storage at 85.8% RH, traces of α-lactose monohydrate crystals were detected. The crystal forms may also be dependent on the crystallization temperature. Würsch et al. (70) reported that lactose in spray-dried whole milk with 3.1% water crystallized at 60°C as anhydrous β-lactose. Saito (66) found that lactose in spray-dried whole milk crystallized as α-lactose monohydrate when stored at 37°C and 75% RH, but as anhydrous β-lactose when relative humidity was less than 20%. Drapier-Beche et al. (59) found that lactose in skim milk powder crystallized mainly as anhydrous β-lactose at 20°C and 43%

Table 2 Crystal Forms of Lactose Produced in Selected Duplicate Lactose Samples Stored at Various RH Conditions[a]

RH (%)	Storage time (h)	α-Lactose monohydrate		Unstable anhydrous α-lactose[b]		Anhydrous β-lactose		Anhydrous crystals with α- and β-lactose in a molar ratio of 5:3		Anhydrous crystals with α- and β-lactose in a molar ratio of 4:1	
44.4[c]	768	++	++			+	+	++	++		
	1440	++	++			+	+	++	++		
	2160	++	++			+	+	++	++		
44.4[d]	768	+	+++			+	+	+++	+	+	
	1440	++	++			+	+	++	++	+	+
	2160	++	++			+	+	++	++	+	+
53.8	48	+++	+++	+				+			
	192	++	++	+				++	++	+	
	384	++	+++	+				++	+	+	
66.2	24	++	++	+				++	++	+	
	72	+++	++	+	+			+	++		+
	120	++	+++	+				++	+	+	
76.4	21	++	+++	+	+			++	+	+	
	36	++	++					++	++	+	+
	72	+	+					+++	+++	+	+

[a] Approximate portions of various crystal forms were marked in the following way: predominant crystal form (approximate portion of the crystal form was more than half of sample) with +++, typical crystal form (approximate portion of the crystal form was less than half of sample but intensity of the typical peaks of crystal form was moderate) with ++, and traces of the crystal form (the typical peaks of crystal form could be detected although intensity of the peaks was low) with +.

[b] Samples containing unstable anhydrous α-lactose may contain also stable anhydrous α-lactose which was extremely difficult to be distinguished from unstable form using X-ray diffraction patterns because of overlapping of some typical peaks.

[c] α-lactose monohydrate was used in preparation of amorphous lactose.

[d] β-lactose was used in preparation of amorphous lactose.

RH, but as α-lactose monohydrate at relative humidity of 59% and 75%. At 53% RH, lactose crystallized as a mixture of anhydrous β-lactose and α-lactose monohydrate (59).

We believe that the differences in the various crystal forms produced from amorphous lactose may result from whether lactose crystallizes alone or in the presence of milk solids or other food components. Differences in the crystallization behavior may also occur when dairy foods are prepared from temperature-treated or preconcentrated liquids, where mutarotation may cause differences in the α:β ratio of lactose to produced powders. Unfortunately, there are no experimental data on the effects of pretreatment and processing conditions on the crystallization behavior of amorphous sugars in dairy or other foods.

B. Crystallization in Amorphous Starch

Starch refers to various glucose polymers, which exist in nature in starch granules of a number of grains, legumes, and potatoes. Starch consists of two types of glucose polymers, amylose and amylopectin. Amylose is formed of the largely linear molecules of starch where glucose units are linked with α-1,4 bonds. Amylopectin is formed of branched molecules with α-1,4- and α-1,6-linked glucose units (e.g., Ref. 82). Native starches are believed to be composed of amorphous amylose and partially crystalline amylopectin (83). Gelatinization of native starch results in the destruction of starch granules and the partial melting of the crystalline regions (84). Cooling and storage of starch gels is known to lead to rapid crystallization of amylose and relatively slow crystallization in the amylopectin fraction (85).

Starch retrogradation or amylopectin crystallization during storage of foods containing gelatinized starch is a well-known phenomenon. Retrogradation, as a phenomenon often considered to be responsible for the staling of bread, has been studied extensively either in starch gels or amorphous starch (e.g., Refs. 12, 15, 86–106) or in bread (93, 95, 106–111), involving numerous methods for the observation of the development of toughening or crystallinity. The methods have included thermoanalytical techniques used to determine the heat of melting of crystallites formed during storage (e.g., Refs. 12, 86–89, 97, 100), X-ray diffractometry detecting the type and extent of crystallinity (e.g., Refs. 15, 99, 100), and rheological measurements detecting an increase in gel firmness, i.e., hardening, and changes in viscoelastic properties (e.g., Refs. 90, 100, 102). Studies of starch retrogradation have also used Raman spectroscopy (96), Fourier transform IR spectroscopy (101, 112), and NMR (104, 105, 113).

However, differences in sample materials (e.g., origin of starch and amylopectin/amylose ratio), preparation techniques probably affecting the degree of starch gelatinization and thermal breakdown of the starch polymers (e.g., heating rate, heating time, and temperature range), and different methods used in evaluation of the development of crystallinity have made comparisons of reported crystallization behavior difficult. We have observed that the development of crystallinity in gelatinized starch during storage is affected by water plasticization and the glass transition.

1. Effect of Physical State

Crystallization in starch has been suggested to occur at temperatures above the glass transition (2, 71). Although starch retrogradation has been considered as a crystallization phenomenon, affected by water, the physical state of the amorphous starch components were not related to the glass transition until the pioneering work of Levine and Slade (2, 71). Our experimental data on crystallization in starch have suggested that crystallization occurs time-dependently at tempera-

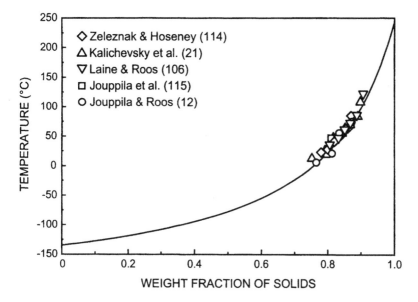

Figure 7 State diagram of amorphous starch. The T_g curve indicates estimated T_g calculated with the Gordon–Taylor equation with constant, k, of 5.6 (Ref. 12) and predicted T_g of 243°C for anhydrous starch (Ref. 27).

tures above 25°C if the water content is higher than 24.3 g/100 g of solids (19.5% of starch–water mixture). The established state diagram (Fig. 7) was used with the sorption isotherm at 25°C to derive the water content and the corresponding storage relative humidity of 86% as the critical values decreasing the glass transition temperature of amorphous corn starch to 25°C (12). Analysis of the crystallization data of previous studies and those of our studies with knowledge of the physical state suggested that starch retrogradation is controlled by the glass transition.

2. Rate of Crystallization

Rates of crystallization in starch gels and bread have been followed by measuring the increase in crystallinity with time using X-ray diffraction. Indirect methods to observe the development in crystallinity, such as the determination of the heat of melting of the crystallites formed using DSC and toughening of the gel or bread structure using rheological measurements, have also been applied. The rates of crystallization have been shown to be dependent on the water content of gelatinized starch and on storage temperature (15, 87, 88, 99), suggesting that the rate may be dependent on the physical state of the amorphous starch.

The rates of changes in crystallinity in starch and bread staling have often been modeled using the Avrami equation [Eq. (12)]. The parameter α has been derived from some crystallinity parameter X, for example, from the observed increase in the latent heat of melting of the crystallites, ΔH_m, the increase in the toughness of the gel or bread, or the increase in the intensity of X-ray diffraction.

$$\ln\left(\frac{X_f - X_t}{X_f - X_0}\right) = \ln(1 - \alpha) = -kt^n \tag{12}$$

where X_f is a leveling-off value of the crystallinity parameter, X_t is the value of the crystallinity parameter at time t, and X_0 is the initial value of the crystallinity parameter.

The studies of rates of crystallization in starch have often used starch gels with 50% or less water as models, because then the water contents have been close to that of white bread. A well-recognized finding is that the rate of crystallization in gelatinized wheat starch containing 50% solids is higher at refrigeration temperatures than at room temperature (e.g., 87, 88, 93, 99, 116). This has suggested that staling of bread may occur faster during storage at +4°C than at room temperature. Colwell et al. (87), Longton and LeGrys (88), and Marsh and Blanshard (99) found that the rate of crystallization in gelatinized wheat starch containing 50% solids decreased wth increasing storage temperature from 2 to 43°C. Colwell et al. (87) suggested that crystal formation occurred more rapidly at low temperatures because the degree of supercooling and, therefore, the driving force for crystallization was greater. However, Nakazawa et al. (94) reported opposite results for gelatinized potato starch containing 50% solids, although the difference in crystallinity after storage at 5 and 23°C was fairly small.

Crystallization of amorphous synthetic polymers is known to be controlled by the glass transition (1, 117). Only a few studies, however, have considered the possible effects of the glass transition on the crystallization behavior of gelatinized starch (12, 15, 71, 99, 106). We have studied the effects of the T_g on crystallization of amorphous corn starch at relatively low water contents when stored at various temperatures and $T - T_g$ conditions (12, 15). We have found that at low water contents, the rate of crystallization at a constant water content increases with increasing temperature, for example, in gelatinized corn starch containing 60–80% solids (20–40% water). The rate of crystallization in corn starch was related to $T - T_g$. The results suggested that the rate of crystallization increased with increasing storage temperature, ranging over the temperature range from 10–70°C. There was also an increase with $T - T_g$ over the same temperature range when the solids content was 60 and 70%. In corn starch containing 80% solids, the rate of crystallization increased with increasing storage temperature when stored at 50–80°C, but the rate decreased when the material was stored at 90°C.

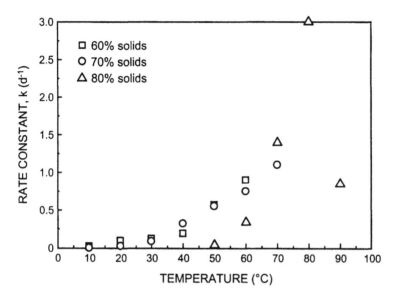

Figure 8   Rate constant of crystallization in gelatinized corn starch containing 60, 70, and 80% solids, as a function of storage temperature. (Data from Ref. 15.)

In our studies (12, 15), the crystallization data followed the Avrami equation, but the rate of crystallization at all water content seemed to be dependent on temperature rather than only on $T - T_g$. The rate constant, k, increased with increasing storage temperature, with concomitant decrease in the half-time of crystallization (Fig. 8 and 9). The data suggested, however, that the extent of crystallinity or the perfection of the crystalline structure increased with temperature. Therefore, the rate of crystallization may also depend on the crystalline structure formed in addition to the temperature and $T - T_g$ effects.

3.   Extent of Crystallization

According to Flory (117), crystallization of a polymer for a long period of time at a temperature close to the melting temperature results in a high degree of crystalline order and a large average size of the crystallites. The degree of perfection of the crystallites may become fairly low if improper annealing or crystallization conditions, e.g., storage at temperatures close to the glass transition, are employed (117).

One of the first studies relating crystallization in starch to the glass transition was that of Laine and Roos (106). They found that the extent of crystallization of gelatinized corn starch containing about 80% solids was a function of

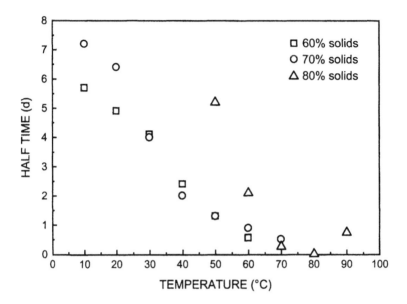

Figure 9 Half-time for crystallization in gelatinized corn starch containing 60, 70, and 80% solids, as a function of storage temperature. (Data from Ref. 15).

$T - T_g$ (Fig. 10). In a subsequent study (12), we observed that the relationship between the extent of crystallinity and $T - T_g$ in crystallized corn starch was parabolic (Fig. 10). The parabolic relationship between the extent of crystallization and $T - T_g$ could be explained using the polymer crystallization theory, as suggested by Slade and Levine (2). We proposed that at low $T - T_g$ conditions, crystallization occurred to a lower extent because molecular mobility was low and the crystal growth was kinetically restricted and slow, although nucleation was likely to be fast. Moreover, the crystals formed at the beginning of storage at low $T - T_g$ conditions act as barriers for molecular rearrangements and crystallization by forming rigid amorphous regions in the vicinity of the crystallites. According to Flory (117), polymer chains in such a stiff region are probably not able to take part in crystallization, which results in a lower extent of crystallization. At high $T - T_g$ conditions, crystal growth occurs rapidly and there is a possibility for the formation of larger and more perfect crystallites.

The parabolic relationship between the extent of crystallinity and $T - T_g$ (12) suggested that the maximum extent of crystallization in gelatinized corn starch occurred at a $T - T_g$ of 87°C (Fig. 10). Therefore, the predicted T_g for gelatinized corn starch containing 60% solids was at −55°C, suggesting that a maximum extent of crystallization was obtained at 32°C. This agreed with the finding of Zeleznak and Hoseney (95) that a maximum extent of crystallization

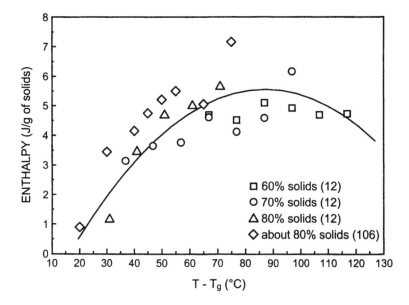

Figure 10 Extent of crystallization in gelatinized corn starch with solids content ranging from 60 to 80%, as a function of temperature difference between storage temperature and glass transition temperature. A second-order polynomial was fitted to experimental data, giving a coefficient of determination of 0.64.

at 25°C occurred in wheat starch gels containing 60% starch, for which we predicted $T - T_g$ to be 80°C. The extent of crystallization in wheat starch containing 50% solids has been found to decrease with increasing storage temperatures above 2°C (e.g., Refs. 87, 88, 93, 99, 116). Our predicted T_g for starch containing 50% solids was −78°C (12), and therefore the predicted $T - T_g$ at 2°C was 80°C, suggesting that a maximum extent of crystallization in wheat starch containing 50% solids occurred at a $T - T_g$ of 80°C or lower.

Starch concentration and the extent of crystallization at a constant temperature may also be related to $T - T_g$ defined by water plasticization. Longton and LeGrys (88) reported that a maximum extent of crystallization at 4°C occurred in wheat starch gel containing 50% solids, for which we found $T - T_g$ to be 82°C (12). However, the temperature at which the maximum extent of crystallinity is produced is likely to be extremely sensitive to differences in composition and water content, which should be taken into account when results of various systems, e.g., amorphous starch, starch gels, and bread, are compared. Starch gels and bread may also be at least partially amorphous and, therefore, more sensitive to water, for water plasticization is likely to occur within the amorphous regions.

It should also be remembered that the crystallinity may depend on the availability of water to hydrate the crystalline structure.

4. Melting Behavior of Crystallites Formed

Melting temperatures of the crystallites formed in gelatinized starch during storage have been found to increase with increasing storage temperature in a number of studies (e.g., 12, 87, 88, 92–94, 98, 103, 111, 116). In most cases, melting behavior was reported for starch with solids contents ranging from 30 to 50% and during storage at refrigeration and room temperatures. Storage of wheat starch gels at higher temperatures was reported to result in the formation of a more perfect crystal structure (87, 88). Zeleznak and Hoseney (111) and Chang and Liu (103) found that the crystals formed in wheat and rice starch gels, respectively, at a higher temperature had a more perfect crystalline structure, due to annealing of the material in the vicinity of the melting temperature during storage. They also found that the melting range of crystallized starch with a more perfect structure was narrower, which was also found by Jouppila and Roos (12).

IV. CONCLUSIONS

Most low-moisture and frozen foods contain amorphous carbohydrates, which exist either in the glassy state or in the more liquidlike supercooled, liquid state. These materials are miscible with water, which is often observed from its plasticizing effect. Water plasticization results in a decrease in the glass transition temperature range. Therefore, at some critical water content or water activity, the glass transition occurs at or below ambient temperature. The depression of the glass transition to below ambient temperature is observed from dramatic changes in mechanical properties, decrease in viscosity, and enhanced flow properties of the material. The increase in molecular mobility occurring above the glass transition may also enhance crystallization of sugars and carbohydrate polymers as they approach the equilibrium crystalline state favored over the nonequilibrium amorphous state. The glass transition seems to control the crystallization of amorphous carbohydrates, for example, crystallization of lactose in dairy powders, crystallization of lactose in frozen desserts, and retrogradation of starch. The crystallization, however, is a complicated process affected by food composition, temperature, and water content, and it seems that the glass transition is not the sole factor contributing to the rate and extent of crystallization of carbohydrates in foods. Crystallization is not likely to occur in glassy carbohydrate systems. Therefore, the knowledge of the glass transition can be used with temperature and water content adjustments to control favorable crystallization in food processing or to reduce detrimental crystallization of amorphous carbohydrates during food processing and storage.

REFERENCES

1. LH Sperling. Introduction to Physical Polymer Science. New York: Wiley, 1986.
2. L Slade, H Levine. Beyond water activity: recent advances based on an alternative approach to the assessment of food quality and safety. CRC Crit Rev Food Sci Nutr 30:115–360, 1991.
3. YH Roos. Phase Transitions in Foods. San Diego: Academic Press, 1995.
4. M Peleg. Mechanical properties of dry brittle cereal products. In: DS Reid, ed. The Properties of Water in Foods ISOPOW 6. London: Chapman and Hall, 1998, pp 233–252.
5. GC Supplee. Humidity equilibria of milk powders. J Dairy Sci 9:50–61, 1926.
6. GS Parks, JD Reaug. Studies on glass. XV. The viscosity and rigidity of glucose glass. J Chem Phys 5:364–367, 1937.
7. GW White, SH Cakebread. The glass state in certain sugar-containing food products. J Food Technol 1:73–82, 1966.
8. L Slade, H Levine. Mono- and disaccharides: Selected physicochemical and functional aspects. In: A-C Eliasson, ed. Carbohydrates in Foods. New York: Marcel Dekker, 1996. pp 41–157.
9. M Yoshioka, BC Hancock, G Zografi. Crystallization of indomethacin from the amorphous state below and above its glass transition temperature. J Pharm Sci 83: 1700–1705, 1994.
10. Y Roos, M Karel. Differential scanning calorimetry study of phase transitions affecting the quality of dehydrated materials. Biotechnol Prog 6:159–163, 1990.
11. Y Roos, M Karel. Crystallization of amorphous lactose. J Food Sci 57:775–777, 1992.
12. K Jouppila, YH Roos. The physical state of amorphous corn starch and its impact on crystallization. Carbohydr Polym 32:95–104, 1997.
13. K Jouppila, J Kansikas, YH Roos. Glass transition, water plasticization, and lactose crystallization in skim milk powder. J Dairy Sci 80:3152–3160, 1997.
14. K Jouppilla, J Kansikas, YH Roos. Crystallization and X-ray diffraction of crystals formed in water-plasticized amorphous lactose. Biotechnol Prog 14:347–350, 1998.
15. K Jouppila, J Kansikas, YH Roos. Factors affecting crystallization and crystallization kinetics in amorphous corn starch. Carbohydr Polym 36:143–149, 1998.
16. S Söderholm, YH Roos, N Meinander. Characterization of effects of molecular mobility in amorphous sucrose using FT-Ramen spectroscopy. In: YH Roos, ed. ISOPOW 7, Proceedings of the Poster Sessions. EKT-series 1143. Helsinki: University of Helsinki, 1998, pp 104–107.
17. IJ van den Dries, PA de Jager, MA Hemminga. Sensitivity of saturation transfer electron spin resonance extended to extremely slow mobility in glassy materials. J Magn Reson 131:241–247, 1998.

18. D Champion, H Hervet, G Blond, D Simatos. Comparison between two methods to measure translational diffusion of a small molecule at subzero temperature. J Agric Food Chem 43:2887–2891, 1995.

19. D Champion, H Hervet, G Blond, M Le Meste, D Simatos. Translational diffusion in sucrose solutions in the vicinity of their glass temperature. J Phys Chem B 101: 10674–10679, 1997.

20. IJ van den Dries, D Dusschoten, M Hemminga. Mobility in glucose glasses. ISOPOW 7 Symposium, Unitas Congress Center, May 30–June 4, 1998, Helsinki, Finland.

21. MT Kalichevsky, EM Jaroszkiewicz, S Ablett, JMV Blanshard, PJ Lillford. The glass transition of amylopectin measured by DSC, DMATA and NMR. Carbohydr Polym 18:77–88, 1992.

22. B Luyet, D Rasmussen. Study by differential thermal analysis of the temperatures of instability of rapidly cooled solutions of glycerol, ethylene glycol, sucrose and glucose. Biodynamica 10(211):167–191, 1968.

23. PD Orford, P Parker, SG Ring, AC Smith. Effect of water as a diluent on the glass transition behavior of malto-oligosaccharides, amylose and amylopectin. Int J Biol Macromol 11:91–96, 1989.

24. Y Roos, M Karel. Plasticizing effect of water on thermal behavior and crystallization of amorphous food models. J Food Sci 56:38–43, 1991.

25. Y Roos. Melting and glass transitions of low molecular weight carbohydrates. Carbohydr Res 238:39–48, 1993.

26. M Karel, S Anglea, P Buera, R Karmas, G Levi, Y Roos. Stability-related transitions of amorphous foods. Thermochim Acta 246:249–269, 1994.

27. Y Roos, M Karel. Water and molecular weight effects on glass transitions in amorphous carbohydrates and carbohydrate solutions. J Food Sci 56:1676–1681, 1991.

28. YH Roos, M Karel, JL Kokini. Glass transitions in low moisture and frozen foods: effects on shelf life and quality. Food Technol 50(11):95–108, 1996.

29. Y Roos, M Karel. Amorphous state and delayed ice formation in sucrose solutions. Int J Food Sci Technol 26:553–556, 1991.

30. RMH Hatley, C van den Berg, F Franks. The unfrozen water content of maximally freeze concentrated carbohydrate solutions: Validity of the methods used for its determination. Cryo-Lett 12:113–124, 1991.

31. MJ Izzard, S Ablett, PJ Lillford. Calorimetric study of the glass transition occurring in sucrose solutions. In: E Dickinson, ed. Food Polymers, Gels, and Colloids. Cambridge: Royal Society of Chemistry, 1991, pp 289–300.

32. A Saleki-Gerhardt, G Zografi. Non-isothermal and isothermal crystallization of sucrose from the amorphous state. Pharm Res 11:1166–1173, 1994.

33. M Sugisaki, H Suga, S Seki. Calorimetric study of the glassy state. IV. Heat capacities of glassy water and cubic ice. Bull Chem Soc Jpn 41:2591–2599, 1968.

34. GP Johari, A Hallbrucker, E Mayer. The glass–liquid transition of hyperquenched water. Nature 330:552–553, 1987.

35. K Jouppila, YH Roos. Glass transitions and crystallization in milk powders. J Dairy Sci 77:2907–2915, 1994.

36. RJ Lloyd, XD Chen, JB Hargreaves. Glass transition and caking of spray-dried lactose. Int J Food Sci Technol 31:305–311, 1996.

37. TG Fox, PJ Flory. Second-order transition temperatures and related properties of polystyrene. I. Influence of molecular weight. J Appl Phys 21:581–591, 1950.
38. K Ueberreiter, G Kanig. Self-plasticization of polymers. J Coll Sci 7:569–583, 1952.
39. Y Roos, M Karel. Phase transitions of mixtures of amorphous polysaccharides and sugars. Biotechnol Prog 7:49–53, 1991.
40. JM Aguilera, JM del Valle, M Karel. Caking phenomena in amorphous food powders. Trends Food Sci Technol 6:149–155, 1995.
41. T Soesanto, MC Williams. Volumetric interpretation of viscosity for concentrated and dilute sugar solutions. J Phys Chem 85:3338–3341, 1981.
42. M Peleg. Mapping the stiffness–temperature–moisture relationship of solid biomaterials at and around their glaass transition. Rheol Acta 32:575–580, 1993.
43. CA Angell, RD Bressel, JL Green, H Kanno, M Oguni, EJ Sare. Liquid fragility and the glass transition in water and aqueous solutions. J Food Eng 22:115–142, 1994.
44. CA Angell. Formation of glasses from liquids and biopolymers. Science 267:1924–1935, 1995.
45. TJ Laaksonen, YH Roos. Dielectric and dynamic-mechanical properties of frozen doughs. In: YH Roos, ed. ISOPOW 7, Proceedings of the Poster Sessions. EKT series 1143. Helsinki: University of Helsinki, 1998, pp 42–46.
46. EE Katz, TP Labuza. Effect of water activity on the secondary crispness and mechanical deformation of snack food products. J Food Sci 46:403–409, 1981.
47. F Sauvageot, G Blond. Effect of water activity on crispness of breakfast cereals. J Text Stud 22:423–442, 1992.
48. M Peleg. A model of mechanical changes in biomaterials at and around their glass transition. Biotechnol Prog 10:385–388, 1994.
49. YH Roos, K Roininen, K Jouppila, H Tuorila. Glass transition and water plasticization effects of crispness of a snack food extrudate. Int J Food Properties 1:163–180, 1998.
50. RJ Bellows, CJ King. Product collapse during freeze drying of liquid foods. AIChE Symp Ser 69(132):33–41, 1973.
51. DP Downton, JL Flores-Luna, CJ King. Mechanism of stickiness in hygroscopic, amorphous powders. Ind Eng Chem Fundam 21:447–451, 1982.
52. BL Herrington. Some physico-chemical properties of lactose. I. The spontaneous crystallization of supersaturated solutions of lactose. J Dairy Sci 17:501–518, 1934.
53. JH Bushill, WB Wright, CHF Fuller, AV Bell. The crystallization of lactose with particular reference to its occurrence in milk powder. J Sci Food Agric 16:622–628, 1965.
54. S Warburton, SW Pixton. The moisture relations of spray dried skimmed milk. J Stored Prod Res 14:143–158, 1978.
55. G Vuataz. Preservation of skim-milk powders: role of water activity and temperature in lactose crystallization and lysine loss. In: CC Seow, ed. Food Preservation by Water Activity Control. Amsterdam: Elsevier Science, 1988, pp 73–101.
56. H-M Lai, SJ Schmidt. Lactose crystallization in skim milk powder observed by hydrodynamic equilibria, scanning electron microscopy and ^{2}H nuclear magnetic resonance. J Food Sci 55:994–999, 1990.

57. K Joupilla, YH Roos. Water sorption and time-dependent phenomena of milk powders. J Dairy Sci 77:1798–1808, 1994.
58. G Buckton, P Darcy. Water mobility in amorphous lactose below and close to the glass transition temperature. Int J Pharm 136:141–146, 1996.
59. N Drapier-Beche, J Fanni, M Parmentier, M Vilasi. Evaluation of lactose crystalline forms by nondestructive analysis. J Dairy Sci 80:457–463, 1997.
60. B Makower, WB Dye. Equilibrium moisture content and crystallization of amorphous sucrose and glucose. J Agric Food Chem 4:72–77, 1956.
61. HA Iglesias, J Chirife. Delayed crystallization of amorphous sucrose in humidified freeze dried model systems. J Food Technol 13:137–144, 1978.
62. A Senoussi, ED Dumoulin, Z Berk. Retention of diacetyl in milk during spray-drying and storage. J Food Sci 60:894–897, 905, 1995.
63. CJ Kedward, W Macnaughtan, JMV Blanshard, JR Mitchell. Crystallization kinetics of lactose and sucrose based on isothermal differential scanning calorimetry. J Food Sci 63:192–197, 1998.
64. L-E Briggner, G Buckton, K Bystrom, P Darcy. The use of isothermal microcalorimetry in the study of changes in crystallinity induced during the processing of powders. Int J Pharm 105:125–135, 1994.
65. T Sebhatu, M Angberg, C Ahlneck. Assessment of the degree of disorder in crystalline solids by isothermal microcalorimetry. Int J Pharm 104:135–144, 1994.
66. Z Saito. Particle structure in spray-dried whole milk and in instant skim milk powder as related to lactose crystallization. Food Microstruct 4:333–340, 1985.
67. YH Roos. Water activity and physical state effects on amorphous food stability. J Food Process Preserv 16:433–447, 1993.
68. MP Buera, K Jouppila, YH Roos, J Chirife. Differential scanning calorimetry glass transition temperatures of white bread and mold growth in the putative glassy state. Cereal Chem 75:64–69, 1998.
69. KJ Palmer, WB Dye, D Black. X-ray diffractometer and microscopic investigation of crystallization of amorphous sucrose. J Agric Food Chem 4:77–81, 1956.
70. P Würsch, J Rosset, B Köllreutter, A Klein. Crystallization of β-lactose under elevated storage temperature in spray-dried milk powder. Milchwissenschaft 39:579–582, 1984.
71. H Levine, L Slade. Influences of the glassy and rubbery states on the thermal, mechanical, and structural properties of doughs and baked products. In: H Faridi, JM Faubion, eds. Dough Rheology and Baked Product Texture. New York: AVI, 1990, pp 157–330.
72. RW Hartel, AV Shastry. Sugar crystallization in food products. Crit Rev Food Sci Nutr 30:49–112, 1991.
73. M Avrami. Kinetics of phase change. I. General theory. J Chem Phys 7:1103–1112, 1939.
74. MN Kim, M Saltmarch, TP Labuza. Non-enzymatic browning of hygroscopic whey powders in open versus sealed pouches. J Food Process Preserv 5:49–57, 1981.
75. P Gabarra, RW Hartel. Corn syrup solids and their saccharide fractions affect crystallization of amorphous sucrose. J Food Sci 63:523–528, 1998.

76. DC Fries, ST Rao, M Sundaralingam. Structural chemistry of carbohydrates. III. Crystal and molecular strucure of 4-*O*-β-D-galactopyranosyl-α-D-glucopyranose monohydrate (α-lactose monohydrate). Acta Cryst B27:994–1005, 1971.
77. TJ Buma, GA Wiegers. X-ray powder patterns of lactose and unit cell dimensions of β-lactose. Neth Milk and Dairy J 21:208–213, 1967.
78. TD Simpson, FW Parrish, ML Nelson. Crystalline forms of lactose produced in acidic alcoholic media. J Food Sci 47:1948–1951, 1954, 1982.
79. RS Shallenberger, GG Birch. Sugar Chemistry. Westport, Conn.: AVI, 1975, pp 62–67.
80. FE Young, FT Jones. Sucrose hydrates. The sucrose–water phase diagram. J Phys Coll Chem 53:1334–1350, 1949.
81. CA Aguilar, GR Ziegler. Physical and microscopic characterization of dry whole milk with altered lactose content. 2. Effect of lactose crystallization. J Dairy Sci 77:1198–1204, 1994.
82. DR Lineback. Current concepts of starch structure and its impact on properties. J Jpn Soc Starch Sci 33(1):80–88, 1986.
83. A-C Eliasson, K Larsson, S Andersson, ST Hyde, R Nesper, H-G von Schnering. On the structure of native starch—an analogue to the quartz structure. Starch/Stärke 39:147–152, 1987.
84. WA Atwell, LF Hood, DR Lineback, E Varriano-Marston, HF Zobel. The terminology and methodology associated with basic starch phenomena. Cereal Foods World 33:306, 308, 310–311, 1988.
85. MJ Miles, VJ Morris, SG Ring. Gelation of amylose. Carbohydr Res 135:257–269, 1985.
86. RG McIver, DWE Axford, KH Colwell, GAH Elton. Kinetic study of the retrogradation of gelatinised starch. J Sci Food Agric 19:560–563, 1968.
87. KH Colwell, DWE Axford, N Chamberlain, GAH Elton. Effect of storage temperature on the ageing of concentrated wheat starch gels. J Sci Food Agric 20:550–555, 1969.
88. J Longton, GA LeGrys. Differential scanning calorimetry studies on the crystallinity of ageing wheat starch gels. Starch/Stärke 33:410–414, 1981.
89. A-C Eliasson. Differential scanning calorimetry studies on wheat starch–gluten mixtures. II. Effect of gluten and socium stearoyl lactylate on starch crystallization during aging of wheat starch gels. J Cereal Sci 1:207–213, 1983.
90. R Germani, CF Ciacco, DB Rodriguez-Amaya. Effect of sugars, lipids and type of starch on the mode of kinetics of retrogradation of concentrated corn starch gels. Starch/Stärke 35:377–381, 1983.
91. RR del Rosario, CR Pontiveros. Retrogradation of some starch mixtures. Starch/Stärke 35:86–92, 1983.
92. F Nakazawa, S Noguchi, J Takahashi, M Takada. Gelatinization and retrogradation of rice starch studied by differential scanning calorimetry. Agric Biol Chem 48:201–203, 1984.
93. A-C Eliasson. Retrogradation of starch as measured by differential scanning calorimetry. In: RD Hill, L Munck, eds. New Approaches to Research on Cereal Carbohydrates. Amsterdam: Elsevier Science, 1985, pp 93–98.
94. F Nakazawa, S Noguchi, J Takahashi, M Takada. Retrogradation of gelatinized

potato starch studied by differential scanning calorimetry. Agric Biol Chem 49: 953–957, 1985.

95. KJ Zeleznak, RC Hoseney. The role of water in the retrogradation of wheat starch gels and break crumb. Cereal Chem 63:407–411, 1986.

96. BJ Bulkin, Y Kwak, ICM Dea. Retrogradation kinetics of waxy-corn and potato starches; a rapid, Raman-spectroscopic study. Carbohydr Res 160:95–112, 1987.

97. PL Russell. The aging of gels from starches of different amylose/amylopectin content studied by differential scanning calorimetry. J Cereal Sci 6:147–158, 1987.

98. A-C Eliasson, G Ljunger. Interactions between amylopectin and lipid additives during retrogradation in a model system. J Sci Food Agric 44:353–361, 1988.

99. RDL Marsh, JMV Blanshard. The application of polymer crystal growth theory to the kinetics of formation of the β-amylose polymorph in a 50% wheat-starch gel. Carbohydr Polym 9:301–317, 1988.

100. Ph Roulet, WM MacInnes, P Würsch, RM Sanchez, A Raemy. A comparative study of the retrogradation kinetics of gelatinized wheat starch in gel and powder form using X-rays, differential scanning calorimetry and dynamic mechanical analysis. Food Hydrocolloids 2:381–396, 1988.

101. RH Wilson, PS Belton. A Fourier-transform infrared study of wheat starch gels. Carbohydr Res 180:339–344, 1988.

102. CG Biliaderis, J Zawistowski. Viscoelastic behavior of aging starch gels: effects of concentration, temperature, and starch hydrolysates on network properties. Cereal Chem 67:240–246, 1990.

103. S-M Chang, L-C Liu. Retrogradation of rice starches studied by differential scanning calorimetry and influence of sugars, NaCl and lipids. J Food Sci 56:564–566, 570, 1991.

104. CH Teo, CC Seow. A pulsed NMR method for the study of starch retrogradation. Starch/Stärke 44:288–292, 1992.

105. JY Wu, TM Eads. Evolution of polymer mobility during aging of gelatinized waxy maize starch: a magnetization transfer ^{1}H NMR study. Carbohydr Polym 20:51–60, 1993.

106. MJK Laine, Y Roos. Water plasticization and recrystallization of starch in relation to glass transition. In: A Argaiz, A López-Malo, E Palou, P Corte, eds. Proceedings of the Poster Session. International Symposium on the Properties of Water. Practicum II. Cholula, Puebla, Mexico: Universidad de las Américas-Puebla, 1994, pp 109–112.

107. T Fearn, PL Russell. A kinetic study of bread staling by differential scanning calorimetry. The effect of loaf specific volume. J Sci Food Agric 33:537–548, 1982.

108. PL Russell. A kinetic study of break staling by differential scanning calorimetry and compressibility measurement. The effect of different grists. J Cereal Sci 1: 285–296, 1983.

109. PL Russell. A kinetic study of bread staling by differential scanning calorimetry and compressibility measurements. The effect of added monoglyceride. J Cereal Sci 1:297–303, 1983.

110. PL Russell. A kinetic study of bread staling by differential scanning calorimetry. The effect of painting loaves with ethanol. Starch/Stärke 35:277–281, 1983.

111. KJ Zeleznak, RC Hoseney. Characterization of starch from bread aged at different temperatures. Starch/Stärke 39:231–233, 1987.
112. BJ Goodfellow, RH Wilson. A fourier transform IR study of the gelation of amylose and amylopectin. Biopolymers 30:1183–1189, 1990.
113. CC Seow, CH Teo. Staling of starch-based products: a comparative study by firmness and pulsed NMR measurements. Starch/Stärke 48:90–93, 1996.
114. KJ Zeleznak, RC Hoseney. The glass transition in starch. Cereal Chem 64:121–124, 1987.
115. K Jouppila, T Ahonen, Y Roos. Water adsorption and plasticization of amylopectin glasses. In: E Dickinson, D Lorient, eds. Food Macromolecules and Colloids. London: Royal Society of Chemistry, 1995, pp 556–559.
116. T Jankowski, CK Rha. Retrogradation of starch in cooked wheat. Starch/Stärke 38:6–9, 1986.
117. PJ Flory. Principles of Polymer Chemistry. Ithaca, New York: Cornell University Press, 1953, pp 563–571.

5

Construction of State Diagrams for Cereal Processing

Gönül Kaletunç
The Ohio State University, Columbus, Ohio, U.S.A.

I. INTRODUCTION

The quality of cereal-based foods can be evaluated by several criteria, including sensory attributes and laboratory-measured physical properties. The development of a fundamental understanding of the influence of processing and storage on quality parameters of food products requires identification and investigation of measurable physical properties of food materials as a function of the conditions that simulate processing and storage environments; for example, texture and stability during storage determine the shelf life of the product. It is important to identify the key physical properties for each product with which quality parameters such as texture and stability can be correlated. Thermal and mechanical properties have been shown to correlate with the sensory textural attributes of cereal-based foods (1–4). Given the variety of available parameters, it is important to develop a mechanism for evaluating the quality of a product in terms of multiple parameters.

Thermodynamic phase diagrams have a long history of application to evaluate the equilibrium properties of a material as a function of composition and environmental parameters. Recently, the concept of a phase diagram has been extended to include properties that are not rigorous thermodynamic parameters and systems away from equilibrium. Phase diagrams in the form of temperature–composition plots were generated using thermal, volumetric, and mechanical properties data for aqueous model systems of carbohydrates and polymers in order to interpret, qualitatively or semiquantitatively, the freezing behavior of cells and tissues (5). These plots showed the boundaries of solid and liquid phases that are in thermodynamic equilibrium. Studies on these systems also showed the existence of kinetically controlled metastable states, especially when they were subjected to rapid heating or cooling (5–7). MacKenzie (5) included properties of nonequilibrium behavior on phase diagrams referring to the total display as "supplemented phase diagrams." These diagrams, also called extended phase diagrams, dynamic phase diagrams, or more commonly state diagrams, permit qualitative or semiquantitative analysis of food and pharmaceutical systems (8, 9).

Thermal and mechanical properties data collected as a function of processing and storage conditions can be used to construct state diagrams, which describe the states of a material prior to and during processing. Similar state diagrams for the corresponding products can be constructed and used to assess the physicochemical changes that result from processing as well as product storage. Given the number of parameters associated with processing and storage, the number of dimensions of multidimensional state diagrams may be large. The axes of such state diagrams may include temperature, pressure, moisture content, concentrations of additives, and a variety of processing parameters. Diagrams of more than three dimensions cannot be represented in space, making interpretation and application of the complete diagram difficult. Therefore, simplified two- or three-dimensional subdiagrams of the parent state diagrams are constructed.

Over the years, a number of investigators have proposed that state diagrams (which map the physical states a material assumes as a function of concentration, temperature, pressure, and time) can be used to develop a fundamental understanding of food products and processing (10–15). As early as 1966, White and Cakebread (16) recognized the importance of characterizing the physical state(s) of sugars in sugar-containing food products to interpret some of the defects associated with changes in storage conditions (e.g., temperature, relative humidity). These investigators discussed the implications of the breakdown of the glassy state on the quality, safety, and storage stability of boiled sweets, milk powder, ice cream, and some freeze-dried products. Since 1966, numerous other researchers have measured fundamental properties (thermodynamic, rheological, kinetic) to define the physical states of food components under conditions that simulate processing and storage. Because the nature of these physical states can be related to the quality, shelf life, and sensory attributes of the end products, physical characterization of food components has received growing attention in the pharmaceutical and food science literature. MacKenzie (5) and Franks et al. (10) developed state diagrams for hydrophilic polymers, in connection with their use as cryoprotectants in the preservation of biological structure. Levine and Slade (11) described the application of an idealized state diagram for a hypothetical small carbohydrate to explain and/or predict the functional behavior of such carbohydrates in frozen foods, in connection with cryostabilization technology. These researchers proposed that the glass transition should be considered the critical reference point, because various food materials are in an amorphous state at temperatures above the glass transition temperature. Roos and Karel (12) discussed the use of a generic, simplified state diagram for a water-soluble food component to show the effect of temperature and moisture content on stability and material characteristics. They also suggested the possibility of using such diagrams to map the path of various food processes, as had Slade and Levine (13).

Slade and Levine (17) demonstrated that a sucrose–water state diagram (which reveals the relative locations of the glass, solidus, liquidus, and vaporus curves) is useful for characterizing various aspects of cookie and cracker manufacturing, including dough mixing, lay time, machining, and baking, as well as for characterizing finished-product texture, shelf life, and storage stability. In a review, Roos (18) reemphasized the use of glass transition temperature (T_g) values to establish state diagrams that describe the effect of composition on stability and which show the effects of temperature and moisture on viscosity, structure, and crystallization.

In this chapter, we review characterization mainly by calorimetry of thermally induced conformational changes and phase transitions in pre- and postprocessed cereal biopolymers and the use of calorimetric data in the construction of state diagrams. The utilization of mechanical properties in expanding our understanding of biopolymers during processing also is discussed. Finally, the benefits of employing state diagrams in development of new products or in the design of new processes or the improvement of existing processes is reviewed through specific illustrations of the applications of state diagrams reported in the literature.

II. DEVELOPMENT OF STATE DIAGRAMS

A. Physial State of a Material

In physical chemistry, the *state of the system* is defined by the condition of the system at a given time. The system is the material being studied at a given moment, and the values of the variables, including temperature, pressure, and composition, specify the state of the system. Processes can change the state of the system because the variables specifying the state are changed. If the two states are in thermodynamic equilibrium such as in melting, in which the chemical potential of any substance in both states has the same value, they are referred to as phases rather than states and the transition is referred to as a phase transition. However, kinetically controlled metastable states frequently occur in biological systems. In such states, the physical properties of a given material can be time, temperature, and composition dependent. Several investigators have observed that nonequilibrium transitions resulting in metastable states are commonly encountered in water–sugar and water–polymer systems due to the effect of cooling rate on crystallization and vitrification (5, 7, 10). It has also been reported that unstable states, such as those formed in sugar glasses, can lead to various defects under unfavorable conditions because of their instability in the presence of moisture (16).

A complete understanding of the behavior of a given system requires the study of phases (states) in equilibrium and the transitions between these states as well as nonequilibrium, time-dependent phenomena. Kinetically controlled states can be investigated under conditions relevant to processing and storage. Rasmussen and MacKenzie (19) demonstrated the effect of heating rate on the glass transition temperature of amorphous water. These investigators emphasized the kinetic nature of the glass transition and stated that a family of glass transition curves can be obtained depending on the heating rate chosen for the measurements. Therefore, it is essential that nonequilibrium transitions such as the glass transition be reported with the experimental time scale.

B. Techniques for Determining the Physical State of a Material

Thermal, rheological, and spectroscopic techniques are employed to characterize the physical state of materials. Thermal analysis techniques, specifically differential scanning calorimetry (DSC), are commonly used to determine the transformations of the equilibrium states as well as of nonequilibirum states for sugar–water, biopolymer–water, and pre- and postprocessed cereal–water systems. The use of differential scanning calorimetry to detect, characterize, and monitor thermally induced transitions in pre- and postprocessed cereals is discussed in detail in Chapter 1.

Thermally induced transitions as a function of moisture content can be determined from specific heat capacity versus temperature curves. Transformations between the states appear as apparent anomalies in the temperature dependence of the specific heat capacity of the material. The transition temperature of a thermally induced transition defines the point at which the two states coexist at a given moisture content. DSC detects both first-order and second-order transitions. In cereal–water systems, first–order transitions, which manifest themselves as endothermic or exothermic peaks, include starch melting, starch gelatinization, ice melting, crystallization, protein denaturation, protein aggregation, and lipid melting. Among these transitions melting is a true phase transition. However, crystallization (5) and aggregation (20) are known to be kinetically controlled. Furthermore, the nonequilibrium nature of crystallization will affect the melting characteristics of the crystals when they are subsequently subjected to warming. This effect becomes important especially during the freezing of solute–water systems. For a sucrose–water system, MacKenzie (5) reports that in low sucrose concentration solutions (0–20%), all of the water is converted into ice during both slow and rapid cooling. In moderately concentrated sucrose solutions (30–60%), ice crystallization occurred during cooling if the solution was cooled slowly. When cooling was rapid, ice crystallization occurred during slow warming following the rapid cooling. In concentrated sucrose solutions (70–80%), ice formation did not occur during either cooling or rewarming.

The glass transition observed in cereal–water systems is a second-order transition, which appears as a discontinuity in the specific heat in a DSC thermogram. The glass transition is associated with amorphous (noncrystalline) regions in the cereal–water systems. The implications of the glassy state and glass transition in sugar-containing food systems were first interpreted in a review by White and Cakebread (16). White and Cakebread (16) stated that defects such as stickiness, caking, sandiness, and partial liquefaction in sugar-containing foods—such as boiled sweets, milk powder, ice cream, and freeze-dried products—are associated with the breakdown of the glassy state due to exposure to moisture. The glass transition has been shown to be kinetically controlled (5). Rasmussen and MacKenzie (19) estimated the time dependence of the glass transition of pure water by measuring the glass transition temperatures for various concentrations of aqueous methanol, ethylene glycol, and glycerol solutions and extrapolating them to pure water. They stated that lower heating rates allow longer times for the material to relax resulting in the detection of the glass transition at lower temperatures.

In addition to the discontinuity in specific heat occurring with the glass transition, several additional physical properties change (21, 22). A change in the slope of the volume expansion detectable by dilatometry is observed. A discontinuity in the thermal expansion coefficient becomes apparent. The glass transition is associated with a dramatic reduction in viscosity of a given system, typically about three-orders of magnitude (23). Therefore, rheological techniques are also utilized to determine the glass transition temperature by measuring the change in viscosity, Young's modulus, and dynamic viscoelastic properties (24). Several studies were conducted using rheological analysis to measure the glass transition temperature of pre-and postprocessed cereals as well as their biopolymer components (25–31). Amemiya and Menjivar (26) determined the Young's modulus of cookies and crackers as a function of moisture content (fat-free dry basis). Their results showed that at ambient temperature, the modulus displays a sharp decrease at various moisture contents depending on the formulation of the cookie or cracker. The critical moisture contents were determined to be 10%, 8%, 6–7%, and 5% for lean cracker, rich cracker, cookie/cracker hybrid, and wire-cut cookie, respectively. These investigators concluded that at ambient temperature the glass transition occurred at the reported critical moisture content specific for each product.

Because molecular mobility of the system changes over the glass transition, techniques sensitive to the changes in the motion of molecules also can be utilized for characterization of glass transition. Nuclear magnetic resonance (NMR) has been used to describe the glass transition of cereal systems by measuring relaxation times after application of a radio frequency pulse. It has been found that for amylopectin, the T_2 relaxation time is independent of temperature below the rigid lattice limit (RLL) temperature and increases with temperature above it (27). It is important to recognize that temperature corresponding to the RLL may not be identical to the glass transition temperature measured by techniques sensitive to the bulk properties of the material, although it is related to the glass transition (24).

Kalichevsky et al. (27) reported that the glass transition temperature of amylopectin differed depending on the technique used. The measured glass transition temperature of amylopectin increased with the technique used in order from NMR to Instron universal texturometer to DSC, to dynamic mechanical thermal analysis. The differences may be due to the sensitivity of a specific observable to the degree of mobility at the level of measurement. While NMR can detect a local mobility transition of a side chain, other techniques measure the glass transition temperature averaged over the entire sample. It should also be emphasized that because the glass transition is kinetically controlled, differences in the frequency of the measurements and the rate of heating are expected to influence the measured glass transition temperature. Furthermore, transitions may or may not be observable depending on the compatibility of the relaxation time of measured transition and the experimental time domain and the sensitivity of the measurement technique.

C. Construction of State Diagram

Phase diagrams for systems under thermodynamic equilibrium have been applied extensively to describe the states of one, two, or three component systems. The Gibbs phase rule specifies the minimum number of variables (intensive and/or extensive) that are sufficient to describe the system (32). The phase diagram describing that system will have the same number of coordinate axes as the number of variables prescribed by the phase rule. Two parameters are sufficient to define the thermodynamic state of a single component system. Pressure (P) and temperature (T) typically are selected and are particularly relevant variables for food applications. This information can be displayed on a two-dimensional (P, T) phase diagram. Where two phases coexist in equilibrium in a single component system either temperature or pressure is sufficient to define the system. For a binary system, the relationships between composition and the phase transition temperatures can be displayed as a two-dimensional phase diagram at constant pressure. The relationship among the components of a ternary system under equilibrium can be described using a triangle at constant temperature and pressure.

MacKenzie (5) and Franks et al. (10) have extended the phase diagram by including nonequilibrium behavior of aqueous systems to construct supplemented phase diagrams, now commonly referred to as state diagrams. MacKenzie (5) constructed state diagrams of sucrose–water and polyvinylpyrrolidone (PVP)–water systems by incorporating equilibrium and nonequilibrium data obtained from first- and second-order transitions observed either during cooling or subsequent heating scans (Fig. 1). These state diagrams display the temperatures of calorimetrically detected transitions of aqueous solutions as a function of solute concentration. These transitions include melting (T_m), devitrification (T_d), recrystallization (T_r), and homogeneous nucleation (T_h). Except for melting, all of the transitions in the state diagram are kinetically controlled and cannot be described as true phase transitions. The curve defined by the calorimetrically measured melting points is known as the liquidus curve. Each of the additional curves in the state diagram is one of a family of curves that differ by the heating or cooling rate used to achieve the initial states and to observe the underlying transition.

Figure 1 reveals the close resemblance between the PVP–water and sucrose–water state diagrams, although a significant difference in their molecular weights. The main difference between the two state diagrams appears to be the limiting concentration of solute above which ice will not separate out (e.g., freezing and melting cannot be observed), 3.3 mol water/(base mol) or PVP and 4.2 mol water/mol of sucrose (5, 10). Franks et al. (10) also reported the limiting

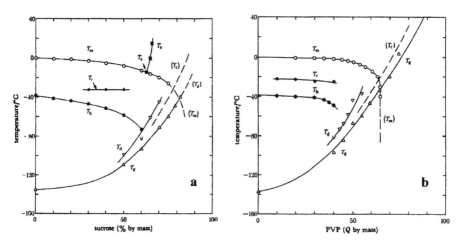

Figure 1 Supplemented phase diagrams for (a) sucrose–water and (b) polyvinylpyrrolidone–water (T_m) melting, (T_d) devitrification, (T_r) recrystallization, (T_h) homogeneous nucleation, (T_g) glass transition. (From Ref. 5.)

concentration for hydroxyethyl starch to be 70% by weight (or 3.6 water/(base mol) of hydroxyethyl starch). Both the studies of MacKenzie (5) and Franks et al. (10) correlated calorimetric data with the structures of solid states observed by electron microscopy.

The state diagram for the sucrose–water system was expanded by Levine and Slade (13) to include solidus and vaporous curves (Fig. 2) to widen the application of state diagrams to the manufacture of sugar containing foods well above 0°C. Slade and Levine (17) refer to the glass transition as the critical reference point on the state diagram because it defines the performance of a given system during processing and storage. The intersection of the extrapolated equilibrium melting curve and the glass curve is defined by a temperature, $T_{g'}$, and by a water content, $C_{g'}$, on the two-dimensional state diagram. This point corresponds to a maximally freeze concentrated solution in which ice crystallization cannot be detectable by DSC, either because the signal is below the threshold sensitivity of the instrument or crystallization cannot be attained on the experimental time scale or both (9). The maximally freeze concentrated state is reached as a result

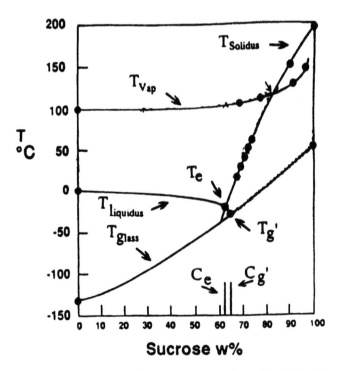

Figure 2 State diagram for sucrose–water, illustrating the locations of the glass, solidus, liquidus, and vaporous curves. (Adapted from Ref. 17.)

of gradual concentration of the aqueous solution due to ice crystallization and separation during slow cooling. The conditions defining this state have been shown to be important for designing processes for frozen-food technology including freeze concentration and freeze drying, as well as, for maintaining the physical stability of food products during frozen storage (9, 11).

During the past two decades, a number of investigators have presented state diagrams for oligomeric carbohydrates (29, 33) and proteins (31, 34) due to the significance of their structure-property relationships in food processing. These state diagrams include the thermally induced transitions (glass, melting, gelatinization, crystallization, denaturation) of individual bipolymer components of cereal flours. Kaletunç and Breslauer (15) warned that the use of thermal data on the individual components to interpret complex systems, such as cereal flours, ignores interactions between the components and the potential alteration of component thermal properties that might occur as a result of the separation process. These investigators stated that because wheat flour is a complex mixture of interacting biopolymers, the flour itself provides the appropriate thermodynamic reference state for processing-induced alterations in the thermal properties of the postprocessed material. They utilized the differential thermal properties of pre- and postextruded wheat flour (a real food system) as a function of plasticizing water to develop a state diagram for wheat flour, with the goal of characterizing the physical state of the wheat flour prior to, and during, and after extrusion processing (Fig. 3). Curves on the state diagram, which define the moisture-content dependence of the calorimetrically detected transition temperatures, form the

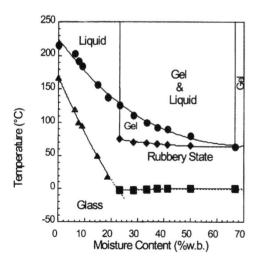

Figure 3   Wheat flour–water state diagram at 30 atm. (From Ref. 15.)

boundaries between regions that correspond to particular states of the wheat flour.

Being a partially crystalline polymer system, wheat flour displays thermally induced transitions typical of both amorphous and crystalline materials. Figure 3 shows the transition temperatures of the calorimetrically detected, thermally induced transitions as a function of moisture content, to construct melting, gelatinization, and glass curves for wheat flour and a melting curve for the freezable water. As seen in the state diagram, at a moisture content lower than 20%, a high temperature endotherm, corresponding to the melting of crystalline regions of wheat flour is observed. The transition temperature of this melting endotherm (T_m) shifts to lower temperatures as the moisture content of the wheat flour increases. At temperatures above the melting curve, defined by T_m values as a function of moisture content, a free flowing, amorphous liquid state is reached. A second endotherm begins to appear, above 23% moisture and temperatures below the high temperature melting endotherm. The transition temperature of this second endotherm (T_{gel}), which frequently is referred to as the gelatinization endotherm, is not as sensitive to the moisture content as is T_m. At about 67% moisture, the two endotherms coalesce into a single endotherm, the transition temperature of which is independent of further increases in moisture content.

Although, wheat flour is a mixture of biopolymers, Kaletunç and Breslauer (15) reported a single apparent T_g for wheat flour as observed in DSC thermograms. They stated that this does not necessarily mean that there is only one glass transition, but if the energy associated with any glass transition, which manifests itself as a heat capacity change, is small enough, that particular transition may not be detectable using DSC. However, components such as proteins present in small amounts relative to carbohydrates contribute to the structure and texture of cereal products such as dough, bread, and extruded products. Due to increased mobility at temepratures above the glass transition, reactions may occur, which manifest themselves in altered rheological properties. State diagrams for the cereal proteins zein, gliadin, glutenin, and gluten were developed using calorimetric and rheological data (31, 34, 35). The various regions on the two-dimensional temperature–water content state diagram defined by changes in the storage modulus (G') and the loss modulus (G'') include "reaction" and "softening" zones which cannot be considered states of the protein because they do not represent merely physical changes but occur as a result of chemical reactions. Toufeili et al. (35) attribute the changes in G' and G'' to cross-linking reactions, including disulfide bond formation. Nevertheless, these zones may have importance in interpretation of structure formation and textural characteristics of cereal products.

Although the major components of cereal flours are starches and proteins, the composition changes among various cereal flours and even among varieties within the same flour. A systematic study of the physical properties of cereal flours as a function of protein, lipid, and starch content and amylose:amylopectin ratio will be valuable in assessing the extent of composition-induced changes in the physical properties of the flour. A number of studies demonstrate that additives influence the thermal properties of cereal products directly by plasticizing the bipolymer systems and indirectly by modifying the process parameters. The curves that define the boundaries of the state diagram may shift in the presence of additives and as a result processing conditions may need to be modified. Barrett et al. (36) reported the plasticization effect of sucrose on corn extrudates in the presence of moisture as well as its effect on reduction of shear (lowering of the specific mechanical energy) in the extruder. These investigators stated that addition of sucrose requires modification of the operating conditions to produce extrudates with desirable attributes. The modification of processing conditions, in a rational fashion, to compensate for the presence of additives requires knowledge of the physical states of the material as a function of additive concentration under conditions relevant to processing and storage.

The glass transition occurs over a temperature range rather than at a well-defined temperature for a given water content. This becomes more apparent for high molecular weight polymers with a high polydispersity ratio (the ratio of weight average molecular weight to number average molecular weight). In a recent review, Angell (37) discussed the possibility of multiple glassy states with different packings in glass-forming proteins and other biopolymers. In addition to the multicomponent, heterogeneous nature of preextruded material, the breadth of the transition may originate from additional heterogeneity created during extrusion due to polymer fragmentation. It is reasonably well established that T_g is related to the molecular weight of a polymer and that the glass transition region can be interpreted qualitatively as the onset of long-range, coordinated molecular motion (21). Different molecular weight fragments present in an extrudate will go through their glass transition at different temperatures. In the case of a broad molecular weight distribution, the glass transition for fragments of increasing molecular weight will occur at successively higher temperatures. Consequently, the glass transition for extruded products may occur as a continuum over a broad temperature range. One should therefore recognize when constructing and using state diagrams that the glass transition occurs over a range of temperature and that regions of the diagram near T_g-defined boundaries may actually represent states that are mixtures of glassy and rubbery material.

III. APPLICATION OF STATE DIAGRAMS IN THE CEREAL PROCESSING INDUSTRY

State diagrams that display equilibrium and nonequilibrium states (under specific conditions) for a given system allow us to identify the kinetic and thermodynamic factors controlling the behavior of the system under processing and storage conditions. Levine and Slade (11) emphasized that time is a critical parameter to describe kinetically controlled phenomena in the rubbery state. The rubbery state frequently is reached during processing and is reached under some storage conditions. Roos and Karel (12) discussed the benefits of applying state diagrams in food processing and product development. In the following three sections, some specific examples of application of state diagrams in the design of processing and storage conditions and in product development are discussed.

A. Process Analysis and Design

State diagrams are important to assess and to optimize the temperatures and water contents used in the processing of cereals. Slade and Levine (17) demonstrated that a sucrose–water state diagram (which reveals the relative locations of the glass, solidus, liquidus, and vaporous curves) could be used to understand the cookie and cracker baking process. These investigators used the sucrose–water state diagram (Fig. 2) for mapping the various steps of cookie and cracker manufacturing, including dough mixing, lay time, and machining, and baking, as well as for characterizing finished-product texture, shelf life, and storage stability (Fig. 4). Slade and Levine (17) chose the sucrose–water state diagram to characterize the cookie and cracker baking process due to the dependence of physical states attained by the glass-forming versus crystallizing behaviors of sucrose on the final product. It is apparent from Figure 4 that, depending on the sugar/fat ratio and on the amount of crystalline sugar dissolved during dough mixing and lay time, the sucrose in dough assumes various physical states prior to baking as a function of dough moisture content and temperature (points labeled LEAN, RICH, A, and B). Prior to baking, all formulations are well above the glass curve so that removal of moisture is facilitated due to mobility in the system. As water in the cookie or cracker is vaporized following the T_{vap} curve, the sucrose concentration increases. Upon cooling, the physial states of products vary, as shown by the box labeled ''products,'' depending on the path of baking and cooling, and final temperature and sucrose concentration in the product. The final product can be in a glassy or a rubbery state upon cooling, during storage, or during distribution. While point E represents a product with optimum initial quality and storage stability, a product described by points F or G is clearly in an unstable rubbery state and is expected to have a shorter shelf life. Depending on the temperature and relative humidity conditions during distribution and storage, the amount of water removed during baking must be adjusted to attain the desired glass transition temperature. Slade and Levine (17) state that more complex behavior is expected for cookie doughs due to the high initial sugar level. Sucrose can be either completely (point A unsaturated solution) or partially (point B supersaturated solution) dissolved in the dough prior to baking. Therefore, two different paths are followed during the baking process. Through both paths, due to water

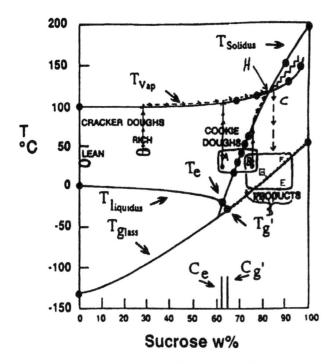

Figure 4   Sucrose–water state diagram on which the paths of cookie and cracker baking processes are superimposed. (From Ref. 17.)

loss, a supersaturated sucrose solution is created within the metastable region, C, from which recrystallization of sucrose may occur during baking, cooling, or storage. The path starting from point B will have additional crystalline sugar in the cookie because during baking, the cookie dough goes through the solidus curve and crystalline sugar forms. In using the sucrose–water state diagram for sucrose containing products, it must be remembered that weight percent of sucrose shown on this diagram does not account for the additional components of the system.

The temperature and moisture content ranges covered by the wheat flour state diagram presented in Figure 3 correspond to those applied to cereal flours as part of many food-processing operations, such as baking, pasta extrusion, and high-temperature extrusion cooking. Kaletunç and Breslauer (15) mapped on this state diagram the path of a high-temperature extrusion cooking process to reveal the physical state of wheat flour during the various steps of processing. Kaletunç and Breslauer (15) also placed the T_g values of wheat flour extrudates on the state diagram to assess the impact of shear, in terms of specific mechanical energy

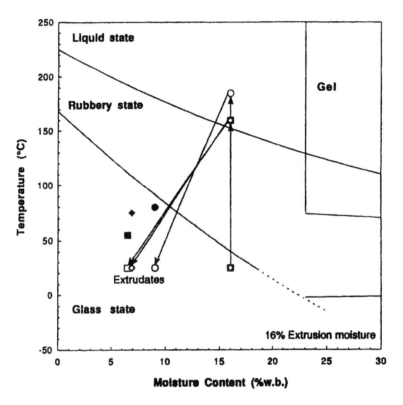

Figure 5   Wheat flour–water state diagram on which the paths of the extrusion process at various SME values are superimposed. T_g values of the extrudates are indicated as the corresponding filled symbols (From Ref. 15.)

(SME), on the wheat flour during extrusion processing (Fig. 5). They evaluated the effects of extrusion processing on wheat flour as a function of SME by comparing the T_g values of wheat flour with the T_g values of extrudates at the same moisture content. Figures 5 shows the path of the extrusion process (open symbols) on the wheat flour state diagram, in terms of the various SME conditions and the T_g values of the extrudates at their measured moisture contents (the corresponding filled symbols). It is apparent from figure 5 that all of the extrudates were in the glassy state at room temperature and moisture contents of 7 to 9.5% (w.b.) (comparison of closed and open symbols). Two important conclusions can be deducted from Figure 5. First, different values of SME for the extrusion process lead to different extents of fragmentation of the wheat flour, as reflected in the difference between the T_g values of the extrudate (closed symbols) and the corresponding temperature on the glass curve of wheat flour at the same moisture content. Second, the difference between the ambient temperature of the T_g values of the extrudates can be used to assess to the storage stability of the extrudate. The greater the difference between the storage temperature and the T_g value of the extrudates, the greater the stability of the extrudate and the longer its shelf life. For highly fragmented extrudates, which display low T_g values, elevated storage temperature may induce the rubbery state. Moreover, an increase in the relative humidity of the storage environment may increase the moisture content of such extrudates, thereby leading to even lower T_g values. The rubbery state will be entered at even lower storage temperatures, which will lead to reduced shelf life.

It is apparent that the T_g values of wheat flour extrudates decrease with increasing SME generated in the extruder. Kaletunç and Breslauer (15) noted that, based on the wheat-flour state diagram and knowledge of the SME dependence of T_g, processing conditions can be adjusted to yield extrudates with desired T_g values. To make these correlations of practical value, a criterion that relates the target T_g to the end-product attributes of significance to the consumer should be defined. In an earlier study on corn flour extrudates, Kaletunç and Breslauer (1) demonstrated that an increase in T_g is related to an increase in the sensory textural attribute of crispness, which is one of the major criteria by which a consumer judges the quality of a breakfast cereal or puffed snack product. Determination of the T_g for extruded amorphous products can be useful for quality control in the manufacture of such products. This approach has the advantages of being inexpensive, rapid, and objective.

The state diagram displayed on Figure 3 also can be used to map the path of the pasta extrusion and postextrusion drying process (Fig. 6). During the pasta extrusion process, the dough is in the rubbery state for ease of hydration and processability. Following extrusion, pasta can be dried following low (40–60°C), high (60–84°C), or very high (>84°C) temperature drying schemes. Drying time decreases with increasing drying temperature. It is apparent from Figure 6 that pasta is in the rubbery state during the drying and part of the cooling process. As the moisture content of the pasta decreases, the difference between the temperature of the pasta and its glass transition temperature at the given moisture content decreases. A greater difference between the drying temperature and the glass transition of the pasta will result in a faster rate of dehydration due to the increased water mobility and diffusion rates in the rubbery state. Zweifel et al. (38) used the state diagram of starch to map the pasta-making process in order to evaluate the thermal modifications of starch during high-temperature drying of pasta.

B. Product Development

State diagrams may be used in product development by evaluating the effect of additives on the physical state of the material during processing. Development of state diagrams provides a rational basis for developing new products without

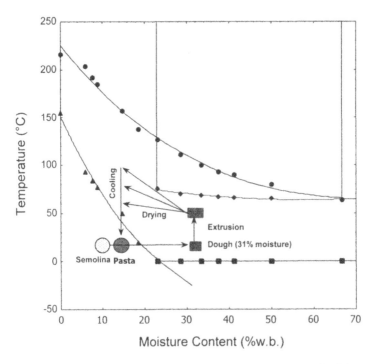

Figure 6 Wheat flour–water state diagram on which the path of the pasta extrusion process with three drying schemes are superimposed.

costly trial and error runs. Knowledge of a product's thermal properties, together with the state diagram of its raw materials, can be used to adjust processing conditions and/or the formulation of raw materials to achieve desired end-product attributes. In addition to improving the quality of an existing product, one can use state diagrams to judge the feasibility of potential products as part of a rapid evaluation procedure in the product development area. For example, sucrose, a common additive in breakfast cereals, is known at act as a plasticizer. Barrett et al. (36) have shown by mechanical and thermal measurements that sucrose plasticizes corn extrudates. The T_g of extrudates will decrease with the addition of sugar, resulting in a loss of crispness. Thus, a state diagram can be used to adjust extrusion-processing conditions (e.g., by decreasing SME) to compensate for the T_g-depressing influence of sucrose. This can be done prior to any test runs, thereby decreasing the number of test runs required to produce a product with desired properties and reducing significantly the time, materials, and overall expenditures required for product development.

First and Kaletunç (39) investigated the efficacy of using semolina pasta processing conditions for composite flour (semoline and soy) pasta processing utilizing a state diagram. The results showed that the T_g of the composite flour dough was lower than the T_g of semolina dough at corresponding water contents. These investigators concluded that the redness developed during drying of soy-added pasta could be avoided by drying at a lower temperature, which is possible due to the lower T_g of the composite flour dough.

C. Design of Storage Conditions

To define storage conditions (temperature, relative humidity) that enhance the stability and, therefore, the shelf life of cereal products, can use moisture sorption isotherms in conjunction with the glass curves of cereal products (18, 40). Roos (18) combined glass transition curves and sorption isotherms for maltodextrins as a function of relative humdity (% RH) of the environment (Fig. 7). In Figure 7, the sorption isotherm shows the water content of a material as a function of relative humidity of the environment in equilibrium with the material at a constant temperature. In Figure 7, the glass curve of the same material is also plotted as a function of % RH. The critical relative humidity level (the relative humidity level sufficient to depress the glass transition temperature to just below ambient

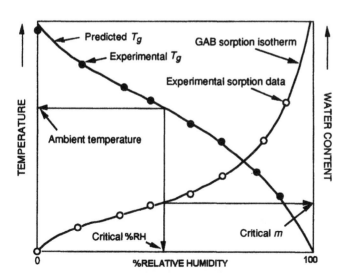

Figure 7 Glass curve and sorption isotherm as a function of relative humidity. Arrows indicate the critical moisture content and critical relative humidity for stability at ambient temperature. (Adapted from Ref 14.)

temperature) can be determined using the glass transition curve. The sorption isotherm can then be used to determine the moisture content corresponding to the critical relative humidity. The complete diagram in Figure 7 is useful for the selection of storage conditions for low- and intermediate-moisture cereal foods.

Nikolaidis and Labuza (41) developed a state diagram for a baked cracker measuring its glass transition temperature as a function of moisture content using dynamic mechanical thermal analysis (DMTA). They used the glass curve for the cracker together with its sorption isotherm to assess the optimum relative humidity and temperature for retention of desirable textural attributes during storage. These investigators emphasize that a comparison of the glass transition curve of the components of a complex food system with that of the product is needed to determine the effect of formulation on the final product attributes.

For storage stability at a constant temperature, it is essential to keep the relative humidity of the surrounding environment sufficiently low so that moisture content of the product will be below the critical moisture content. Van den Berg et al. (42) constructed a state diagram for amorphous sucrose by plotting glass transition temperature versus relative humidity in order to assess the relative humidity and temperature conditions for maintaining the stability of amorphous sucrose.

The combined sorption isotherm and state diagram also can be used in the design of suitable packaging to enhance the shelf life of products. Roos (14) notes that accelerated shelf life studies should be interpreted cautiously because at temperatures above the glass transition, the rates of change for various quality parameters are markedly different from those below the glass transition. Therefore, sorption isotherms in conjunction with state diagrams should be used to determine critical water activity levels to evaluate the extrapolation of shelf life data.

IV. CONCLUSION

Improvements of the quality of existing products or processes, as well as development of new products and design of novel processes requires an understanding of the impact of processing and storage conditions on the physical properties and the structures of pre- and postprocessed materials. To achieve this fundamental understanding, one needs to study the physical properties of such materials under conditions simulating processing and storage. The database generated from such studies can be used to construct state diagrams. Quantitatively accurate state diagrams, and kinetic data for the physical and chemical phenomena that occur in the rubbery state, are essential in order to assess the impact of processing on the properties and storage stability of products.

By superimposing a processing path on a temperature versus moisture state diagram, one can determine the physical state of the material by locating the intersection of the temperature and moisture content values corresponding to a given state of the process. For any raw material, such a diagram can be used as a predictive tool for evaluating the performance of that material during processing. Furthermore, a state diagram for pre- and postprocessed cereal can be used in conjunction with the corresponding moisture sorption isotherms to define and adjust processing conditions, so as to obtain products with enhanced stability. This predictive ability, prior to processing, enables one to improve performance, by changing processing conditions and/or changing the formulation, to favor desired end-product attributes. Because they can provide a rational basis for designing processing conditions and/or raw material formulations, state diagrams have great potential value in analysis of food manufacturing processes, product development, and in the design of processes and storage conditions.

REFERENCES

1. G Kaletunç, KJ Breslauer. Glass transitions of extrudates: relationship with processing-induced fragmentation and end product attributes. Cereal Chem 70:548–552, 1993.
2. AM Barrett, AV Cardello, LL Lesher, IA Taub. Cellularity, mechanical failure and textural perception of corn meal extrudates. J of Text Stud 25:77–95, 1994.
3. T Suwonsichon, M Peleg. Instrumental and sensory detection of simultaneous brittleness loss and moisture toughening in three puffed cereals. J of Text Stud 29:255–274, 1998.
4. B Valles-Pamies, G Roudaut, C Dacremont, M Le Meste, JR Mitchell. Understanding the texture of low moisture cereal products: mechanical and sensory measurements of crispness. J Sci Food Agric 80:1679–1685, 2000.
5. AP MacKenzie. Non-equilibrium behavior of aqueous systems. Phil Trans R Soc Lond. B. 278:167–189, 1977.
6. H Levine, L Slade. Non-equilibrium behavior of small carbohydrate-water systems. Pure and Appl. Chem 60:1841–1864, 1988.
7. F Franks, SF Mathias, RHM Hatley. Water, temperature and life. Phil Trans R Soc Lond. B 326:517–533, 1990.
8. AP MacKenzie. The physico-chemical basis for the freeze drying process. Dev Biol Stand 36:51–67, 1977.
9. F Franks. Freeze drying. From empricism to predictability. Cryo-Lett. 11:93–110, 1990.
10. F Franks, MH Asquith, CC Hammond, HB. Skaer, P Echlin. Polymeric cryoprotectants in the preservation of biological ultrastructure. J. Microscopy, 110:233–238, 1977.
11. H Levine, L Slade. Principles of "cryostabilization" technology from structure/property relationships of carbohydrate/water systems. Cyro-Lett 9:21–63, 1988.
12. Y Roos, M Karel. Applying state diagrams to food processing and development. Food Technol 45:66–71, 1991.
13. L Slade, H Levine. The glassy state phenomenon in food molecules. In: JMV Blansh-

ard, PJ Lillford, eds. The Glassy State in Foods. Loughborough: Nottingham University Press, 1993, pp 35–101.

14. YH Roos. Phase Transitions in Foods. San Diego: Academic Press, 1995.

15. G Kaletunç, KJ Breslauer. Construction of a wheat-flour diagram: application to extrusion processing. J Thermal Analysis 47:1267–1288, 1996.

16. GW White, SH Cakebread. The glassy state in certain sugar-containing food products. J. Food Technol 1:73–82, 1966.

17. L Slade, H Levine. Water and the glass transition. Dependence of the glass transition on composition and chemical structure: special implications for flour functionality in cookie baking. J Food Eng 24:431–509, 1995.

18. Y Roos. Characterization of food polymers using state diagrams. J Food Eng 24: 339–360, 1995.

19. DH Rasmussen, AP MacKenzie. The glass transition in amorphous water. Application of the measurements to problems arising in cryobiology. J Phys Chem 75:967–973, 1971.

20. G Kaletunç, KJ Breslauer. Heat denaturation studies of zein by optical and calorimetric methods. IFT Annual Meeting, June 20–24, New Orleans, Louisiana, abs. 289, 1992.

21. LH Sperling. Introduction to Physical Polymer Science. 2nd ed. New York: Wiley, 1992.

22. B Wunderlich. Thermal Analysis. Boston: Academic Press, 1990.

23. A Eisenberg. The glassy state and the glass transition. In: JE Mark, A Eisenberg, WW Graessley, L Mandelkern, JL Koenig, eds. Physical Properties of Polymers. Washington, DC: Am Chem Soc, 1993, pp 61–97.

24. JMV Blanshard. The glass transition, its nature and significance in food processing. In: ST Beckett, ed. Physico-chemical Aspects of Food Processing. New York: Blackie Academic & Professional, 1995, pp 17–49.

25. AM Cocero, JL Kokini. The study of the glass transition of glutenin using small amplitude oscillatory rheological measurements and differential scanning calorimetry. J Rheol 35:257–270, 1991.

26. J Amemiya, JA Menjivar. Mechanical properties of cereal-based food cellular systems. American Association of Cereal Chemists 77th Annual Meetings, abs. 207, 1992.

27. MT Kalichevsky, EM Jaroszkiewicz, S Ablett, JMV Blanshard, PJ Lillford. The glass transition of amylopectin measured by DSC, DMTA, and NMR. Carbohydr Polym 18:77–88, 1992.

28. EM Degraaf, H Madeka, AM Cocero, JL Kokini. Determination of the effect of moisture on glass transition of gliadin using mechanical spectrometer and differential scanning calorimeter. Biotechnol Prog 9:210–213, 1993.

29. S Ablett, AH Darke, MJ Izzard, PJ Lillford. Studies of the glass transition in malto-oligomers. In: JMV Blanshard, PJ Lillford, eds. The Glassy State in Foods. Loughborough: Nottingham University Press, 1993, pp 189–206.

30. TR Noel, R Parker, SG Ring, AS Tatham. The glass transition behavior of wheat gluten proteins. Int J Bio Macromo 17:81–85, 1995.

31. H Madeka, JL Kokini. Effect of glass transition and crosslinking on rheological

properties of zein: development of a preliminary state diagram. Cereal Chem 73: 433–438, 1996.

32. RG Mortimer. Physical chemistry. San Diego: The Benjamin/Cummings Publishing Co., Inc., 1993.

33. K Kajiwara, F Franks. Crystalline and amorphous phases in the binary system water–raffinose. J Chem Soc Faraday Trans 93:1779–1783, 1997.

34. JL Kokini, AM Cocero, H Madeka. State diagrams help predict rheology of cereal proteins. Food Technol 49(3):74–82, 1995.

35. I Toufeili, IA Lambert, JL Kokini. Effect of glass transition and cross-linking on rheological properties of gluten: development of a preliminary state diagram. Cereal Chem 79:138–142, 2002.

36. A Barrett, G Kaletunç, S Rosenberg, K Breslauer. Effect of sucrose and moisture on the mechanical, thermal, and structural properties of corn extrudates. Carbohydr Polym 26:261–269, 1995.

37. CA Angell. Formation of glasses from liquids and biopolymer. Science 267:1924–1935, 1995.

38. C Zweifel, B Conde-Petit, F Escher. Thermal modification of starch during high-temperature drying of pasta. Cereal Chem 77:645–651, 2000.

39. L First, G Kaletunç. Application of state diagram to pasta processing. IFT Annual Meeting, June 23–27, New Orleans, Louisiana, abs. 88C-5, 2001.

40. G Kaletunç, KJ Breslauer. Influence of storage conditions (temperature, relative humidity) on the moisture sorption characteristics of wheat flour extrudates. IFT Annual Meeting, June 14–18, Orlando, Florida, abs. 18-3, 1997.

41. A Nikolaidis, TP Labuza. Glass transition state diagram of a baked cracker and its relationship to gluten. J Food Sci 61:803–806, 1996.

42. C Van Den Berg, F Franks, P Echlin. The ultrastructure and stability of amorphous sugars. In: JMV Blanshard, PJ Lillford, eds. The Glassy State in Foods. Loughborough: Nottingham University Press, 1993, pp. 249–267.

6

Powder Characteristics of Preprocessed Cereal Flours

G. V. Barbosa-Cánovas and H. Yan
Washington State University, Pullman, Washington, U.S.A.

I. INTRODUCTION

Fruits of cultivated grasses, which are members of the monocotyledonous family *Gramineae*, are called cereals. Wheat, barley, oats, rice, corn, sorghums, millets, and rye are principal cereal crops, though few are consumed in their original form. Milling is an ancient and the most common method used to produce intermediate products of various particle sizes from cereal crops. Such intermediate products can then be used to make more palatable, digestible, and desirable final food products (1). The milling process changes the appearance and nutrition value of cereal crops by separating the bran and or/germ from the endosperm and reducing the particle size.

After milling, preprocessed cereal flours are often in the form of powders during further processing, transportation, and even marketing. Like many other food powders, preprocessed cereal flours represent a wide range of powders different in their chemical and physical properties. The only characteristic they have in common is that they will develop physical and chemical changes depending on their temperature–moisture history (2). This chapter presents general descriptions of some physical properties that are essential to the processing and handling of preprocessed cereal flours. These include particle size, particle size distribution, bulk density, compressibility, flowability, moisture content, mixture characteristics, caking and anticaking agents, as well as dust explosion concerns.

II. PARTICLE SIZE

Particle size is one of the most essential product properties. For a spherical particle, its size is directly defined as its diameters (the distance between two points on the particle) or calculated from the two-dimensional projected area or three-dimensional volume. All three of these estimates will agree very well (3). For some nonspherical regular-shaped particles, such as cone and cuboid, more than one particle size dimension must be specified, while irregularly shaped particles require derived diameters to define their particle sizes.

Derived diameters are usually determined by measuring a size-dependent property of the particle and relating it to a linear dimension (4). As concluded by Schubert (5), attributes often used to characterize particle sizes may be classified as geometrical properties (linear dimension, area, and volume), mass, settling rate in a fluid, and field interference (electrical field interference and light scattering or diffraction). It is obvious that when different physical principles are used in particle size determination, it cannot be assumed that they will produce identical results.

An irregular particle may be described by a number of sizes, depending on what dimension is measured or method used. Some definitions of particle sizes are listed in Table 1. These sizes may be divided into three basic groups: statistical diameter (Martin, Feret, and shear diameters), equivalent circle diameter (projected area and perimeter diameters), and equivalent spherical diameter (volume, surface, surface volume, Stokes, and sieve diameters) (6). The last group is the most widely used.

Among all the different particle sizes, the sieve diameter is the one most often used to characterize flour sizes in the wheat milling process. It is the side length of the minimum square aperture through which the particles will pass if square-mesh sieves are used. In the wheat milling process illustrated by Hoseney (7), the stock is sifted and the flour removed after each grinding pass, and the coarser particles are sent to the appropriate reduction rolls. Purifiers (essentially, inclined sieves) are used to classify the middlings according to sizes after reduction rolls. Finally, the flour is sieved through a 10XX flour cloth with 136-µm openings. In North America, flour is generally defined as stock passing through a sieve screen of 112-µm opening size, a dressed flour 132-µm, and cake flour 93-µm (8).

For preprocessed cereal flours, particle size is one of the most important physical properties because of its key role in unit operations such as milling, mixing, hydrating, extruding, and pneumatic handling. Particle size measurements are often made to control final product quality (especially in the milling process) because the latter may be correlated with a certain particle size. Particle size is important in flour quality evaluation, flour behavior in processing, and even the appearance and acceptance of final baked products. Take wheat flour,

Table 1 Particle Size Definitions

Symbol	Name	Definition
d_p	Projected diameter	Diameter of a circle having the same projected area as a particle resting in its most stable position
d_c	Perimeter diameter	Diameter of a circle having the same perimeter as the projected outline of the particle
d_M	Martin's diameter	Mean length of a line intercepted by the profile boundary that approximately bisects the projected area of a particle
d_F	Feret's diameter	Mean length of the distance between two parallel tangents on opposite sides of a particle's projected outline
d_{sh}	Shear diameter	Width of a particle obtained with an image-shearing eyepiece
d_V	Volume diameter	Diameter of a sphere having the same volume as a particle
d_s	Surface diameter	Diameter of a sphere having the same surface area as a particle
d_{SV}	Specific surface diameter	Diameter of a sphere having the same ratio of surface area to volume as a particle
d_{SIE}	Sieve diameter	Width of the minimum square aperture on a sieve through which a particle will pass
d_{ST}	Stock's diameter	Diameter of a sphere having the same free-flowing velocity as a particle in the laminar flow region
T	Average thickness	Average distance between the upper and lower surfaces of a particle in its most stable resting position
L	Average length	Average distance of the longest chords measured along the upper surface of a particle in its stable position

Source: Modified from Refs. 3, 4, and 6.

for example: Flour protein contents, usage, and even starch damage after milling are all related to flour size. Rice flour particle size greatly influences pasting characteristics, gel consistency after cooking, and the palatability of rice flour butter cakes (9). Similarly, the particle size of sorghum has been shown to affect the grittiness of baked products with wheat flour (10).

Wheat flour may be classified into three main fractions according to their different sizes: (a) whole endosperm cells, segments of endosperm cells, and clusters of starch granules and protein (>35 μm in diameter) in which the protein content is similar or higher than that of the parent flour; (b) large and medium-sized starch granules, some with protein attached (15–35 μm in diameter); and (c) small chips of protein and detached starch granules (<15 μm in diameter). The protein contents in groups (b) and (c) vary from 0.5 to 2 times that of the parent flour. Thus, the particle size classification permits the concentration of protein and starch into different fractions, making it possible to get a range of flours with varying properties that is suitable for various bakery products from one parent flour (1, 11).

Particle size also plays an important role in wheat classification during milling. Normally, milled wheat flour is a mixture of particles of different sizes and compositions. Because the content of flour protein varies with particle size, fine grinding and air classification methods are used to alter or improve certain properties. After fine grinding, the flour is channeled to an air classifier, where larger particles are funneled down and away and the smaller ones lifted and separated, because the effect of centrifugal force is opposed to the effect of air drag on individual particles in the classifier.

The relationships between flour particle size, milling yield, and protein content for one kind of pinmilled wheat flour after air classification are shown in Figure 1. It is clear that the finer particle size material has an increased protein content, with shifts from 13.4% protein content in the parent flour to 24% in the classified flour. Based on their different protein contents and particle sizes, the

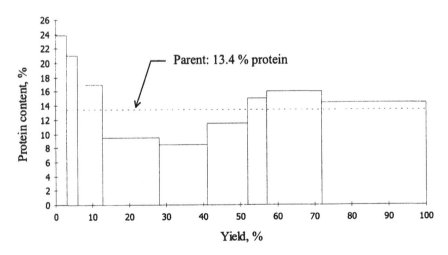

Figure 1   Results of air classification of wheat flour into nine fractions of different sizes. From left to right, the eight rectangles represent pinmilled wheat flour classified at 10, 13, 17, 22, 28, 35, 44, and 55 μm nominal cut sizes. (Modified from Ref. 1.)

Table 2 Flour Particle Size Criteria for Different Uses

Particle size range (μm)	Use	Particle size range (μm)	Use
1–150	Pan bread	0–150	Soups and gravies
1–150	Hearth bread	0–125	Crackers
1–150	Variety bread	0–125	Cookies
1–150	Soft roll	0–125	Layer cakes
1–150	Sweet goods	20–60	Form cakes
1–150	Home baking	0–90	Biscuits

Source: Modified from Ref. 12.

fine fraction can be used to increase the protein content of bread flour milled from low- or medium-protein wheat, gluten-enriched bread, and starch-reduced products. The chlorinated-medium fraction is recommended for sponge cakes and premixed flours, while the coarse fraction is said to be good for biscuit manufacturing because it has a uniform particle size and granular structure (1). General wheat flour particle size criteria for different uses (12) are listed in Table 2. The nomination of dry corn milling products based mainly on particle size range (13) are listed in Table 3.

Starch damage in wheat flour is also said to be correlated with flour particle size. The starch damage level in flour is one of the concerns in flour milling because it influences the flour's water-absorbing ability. Damaged starch in gran-

Table 3 Dry Corn Milling Product Nomination with Respect to Particle Size Ranges

	Particle size range				
	U.S. Standard sieve number		Sieve diameter (μm)		
Product	Pass through	Retained on	Less than	More than	Yield (%)
Hominy grits	3.5	6	5,660	3,360	12
Coarse grits	10	14	2,000	1,410	15
Brewers grits	12	30	1,680	590	30
Regular grits	14	28	1,410	638	23
Coarse meal	28	50	638	297	3
Dusted meal	50	75	297	194	3
100% meal	28	pan	638	pan	10
Fine meal	50	80	297	177	7
Flour	75	pan	194	pan	4

Source: Modified from Ref. 13.

ules results in a susceptibility to fungal α-amylase and produces weak side walls and sticky crumbs if sufficient amylolytic enzymes are available (7). The higher the damaged starch content, the higher the water absorption. This can be advantageous, but excessive dextrin production from the amylolytic breakdown of damaged starch is undesirable (1). When Scanlon et al. (14) studied some particle size–related physical properties of flour produced by smooth-roll reduction of hard red spring wheat farina, they found that for the flour produced at different milling conditions in their experiments, the under-53-μm fraction exhibited much greater starch damage than the coarser fractions with size ranges from 91 to 136 and 53 to 91 μm.

Flour color is dependent on particle size, because the latter affects light reflection, causing coarser particles to appear darker due to their larger shadows. Among the many factors influencing bread whiteness, such as wheat pigment, grain content, and grain fineness, flour particle size is the most important. As explained by Kruger and Reed (15), the finer the grain, the better reflection on pore structure under incident light and hence the whiter the crumb appears. They also noticed that consumers prefer a very white bread crumb in white bread, rolls, and buns.

Particle size analysis of cereal and cereal products may be made by standard sieving tests, the Simon funnel method, a Coulter counter, the Andreassen pipette method, laser light scattering, or the AACC Methods 50–10 based on the principles of centrifugation or sedimentation (1, 12). A detailed description of these methods is beyond the scope of this chapter. Interested readers are referred to Allen's book about particle size measurement (4), in which most of these methods are well illustrated.

III. PARTICLE SIZE DISTRIBUTION

The measurement of particle size and particle size distribution is one of the most widely used methods in the industry (16), because, in combination, they affect such other physical properties of a powder system as flowability, bulk density, and compressibility (2, 17). Even segregation will happen in a free-flowing powder mixture because of differences in particle size (18). Because preprocessed cereal flours are rarely of uniform particle size, it is necessary to measure a great many particles and produce a description of their size distribution. For quality control or system property descriptions, the usefulness of representing the particle size distribution of a preprocessed cereal flour is apparent, and proper descriptors of the size distribution are essential in the analysis of their handling, processing, and functionality.

Methods of presenting particle size distribution data for a given powder include tabular form, the cumulative percentage frequency curve, and the relative

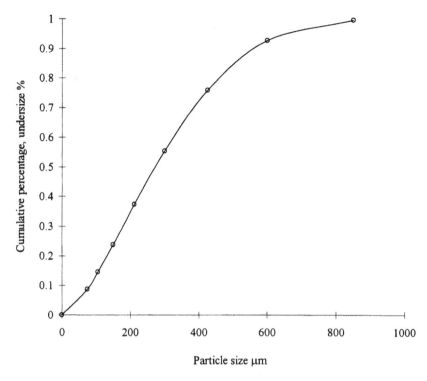

Figure 2 Schematic of cumulative percentage frequency distribution curve.

percentage frequency curve. In the cumulative percentage frequency distribution curve, as shown in Figure 2, the abscissa is particle size while the ordinate is the percentage smaller or larger than a given particle size, and the value range for cumulative frequency is from 0 to 1 (or 0 to 100%). The relative percentage frequency distribution curve is shown in Figure 3. By its definition, the area under the curve of frequency against particle size should be equal to 1 (6). The characteristics of a powder distribution (either in cumulative or relative frequency form) may be expressed by the number, length, area, or volume of its particles (4).

Representing particle size distribution curves by some mathematical functions is another way to characterize a powder system. There are many different types of size distribution functions, either in cumulative or relative percentage frequency form. These include the normal distribution, log-normal distribution, and Rosin–Rammler functions (4); Gates–Gaudin–Schuhmann, Bennett's form, Gaudin–Meloy, and modified Gaudin–Meloy functions (19, 20); Roller and Svenson functions (21); the error function (3); modified beta function (22, 23):

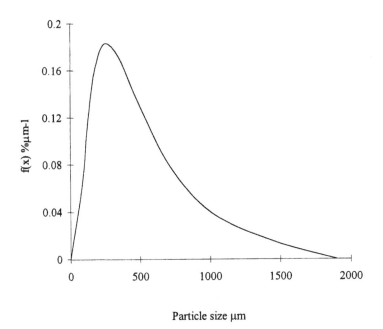

Particle size μm

Figure 3 Schematic of relative percentage frequency distribution curve.

and Griffith and Johnson's S_B function (24). Among these particle size distribution functions, the five most commonly used were selected for detailed explanation here. They are the Gates–Gaudin–Schuhmann, modified Gaudin–Meloy, Rosin–Rammler, log-normal, and modified beta functions.

The Gates–Gaudin–Schuhmann function is expressed as (20):

$$Y = \left(\frac{x}{k}\right)^m \tag{1}$$

where Y is the cumulative weight fraction under size x, k is the characteristic size of the distribution (also called the Schuhmann size modulus), and m is the measurement of the distribution spread (i.e., the Schuhmann slope).

The Rosin–Rammler function was introduced in comminution studies in 1933 (25) and also used to describe the particle size distribution of moon dust (4). Usually, it is a two-parameter function given as an undersize cumulative percentage (6):

$$Y = 1 - \exp\left[-\left(\frac{x}{x_R}\right)^n\right] \tag{2}$$

where Y is the weight fraction of material finer than size x, x_R is a constant giving a measure of the present particle size range, and n is a constant characteristic of the analyzed material that gives a measure of the steepness of the cumulative curve. The x_R can easily be found from the plot in the Rosin–Rammler graph because it is the size corresponding to $100/e = 36.8\%$, where n is the slope of the line. Lower values of n are associated with a more scattered distribution, while higher values of n imply an increasingly uniform particle size distribution (21).

The modified Gaudin–Meloy function is expressed as (19, 20):

$$Y = \left[1 - \left(1 - \frac{x}{x_0} \right)^r \right]^m \tag{3}$$

where Y is the cumulative weight fraction under size x, x_0 is the parameter related to the maximum particle size (100% passing size), m is the Schuhmann slope, and r is the ratio of x_0 to the size modulus, which is related to the Schuhmann size modulus in Eq. (1).

Among all the different types of distribution functions, the most useful is the log-normal (26). It can be given in this form:

$$f(x) = \frac{1}{x \ln \sigma_g \sqrt{2\pi}} \exp \left[- \frac{(\ln x - \ln x_g)^2}{2 \ln^2 \sigma_g} \right] \tag{4}$$

where $f(x)$ is the size distribution function for particle size x, x_g is the geometric mean of the distribution, and σ_g is the geometric standard deviation of $\ln x$ (22).

For many processes in which the population mode and spread vary independently and the size distributions have a finite range, the modified beta distribution function is more appropriate than the log-normal function, because it has a finite range and the ability to describe symmetric as well as asymmetric distributions skewed to the right or left (23). The modified beta distribution $f_{am}(x)$ is defined as:

$$f_{am}(x) = \frac{x^{am} (1 - x)^m}{\displaystyle\int_0^1 x^{am} (1 - x)^m \, dx} \tag{5}$$

where a and m are constant and x is the normalized length given by:

$$x = \frac{X - X_{\min}}{X_{\max} - X_{\min}} \tag{6}$$

where $X_{\min}$ and $X_{\max}$ are the smallest and largest particle sizes, respectively, and therefore $X_{\min} < X < X_{\max}$, $0 < x < 1$.

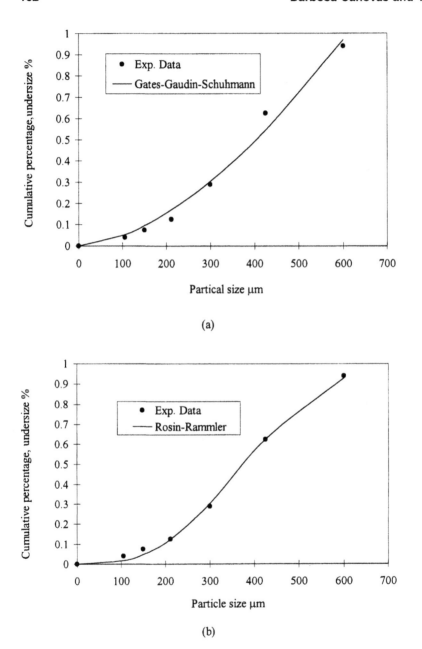

(a)

(b)

Figure 4 Experimental particle size distribution data fit by five distribution functions for corn meal. (Modified from Ref. 27.)

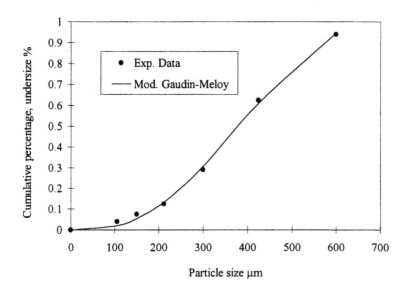

(c)

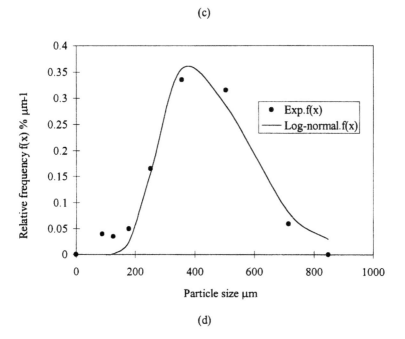

(d)

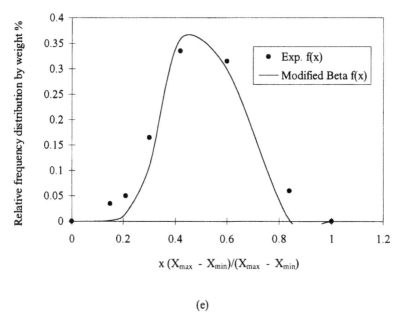

$$x \, (X_{max} - X_{min})/(X_{max} - X_{min})$$

(e)

Figure 4 Continued

In Yan and Barbosa-Cánovas's recent study (27), the particle size distributions of selected food powders were analyzed by sieving methods and characterized by the five size distribution functions mentioned earlier because they are well accepted for nonfood powder systems and no such work had been done in the food powder field before. One of the materials they used was corn meal, the milling product of corn. The experimental particle size distribution data fit by the five distribution functions for corn meal are diagrammatically shown in Figure 4. Using the curve-fitting correlation coefficient R^2 for each function as the overall "goodness of fit" evaluation, they found that the modified Gaudin–Meloy and Rosin–Rammler functions were the best to characterize the size distributions of all the selected food powders studied and that the Gates–Gaudin–Schuhmann function was also suitable for corn meal.

By using size distribution functions to describe the size distribution of a given powder, it is possible to get such useful information about the powder system as the largest particle size (by using the modified Gaudin–Meloy function), as well as the mean, median, and mode (by using the log-normal function), all of which are important to the on-line quality control of preprocessed cereal flours during milling processes. Flour production conditions must also be considered in choosing the most appropriate equation among many size distribution functions to describe the size distribution of a selected cereal flour.

After grains are ground in a burr-type mill, measuring the particle size distribution of the ground material has proven to be a very successful and practical method to evaluate cereal grain *hardness*, which is defined as the resistance to breakage or reduction in particle size. Hardness is an important attribute of grains because it affects many other properties, such as susceptibility to insect attack, breakage during handling, starch damage during dry milling, and the ability to produce certain products. It is difficult to measure grain hardness based on the known kernel geometry, because of the various grain shapes and sizes. After grinding and measurement of particle size distribution, the resulting particle size index is widely used as a hardness measurement: Under the same grinding conditions, the harder the grains, the larger the particles produced. And by using a burr-type grinder, a wider particle size distribution is produced from grains with varied hardness (28).

IV. BULK DENSITY, COMPRESSIBILITY, AND FLOWABILITY

A. Bulk Density

Bulk density is an important physical property of preprocessed cereal flours because it plays an important role in storage, transportation, and marketing. Bulk density is of great concern in grocery stores because consumers expect that the indicated mass on the package fulfills the package volume completely, while in many cases producers wish to offer powders with a small mass but large volume (5). These problems and difficulties are also associated with the bulk density of flours.

The *bulk density* of powders is defined as the mass of particles that occupies a unit volume of the container. It is usually determined by dividing the powder's net weight by the volume it occupied in a container. Even the particles that make up the powder system have their own *particle density*, which is defined as the particle's actual mass divided by its actual volume. The relationship between bulk density ρ_b and particle density ρ_s is expressed as:

$$\rho_b = (1 - \varepsilon_p)(1 - \varepsilon_b)\rho_s = (1 - \varepsilon)\rho_s \qquad (7)$$

where ε_p is the particle porosity, ε_b is the bulk porosity (the ratio of void volume between particles to the total volume), and ε is the porosity, defined as the ratio of total void volume (inter- and intra-particle) to total powder volume (5, 29). The solid and bulk densities of some cereal grains are listed in Table 4, which

Table 4 Solid and Bulk Densities of Some Preprocessed Cereal
Flours

Name	Solid density ρ_s (g/cm^3)	Bulk density ρ_b (g/cm^3)
Wheat flour	1.45–1.49	0.4–0.75
Rye flour	1.45	0.45–0.7
Corn flour	1.54	0.5–0.7
Corn starch	1.62	0.55
Polished rice	1.37–1.39	0.7–0.8

Source: Modified from Ref. 5.

makes it clear that the bulk densities of preprocessed cereal flours are totally
different from those of their crops, because of the milling process.

Because powders are compressible, their bulk density is usually given with
additional specifications, such as loose bulk density (after powders are poured
onto a bed), tapped bulk density (after vibration or tapping), or compacted bulk
density (after compression) (29). The relationship between the normalized volume changes $\gamma(n)$ after n number of tapping is given by Sone's model (30):

$$\gamma(n) = \frac{V_0 - V_n}{V_0} = \frac{\alpha\beta n}{1 + \beta n} \tag{8}$$

V_0 is the initial bulk volume, V_n is the tapped volume after n taps, and α and β
are constants.

In Eq. (8), when $n \to \infty$, $\gamma(n) \to$ a constant (i.e., its asymptotic value). If
that value is denoted as ϕ, the ratio of $1/(1 - \phi)$ is known as the very useful
Hausner ratio (the same ratio of its loose bulk density to its tapped bulk density).
The Hausner ratio may be used as an internal friction index for relatively nonco-
hesive powders, to evaluate powder flowability (31), or as a practical parameter
quantifying the maximum compressibility under vibration for a given powder,
with implications for powder-handling and -filling operations (30).

B. Compressibility

A considerable increase in the bulk density of preprocessed cereal flours may be
caused by vibration or tapping during transportation or handling, static pressure
from storage in high bins, and mechanical compression. In powder technology,
great attention has been paid to the general behavior of powders under compressive stress (2). Compression tests have been widely used in pharmaceutics, ceramics, metallurgy, civil engineering, as well as the food powder field as a simple

and convenient technique to measure such physical properties as powder compressibility and flowability.

In order to get the pressure–density relationship for a given powder, a set of compression cells (usually a piston in a cylinder) is used. The tested powder is poured into the cylinder and compressed with a piston attached to the crosshead of a TA-XT2 Texture Analyzer or Instron Universal Testing Machine. Normally, a force–deformation relationship during a compression test will be recorded by the instrument. Figure 5 shows typical force–deformation relationships for agglomerated low-fat milk powders of different particle sizes under compression tests (32). It is easy to change this relationship into a pressure–density relationship to get the compressibility after data treatment if the cross-sectional area of the cell and the initial powder weight are known.

The compression mechanisms for fine powders have been well studied by many researchers. The compression process takes place in two stages. The first involves filling voids with particles of the same or smaller size than the voids

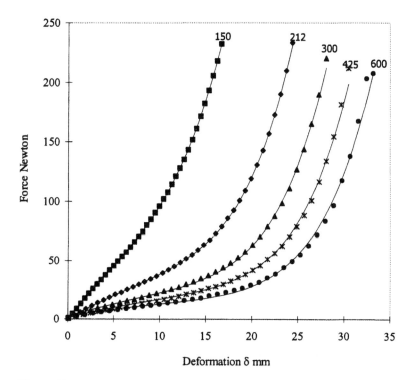

Figure 5 Force deformation relationships for agglomerated low-fat milk powders of different particle sizes under compression tests. (Modified from Ref. 32.)

by particle movement. The second stage comprises the filling of smaller voids by the particles' elastic, plastic deformation and/or fragmentation (33–35). A number of authors have suggested empirical equations to describe the pressure–density relationship during compression processes: Athy (36), Heckel (37), Kawakita (38), and Sone (39). As mentioned by Georget et al. (38), these equations usually concentrate on the compaction of particle sizes less than 1 mm. The last three are highlighted next.

The Heckel model is expressed as (35, 40):

$$\ln \frac{1}{1 - D} = aP + b \tag{9}$$

where D is the relative density, which equals ρ/ρ_p, ρ is the apparent density, ρ_p is the true density of particles, P is the applied pressure, and a and b are constants. The two constants can be determined from the slope and intercept, respectively, of the extrapolated linear portion of the plot of $\ln[1/(1 - D)]$ versus P. They have been respectively identified with the reciprocal of the material yield pressure and particle movements during the initial stages of compression. The slope of the linear portion is the constant a, which is related by Heckel to the yield stress Y of the material by the expression $a = Y/3$. Thus, the constant a is regarded as a material constant for determining the deformation mechanisms of materials (33).

The widely used Kawakita and Ludde equation may be written as (38):

$$\frac{P}{C'} = \frac{1}{A'\,B'} + \frac{P}{A'} \tag{10}$$

where P is the applied pressure, C' is the relative volume change, equal to $(V_0 - V)/V_0$, V_0 is the initial volume, V is the powder volume under applied pressure, and A' and B' are constants. As the pressure tends to infinity, C' tends to equal A', which can be written as:

$$A' = \frac{V_0 - V_\infty}{V_0} = 1 - \frac{\rho_0}{\rho_p} \tag{11}$$

where V_∞ is the volume at infinitely large pressure, ρ_0 is the initial density, and ρ_p is the particle density. Thus, A' can be identified as the initial porosity. Adams et al. (41) found that the term $1/B'$ is qualitatively related to the yield stress of compact particles.

The pressure–density relationship for powders in a compression test at low-pressure range can be described by the following equation (17):

$$\frac{\rho(\sigma) - \rho_0}{\rho_0} = a + b \log \sigma \tag{12}$$

where $\rho(\sigma)$ is the bulk density under the applied normal stress σ, ρ_0 is the initial bulk density, and a and b are constants. The constant b represents the compressibility of a powder specifically. Because the interparticle forces that enable open structure in a powder bed succumb under relatively low pressure, compressibility has been found to correlate with the cohesion of powders and could be a simple parameter to indicate flowability changes. It is believed that powders showing high compressibility have poor flowing properties (5)

C. Flowability

Powder flow is defined as the relative movement of a bulk of particles among neighboring particles or along a container wall surface (2). The flow characteristics of powders are of great importance in many problems encountered in bulk material-handling processes in the agricultural, ceramic, food, mineral, mining, and pharmaceutical industries, because the ease of powder conveying, blending, and packaging depends on them (42). To ensure steady and reliable flow, it is crucial to accurately characterize the flow behavior of powders (43).

The forces involved in powder flow are gravity, friction, cohesion (interparticle attraction), and adhesion (particle–wall attraction). Particle surface properties, particle shape and size distribution, and the geometry of the system are additional factors that affect the flowability of a given powder. Therefore, it is obvious that it is hard to have a general theory applicable to all food powders in all possible conditions that may develop in practice (2). Among the many methods and tests used to measure powder flowability are the flow test, repose angle measurement, compression, shear cell test, and tapping test.

In the flow test, a powder is allowed to flow through laboratory bins or a conical funnel of different shapes with or without the aid of controlled vibrations. The flowability of a powder is evaluated based on the mass flow rate or the flow time for a standard material mass. In general, the mass flow rate V_m of a powder through a funnel could be described as follows:

$$V_m = \alpha\rho_b(D - \varphi d) \tag{13}$$

where π_b is the bulk density of the powder, D is the orifice diameter, d is the mean diameter of the constituent particles, and α and φ are empirical coefficients. If the orifice diameter D is fixed, the volumetric flow rate of the powder can be used as an indication of its flow properties (44). Obviously this method is applicable only for free-flowing powders, but it can provide useful guidelines for flowability analyses under some limited conditions (2).

From a technical point of view, perhaps the simplest test for determining the flowability of a powder is measuring the *repose angle*, which is the angle with the horizontal plane formed by a pile of powder when it is gently poured from a fixed height (44). Both frictional and interparticle cohesive forces are involved in the formation of the cone and impact effects that may lead to segregation, so the actual measurements of the repose angle depend on the experimental method. It is clear that the results obtained by these techniques are significantly different and therefore not comparable. But regardless of the methods by which the cone (or the powder shape) is formed, it can be assumed that the smaller the angle, the more free-flowing the powder. The rule to remember is: Powders with repose angles less than about 40° are free-flowing, while these with 50° or more will have flow problems (2).

Compression tests are also useful in characterizing the flowability of powders because the interparticle forces that enable open structures in powder beds succumb under relatively low pressures. Thus, the compressive behavior of powders is presented in the form of bulk density–pressure (compressive stress) relationships (39). As shown in Eq. (12), the constant b, representing the change in bulk density by the applied stress, is referred to as the powder *compressibility*. It has been found that b can also be correlated with the cohesion of a variety of powders and therefore could be a simple parameter for indicating flowability changes. Generally, the higher the compressibility, the poorer the flowability. But if quantitative information about flowability is required, shear tests are necessary (5).

The shear cell developed in the 1960s by A. W. Jenike is the most common instrument for flowability evaluation (2). A schematic diagram of a Jenike shear cell is shown in Figure 6. It contains three main parts: mobile base A, ring B resting on the base, and lid C. When a sample is put in the shear cell and compressed by a normal force from the lid C, the base A can be placed in motion by a horizontal shearing force. Running the tests with identical preconsolidated samples under different normal forces gives maximum shearing forces for every normal force.

The curve representing the relation between the maximum shear and normal stresses is called the *yield locus*, which is shown in Figure 7. For many powders,

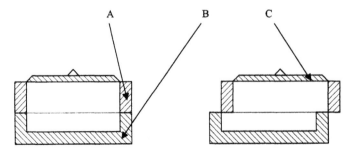

Figure 6 Schematic diagram of a Jenike shear cell. (Modified from Ref. 2.)

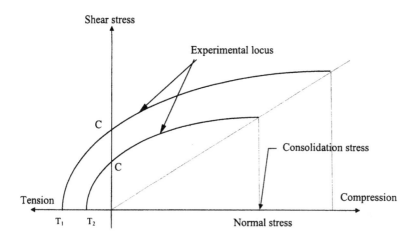

Figure 7 Yield locus curves showing the relationship of normal stress versus shear stress for cohesive powders. (Modified from Ref. 2.)

yield locus curves can be described by the empirical Warren–Spring equation (45):

$$\left(\frac{\tau}{c}\right)^{n} = \frac{\sigma}{t} + 1 \tag{14}$$

where τ is the shear stress, c is the material's cohesion, σ is the normal stress, t is the tensile stress, and n is the shear index ($1 < n < 2$). Wheat flour is usually stored in large quantities in bins, so reliable flow from these bins is desirable. The formation of a stable solid arch above the aperture of a container is also possible. The flow properties measured by using the Jenike shear cell (i.e., cohesion c and the slope of yield locus) along with other parameters, such as bin wall friction angle and the effective angle of internal friction, are very useful for bin design (43).

Each yield locus also gives one pair of values for the unconfined yield strength and major consolidation stress. The ratio of the consolidation stress to the unconfined yield strength, called the *flow function* ff_c, can be used to characterize the flowability of powders. According to the different flow function values, powders can be classified into different groups with similar flowability. Powders with $ff_c < 2$ belong to a very cohesive and nonflowing group; $2 < ff_c < 4$, cohesive; $4 < ff_c < 10$, easy-flowing; and $10 < ff_c$, free-flowing (5).

A general flowability index was proposed by Stainforth and Berry in 1973 (46) in which the slope value of the function linking Jenike's flow function ff_c was used with inverse specific tension (the reciprocal of tensile strength divided by its corresponding value of steady-state normal stress). Depending on the general flowability values, the flowability of about 60 regular powders was classified.

One of the standard methods to evaluate the flowability of a particulate system is to calculate the *Hausner ratio after tapping*, which is defined as the ratio of a powder system's tapped bulk density to its initial (loose) bulk density (i.e., the ratio of loose volume to tapped volume). In such a test, a sample of loose powder is placed into a graduated cylinder and jarred in a tap-density instrument for a specified number of taps. Each tap is the result of rotation of a cam that lifts the cylinder 0.3 cm and allows it to drop the same distance vertically. The Hausner ratio can then be calculated and powder flowability assessed when the loose and tapped volumes of the test material are known. For a Hausner ratio of 1.0–1.1, the powder is classified as free-flowing; 1.1–1.25, medium-flowing; 1.25–1.4, difficult-flowing; and >1.4, very-difficult-flowing (31).

Powder flow behavior in relation to repose angle, flow function, and Hausner ratio is listed in Table 5. All these test methods are simple to perform, and the equipment required is inexpensive. However, it should be mentioned that no flowability test is universally applicable, and the state of the powder in the actual process should be reflected in the chosen flowability test (44).

One way to improve flour flowability is by using the agglomeration process, which is accomplished by wetting flour particles in an atmosphere of water or suitable solvent droplets, causing the particles to collide and stick together, and then drying the agglomerated material in an air stream. Thus, the agglomerated flour has a controlled bulk density and particle size distribution, improved flowability, and better wettability and dispersibility in liquids and is dust free (11, 7).

Table 5 Flow Behavior in Relation to the Repose Angle, Flow Function, and Hausner Ratio

Flowability	Repose angle	Flow function	Hausner ratio
Nonflowing	>60°	<2	>1.4
Cohesive	>60°	2–4	>1.4
Fairly free-flowing	45–60°	4–10	1.25–1.4
Free-flowing	30–45°	>10	1–1.25
Excellent-flowing	10–30°	>10	1–1.25
Aerated	10°	>10	1–1.25

Source: Modified from Ref. 44.

V. MIXTURE CHARACTERISTICS

Mixing ingredients is a common and ancient operation and still very important in many industrial fields, such as food, pharmaceutics, paper, plastics, and rubbery. The objective of mixing is homogenization (i.e., reducing nonuniformities or gradients in the composition, concentration, properties, or temperature of bulk materials), which requires motion of bulk materials. In food and agriculture processing, mixing operations are often used to blend ingredients ranging from cohesive powders to viscous non-Newtonian fluids to dispersed particles suspended in Newtonian liquids (47–49). Mixing of solid particulate food material is of the most concern in this section.

In the milling industry, the mixing process becomes necessary at three different levels, as specified by Melcion (50):

1. *Raw material level*: mixing wheat with different characteristics, origins, and prices to obtain a mixture of ground materials with a constant composition
2. *Half-finished product level*: mixing flour in variable proportions to their ash contents and destination to get a targeted product
3. *Commercialized product level*: adding various additives, whitening agents, gluten improvers, mineral salts, and vitamins into flours in small percentages

In most cases, mixing preprocessed cereal flours belongs to the last two levels, but sometimes other ingredients or materials with specific functionality, such as sugars, yeasts, and flavors, are also required to make final desirable food products. It is necessary to ensure that final powder mixtures are homogeneous because of nutritional, product quality control, and processing design considerations. In addition, consumers expect all containers of such mixed food products as soups, breakfast cereals, and fruit to have the same amount of each ingredient (47).

In solid mixing operations, two or more particulate solid materials are scattered randomly in a mixer. By definition, a solid mixture is said to be homogeneous when the composition is uniform throughout the whole mixture. To determine the homogeneity of a mixture, some criteria for the degree of homogeneity should be defined; more than 30 types of standards/tests have been proposed by various investigators (51). the simplest degree of homogeneity M is given by Lacey (50, 51):

$$M = \frac{\sigma_0^2 - \sigma^2}{\sigma_0^2 - \sigma_r^2} \tag{15}$$

with

$$\sigma_0 = P(1 - P) \tag{16}$$

$$\sigma_r = \frac{P(1 - P)}{N} \tag{17}$$

$$\sigma = \frac{1}{n} \sum_{i=1}^{n} (x_i - X)^2 \tag{18}$$

$$X = \frac{1}{n} \sum_{i=1}^{n} x_i \tag{19}$$

where σ_0 is the standard deviation of the mixture before mixing, σ_r is the standard deviation of the complete random mixture, σ is the measured standard deviation in the sample, P is the proportion of the considered component in the mixture, N is the total number of particles in a sample, n is the number of spot samples, x_i is the ith value of x, which represents a spot sample's characteristics (such as composition), and X is the arithmetic mean of the spot samples. The M value is between 0 and 1.

The tendency of particles to separate according to size and/or density differences (known as *segregation*) is generally found in free-flowing solid mixtures. For particles with different densities but similar size ranges, heavier particles tend to remain near the bottom of containers, while round or small ones stay more toward the top. However, for particles of the same density but different sizes, the smaller particles go to the bottom. Avoiding segregation is a challenge in the food industry, where materials with a wide range of properties are mixed. Lindley (48) noticed that segregation is unlikely for particles of less than 10 μm. Adding small quantities of moisture can transfer a mixture with high segregating tendencies into a nonsegregating one, and a mixture can be free of segregation if its coarse particles have a rough or fibrous shape.

Segregation can be minimized by the following methods: (a) using rational equipment in such processes as mixing or transportation, (b) reducing particle mobility via close packing, (c) reducing particle size by grinding, and (d) wet-mixing and spray-drying of mixture gradients. Some special segregation indices have been introduced to measure this kind of phenomenon kinetics. A noteworthy example is William's segregation coefficient. The William segregation test cell can be separated into two halves from the middle, with one called the up cell and the other the bottom cell. The segregation for a binary powder mixture can be evaluated by monitoring the segregation intensity in the William cell, subjected to tapping. Based on the coarse fractions in the upper and lower cells, the potential segregation can be evaluated as the S_{index} (18):

$$S_{index} = \frac{X_{CT} - X_{CB}}{X_{CT} + X_{CB}} \tag{20}$$

where X_{CT} and X_{CB} are the weight fractions of the coarse material at the top half and at the bottom half of the cell, respectively.

Instead of a cell split only at the middle, a multiring split cell, as shown in Figure 8, may be used for segregation studies, which makes it possible to see the overall distribution of fines and coarse fractions along the cell height after the mixture is subjected to vibration or tapping (i.e., undergone segregation). Since the contents of each ring can be weighed separately, density and content changes along the cell height can easily be detected. The segregation index S_{index} for the multisplit cell is calculated by (18, 52):

$$S_{index} = \sqrt{\frac{\sum_{i=1}^{n} W_i(X_i - X)^2}{\sum_{i=1}^{n} W_i}} \tag{21}$$

where W_i is the weight of powder in the ith ring, X_i is the concentration of a given component, and X is the mean concentration of the component in the mixture.

It can be shown that S_{index} can theoretically vary between 0 (total mixing) and 0.5 (total segregation) if the mixture of two components with a $1:1$ density ratio is uniform along the cell's vertical axis after vibration and the split cell number is even. The maximum theoretical S_{index} values for binary mixture components with the same bulk density but different weight ratios in a multisplit cell with different rings are listed in Table 6. If the densities of the binary components are different, the height ratio at complete separation H is given by:

$$H = \frac{L_A}{L_B} = \frac{X_A \rho_B}{X_B \rho_A} \tag{22}$$

where L_A and L_B are the heights, X_A and X_B and the weight fractions, and ρ_A and ρ_B are the densities for components A and B, respectively (18).

It is very helpful to know what type of mixture is created after the mixing process. Basically, there are four types of mixtures: random, partially randomized, ordered, and partially ordered random. In random mixtures, "the probability of finding a particle of a given type A at any point in the mixture is a constant equal to the proportion of that kind of particle in the whole mixture" (53). Only free-flowing particles with equal size, shape, and density have such characteristics. The $1:1$ (w/w) dry sucrose and citric acid (both have the same particle size

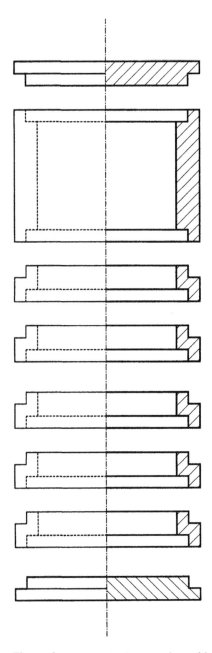

Figure 8   Schematic diagram of a multisplit cell for segregation kinetic studies. (Modified from Ref. 18.)

Table 6 Maximum Theoretical Segregation Index S_{index} Values for a Binary Mixture with the Same Bulk Densities in a Multisplit Cell with Different Ring Numbers

	Number of rings				
A:B (weight ratio)	4	5	6	7	20
5:95	0.087	0.100	0.112	0.122	0.218
10:90	0.173	0.200	0.224	0.245	0.300
25:75	0.433	0.387	0.382	0.401	0.433
40:60	0.424	0.490	0.447	0.407	0.490
50:50	0.500	0.447	0.500	0.463	0.500

Source: Modified from Ref. 18.

ranges and similar densities) mixture and 1:1 (w/w) dry sucrose and malic acid granule mixture belong to these kind of mixtures. If the particles are not identical for a random mixture of freely flowing particles and there is no interaction among components, then a partially randomized mixture will be formed. A 1:1 (w/w) dry granulated sucrose and cornstarch mixture is an example of this kind of mixture.

When two types of particles have different particle sizes and interactions, an ordered mixture is formed, because fine particles adhere onto large carrier particles but do not distribute randomly in the mixture. Ordered mixtures require particle interaction through adsorption, surface tension, chemisorption, friction, electrostatic or other type of adhesion. A typical ordered mixture is the 9:1 (w/w) granular sucrose and cornstarch mixture after being exposed to 100% RH for 3 hours and dried. For an ordered mixture, if the number of fine particles exceeds the limited number that can adhere to the larger particles, a partially ordered randomized mixture will be formed. A 1:1 (w/w) dry soy protein and cornstarch mixture is an example of this, because some cornstarch particles are adhering on the protein surface and others distributed in the mixture randomly (53).

The density and compressibility of various food gradient mixtures were determined and compared to those of pure components by Barbosa-Cánovas et al. in 1987 (17). It was found that the density and/or compressibility of the mixture could not be deduced from those of the pure gradients unless the particles were of similar sizes and properties. This implied that the mixture should be treated as a new powder and that some of its physical properties were changed by mixing.

A satisfactory mixing process produces a uniform mixture in a minimum time with a minimum cost of overhead, power, and labor (48). Making a mixture is associated with the physical characteristics of the components, the type of mixer used, and the final use of the mixture. Mixers for both research and commercial uses are usually classified according to mixing mechanisms or mixer movement. For both types the mixing mechanisms can be systematically classified according to motion types: within the bulk material, in the centrifugal field (gravitational or centrifugal), in a fluidized bed, in a suspended condition, in the free-fall process, or in several streams of materials (54).

Mixing operations may be batch or continuous. The batch-type mixer is commonly used since it is more flexible. There are two types of batch mixers. The first type has rotational moving vats shaped like a V, cone, drum, or cube that can be used for the premixing of additives or minerals for the ultimate concentrate preparation and are satisfactory for small agricultural operations. These types of mixers are affected by particle segregation. The second type of batch mixer has a stationary container (often U-shaped) with a rotating shaker that creates shearing and circulating actions at the center of the mass. The shaker may be paddles, ploughs, turbines, or ribbons. These mixers can be used for larger or more difficult operations in the livestock feed, baking, and cereal-processing industries (48, 50).

Static or dynamic types of continuous mixing procedures are most suitable for large and extensive operations. With static continuous mixers, mixing is done by air or gravity. Dynamic mixers are derived from batch mixers. The ingredients are usually added volumetrically, by auger, star wheel, or other device, to a screw conveyor (50, 55). In some cases, mixing is accomplished during conveying, so no additional mixer is required (48).

Mixer selection requires care but is primarily a trial-and-error process (47); some typical criteria are: (a) mixing characteristics, such as mixing quality, time, type, and capacity, (b) physical construction, which includes dimension, horsepower requirement, ease of loading, unloading, and cleaning, and safety, (c) cost factors, such as the capital cost of the mixer and necessary auxiliary equipment, labor, and other operating costs (51).

VI. MOISTURE CONTENT

Determining the moisture content of preprocessed cereal flours is an essential step in their quality evaluation, because it greatly influences the behavior of cereal grains and their flour products during milling and storage (12). Among all the storage environment factors (moisture content, temperature, and time), moisture content is the most important, because it influences the growth of fungi, which is a major cause of spoilage and quality reduction in grains and their products. Moisture also plays an important role in the merchandising of cereal grains and their products because it demands the same price as the main components. Both buyers and sellers are interested in knowing how much of this costly water they are dealing with. Accordingly, moisture limits are generally specified in purchasing contracts (56).

Moisture content is often expressed either in wet- or dry-basis percentages. Grain-marketing institutions prefer a wet basis because it directly expresses the amount of water in the grain as a percent of the total grain weight, while engineering and specific researchers prefer to express moisture content in dry basis because the dry matter in the grains remains constant for all ranges of moisture content. The wet-basis moisture content C_W may be calculated from the following equation if a drying process is used to measure it:

$$C_W = \frac{\text{weight before drying} - \text{weight after drying}}{\text{weight before drying}} \times 100 \qquad (23)$$

The dry basis C_D moisture content is calculated as:

$$C_D = \frac{\text{weight before drying} - \text{weight after drying}}{\text{weight after drying}} \times 100 \qquad (24)$$

Conversion between dry- and wet-basis moisture contents is fairly simple. Moisture contents from a dry basis to wet basis, or vice versa, may be expressed by the following equations (57):

$$C_W = 100 \times \frac{C_D}{100 + C_D} \qquad (25)$$

$$C_D = 100 \times \frac{C_W}{100 - C_W} \qquad (26)$$

Cereal grains and products are hygroscopic, so they will gain or lose moisture when exposed to ambient air of a certain relative humidity (56). More importantly, they tend to develop physical and chemical changes with a strong dependency on their temperature–moisture history (2). The relative absence of cohesive forces in particulate materials is related to their particle sizes and moisture contents.

In general, moisture sorption is associated with increased cohesiveness, mainly due to the formation of interparticle liquid bridges. Even for the same moisture sorption, the results are different if the original particle properties are different. Higher moisture will result in lower loose bulk density in fresh-sieved or flowing powders, but in liquefaction, and hence increased density (i.e., caking), for powders containing soluble crystalline compounds (29, 48). Moreyra and Peleg (39) found that moist or cohesive powders showed low loose bulk density and that the compressibility of moist powders will be greater than that of dry or less cohesive powders.

The moisture content of cereal flours also influences the final quality of bakery products. The effect of flour moisture content on cookie dough stickiness and consistency was studied by Gaines and Kwolek (58), who found that the dough made from low-moisture-content flour was more sensitive to dough water absorption changes than that of dough made from high-moisture-content flour.

The methods developed to measure the moisture content of cereal grains and their products are classified into three groups: (a) fundamental reference, (b) routine reference, and (c) practical based on background, ease of use, and/or cost. The Karl Fischer method belongs to the fundamental reference group; distillation, vacuum oven and air oven methods belong to the routine reference group; and electronic meter, nuclear magnetic resonance, and near-infrared spectrophotometer belong to the practical. These methods were well described by Christensen et al. (56) and will not be repeated here. By using these methods, moisture content results may vary considerably. Therefore, it is important that the method used be specified in the report and that the same method be used for all moisture tests if the results are used for the same purpose [12].

VII. CAKING AND ANTICAKING AGENTS

A simple definition of *caking* is ''when two or more macroparticles, each capable of independent translational modes, contact and interact to form an assemblage in which the particles are incapable of independent translations.'' Caking is such a common phenomena that almost every industry dealing with powders must deal with the problems that caking can cause. For many powdered products, such as foods, detergents, pigments, fertilizers, and chemicals, one important quality criterion is whether or not they cake under normal use conditions. Furthermore, lumped products will be considered of poor quality by consumers (59).

Caking is characterized by soft-lump formation or total solidification caused by interparticle forces developed under moisture absorption, elevated temperature, or static pressure. It is associated with flowability reduction and stable bridge formation (60). While flow problems exist for grains in silos, caking also happens in some preprocessed cereal products, such as flour, oatmeal, tapioca, and starch (59). Because the preprocessed cereal products are very often the main ingredients in numerous food formulations, the effect of caking on the flowability and physical properties of these powders is of prime concern in many processes.

There is a wide variety of reasons causing solids to cake or form lumps, but the main factors include moisture content, composition, pressure, crystal size and shape, temperature, and humidity variations (61). Most caking phenomena can be classified as one of four major types (59, 62):

1. *Mechanical caking*: This is particle-shape related and caused by particle interlocking or ''bird nesting.'' It usually occurs with fibrous or plate-shaped particle but not with spheres.
2. *Plastic-flow caking*: This occurs with highly viscous materials, such

as tars, gels, and waxes, whose soft crystalline substances stick together when subjected to either pressure or higher temperatures.

3. *Chemical caking*: This is the most common caking type and may be caused by chemical reactions in which a new compound or no new compound has been generated, such as decomposition, hydration, dehydration, recrystallization, and sublimation.

4. *Electrical caking*: As the name implies, this is caused by electrical charges on powders. Except for static electrical charges, most other electrical charges are the results of unsymmetrical properties (either physically or chemically) in the particle's crystal structure.

These four types of caking are remarkably different, so it is not difficult to distinguish them, even though there might be more than one type of caking in a given system. Once the type of caking has been identified and its cause understood, the work to eliminate the problem can be started in a predictable and organized way with greater confidence [59]. The caking tendency can be minimized by taking the following precautionary measures: drying, granulation, mixing with an insoluble fine powder, using crystal habit modifiers to change crystal shape, and coating surfactants [61]; using anticaking agents achieves most of these.

Anticaking agents, also called flow conditioners, glidants, antiagglomerants, lubricants, and free-flowing agents, are defined as substances added to finely powdered or crystalline food powders to prevent caking, lumping, or aggregation (i.e., to improve the latter's flowability and/or inhibit their tendency to cake). usually, the anticaking agents used in food powders are finely divided solids (particle size on the order of a few microns) made of chemically or practically inert substances. Their legally permitted concentration level is on the order of 1% or less (63, 64). Common anticaking agents include silicon dioxides, silicates, insoluble phosphates, the bi-or trivalent salts of stearic acid, talcum, starches, or modified carbohydrates, such as aluminum calcium silicate, silicon dioxide, and magnesium phosphate [63]. Rice flour produced by grinding is also used as a dusting or anticaking agent, for refrigerated biscuit dough (9).

In general, anticaking agents must be able to adhere to host powder particles to affect the latter's surface properties effectively. The adherence patterns may be complete surface coverage, scarce coverage, or scattered coverage. The mechanisms by which anticaking agents can affect powder flowability and caking tendency include [63]:

1. *Physical separation*: Host particles are physically separated when coated with a layer of anticaking agents, resulting in the lubrication (internal friction decreasing) or interruption of liquid bridging (humidity caking). This lubricant effect was proved by conditioning powdered

sucrose with calcium stearate. In this case, the former's friction angle was decreased by 2–5°.

2. *Competition for adsorbed water*: Usually, anticaking agents have comparatively large water adsorptive ability so that they can compete with the host particles for the available water, thus blocking the formation of liquid bridges and decreasing the consequent interparticle attraction. This mechanism works well if the available moisture is limited and the host powders are not highly hygroscopic. For example, when water supply is unlimited, using 1% (by weight) aluminum calcium silicate or calcium stearate has no anticaking effect at all on highly hygroscopic powdered onion.

3. *Cancellation of electrostatic charges and molecular forces*: The addition of anticaking agents can either neutralize electrostatic charges or reduce superficial molecular attractive forces on host particles to affect the latter's flowability. This effect was demonstrated in silica-conditioned carbowax. Because of the few published data on the electrostatic charge patterns of food powders, it is difficult to quantitatively analyze the role of molecular forces. However, the selective surface affinity between different food ingredients and conditioners serves as a strong indication that molecular forces do affect the flowability of powders.

4. *Modification of the crystalline lattice*: When an appropriate anticaking agent is used, it will inhibit crystal growth and change its lattice pattern. And when it is present at the surface of otherwise-normal crystalline particles, the recrystallization of any liquid bridges will be in the form of "dendritic" solid bridges, resulting in easily crumbed or disintegrated lumps or aggregates. For example, by adding a small quantity of urea to the crystallizer, some commercial table salt is crystallized as octahedral particles instead of its characteristic hard cubic shape (59).

VIII. DUST EXPLOSION

The first recorded flour dust explosion happened in a Turin flour mill in 1785. Since then, continued explosions have drawn scientific, economic, and even political attention. During the 62-year period from 1860 to 1922 in the United States, there were 119 explosions in the grain and grain-processing industry, which caused 215 deaths and 271 injuries. A series of United States grain elevator explosions that happened within eight days in 1979 killed 59 and injured 48, destroyed 2.5% of the nation's export elevators, and led to the initiation of extensive programs for researching the causes and prevention of grain elevator and mill dust explosions [1, 65]. While modern technology has reduced dust explosion hazards, it still happens, causing immense property damage and even loss of life.

As defined by Hertzberg and Cashdollar [66], a dust explosion is a rapid chemical oxidation of dust particles dispersed in air that leads to a rapid energy release that increases the system temperature so rapidly that a pressure increase follows. Industries concerned with the manufacture or handling of exlosive dusts include those dealing with agriculture, grain, milling, foods, chemicals, mining, metals, pharmaceuticals, plastics, and woodworking. In these areas, dusts may be produced as either end products or by-products. Dust clouds will often burn with explosive violence while the parent bulk material has no remarkable flammable hazard.

Dust clouds can be created during processes such as grinding and fluidized drying, as well as product handling such as elevator emptying, pneumatic transportation, and filter vibration [67]. The dust problem caused by mechanical attrition may also develop into a dust explosion hazard. For some agricultural products, the explosibility indices of their dust are ranked by Carr as [29]: starches 50, sugar 13.2, grains 9.8, wheat flour 3.8, wheat 2.5, skim milk 1.4, cocoa 1.4, and coffee <0.1 (a severe hazard is ranked by an index of 10 and above, strong by 1–10, moderately by 0.1–1, and weak by <0.1).

It is found that three conditions, often called the *triangle of fire*, must be satisfied before a dust explosion can occur. They are: (a) the combustible dust must be dispersed and mixed with air while contained within a volume, (b) the concentration of dispersed dust must be above the minimum explosive concentration, and (c) an ignition source with sufficient density and total energy to initiate the combustion wave must be present (66). Generally, an explosion occurs only where dust is dispersed in the air and a source of ignition is present. Thus, dust explosion can be prohibited with certainty if one of those three requirements is reliably eliminated (68).

Suppression of dust and avoidance of ignition sources are direct methods to avoid dust explosion. Dust formation can be suppressed by light damping with water (about 1% by wt) in the flour-milling process. Dust water content is a significant factor because the evaporation of water can take up some reaction heat and prevent an explosion from rapidly proceeding. Additionally, a dry dust is less cohesive and hence more likely to form a flammable cloud (1).

There are several factors related to dust dispersion into the air: the individual dust particle density, particle diameter and shape, cohesive properties with respect to each other, and particle's adhesive properties with respect to support surfaces. Dust dispersibility is difficult to characterize, but in principle it depends on humidity and particle shape. The minimal explosive concentrations for dust particles smaller than 100 μm are generally between 20 and 100 g/cm^3. However, because of the difficulty in getting a homogenous dust suspension and the variation in ignition energy and closure volume in explosion experiments, the figures concerning the minimal explosive concentration should be taken with skepticism (67).

When dust clouds are present, keeping the dust concentration outside the explosive range is rarely possible because of the sedimentation or whirling up of fine particles. Thus, an explosive atmosphere can be avoided only by reducing the concentration of oxygen (i.e., using inerting gas or working in a vacuum). In most situations where neither of those two alternative methods can be used, the only choice is either to avoid ignition sources or to use explosive-proof equipment (68).

Ignition sources may be presented in many ways, but are usually characterized according to the type of energy they introduce to the system. The most common types of ignition sources are electrical, chemical, and purely thermal (66). Welding and hand lamps are the most frequently responsible ignition sources in the flour-milling industry. Other sources are flames, hot surfaces, bearings, friction sparks, electric appliances, spontaneous heating, static electricity, magnets, bins, and bucket elevators (1).

Avoiding the presence of ignition sources is the most important prevention measure, but is still not sufficient by itself. Other methods also necessary for dealing with explosion protection include minimizing dust cloud formation, containment, proper plant layout, venting, inerting by gas or dust, flame traps, and automatic barriers (69). The flour agglomeration process may also be used to decrease the dust explosion potential because its final agglomerated products are almost dust-free.

IX. CONCLUSIONS

The physical properties of preprocessed cereal powders are very important in the grain, milling, and food-processing industries. Properties such as bulk density, compressibility, and flowability are essential to powder handling, processing, and transportation, while particle size and moisture content are related to final product quality and consumer acceptance. It is thus clear that a better understanding of these physical properties will enable greater control of preprocessed cereal powder processing and final product quality.

REFERENCES

1. NL Kent. Technology of Cereals. 3rd ed. New York: Pergamon Press, 1983, pp 1–196.

2. M Peleg. Flowability of food powders and methods for its evaluation—a review. J Food Pro Eng 1:303–328, 1977.
3. G Herdan. Small Particle Statistic. 1st ed. London: Butterworths, 1960, pp 73–89.
4. T Allen. Particle Size Measurement. 3rd ed. London: Chapman & Hall, 1981, pp 103–164.
5. Schubert. Food particle technology. Part I: Properties of particles and particulate food systems. J Food Eng 6:1–32, 1987.
6. L Svarovsky. Solid–Liquid Separation. 1st ed. London: Butterworth, 1981, pp 9–32.
7. RC Hoseney. Principles of Cereal Science and Technology. 2nd ed. St. Paul, MN: American Association of Cereal Chemists, 1994, pp 125–146.
8. CF Morris, SP Rose. Wheat. In: RJ Henry, PS Kettlewell, eds. Cereal Grain Quality. London: Chapman & Hall, 1996, pp 3–54.
9. KD Nishita, MM Bean. Grinding methods: their impact on rice flour properties. Cereal Chem 59(1): 46–49, 1982.
10. LW Rooney, SO Serna-Saldivar. Sorghum. In: KJ Lorenz, K Kulp, eds. Handbook of Cereal Science and Technology. New York: Marcel Dekker, 1991, pp 233–270.
11. GE Inglett, RA Anderson. Flour milling. In: GE Inglett, ed. Wheat: Production and Utilization. Westport, CN: AVI, 1974, pp 186–198.
12. VF Rasper. Quality evaluation of cereals and cereal products. In: KJ Lorenz, K Kulp, eds. Handbook of Cereal Science and Technology. New York: Marcel Dekker, 1991, pp 595–638.
13. LA Johnson. Corn: production, processing, and utilization. In KJ Lorenz, K Kulp, eds. Handbook of Cereal Science and Technology. New York: Marcel Dekker, 1991, pp 55–132.
14. MG Scanlon, JE Dexter, CG Biliaderis. Particle size related physical properties of flour produced by smooth roll reduction of hard red spring wheat farina. Cereal Chem 65(6):486–492, 1988.
15. JE Kruger, G Reed. Enzyme and color. In: Y Pomeranz, ed. Wheat: Chemistry and Technology. St. Paul, MN: American Association of Cereal Chemists, 1988, pp 441–480.
16. TP Meloy, NN Clark. Introduction to modern particle size and shape characterization. Powder and Bulk Solids Conference Exhibition, Rosemont, IL, 1987, pp 1–28.
17. GV Barbosa-Cánovas, J Malavé-López, M Peleg. Density and compressibility of selected food powder mixtures. J Food Pro Eng 10:1–19, 1987.
18. GV Barbosa-Cánovas, J Malavé-López, M. Peleg. Segregation in food powders. Biotech Prog 1(2):140–146, 1985.
19. BH Bergstrom. Empirical modification of the Gaudin–Meloy equation. AIME Trans 235:45–46, 1966.
20. CC Harris. The application of size distribution equations to multi-event comminution processes. Trans SME 241:343–358, 1968.
21. B Beke. Principles of Comminution. 1st ed. Budapest: Publishing House of the Hungarian Academy of Science, 1964, pp 27–62.
22. M Peleg, MD Normand, JR Rosenau. A distribution function for particle populations

having a finite size range and a mode independent of the spread. Powder Tech 46: 209–214, 1986.

23. M Peleg, MD Normand. Simulation of size reduction and enlargement processes by a modified version of the beta distribution function. AICHE J 32(1):1928–1930, 1986.

24. AB Yu, N Standish. A study of particle size distribution. Powder Tech 62:101–118, 1990.

25. CC Harris. A multi-purpose Alyavdin–Rosin–Rammler Weibull chart. Powder Tech 5:39–42, 1971/1972.

26. JK Beddow, JK Meloy. Testing and Characterization of Powders and Fine Particles. 1st ed. London: Heyden, 1980, pp 144–171.

27. H Yan, GV Barbosa-Cánovas. Size classification of selected food powders by five particle size distribution functions. Food Sci Tech Intern 3(5):361–369, 1997.

28. RC Hoseney, JM Faubion. Physical properties of cereal grains. In: DB Sauer, ed. Storage of Cereal Grains and Their Products. St. Paul, MN: American Association of Cereal Chemists, 1992, pp 1–38.

29. M Peleg. Physical characteristics of food powders. In: M Peleg, E Bagley, eds. Physical Properties of Foods. New York: AVI, 1983, pp 293–323.

30. J Malave, GV Barbosa-Cánovas, M Peleg. Comparison of the compacting characteristics of selected food powders by vibration, tapping and mechanical compression. J Food Sci 50:1473–1476, 1985.

31. GD Hayes. Food Engineering Data Handbook. 1st ed. New York: Wiley, 1987, p 83.

32. H Yan, GV Barbosa-Cánovas. Compression characteristics of agglomerated food powders: Effect of agglomerate size and water activity. Food Sci Tech Intern 3(5): 351–359, 1997.

33. M Duberg, C Nystrom. Studies of direct compression of tablets: XVII. Porosity–pressure curves for the characterization of volume reduction mechanisms in powder compression. Powder Tech 46:67–75, 1986.

34. JT Carstensen, XP Hou. The Athy–Heckel equation applied to granular agglomerates of basic tricalcium phosphate [(3 CA_3PO_4)2 · $Ca(OH)_2$]. Powder Tech 42:153–157, 1985.

35. TRR Kurup, N Pilpel. Compression characteristics of pharmaceutical powder mixtures. Powder Tech 19:147–155, 1978.

36. M Chen, SG Malghan. Investigation of compaction equations for powder. Powder Tech 81:75–81, 1994.

37. JM Geoffroy, JT Carstensen. Effects of measurement methods on the properties of materials. Powder Tech 68:91–96, 1991.

38. DMR Georget, R Parker, AC Smith. A study of the effects of water content on the compaction behavior of breakfast cereal flakes. Powder Tech 81:189–195, 1994.

39. R Moreyra, M Peleg. Compressive deformation patterns of selected food powders. J Food Sci 45:864–868, 1980.

40. R Ramberger, A Burger. On the application of the Heckel and Kawakita equations on powder compaction. 43:1–9, 1985.

41. MJ Adams, MA Mullier, JPK Seville. Agglomerate strength measurement using a uniaxial confined compression test. Powder Tech 78:5–13, 1994.

42. XD Chen. Mathematical analysis of powder discharge through longitudinal slits in a slow rotating drum: objective measurements of powder flowability. J Food Eng 21:421–437, 1994.
43. S Kamath, VM Puri, HB Manbeck. Flow property measurement using the Jenike cell for wheat flour at various moisture contents and consolidation times. Powder Tech 81: 293–297, 1994.
44. JAH de Jong, AC Hoffmann, HJ Finkers. Properly determine powder flowability to maximize plant output. Chem Eng Prog 95(4):25–34, 1999.
45. P Chasseray. Physical characteristics of grains and their byproducts. In: B Godon, C Willm, eds. Primary Cereal Processing. New York: VCH, 1994, pp 85–142.
46. PT Stainforth, RER Berry. A general flowability index for powders. Powder Tech 8:243–251, 1973.
47. JA Lindley. Mixing processes for agricultural and food materials. 1. Fundamentals of mixing. J Agric Eng Res 48:153–170, 1991.
48. JA Lindley. Mixing processes for agricultural and food materials: 3. Powders and particulates. J Agric Eng Res 49:1–19, 1991.
49. JA Lindley. Mixing processes for agricultural and food materials: Part 2. Highly viscous liquids and cohesive materials. J Agric Eng Res 48:229–247, 1991.
50. JP Melcion. The blending of powdery matter. In: B Godon, C Willm, eds. Primary Cereal Processing. New York: VCH, 1994, pp. 255–268.
51. LT Fan, SJ Chen, CA Watson. Solid mixing. Ind Eng Chem 62(7): 53–69, 1970.
52. JL Olson, EG Rippie. Segregation kinetics of particulate solids systems I: Influence of particle size and particle size distribution. J Pharm Sci 53(2):147–150, 1964.
53. GV Barbosa-Cánovas, R Rufner, M Peleg. Microstructure of selected binary food powder mixtures. J Food Sci 50:473–477 & 481, 1985.
54. K Sommer. Powder mixing mechanisms. J Powder Bulk Solids Tech 3(4):2–9, 1979.
55. SS Widenbaum. Mixing of Powders. In: ME Fayed, L Otten, eds. Handbook of Powder Science and Technology. New York: Van Nostrand Reinhold, 1984, pp 345–364.
56. CM Christensen, BS Miller, JA Johnston. Moisture and its measurement. In: DB Sauer, ed. Storage of Cereal Grains and Their Products. St. Paul, MN: American Association of Cereal Chemists, 1992, pp 39–54.
57. RC Brook. Drying cereal grains. In: DB Sauer, ed. Storage of Cereal Grains and Their Products. St. Paul, MN: American Association of Cereal Chemists, 1992, pp 183–218.
58. CS Gaines, WF Kwolek. Influence of ambient temperature, humidity, and moisture content on stickiness and consistency in sugar-snap cookie doughs. Cereal Chem 59(6):507–509, 1982.
59. EJ Griffith. Cake Formation in Particulate Systems. New York: VCH, 1991, pp 1–33.
60. MA Rao. Transport and storage of food products. In DR Heldman, DB Lund, eds. Handbook of Food Engineering. New York: Marcel Dekker, 1992, pp 199–246.
61. YL Chen, JY Chou. Selection of anti-caking agents through crystallization. Powder Tech 77:1–6.
62. H Schubert. Principles of agglomeration. Intern Chem Eng 21(3):363–377, 1981.

63. M Peleg, AM Hollenbach. Flow conditioners and anticaking agent. Food Tech (Mar): 93–102, 1984.
64. AM Hollenbach, M Peleg, R Rufner. Effects of four anticaking agents on the bulk characteristics of ground sugar. J Food Sci 47:583–544, 1982.
65. CW Kauffman. Recent dust explosion experiences in the U.S. grain industry. In: KL, Cashdollar, M Hertzberg, eds. Industrial Dust Explosions. Pittsburgh: ASTM, 1987, pp 243–264.
66. M Hertzberg, KL Cashdollar. Introduction to dust explosions. In: KL Cashdollar, M Hertzberg, eds. Industrial Dust Explosions. Pittsburgh: ASTM, 1987, pp 5–32.
67. JP Pineau. The risk of dust explosion. In: B. Godon, C. Willm eds. Primary Cereal Processing. New York: VCH, 1994, pp 377–400.
68. N Jaeger, R Siwek. Prevent explosions of combustible dusts. Chem Eng Prog 95(6): 25–37, 1999.
69. J Cross, D Farrer. Dust Explosion. New York: Plenum Press, 1982, pp 115–164.

7
Rheological Properties of Biopolymers and Applications to Cereal Processing

Bruno Vergnes
Ecole des Mines de Paris, Sophia-Antipolis, France

Guy Della Valle and Paul Colonna
Institut National de la Recherche Agronomique, Nantes, France

I. INTRODUCTION: DOUGH RHEOLOGY AND BIOPOLYMER PROCESSING

Rheology is the science of deformation and flow of matter. The rheological properties of a material provide relevant information on its structure, its behavior during processing, and its end-use properties. These properties influence the flow process and are themselves influenced by the structural changes generated during the process. Initially a branch of mechanics, this science has developed considerably and, for example, in the field of synthetic polymers, aims to compare results from continuum mechanics, molecular theories, and computer simulations. Unfortunately, application of these approaches to biopolymers and cereals is more difficult, for the following reasons:

> In contrast to synthetic polymers, cereal grains are variable products, whose suitability for a given process/product goal is affected by the genotype, the environmental conditions encountered during grain development and finally the milling process; these difficulties have been reinforced by progress in plant breeding and by the large offers created by the international cereals trade.

Cereal products very often have a complex formulation, with several components (starch, proteins, water, sugars, lipids) that can interact and lead to more or less organized structures; starch itself is made of two types of macromolecules, the linear amylose and the branched amylopectin. It can thus be considered as a multiphasic, rheologically complex material.

Doughs and melts from cereal products are highly non-Newtonian, with a high level of elasticity, and are very sensitive to the temperature, the water content, and, more generally, the composition (starch origin, presence of lipids).

Some components, even in small quantities, like lipids, may induce slip at the wall and thus totally modify the flow conditions.

Biopolymers are not as thermally stable as casual synthetic polymers. They start to decompose at 200–220°C. But before this threshold, chemical interactions can occur, leading to intra- and intermolecular covalent cross-linking. The kinetics of these reactions are highly determined by temperature and also by water content: Material processing itself induces important changes to the structure.

All of these reasons imply that any investigation should be based upon samples whose biological origin and conditions of preparation are not only known but also reproducible. The variability induced by these various factors leads to a broad range of behavior during the three basic operations in cereal technology: mixing, dough processing (including shaping), and baking or cooking. These operations commonly lead to the following phenomenological changes:

Obtaining a macroscopic homogeneous phase: from a powder (flour) to a dough or a melt

Flow and dough forming (laminating, sheeting)

Foam creation by bubble nucleation and growth during fermentation and heating, vaporization: from a dough or a melt to a solid foam

Various states or types of organization (suspension, network, melt) may thus be encountered under many different flow conditions, as expressed in Table 1. To face the complexity of these changes, empirical rheological tools have first been used in the laboratories, to predict cereal and flour performances. Their goal is to provide qualitative information related to process adequacy. Section II.A is devoted to these tests, which are still widely employed, not only in industry but also in research laboratories. However, no device has ever been developed to follow the on-line rheological properties of dough during baking: This would explain, for instance, why all final testing procedures for bread making include a small baking test. This underlines that this last step of bread technology can reveal unusual behaviors, unexpected on the basis of all empirical tests devoted to only the two first steps, mixing and dough processing.

Table 1 Examples of Products, Behaviors, and Processing Conditions Encountered During Overall Processing

	Bread baking	Biscuit dough forming	Dough development	Breakfast cereals and snacks
Application and/or end product	Bread baking	Biscuit dough forming	Dough development	Breakfast cereals and snacks
Process and/or phenomenon	Fermentation and bubble growth	Laminating, sheeting	Mixing and kneading	Extrusion and die flow
Typical rheological behavior	Viscoelastic solid	Concentrated suspension	Transient network	Polymer melt, viscous liquid
Range of moisture content and temperature	10–40% 50–200°C	10–30% 10–60°C	30–50% 10–60°C	10–30% 100–200°C
Range of shear rate (s^{-1})	$10^{-3}–10^{-2}$	$1–10^2$	$10–10^2$	$10–10^3$

More objective experimental approaches are now available. They are described in Section II.B. The adaptations that have to be made to classical methods to study biopolymers are also emphasized in Section II.C. Since the pioneering work of Schofield and Scott-Blair (1), many applications of these approaches to cereal dough characterization can be found in the literature. The variety of products, processes, and methods encountered explain why no global literature synthesis is available on this topic. Thus, rather than trying to be exhaustive, we have chosen, in Section III, to present some recent examples of the use of such methods for different types of products encountered during different processing steps, such as those mentioned in Table 1. This chapter also aims to show the point up to which rheological methods can actually be useful for cereal processing and which improvements they eventually require.

II. OVERVIEW OF RHEOLOGICAL MEASUREMENTS

Rheological properties create the link between the stresses applied to a material and the way this material will flow or deform. All experiments associating on one hand a deformation and on the other hand a stress or a force can be considered a rheological test. However, some conditions are indispensable to really deducing an objective value characteristic of the material from such a test. In other words, rheological measurements can be split into two categories:

Empirical measurements, widely used in the food industry, which are of real importance to comparing different raw materials and to assessing their aptitude to processing but cannot provide constitutive equations.

Objective measurements, in which strain and stresses are usually simple but perfectly determined: They allow one to define intrinsic parameters of the rheological behavior.

A. Empirical Measurements and Industrial Tests

Empirical measurements have been developed for approaching the quality of the raw ingredients and following the first two steps of cereal processing (mixing and dough processing) over a limited domain. The basic principle of these empirical tests is to mimic each step of the industrial process at a laboratory scale. One advantage of these methods is that they are based upon samples that have been submitted to the same well-defined deformation history (2). Conversely, no intrinsic rheological property is directly accessible. In their design, these measurements are easy, quick to carry out, and inexpensive, which explains their great success on a world scale. They deliver data expressed in arbitrary units, but they are included in international procedures and their uses are completely described by official specifications (ISO, International Standardization Organization; AACC, American Association of Cereal Chemistry). Thus, they are highly reproducible. Their technological value relies on the existence of correlations with values observed at pilot or industrial scales: Their usefulness is based upon a large body of knowledge, accumulated over the past 50 years (3).

The major drawback of these empirical methods is that it is difficult to define the strains and stresses applied to the dough during the experiments. Another question is the adequacy of the shear and extensional rates involved in physical testing to those encountered in real cereal-processing operations (Table 2). Depending on the processing step being considered, laboratory tests are carried out at rates at least 10- to 100-fold higher (for baking, for example) or lower (for mixing) than for the actual industrial application (4). Conversely, objective measurements are often performed at low strain (0.1–5%), in the linear viscoelastic domain, whereas empirical tests are carried out for strain values for which no linear domain is observed (5, 6). Strain may reach 100% during sheeting, 1000% during fermentation, and up to 500,000% during mixing (7). The narrowness of the linear domain is also due to the high reactivity of dough components, the structure of which continuously changes during processing. For instance, Amemiya and Menjivar (8) considered three different zones for the variation of the shear stress of wheat dough as a function of strain γ:

For $\gamma < 3\%$, weak starch–starch and starch–proteins interactions are predominant, leading to a power law relationship, with a constant exponent.
For $3\% < \gamma < 25\%$, these weak interactions diminish, giving a decrease in the exponent of the power law relationship.
For $25\% < \gamma < 1300\%$, a strain-hardening behavior is obtained, probably due to interactions between proteins fibrils.

An important outcome of these tests is that they are also used for determining in which processing conditions each flour sample will give the best final product. So water volume and mixing time can be adjusted for each flour batch to an ideal behavior. Other dough additives, such as oxidants, enzymes, and emulsifiers, can be studied in the same way.

1. Recording Dough Mixers

The first family of empirical tests concerns recorders of dough mixers that mimic the mixing step. The basic idea is to follow the formation of a dough in a unique procedure. Starting from flour and water, a mixing step is imposed on this particle suspension, during which torque developed by the mixer is measured and recorded throughout mixing. Flow conditions are very complex in these geometries, so both shear and elongational deformations are encountered during mixing. Dough sticking also influences the measurements. In addition, due to unfilled

Table 2　Basic Events Occurring During Wheat Breadmaking

Process step	Technological objectives	Estimated shear and extensional rate (s^{-1})	Mechanistic events, including rheology
Mixing	Optimal consistancy Hydration level Duration, intensity, and type of mixer Overmixing tolerance	10–10^2	Shear and elongation deformations beyond the rupture limits Adhesion Hydration Solubilization and swelling of albumins, globulins, damaged starch, and arabinoxylans Gluten formation and breakdown, including enzymatic reactions Gas-cell formation
Dough processing			
Sheeting and molding	Shaping	10	Shear and elongation deformations beyond the rupture limits Flavor formation
Fermentation	Size stability Gas retention	10^{-4}–10^{-3}	Elongation deformations below the rupture limits Permeability, diffusivity Gas-cell formation, disproportionation, and coalescence
Baking	Oven rising Thermal setting	10^{-3}–10^{-2}	Flavor formation Elongation deformations below the rupture limits Permeability, diffusivity Biopolymers cross-linking and phase transition Gas cell: disproportionation and coalescence Water migration

Source: Estimations of deformation rates from Ref. 4.

conditions, flow geometry is not constant over time. Consequently, rheological modeling is currently completely impossible, and these devices are strictly representative of empirical measurements.

The dough-making process encompasses various phases, which in practice overlap (9). The first phase is the moistening of the flour particles, where adsorption of water onto particle surfaces induces high adhesion forces (10). Then solubilization and swelling of albumins, globulins, arabinoxylans, and damaged starch granules are followed by the restructuring of a gliadins and glutenins network. This last step involves a continuous overlap of molecular buildup and breakdown, proceeding simultaneously during mixing. Therefore, consistency increases up to a maximum and then falls off. This continuous degradation explains why energy expended during mixing is involved not only in dough structural buildup, but also in heat and irreversible structural breakdown (11). In bread doughs, gas-holding ability is generally associated with the formation of a continuous gluten network (12). The most critical factors in the mixing stage are flour quality, amount of water added, and magnitude of work provided the developing dough. Interpretation of the empirical tests elucidates three major parameters: the development time of a flour, its tolerance to overmixing, and its optimum water absorption.

All empirical tests can be used to predict dough water absorption, dough stiffness, and mixing requirements, including the flour's ability to support overmixing. For a given flour sample, there is a specific mixing intensity that will ensure optimal dough development. It is reached by adjusting mixing time rather than mixing speed. These relationships are often valid only within a limited range of flours and mixing conditions. Outside this range, the relationships often break down or are misleading. Hence, correlations between dough rheology and baking performance have often produced inconsistent and conflicting results (13).

The majority of the literature on recording dough mixers concerns bread. Nevertheless, this technique is also used for testing the suitability of flours to cookie and cracker technology (14) and to pasta technology (15, 16).

 a. Farinograph from Brabender. Developed initially by Hankoczy and Brabender, it has remained unchanged in its principle and geometry (17). It works by measuring the resistance of a dough against sigmoid-shaped mixing paddles, turning at a 1.5/1 differential speed (93 and 62 rpm). The paddles hold a flour–water dough (constant flour weight: 300, 50, or 10 g of dry flour) to a prolonged, relatively gentle kneading action, at a constant temperature (30°C). This feature is important, mainly for North American flours (18). Shear rates are estimated to be around 10 s^{-1} for a dough consistency of 500 BU (Brabender units) (4). Designed before high-intensive mixers were widely used, the farinograph does not reflect the mixing requirements of commercial production. The variation of the torque during mixing has a characteristic shape and is known as a *farinogram*

Torque or consistancy (BU)

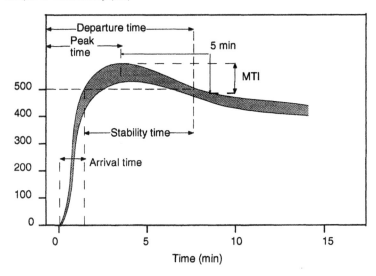

Figure 1 Schematic representation of a farinogram and associated measurements: departure time, peak time, stability time, arrival time, mixing tolerance index MTI.

(Fig. 1). It provides information on the short-term transient changes in dough rheology during mixing. Water absorption by flour particles is supposed to be one basic reaction used in generating a farinogram.

The following characteristic values are extracted from each farinogram:

The main farinogram value is the *arrival time*, corresponding to the time necessary for the top of the curve to first intersect the 500-Brabender-units (BU) line, as the water is being rapidly absorbed.

The time required to reach a point of maximal dough consistency before any indication of dough breakdown is considered the dough's development or *peak time*.

The *departure time* is the time at which the top of the curve drops below the 500-BU line. A long departure time suggests a strong flour.

Stability or tolerance *time* is the difference (min) between the arrival and departure times: It reflects the flour's tolerance to mixing.

The *time to breakdown* is defined as the time from the start of mixing to the time at which the curve has dropped by 30 BU from the peak point.

The *torque mixing peak* has been traditionally related to optimum dough development for baking. Occasionally, farinograms may present two peaks: The first one is related to hydration, whereas the second one is considered the true one.

The *mixing tolerance index*, MTI, is the difference (in BU) between the heights at peak time and 5 min later: It indicates the dough breakdown rate.

The *20-min drop* is the distance in BU between the development peak and the point 20 min after the peak time: It indicates the rate of breakdown in dough strength. High values indicate weak flours.

Water absorption inferred from global behavior is certainly the most widely accepted measurement (17). *Farinograph optimum absorption* is defined as the amount of water required to locate the peak area of a farinograph curve on the 500-BU line for a flour–water dough, water being the adjusting variable.

The use of the farinograph test has been extended to rye-grain quality, owing to large and unexpected variations from year to year for this crop.

b. Mixograph. Developed initially by Swansson and Working (19), this device was built to mimic the powerful mixing (pull, fold, pull) of U.S. commercial bread dough mixers. Its originality is in the planetary head, which was designed to mix flour/water doughs (constant flour weight: 2, 10, and 35 g of flour) at 88 rpm. The mixograph uses a harsher pin mixing method.

The recorded response, called a *mixogram* (Fig. 2), is a two-part curve, with ascending and descending arms. Interpretation is also manual and gives five main values:

Peak time t, similar to arrival time in the farinogram
Peak height H, which is a measure of the dough's resistance to the extension caused by the passage of the pins
Developing or ascending angle D (rate of dough development)
Weakening or descending angle W (rate of dough breakdown)
Tolerance T, which is the angle given by the difference between the developing and weakening slopes

The curve length is related to the time the dough has been mixed. The curve width is related to the cohesiveness and elasticity of the dough. This latter indicates a dough's mixing tolerance. Curve peak time and height are determined on one hand by the quality and the protein content of the flour and on the other hand by the water absorption (20).

In contrast to the farinogram, the mixogram is more complex for obtaining the amount of water needed to produce a dough of optimum absorption; additional measurements are needed (21). But it gives the best prediction of the mixing time for optimum bread quality in the bakery mixer test (22).

Torque

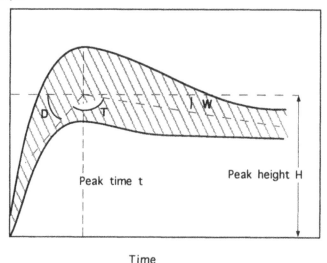

Time

Figure 2 Schematic representation of a mixogram and associated measurements: development angle D, tolerance angle T, weakening angle W, peak time t, peak height H.

Other outstanding recording dough mixers are:

The *alveograph* from Chopin, which uses a sigmoid blade first to mix and then, by reversing the rotating movement of the blade, to extrude the dough into a uniform sheet. Extensional rates are estimated to be in the range 10^{-1}–1 s^{-1} (4). It does not measure dough rheology, but can be used to prepare dough samples for rheological measurements.

The *Do-corder* from Brabender, which is based upon the continuous recording of the torque exerted by different types of blades upon flour–water blends. It is a very versatile piece of equipment, able to work in large ranges of temperature (40–300°C) and mixing speed (5–250 rpm). According to Nagao (23), typical results of torque/temperature curves present two peaks at about 75 and 85°C, associated with the formation of disulphide bonds and the initiation of starch gelatinization, respectively.

2. Stretching Devices

After the mixing, shaping, and fermentation steps, the dough is submitted to stretching. This is expected to reveal the more permanent structural changes occurring in doughs as a result of mixing. Dough has to be considered a filled-polymer system with a combination of three phases: a solid phase, made of gluten with starch granules embedded in it; a liquid phase, with water and water-soluble components, such as globulins, albumins, and arabinoxylans; and a gaseous phase, with gases entrapped during mixing or generated by yeast or chemicals. The main point is the difficulty of working with fermenting doughs, where dimensions and physical properties are continuously changing. In these tests, a dough sample is submitted to large deformations until it breaks. The resistance curve of the dough sample during stretching is recorded and two main parameters are extracted: resistance to large deformations, and stretching suitability. The alveograph and the extensograph provide deformations similar to those that take place during fermentation and oven rise, with higher rates of deformation.

All devices reproduce more or less the mechanical actions encountered during shaping. However, Janssen et al. (24) found no clear relationship between the extensibilities determined by the empirical methods and loaf volume.

a. Alveograph from Chopin. This device is specifically designed to measure the resistance to biaxial extension of a thin sheet of flour/water/salt dough, generally at a constant hydration level. This process is similar to sheeting, rounding, and molding in the baking process. From the dough sheet obtained after the mixing-extruding device, five individual disks are cut and allowed to relax for 20 min. Then each disk is clamped above a valve mechanism, and air is blown under the disk at a constant rate, thus creating a bubble. The pressure inside the bubble is recorded until rupture occurs, giving the final information on the dough's resistance to deformation (Fig. 3). Maximum strain rate is of approximately 0.5 s^{-1}. The alveograph is considered to operate at strains close to those observed during baking expansion. The shape of the alveogram can be calculated for rheological model materials (25). With modification, it can be used to obtain fundamental tensile rheological properties (12, 26). Wheat flour doughs exhibit a clear power law strain-hardening under a large extension rate, which is an important feature for bubble stabilization (24).

Measurements made on the average curve for the five replicates include:

Adjusted peak height (H, mm). The H value (overpressure, mm) is related to the dough's tenacity.

Curve length (L, mm). This is proportional to the volume obtained before rupture. The L value is generally related to dough extensibility and predicts the handling characteristics of the dough.

Work input (W, J/g). The W value is the amount of work required for the deformation of the dough and is related to the baking "strength" of the flour.

Pressure

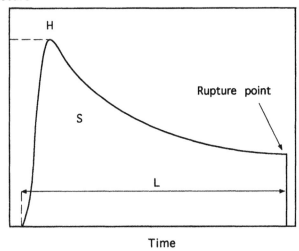

Figure 3 Schematic representation of an alveogram and associated measurements: peak height *H*, curve length *L*, area under the curve *S*.

Strong flours are characterized by high *H* and *W* and low/medium *L* values. However, damaged starch content will considerably affect the response when working at a constant hydration level. Therefore, Rasper et al. (27) and Chen and d'Appolonia (28) used the Farinograph Brabender to determine which hydration level has to be used in the stretching analysis. With alveograms, as with extensograms, good baking flours exhibited stronger resistance to extension and a greater extensibility, but the differences found are not always directly related to the results of the bakery test (29). Final bread volume is predicted with a better precision, whereas this procedure fails for cookie development (27, 28).

b. Extensograph from Brabender. The extensograph measures dough extensibility and dough relaxation behaviors. Flour/water/salt doughs are first prepared in a farinograph at a water content slightly higher than normal absorption (to compensate for the salt). A piece of dough is molded into a cylinder and clamped into a saddle. After a rest time, a hook stretches the dough. Extensional rates are estimated to be in the range 10^{-1}–1 s^{-1} (4). The graphical output (Fig. 4) has time, representing extension, on the *x*-axis and resistance to extension on the *y*-axis.

Averaged values, based upon four measurements, are:

Torque (BU)

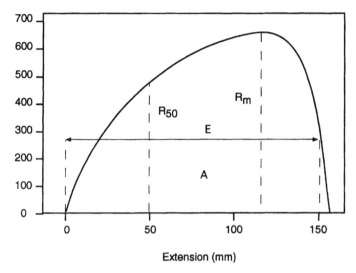

Extension (mm)

Figure 4 Schematic representation of an extensogram and associated measurements: resistance to extension R_{50}, maximal resistance to extension R_m, length of the curve E, area under the curve A.

Resistance to extension (R_{50}, BU), measured 50 mm after the curve has started. It is related to the elastic properties.

Maximal resistance to extension (R_m, BU).

Extensibility, which is the length of the curve (E, mm).

Strength value, given by the area A (cm^2) under the curve.

Resistance R_m and extensibility E are the most used parameters (21). In practice, the measurement is repeated for four different rest times (10, 20, 30, and 40 min), but there is no general rule for interpretation.

    c. Extensometer and Extrudometer Simon. After a mixing-extrusion step in the Extrudometer Simon, a cylinder of dough is submitted to a uniaxial extension. Similar information to that gained with the extensograph Brabender is obtained. The Extrudometer Simon measures the time necessary for a dough to flow from a reservoir through a cylindrical orifice under a constant piston load, the motion of which is measured by a micrometer. This principle is very close to that of the melt-indexer, widely used in the synthetic thermoplastic industry. It has been used to determine the optimum hydration of the dough. The result is a combination of extension (at the die entry), shear (into the die), and sticking properties of the dough.

3. Special Case of Biscuits

In contrast to bread doughs, where water content is about 45%, biscuits are more complex, for they contain about 20% water, with the presence of sugars and fats. Thus, all aforementioned tests are no longer useful. No specific test has been developed for assessing the mixing step, probably because very little information is available on the structure of biscuit doughs.

Miller (30) has proposed a simple test based upon the penetration of a set of needles into the dough. The force necessary to obtain a given penetration speed is used as a measure of consistency. Fair relationships between consistency and hydration support the idea that consistency could be used to predict the final dimensions and weight of cooked biscuits. In a more recent approach, the suitability of flour to biscuit processing has been approached with a test similar to the Extrudometer Simon (31).

4. Stickiness of Doughs

Stickiness has been recognized for a long time as a key factor that may considerably reduce the production rate in most cases. Unfortunately, the basic physical chemistry is still lacking. The easiest way to estimate sticking behavior is to apply a compression step over a dough sample. Afterwards, the maximal force necessary to remove the piston from the surface of the dough sample is an indirect measurement of stickiness. The energy necessary to unstick the piston from the dough surface can also be used. These measurements depend highly on the procedure employed to set up the sample. Several proposals can be found in the works of Hahn (32) and Chen and Hoseney (33). No normative procedure has been accepted worldwide at the present time. However, this procedure is important for predicting the end-use quality of pasta, once cooked. But the results remain poorly correlated with the stickiness of the pasta in the mouth during chewing, evaluated by a sensory panel.

There is still a need to use basic physical concepts to obtain sound insights into the rheological behavior of dough, particularly the surface rheology, and thereby on the micro- and mesostructure of doughs. All the information provided by empirical measurements has been extensively evaluated in relation to wheat variety and growing and harvesting conditions. Correlations with flour protein content are the first relevant links most often reported in the literature (34, 35). This does not intrinsically reflect any mechanistic link, but rather a pure correlation based upon the simplest chemical determination to carry out in present cereal chemistry.

B. Objective Measurements

In this section, we will first make a distinction between measurements in large deformations and in small deformations. The first kind of measurements covers the habitual conditions of processing, useful to define flow curves and viscous laws, which are necessary for the computation of processes such as extrusion (see Sec. III.B). The second kind, usually made in the range of linear viscoelasticity, is relevant for providing information on the structure of the material (see Sec. III.A).

1. Large Deformations

The materials of interest in this chapter (molten starches and doughs) are highly viscous fluids. During the processes (extrusion, mixing, rolling), they are submitted to high shear rates and large deformations. The classical way to characterize the shear viscosity of such products is to use either a capillary rheometer or a slit die rheometer.

a. The Capillary Rheometer. This rheometer consists of a heated cylindrical reservoir in which a piston can axially move. At the lower part of the reservoir a capillary tube, of length L_c and radius R_c, is fixed. The following conditions are necessary for obtaining reliable measurements:

The flow is incompressible, isothermal, and steady state.
The product is homogeneous.
There is no slip at the wall.
The pressure has no influence on the viscosity.

The displacement of the piston into the reservoir at a fixed speed imposes a volumetric throughput Q_v through the capillary. The measurement of the pressure drop Δp through the capillary allows one to define the apparent viscosity as follows:

The wall shear stress is expressed as:

$$\tau_w = R_c \frac{\Delta p}{2L_c} \tag{1}$$

The apparent wall shear rate is given by:

$$\dot{\gamma}_{app} = 4 \frac{Q_v}{\pi R_c^3} \tag{2}$$

The ratio $\tau_w/\dot{\gamma}_{app}$ provides the apparent viscosity η. This would be the viscosity of a Newtonian fluid. In fact, all the products we are dealing with are non-Newtonian. This means that their viscosity is not a constant, but depends on the rate of deformation $\dot{\gamma}$. The simplest way to describe this behavior is the so-called power law:

$$\eta = K|\dot{\gamma}|^{n-1} \tag{3}$$

where K is the consistency and n is the power law index. The value $n = 1$ corresponds to a Newtonian fluid with a constant viscosity; $n > 1$ corresponds to a shear-thickening fluid, for which the viscosity increases with the shear rate. This is the case, for example, for some concentrated suspensions (36). The value $0 < n < 1$ corresponds to a shear-thinning fluid, whose viscosity decreases when increasing the shear rate. This will be the general case for the products studied in this chapter.

To obtain the viscosity curve for a non-Newtonian fluid, the following procedure has to be respected:

Pressure drops are measured for several values of throughput Q_v and thus for several values of shear stress τ_w and shear rate $\dot{\gamma}_{app}$.
The changes of shear stress with shear rate are plotted in a log-log scale. The slope of this curve allows one to define the power law index:

$$n = \frac{d \log \tau_w}{d \log \dot{\gamma}_{app}} \tag{4}$$

Because the velocity profile in a tube is different between a Newtonian and a non-Newtonian fluid, the apparent wall shear rate [Eq. (2)] must be corrected, to define its true value $\dot{\gamma}_c$. This is known as the Rabinowitsch correction (37):

$$\dot{\gamma}_c = \frac{3n + 1}{4n} \dot{\gamma}_{app} \tag{5}$$

The apparent viscosity at the shear rate $\dot{\gamma}_c$ is defined by the ratio:

$$\eta(\dot{\gamma}_c) = \frac{\tau_w}{\dot{\gamma}_c} \tag{6}$$

This procedure assumes that the pressure drop Δp through the capillary is known. In fact, in a capillary rheometer the pressure is generally measured at the bottom of the reservoir, upstream of the capillary. It is thus necessary to correct the measured value p_m to account for the entry and exit pressure losses. This is usually done by using several capillaries with the same radius but different lengths. For a fixed throughput (which means a fixed apparent shear rate), the measured pressures are plotted as a function of the length L_c. If the experiments are made under the correct conditions, one obtains a linear relationship. The intercept with the ordinate axis provides the entrance pressure drop Δp_0. The pressure drop through the capillary is thus:

$$\Delta p_c = p_m - \Delta p_0 \tag{7}$$

Obviously, this correction, known as the Bagley correction (38), must be done for each value of the apparent shear rate. Another way to perform these corrections is to use an orifice die (of length $L_c \approx 0$) and to subtract the pressure drops through this orifice from the measured values p_m (39, 40).

b. The Slit Die Rheometer. Instead of using a capillary flow for measuring the viscosity, we can use a flow between two parallel plates, or more precisely in a rectangular channel whose width W is large compared to its thickness h (41, 42). The major advantage of this system is the possibility for directly measuring the pressure drop using flush-mounted transducers and thus avoiding the time-consuming Bagley corrections. To obtain reliable measurements, the preceding conditions made for capillary rheometry must obviously be verified. Moreover, two complementary conditions must be fulfilled:

The influence of lateral walls must be negligible, which necessitates that we use a very low h/W ratio, less than 0.1 (43).
The pressure profile all along the slit must be linear. To check this condition, it is necessary to use at least three pressure transducers.

Practically, the regulation temperature of the die is adjusted to the product temperature at the entry. The way to obtain a viscosity curve is then similar to that presented for the capillary rheometer:

The imposed volumetric flow rate defines the apparent wall shear rate:

$$\dot{\gamma}_{app} = \frac{6Q_v}{Wh^2} \tag{8}$$

The pressure drop $\Delta p/L$ provides the wall shear stress:

$$\tau_w = \frac{h}{2}\frac{\Delta p}{L} \tag{9}$$

The corrected wall shear rate is obtained by:

$$\dot{\gamma}_c = \frac{2n+1}{3n}\dot{\gamma}_{app} \tag{10}$$

where n is defined by Eq. (4).

Another point of interest of the slit die rheometer is its potential to provide information about some viscoelastic properties. Two methods have been proposed in the literature, but it is important to emphasize that, even for simple fluids, these methods are difficult to apply and are still the object of controversy:

The exit pressure (44–46): The extrapolation of the pressure profile up to the die exit is generally not zero. An exit pressure p_{ex} may be measured, which can lead to an estimation of the first normal stress difference N_1:

$$N_1 = p_{ex}\left[1 + \frac{d\log p_{ex}}{d\log \tau_w}\right] \tag{11}$$

The "hole" pressure (47–50): This is the difference measured between a flush-mounted transducer p_1 and a second one on the opposite wall, which is at a small distance from the wall surface p_2: $p^* = p_1 - p_2$. For circular dies, this value may be correlated to the normal stress differences by the following expression:

$$N_1 - N_2 = 3p^* \frac{d \log p^*}{d \log \tau_w} \tag{12}$$

There have been some tentative attempts to apply these techniques to cereal products (51–54). The results have been generally much too scattered to be really interesting, even if some recent trials seem more promising (55).

c. *Elongational Deformations.* The aforementioned measurements were focused on rheological properties in shear. However, elongational properties play an important role in cereal processing. Unfortunately, the measurement of such properties is much more difficult. The main problem is to deform homogeneously a molten sample at a constant elongational rate. In uniaxial extension, the length of the sample should increase exponentially with time, which usually limits the experimental conditions to very low deformation rates (less than a few per second) and short times. For these reasons, the characterization of the true elongational behavior of cereal products is a real challenge, and very few studies can be found in the literature. Usually, experiments are performed on universal testing machines, for which stresses, deformations, and deformation rates vary during the tests, which limits the interpretation of the results. Recent improvements in the technology of measurement, supported by the studies of Meissner (56), now offer new possibilities in uniaxial extension, and an application of this technique to doughs has recently been proposed (57). The alveograph presented in Sec. II.A generates biaxial extensional deformations, but the rheological interpretation of this test is not straightforward (12, 26). Rheological behavior in biaxial extension can be estimated more easily using lubricated squeezing flow, first developed by Chatraei et al. (58). This consists of the compression of a circular sample between two parallel plates, the surfaces of which are perfectly lubricated. The rate of deformation is given by the variation of the gap h between the plates:

$$\dot{\varepsilon}_B = -\frac{1}{2h(t)} \frac{dh}{dt} \tag{13}$$

If the sample radius is greater than the plate radius, the lubricated surface is constant and the stress is then:

$$\sigma = \frac{F}{S} \tag{14}$$

where F is the applied force. In this type of experiment, one can impose either a constant compression speed (thus the rate of deformation and the stress are not constant) or a constant rate of deformation [with an exponential displacement $h(t)$] or a constant force. Depending on the choice made, the interpretation is different, but all the results lead to a biaxial extensional viscosity:

$$\eta_B = \frac{\sigma}{\dot{\epsilon}_B} \tag{15}$$

Such characterizations have been proposed, for example, by Huang and Kokini (59) and Bagley (60) on wheat flour doughs and by Janssen et al. (24) and Kokelaar et al. (29) on flour and gluten doughs.

Because these methods are restricted to deformation rates much lower than those encountered in forming processes (mixing, sheeting), other techniques have been proposed, approximating a rheological point of view but closer to real processing conditions. Most is known about converging flows (61, 62). In convergent flow (for example, at the entry of a capillary), both shear and elongational deformations are present. Cogswell (63) was the first one to propose expressing the entrance-pressure drop as the sum of two parts, one due to the shear, the other to the elongation:

$$\Delta p_0 = \Delta p_{\text{shear}} + \Delta p_{\text{el}} \tag{16}$$

Δp_{shear} is computed using the lubrication approximation (which means neglecting elongation terms):

$$\Delta p_{\text{shear}} = \frac{2K}{3n \tan \alpha} \left(\frac{1 + 3n}{4n} \right)^n \dot{\gamma}_{\text{app}}^n \left[1 - \left(\frac{R_1}{R_0} \right)^{3n} \right] \tag{17}$$

where R_0 and R_1 are the radius at the entry and the exit of the convergent, respectively, α is the half-angle, n and K are the power law parameters [Eq. (3)], and $\dot{\gamma}_{\text{app}}$ is the apparent wall shear rate computed at the exit [using Eq. (2)]. The elongational viscosity is thus approximated by:

$$\eta_E = \frac{3n(\Delta p_0 - \Delta p_{\text{shear}})}{\dot{\gamma}_{\text{app}} \tan \alpha} \tag{18}$$

The elongational rate of deformation is not constant in this type of flow, and it increases from the entry to the exit. An average value can be estimated by:

$$\dot{\epsilon} = \frac{Q_v}{L} \frac{S_0 - S_1}{S_0 S_1} \tag{19}$$

where S_0 and S_1 are the entry and exit sections, respectively, and L is the length of the convergent. This method has been criticized by the arbitrary split of pressure drop into two independent terms [Eq. (16)]. However, in the field of synthetic polymers, some authors have shown that its results were in general agreement with those obtained on true elongational rheometers (40, 64). In cereal processing, this method was used by Battacharya et al. (52, 55) and Senouci et al. (65), who have shown well the difficulty of adapting such measurements in the case of cereal products.

2. Small Deformations

Although they may also be used for large-deformation studies, cone-and-plate or parallel-plate rheometers are frequently used to characterize the viscoelastic behavior in small-amplitude oscillatory shear (66). In these geometries, either the deformation (or the rate of deformation) or the stress may be imposed. In a controlled-strain rheometer, a plate oscillates with a frequency ω and an amplitude γ_0, and the resulting torque C_0 (and phase lag φ) is measured on the other plate. From these measurements, one can deduce the values of the viscous and elastic parts of the complex modulus G^*:

$$G^* = G' + iG'' \tag{20}$$

where

$$G' = \frac{\sigma_0}{\gamma_0} \cos \varphi, \qquad G'' = \frac{\sigma_0}{\gamma_0} \sin \varphi, \qquad \sigma_0 = \frac{3C_0}{2\pi R^3}$$

The loss angle, usually expressed through the loss tangent, $\tan \varphi = G''/G'$, characterizes the general behavior: $\varphi = 0$ corresponds to an elastic solid, whereas $\varphi = \pi/2$ corresponds to a viscous fluid.

All the preceding equations are defined only in the linear domain, where the moduli G' and G'' are independent of the deformation. The determination of the linear domain is the first step in this type of rheological characterization. Afterwards, it is usual to perform, in this linear domain, frequency sweep measurements, by imposing a range of frequencies, generally between 10^{-1} and 100 rad/s. The mechanical spectrum of a viscoelastic molten polymer over a wide range of frequencies is presented in Figure 5. At low frequencies (terminal zone), G'' is higher than G' and the viscous liquid behavior is dominant. In this portion, the theoretical slopes are 1 for G'' and 2 for G'. At intermediate frequencies, we observe a crossover between G' and G'', which shows a shift from a "viscous fluid" toward an "elastic solid." Afterward, we arrive in the rubbery region, where the storage modulus is quite independent of the frequency. At the highest frequency, glassy behavior is exhibited (67).

For thermorheological simple fluids, measurements at different temperatures may be superimposed to obtain a master curve. This principle of time–

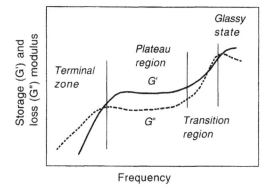

Figure 5   Typical mechanical spectrum of a molten polymer on the whole range of frequency, in a log-log scale.

temperature superposition may be applied to all the material functions that depend on time. For example, the complex modulus at temperature T may be deduced from the values at T_{ref} by:

$$G^*(\omega a_T, T) = G^*(\omega, T_{ref}) \tag{21}$$

where a_T is the shift factor, depending on the temperature. This principle also holds for large deformations. For example, the shear viscosity at any temperature may be obtained by:

$$\eta(\dot{\gamma}, T_{ref}) = \frac{\eta}{a_T}(\dot{\gamma}a_T, T) \tag{22}$$

The value a_T is usually described through an Arrhenius law or a WLF (Williams–Landel–Ferry) law (67). Other shift factors may be defined, for example, to account for a solvent or plasticizer concentration (68). These superposition principles cannot be applied any more when the temperature or the plasticizer modify the structure of the product (69, 70).

From the mechanical spectrum, different parameters can be obtained. For example, the complex viscosity is defined as:

$$\eta^* = \eta' - i\eta'' \tag{23}$$

with

$$\eta' = \frac{G''}{\omega} \quad \text{and} \quad \eta'' = \frac{G'}{\omega}$$

The Cox–Merz rule (71) implies that the variation of complex viscosity η^* as a function of frequency ω is similar to that of apparent viscosity η with the shear rate $\dot{\gamma}$ (measured in large deformations). This rule is verified well by many synthetic polymers, but much less by biopolymers (69, 72, 73), generally because large deformations modify the structure of the products.

Other viscoelastic parameters, such as complex compliance (the reciprocal of the complex modulus) or the spectrum of relaxation times can be deduced from oscillatory shear experiments. Similarly, creep-recovery experiments and the study of nonlinear viscoelastic properties can be of interest for characterizing cereal products. We refer to the literature (74–76) for complementary information on these topics.

C. Specificity of Biopolymers

All the preceding techniques were generally developed for applications involving synthetic polymers. As explained in Sec. I, applications to biopolymers and cereals are usually more difficult. In this section, we just want to illustrate the difficulties encountered in measuring the rheological behavior of cereal products, in both large and small deformations, by examples taken in the domains of starch extrusion and biscuit dough mixing.

1. Large Deformations

Knowledge of the viscous behavior of a molten starch is of vital importance in mastering the extrusion process and controlling starch transformation, bubble expansion, and product quality. However, in a capillary rheometer, there is no mechanical treatment, and thus in this type of rheometer a starchy product cannot be in the same state as in an extruder. Consequently, a classical capillary rheometer is not appropriate for studying the viscosity of molten starches. Two alternatives can be found. The first one is to use a preshearing rheometer, in which a controlled thermomechanical treatment can be applied to the product before the rheological measurement. The Rheoplast (Courbon, Saint Etienne, France) combines a cylindrical shearing chamber (in which the product can be treated at a fixed shear rate, defined by the rotation of the inner piston, during a fixed time duration) and a classical capillary rheometer. It was used for studying the viscous behavior of molten starches (77) and establishing the relationships between starch transformation and specific mechanical energy (SME) (78). The second alternative is to fix a slit rheometer or a capillary rheometer at the exit of an extruder and to characterize directly the extruded product. This was the method largely used in the literature (79–82). However, if it is very easy to define an apparent viscosity at a fixed shear rate, it is more painful to obtain a whole flow curve. Effectively, in this case the flow rate must be varied over a wide range. Unfortunately, by varying the flow rate, the thermomechanical history along the extruder is also modified, and thus the different measurement points would correspond to products with different structures. Consequently, the measurements obtained in this way are hardly reliable. It is possible to improve the technique by simultaneously modifying flow rate and screw speed, to maintain the specific mechanical energy constant and likewise the starch transformation (83). But this leads to long and tedious experiments.

To overcome this problem, we have proposed use of a slit die with twin channels (68). One channel is used for the measurement, the second one to bypass a part of the total flow rate. This allows one to vary the flow rate in the measuring section and to keep the extrusion conditions identical, likewise the product transformation. This system, called Rheopac, was extensively used for characterizing the behavior of different starches and starchy products (54, 69, 84, 85). An example of flow curves obtained in the case of maize starch is presented in Figure 6. For two SME levels leading to different levels of macromolecular degradation, as assessed by intrinsic viscosities values, we clearly obtain two different flow curves. Operating with a single channel at constant screw speed would have led to the "apparent" flow curve represented by the dotted line in the figure, because of the corresponding variation of SME and thus of degradation (SME decreases when increasing feed rate, thus decreasing degradation and increasing viscosity). The main part of the results presented in Sec. III.B have been obtained using

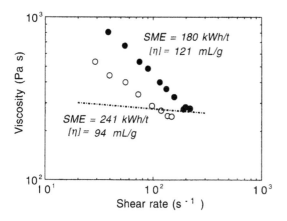

Figure 6   Shear viscosity of maize starch obtained by the Rheopac system for two processing conditions [$MC = 0.17$, $N = 200$ rpm, $Q = 36$ (●) and 29 (○) kg/h] leading to different levels of degradation. The dotted line represents the results obtained by varying the feed rate at constant screw speed without any energy control. (Adapted from Ref. 68.)

this technique. The use of a sidestream valve at the end of the extruder (55) can be considered an equivalent technique.

When the processing conditions are not constant or not accounted for in the measurements, inadequate results are obtained. For example:

Negative values of the power law index (51)
Inconsistency between results obtained with different rheometers (86, 87)
More generally, a wide dispersion of the values reported in the literature (88)

2. Small Deformations

The characterization of a dough in small-amplitude oscillatory shear presents serious experimental problems. For example, the presence of fat (up to 20% for some biscuit doughs) tends to induce slip at the wall. To avoid this phenomenon, it may be necessary to use grooved plates. The lateral free surface of the sample must be covered by a thin layer of silicon oil (89) or vaseline (73), to limit water losses during the measurements. Rest times after mixing play an important role. The complex modulus generally increases with rest time, which is usually explained by water reorganization (89, 90). This necessitates that one strictly control the measuring times and follow an accurate procedure, from the beginning of mixing to the end of measurement. Despite these precautions, the reproducibility is not always perfect (for example, on the absolute values of the moduli G' and G''), and it is difficult to know whether this problem originates from some heterogeneity in the mixer, from the reproducibility of the mixing process, or from the reproducibility of the measurement itself. Phan-Tien et al. (73) reported variations of 10% for wheat flour samples coming from different batches, for the same mixing conditions. Baltsavias et al. (90) found similar results on short doughs and insist on the strong influence of measurement temperature, which must be kept constant at $\pm 1\%$. To overcome these difficulties, the results are sometimes expressed as a ratio of moduli (tan $\varphi = G''/G'$) or using normalized values [for example, $G'(\omega)/G'$ (100), where the storage modulus in a frequency sweep is normalized by its value at 100 rad/s].

III. ILLUSTRATIVE RESULTS

The purpose of this chapter is not to give an overview of the role of rheology in cereal processing but to focus on some examples that show how the knowledge of rheological behavior may help one to understand the mechanisms involved in the processes and to control the quality of the final product.

A. Viscoelastic Properties of Biscuit Doughs

1. Introduction

Biscuit dough is a complex material resulting from the mixing of various ingredients, such as wheat flour, water, sugar, and fat. During the mixing process, specific interactions between these constituents (and especially gluten proteins) develop, giving their structure to the dough. In the following, the viscoelastic behavior of a biscuit dough is characterized in small-amplitude oscillatory shear experiments. These measurements permit one to follow the development of the structure during the kneading process and to check the influence of the process parameters on dough quality.

2. Materials and Methods

The dough formula we studied corresponds to a semisweet biscuit. On a flour weight basis, it contains 18% fat, 35% sugar, and 2.4% leavening agents. Water content is 23%. All the doughs were made using an experimental horizontal-type mixer with a mixing capacity of 6 kg of dough. The various ingredients of the dough formula are preliminarily homogenized at low speed for 2 minutes, then mixed at high speed for a fixed time. Process parameters are the rotation speed (50–120 rpm), the mixing time (200–700 s), and the regulation temperature of the double jacket. After kneading, the dough is left at rest for 30 minutes before rheological measurement.

The rheological measurements were carried out at 20°C using a parallel-plate rheometer in oscillatory mode. To avoid slip at the wall, which may arise from the presence of fat in the formulation, grooved plates were used. The following tests were made:

Deformation sweeps, at controlled frequency, to quantify the linear domain
Frequency sweeps, in the linear domain, between 0.1 and 100 rad/s
Deformation loops, to characterize the dough structure (89)

At a fixed frequency, oscillatory measurements are made for increasing deformations, until a maximum value, then for decreasing deformations. The nonreversibility of the trajectories is characteristic of the structural changes experienced by the dough.

3. Results and Discussion

a. Linear Domain. As for bread doughs (7, 72, 91, 92), the linear viscoelasticity domain is extremely reduced. It can be seen in Figure 7 that this domain is below 0.02% deformation. Above this value, G' and G'' decrease rapidly, indicating a high sensitivity of the structure to the deformation. However, the torque

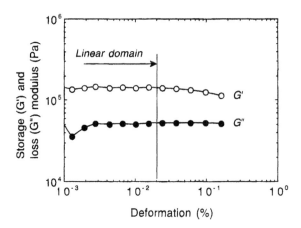

Figure 7   Linear viscoelasticity domain of a biscuit dough ($\omega = 1$ rad/s).

signal remains sinusoidal up to 0.2% and thus allows a correct quantification in this range of deformation, as explained by Berland and Launay (72).

b. Frequency Sweeps.   These are performed at 0.2% deformation, starting from the highest frequencies. An example of a result is presented in Figure 8. Over the whole range of frequencies, the storage modulus G' is higher than the loss modulus G''. Moreover, at low frequencies, both moduli tend toward an equilibrium value, independent of frequency. This behavior is characteristic of a weak physical gel (93) and is evidence of the structured nature of the dough.

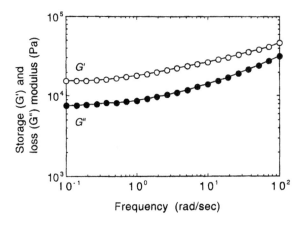

Figure 8   Mechanical spectrum of a biscuit dough ($\gamma = 0.2\%$).

The complex viscosity η^* obeys a power law without any Newtonian plateau at low frequency, which is characteristic of a yield stress fluid (94). From the data presented in Figure 8, we can deduce the following values: consistency $K \approx 15,000$ Pa-s, power law index $n \approx 0.29$.

    c. Deformation Loops. In a previous study (95), Contamine et al. proposed considering the crossover of G' and G'' in a deformation sweep as an index of the structural properties of the dough. Because this crossover occurs in the nonlinear domain, it seems more accurate to perform deformation loops (which means an increase in deformation, followed by a decrease and a return to initial conditions) and to characterize, close to the linear domain, the drop of modulus resulting from the imposed deformations.

    On Figure 9, trajectories during increasing and decreasing deformation are different, and the moduli have undergone a drop of approximately 60% (for a maximum deformation of 80%). This means that the interactions between the dough constituents are weak (as van der Waals or hydrogen bond interactions) and can easily be destroyed by small deformations. In fact, the drop in moduli directly depends on the maximum deformation experienced by the dough and increases regularly with the deformation.

    Figure 10 shows more explicitly the change in the drop of the moduli (that is, $\Delta G'/G'_0$ and $\Delta G''/G''_0$, where G'_0 and G''_0 are the initial values at 0.2% deformation) with the maximum deformation. Even for limited deformations, the drop is important (for example, 8% on G' for 2.5% deformation), which is evidence of the "weakness" of the structure. The drop continues to increase progressively and seems to reach a stabilized value around 60%, for deformations higher than 60–70%. We may observe that these changes are similar for G' and G''. If the

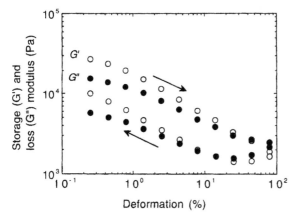

Figure 9   Deformation loop (up to 80%) of a biscuit dough ($\omega = 1$ rad/s).

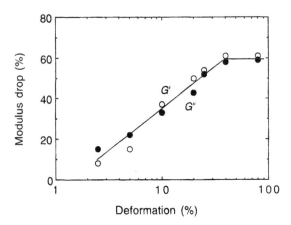

Figure 10 Drop of G' (○) and G'' (●) moduli as a function of the maximum applied deformation.

dough is allowed to rest for a certain time after the return trajectory, we observe a partial recovery of the viscoelastic moduli, which proves that the structural modifications are not totally irreversible (89). Similar changes have also been reported by Berland and Launay (72) on wheat flour doughs.

  d. Influence of Processing Conditions.   Such amplitude oscillatory shear measurements allow one to characterize the viscoelastic behavior of dough and its structural properties. We can apply these techniques for understanding the mixing process and the influence of the main parameters on the dough behavior. For example, we present the influence of kneading time after a preliminary mixing of the different ingredients during 3 minutes at low speed. Figure 11 shows the evolution of the loss tangent (tan φ = G''/G') with frequency. It may be observed that the principal modifications concern the behavior at low frequency. Just after premixing (mixing time = 0), we note an increase of tan φ below 1 rad/s. During the mixing (after, respectively, 1.5, 6, and 12 minutes), a plateau appears, which indicates the development of the viscoelastic structure of the dough. Similar results were reported by Amemiya and Menjivar (8) on a wheat flour/water dough. It has been shown that an increase in rotation speed or in mixer regulation temperature has the same qualitative effect as an increase in mixing time (96).

B. Shear Viscosity of Starchy Product Melts

By submitting starchy products (starch, flour, meal) to high shear stresses and temperatures, a macroscopic homogeneous molten phase is obtained due to starch

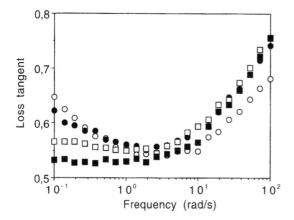

Figure 11 Change in loss tangent with mixing time (N = 120 rpm, T_r = 27°C): O = 0 min (start), ● = 1.5 min, □ = 6 min, ■ = 12 min.

melting. The term *starch melting* includes losses of both native crystallinity and granular structures (97), which may be observed by differential scanning calorimetry (DSC) and optical microscopy, respectively. Such phenomena are achieved on extruders, and their extent may vary according to the values of the many parameters from which results the versatility of this process: die geometry and screw profile, barrel temperature, screw speed (and feed rate in the case of twin-screw extruders), water addition. The aim of this section is not to make a survey of cereal extrusion-cooking (see general textbooks: Refs. 98, 99) but rather to show how the knowledge of rheological properties, mainly shear viscosity, may be helpful in better controlling the extrusion process and product quality. The following issues will be addressed in this section:

Determination of melt shear viscosity in relationship to molecular features and composition
Influence of melt viscosity on extrusion process variables
Role of rheological properties in expansion phenomena and texture acquisition

1. Molecular Features and Shear Viscosity Measurement

The difficulty of such measurements has already been emphasized, and we have explained why it is difficult to compare results obtained using various procedures. Thus, except as mentioned otherwise, results will be referred to as those obtained by the Rheopac system (68). Two major findings can be presented as a link with polymer science: the influence of molecular weight and the influence of chain branching.

a. Influence of Molecular Weight. This can be addressed by comparing viscosity curves for starches from different botanical origins having the same amylose/amylopectin ratio (amylose, linear, and amylopectin, highly branched, are the two main components of starch macromolecules). For this purpose, potato starch and a blend of maize starches of different botanical origin (2/3 waxy maize, with amylose content less than 1%, and 1/3 high-amylose maize, with amylose content around 70%) are extruded under different thermomechanical conditions, yielding either the same specific mechanical energy SME (case 1) or the same macromolecular degradation (case 2). The macromolecular degradation, and thus the molecular weight, are assumed to be described by the intrinsic viscosity [η]. Details on experimental procedure may be found in Della Valle et al. (84, 85). For the same value of SME (125 kWh/t), macromolecular degradation leads to a value of intrinsic viscosity lower for the maize starch blend ([η] = 95 mL/g) than for potato starch ([η] = 180 mL/g). This difference in molecular weight is directly reflected in the larger value of melt viscosity for potato (Fig. 12). Conversely, if mechanical treatments are performed in such a way that the same value of intrinsic viscosity may be obtained ([η] ≈ 100 mL/g; such result is obtained with an SME much higher for potato), then the viscous behaviors of both materials are very close. These results emphasize the importance of an accurate control of the SME when processing cereal products in an extrusion cooker. However, agricultural variability may lead to important differences of behavior when changing the supply of raw material. On the other hand, these results agree with

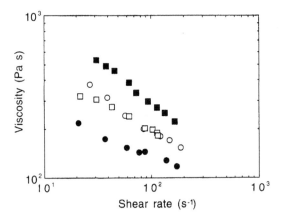

Figure 12 Shear viscosity of molten starches from maize blend (●, ○) and potato (■, □), either for the same specific energy (SME = 125 kWh/t, filled symbols) or the same molecular degradation ([η] ≈ 100 mL/g, open symbols).

the common idea of the influence of average molecular weight on shear viscosity (76).

   b.   Influence of Chain Branching.   In starch macromolecules, chain branching is reflected in the ratio of linear amylose and highly branched amylopectin. This influence may be analyzed by comparing the viscous behavior of waxy (less than 1% amylose) and high-amylose maize starch ($\approx$70% amylose). This was first done by Lai and Kokini (86), who found that higher shear stresses were generally generated, at the same shear rate, by the flow of molten starches containing large amounts of amylose. However, the conclusions and interpretation of their experimental viscous behaviors are not straightforward, since thermomechanical processing conditions were not always the same. A term called *degree of cook* had to be introduced to adjust a rheological model, which can reflect phenomena having antagonistic effects on viscosity. This corresponds to either granular disruption (increasing viscosity) or macromolecular degradation (decreasing viscosity). By using the Rheopac system under controlled thermomechanical conditions leading to complete granule disruption and crystal melting before the die, Della Valle et al. (84) have obtained flow curves of starchy materials with different amylose contents. They found satisfactory adjustment ($r^2 \geq$ 0.84) of the viscous behavior of these starches to the following empirical power law model:

$$\eta = K_0 \exp \left(\frac{E}{RT_a} - \alpha MC - \beta SME \right) \dot{\gamma}^{n-1} \tag{24}$$

where E is the activation energy, R is the gas constant, T_a is the absolute melt temperature, MC is the moisture content, SME is the energy provided to the product, and n is a linear function of moisture and temperature, mainly.

   Variations of the α and β coefficients as a function of composition chiefly show an increasing sensitivity to water content and mechanical treatment when amylopectin content increases. The higher influence of water on highly branched macromolecules is not clearly explained currently. This diluting action could perhaps be connected to a most significant plasticizing action of water, noticed by Lourdin et al. (100). The second trend may be related to the larger sensitivity of amylopectin to macromolecular degradation, due to its higher molecular weight (101, 102). As mentioned before, comparison between starches of different amylose contents is not easy because different thermomechanical histories have to be applied for reaching the same extent of disorganization. Therefore, it may be advisable to use the time/temperature superposition principle, extended to moisture (68). Such a method allows one to superimpose various flow curves obtained under different conditions of moisture and temperature, provided that mechanical degradation is not too different from one to another. By applying this principle to various measurements, it is clearly shown in Figure 13 that amylose leads to higher values of viscosity and a more pronounced shear-thinning behavior. This behavior may be due to a larger ability of linear molecules (compared to the compact conformation of amylopectin) to create entanglements, or to remaining single-helix associations, which could be unwound at high shear rates. Amylose molecules are also more likely to align in the flow at higher shear rates, compared to the highly branched molecules of amylopectin. This trend is all the most noteworthy because these curves are obtained for conditions under which weight average molecular weights M_w, measured by high-pressure size exclusion chromatography (HPSEC-MALLS), are very close (0.8 10^8 g/mole and 1.1 10^8 g/mole, for extruded high-amylose maize and waxy starches, respectively). Thus, the effect of average molecular weight differences may be discarded, and the difference in viscous behavior may actually be considered determined mainly as a consequence of differences in linear/branched structures.

Another factor classically studied in polymer science is polydispersity, which is supposed to decrease the shear viscosity at high shear rates (103). However, no literature data are actually available to confirm that starchy products may follow this general rule. This is due to the practical difficulties encountered in the measurement of molecular weight distribution and to the fact that mechanical treatment may influence polydispersity, concomitant with average molecular weight.

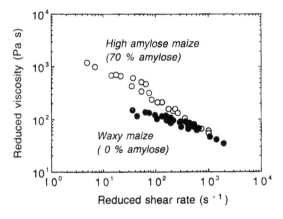

Figure 13 Reduced viscosity of molten starches with different amylose content as a function of reduced shear rate (T_p = 165°C, MC = 0.24, [η] = 95 mL/g). (Adapted from Ref. 84.)

Although the rheological behavior of complex blends during extrusion cannot be understood solely by some concepts of polymer science, those are helpful in showing the most important trends. The addition of sugar, salt, and other ingredients will usually slightly modify the shear viscosity (55). Generally, the influence of a minor component on melt shear viscosity can be taken into account by coefficients similar to α in Eq. (24), provided that its presence does not induce important structural changes during processing. Such changes have been inferred by Willett et al. (104), in the case of glycerol monostearate, the addition of which did not lead to a viscosity decrease. This behavior may be attributed to the creation of semicrystalline structures, including single helical arrangements of amylose, as suggested by Della Valle et al. (69), but the proof of their existence requires careful experiments, including heavy physical methods.

Another explanation could be the antagonistic effect of lubrication, i.e., decreasing viscosity and having protective effects on macromolecules, thus reducing chain splitting. The presence of proteins can hardly be treated straightforwardly. Experience teaches that starches are generally less easily processed than the flours from which they are extracted, which could be interpreted as a diluting effect of proteins. However, under specific conditions, at higher moisture content, for instance, a protein network can be formed by intermolecular cross-linking after denaturation. Under such conditions, significant viscosity increase may be expected, and rheological behavior may be considerably modified. Morgan et al. (105) have proposed to account for these changes by introducing a thermal-history term in the rheological model. However, this generalized model needs more background on structural modifications at the biochemical level to be more widely used (106). Food recipes also include enzymes, often used as processing aids, which lead to partial starch depolymerization. Such effects could be reflected by a term similar to the energetic one (βSME) in Eq. (24), as suggested by Tomas et al. (107).

More experiments are obviously needed to establish detailed rheological models of complex cereal products. But, as stated before, objective and accurate methods are now available to achieve them.

2. Influence of Shear Viscosity on Extrusion Variables

The question of how a change in rheological behavior affects the working conditions of the extruder is raised as soon as a mere change of water addition, for instance, occurs or, more generally, when a modification of the recipe is envisaged. However, a systematic study of the influence of the main rheological features on pressure, energy, and temperature would involve tedious experimental work. To overcome this difficulty, one might simulate the working conditions of the extruders using theoretical models of heat and mass transfers, which are now available for both single- and twin-screw corotating extruders (TSEs), the last being the one most employed in the food industry for cereal extrusion cooking. Such models also offer the possibility of simulating cases that cannot be experimentally achieved or of predicting the variations in variables that cannot be easily measured. All simulations are generally made for products having a viscosity described by the power law [Eq. (3) or (24)].

This is illustrated by Levine et al. (108), who simulated the transient behavior of TSEs, or by Mohamed and Ofoli (109), who predicted temperature profiles via a simple analysis of heat flow in the extruder. A similar approach was used by Chang and Halek (110) to compute temperature and the filling ratio for corn meal. In both cases, the influence of temperature on rheological behavior was taken into account only by an exponential term in the expression of the consistency K [Eq. (24)]. These authors also underlined the importance of shear viscous dissipation in temperature increase. The corresponding volumetric power can be expressed as:

$$\dot{W} = K\dot{\gamma}^{n+1} \tag{25}$$

This term is all the more significant because it can be related to properties reflecting product transformation, such as starch intrinsic viscosity and solubility (78). Tayeb et al. (111) proposed a model that allows one to compute the dissipated power for a TSE, in the case of power law fluids. Della Valle et al. (112) and Barrès et al. (113) have extended its validity to various starchy materials by relating the solubility of extruded products (starch and proteins) to the specific energy delivered by viscous dissipation, assuming that the viscous behavior of the product could be described by a power law.

This first model was further developed and improved to lead to commercial software, called Ludovic®. The theoretical basis has been fully described by Vergnes et al. (114), who also give examples of applications in the field of polymer blending (115) and reactive extrusion (116). Based on a local one-dimensional approach, this software allows one to compute, all along the screws, the change in the main flow parameters, such as pressure, temperature, residence time, shear rate, and filling ratio. Due to its wide range of uses and to the importance of rheological features in the extrusion processing of cereals, we have chosen as an example the prediction of the influence of the variations of the power law parameters on the extrusion variables, especially the computed specific energy.

The following results have been obtained for a power law fluid, following Eq. (24). Standard values are the following: $K_0 = 3.25 \times 10^5$ Pa s^n, $E/R = 4,500$ K, $\alpha = 0.15$, $\beta = 10^{-9}$ (J/m^3)$^{-1}$. These values are close to the ones found by Vergnes and Villemaire (77) for maize starch, and they are also representative of values given by other authors (84, 102, 117) for various low-hydrated starchy materials. Simulations were made for a Clextral BC45 twin-screw extruder of 50-cm length, with a screw arrangement including a 5-cm-long reverse-screw element (RSE), situated 5 cm before the Rheopac system (68). Processing conditions are the following: screw speed: $N = 200$ rpm; feed rate: $Q = 30$ kg/h; total moisture content: MC = 20%; barrel and die temperature: $T_b = 155°$C. We consider adiabatic conditions toward the screws and heat transfer toward the barrel, with a heat transfer coefficient equal to 900 W = m^{-2}K^{-1}. Only one rheological parameter is modified at a time.

By increasing the consistency K_0 from 0.22×10^5 to 10.8×10^5 Pa s^n, the pressure level is considerably increased, from 0.5 to 2.2 MPa at the entrance of the RSE and from 1.7 to 12.1 MPa at the die entry, without significantly changing the pressure drop in the reverse-screw element (Fig. 14a). Pressure

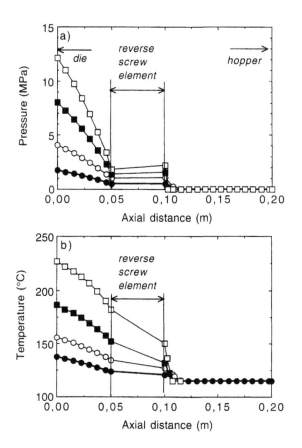

Figure 14 Variations of computed (a) pressure and (b) temperature profiles with the consistency K_0 of a power law fluid: ● $K_0 = 0.22 \times 10^5$, ○ $K_0 = 0.81 \times 10^5$, ■ $K_0 = 3.25 \times 10^5$, □ $K_0 = 10.8 \times 10^5$.

buildup before the RSE is steeper for more viscous fluids. Concomitantly, temperature profiles show a nearly linear increase from the RSE (before which product temperature is 115°C, i.e., the chosen melting temperature) to the die, where values range from 138 to 229°C (Fig. 14b). The increase in consistency leads to an increase in viscous dissipation, as reflected by temperature profiles largely above the barrel temperature for higher values of K. At such high temperatures, viscosity is decreased, which limits the values of die pressure. For lower consistencies, viscous dissipation is not large enough for the material temperature to reach barrel temperature, which also underlines the slight influence of thermal conduction. It also may be shown that an increase in thermal, water, or mechanical sensitivities [increase in E/R, α, or β in Eq. (24)] would lead to the same variations as a drop in K_0.

The role of the power law index on pressure and temperature profiles is illustrated in Figure 15. The increase in the flow index from 0.2 to 0.8 leads to an increase in the die pressure from 5.5 to 13.8 MPa but to a decrease at the RSE entrance from 2 to 0.5 MPa. The first result is due to the increase of viscosity with flow index, whereas the second one is explained by the larger pumping efficiency of screws in the case of fluids having higher flow indices (114). The direct consequence is to modify the length of filled screw channels. The temperature at the die entrance varies from 168°C to 252°C, which is due to the influence of viscous dissipation on the temperature profile, this phenomenon being more important as flow index increases [Eq. (25)].

Ludovic® is also able to predict variations of the length filled with material before the RSE (L_f) and of the dissipated specific energy (CSE). The knowledge of L_f and CSE is useful since filled length affects power requirements, and specific energy is linked to starch destructurization. As a matter of current interest, average residence time may also be computed, but its value in the standard case is affected more by process parameters (screw speed and geometry, feed rate) than by the sole change of n or K values (118, 119). For comparison purposes, we also report in Figure 16 experimental results obtained for starches with various amylose contents, for which detailed results are given by Della Valle et al. (84). Such experimental cases are not easy to find since, for molten starches, K and n generally vary antagonistically with process variables such as temperature, energy, and even moisture content. In this case, changes in n and K values were obtained by modifying the amylose content of starch, so that one parameter (n or K) may vary, with the others remaining approximately constant. Numerical values have been normalized, taking the basic case as reference ($n = 0.4$, $K_0 = 3.25 \times 10^5$ Pa s^n). Normalized values of die pressure and specific energy follow exactly the same trend, for all K and n values tested, which means that shear viscosity changes affect these variables in the same way. Therefore, for clarity's sake, only one of the two variables is represented.

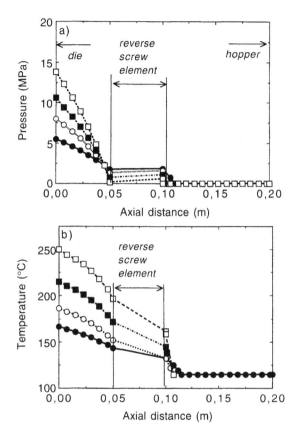

Figure 15 Variations of computed (a) pressure and (b) temperature profiles with the flow index n of a power law fluid: ● $n = 0.2$, ○ $n = 0.4$, ■ $n = 0.6$, □ $n = 0.8$.

The decrease in L_f with the increase in consistency confirms the preceding observed trend. As expected, specific energy increases with consistency, due to viscous dissipation (Fig. 16a). The agreement between predicted and experimental values is remarkable, which suggests a path for the use of Ludovic® to help in predicting product transformation when varying recipes, provided that viscous behavior may be determined beforehand. When increasing the power law index, filled length decreases in agreement with previously observed results. Experimental values of die pressure P_d also follow the same trend as the computed ones.

These results show that modifications of rheological behavior may significantly affect extruder working conditions, through pressure and temperature changes, for instance. Although the rheological behavior of the product is chang-

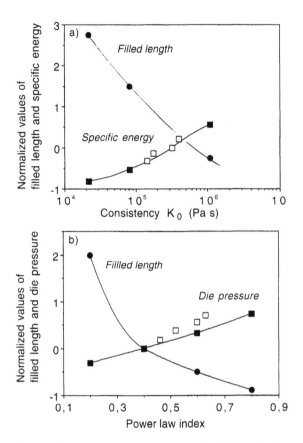

Figure 16 Variations of (a) computed filled length and specific energy with consistency and (b) of computed filled length and die pressure with power law index. Open symbols are experimental values of SME and die pressure.

ing largely during the process and can hardly be defined by a simple power law, it also enhances the efficiency of theoretical tools, such as Ludovic® software, to estimate these changes. In particular, the agreement between measured and predicted values of variables defining the final state of the product at the die offers an opportunity to improve the control of the final texture of the extruded product.

3. Expansion and Texture Formation

Expansion, sometimes called *puffing*, is the phenomenon by which foods, mainly cereals, will acquire a porous structure, like a solid foam, due to transient heat

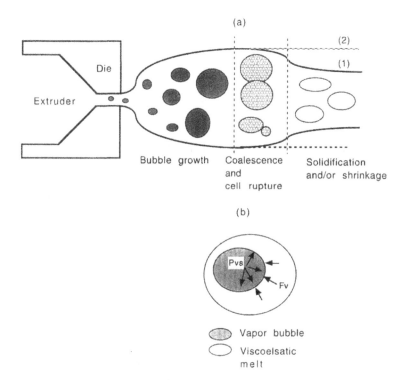

Figure 17 Schematic representation of (a) the flash expansion phenomenon at the die exit of an extruder, (1) with or (2) without shrinkage, and (b) the bubble growth model. (Adapted from Ref. 54.)

and vapor transfers. Expansion can be due to the vaporization of the water contained in the product, but also sometimes to injected gases, carbon dioxide, for example. Pore size and distribution, thickness, and mechanical properties of the wall material will define, in turn, the texture of the product. This texture is strongly influenced by the way the pores are generated. Although it is not the only process leading to such products, let us focus on the direct (or flash) expansion of a starch melt at the die outlet of an extruder. This attractive phenomenon was first studied like a black box, and the influence of many extrusion parameters (barrel temperature, moisture, die geometry, recipe), sometimes contradictory, have been summarized by Colonna et al. (88). However, in the last ten years, more attention has been paid to the basic phenomena that control expansion: nucleation, bubble growth and water evaporation, coalescence, and shrinkage (54, 120–123) (Fig. 17). The aim of this discussion is to understand the influence of rheological properties on the expansion process and to suggest experimental routes to better control it.

The acquisition of a porous structure is generally evaluated by the volumetric expansion index VEI (124), which is related to the porosity ε by:

$$\text{VEI} = \frac{\rho_m}{\rho_e} = \frac{v_m}{v_e} = \frac{1}{1 + \varepsilon} \tag{26}$$

where ρ_m, v_m and ρ_e, v_e are the densities and specific volumes of the melt and the extrudate, respectively. As suggested by Park (125) and Launay and Lisch (126), expansion has two components, defined by radial (SEI) and longitudinal (LEI) indices:

$$\text{VEI} = \text{SEI LEI} \tag{27}$$

where SEI is defined as the ratio of the cross sections of extrudate and die. Each index may be defined at the stage of development of expansion featured in Figure 17a, depending on water content and the storage process.

Very little information is available about nucleation as applied to texturization, but there is common agreement that every inhomogeneity in the melt (solid particle, nondisrupted starch granule), but also injected gas (127), provides a starting point for bubble growth, above a critical radius value (128).

The next stage, i.e., bubble growth in a viscoelastic matrix, is still being widely studied due to the importance of foam production in many other industrial areas. Confusion between expansion and extrudate swell, due to the elastic properties of the melt, is sometimes made, but the latter phenomenon is of interest only when no expansion takes place, i.e., for a temperature low enough to avoid water vaporization, in the case of pellet production, for instance. Otherwise, the recovery of elastic strains usually leads to a sectional index lower than 2 (for length-to-diameter ratios greater than 10) (76). This value is low enough for extrudate swell to be discarded at this stage. However, there is a general agreement that rheological properties play an important role in bubble growth. Kokini et al. (120) first suggested the following relationship:

$$\text{VEI} \approx A \frac{p_v(T)}{\eta} \tag{28}$$

where A is a constant, p_v is the saturating vapor pressure at the extrudate temperature T, and η is the shear viscosity of the melt leaving the die (Fig. 17b). The antagonistic role of shear viscosity on expansion was first suggested by Vergnes et al. (78): Reduction of starch expansion at the die exit of a preshearing capillary rheometer was observed when shear rate decreased, thereby increasing shear viscosity.

This relationship is illustrated by some experimental results obtained for various moisture contents (Fig. 18a) and various amylose contents of the melt (Fig. 18b). These results suggest that, if Eq. (28) allows one to take into account the effects of temperature and melt viscosity, the influence of water content and amylose content cannot be reflected merely by shear viscosity. This discrepancy may be explained in different ways. First, Eq. (28) is adapted from a bubble-growth model, proposed by Amon and Denson (129), who considered a Newtonian matrix. However, we have seen that starchy melts exhibit highly non-Newtonian behaviors. Moreover, it is probable that elongational viscosity should be taken into account, which may not rank in the same way as shear viscosity,

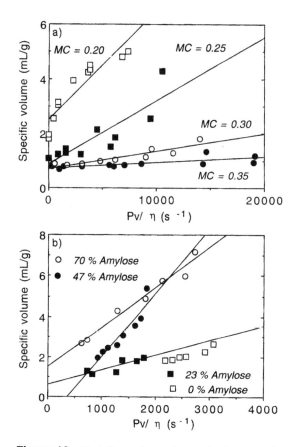

Figure 18 Variations of extrudate volumic expansion with p_v/η ratio for (a) various moisture contents (data from Ref. 120) and (b) various amylose contents (data from Ref. 84).

chiefly according to amylose content. Secondly, VEI is measured at the end of the expansion phenomenon. Consequently, Eq. (28) discards the stages of disproportionation, coalescence, and wall rupture, in which water content may still play an active role.

Disproportionation, coalescence, and cell wall rupture are not well understood presently. However, some studies from the field of synthetic polymers suggest the importance of rheological properties in these phenomena. Neff and Macosko (130) showed that the drop of the normal force, measured on a rotational parallel-plate rheometer equipped with an adapted cup, is due to the cell ruptures of the polyurethane-foaming blend. This is certainly the part of the whole expansion process where elastic properties play the most active role.

The final stages, including shrinkage and solidification, both involve expansion reduction. The reduction of expansion after cooling was first noticed by Guy and Horne (131), who reported variations of SEI by a factor of 2 for moisture content higher than 20% in the case of extruded maize grits. More recently, Pasquet and Arhaliass (132), using an online video system, showed that SEI could decrease by the same order of magnitude, using various die geometries. Shrinkage also explains why longitudinal (LEI) and radial (SEI) indices are often negatively correlated (123), since for constant volumetric expansion, according to Eq. (27), the decrease in SEI has to be balanced by an increase in LEI. Fan et al. (122) suggested two interpretations to explain shrinkage: (a) contraction occurs by release of stored elastic energy and (b) bubbles shrink as the vapor pressure inside the bubbles decreases below atmospheric pressure. Same authors suggested that both mechanisms cease when the extrudate temperature gets lower than $T_g +$ 30°C, T_g being the calorimetric glass transition of the material. The higher the glass transition, the less pronounced is shrinkage. Thus, the strong dependency of T_g on water content explains the negative role of water content on volumetric expansion. The existence of such a limit temperature is confirmed by the results of Della Valle et al. (54), who measured a critical temperature T_{cr} for bubble-growth start (or stop) in a thermostated oil bath (Fig. 19). The difference $T_{cr} - T_g$ is also a function of flow conditions (shear rate) and macromolecular structure (amylose content), which, once again, underlines the role of rheological properties in this phenomenon.

These assumptions are well taken into account in a model for bubble growth and shrinkage proposed by Fan et al. (122), a simplified schema for which is presented in Figure 20. It includes the influence of rheological parameters appearing in Eq. (24). Fan et al. have shown that the most significant influence was due to the thermal dependency of shear viscosity (E/R), whereas flow index n was not found very influential.

Although methodological improvements are necessary to throw light on the early stage of nucleation, this brief survey shows that the knowledge of viscous behavior and transition temperatures helps much in understanding expan-

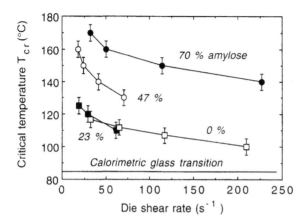

Figure 19 Variations of bubble-growth critical temperature with flow conditions for extrudate with various amylose content. (Adapted from Ref. 54.)

sion. Further progress in this field, like the determination of elongational properties, would give additional opportunity to better control this phenomenon, of crucial importance for final product texture.

C. Extensional Behavior of Bread Dough

Rheology has been used for a long time to study the behavior of wheat flour dough and the breadmaking process. The need to predict wheat performance and formulation parameters first led to the development of empirical methods, de-

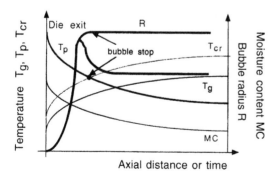

Figure 20 Basic mechanism for expansion and shrinkage of bubbles in extruded starchy products. (Adapted from Refs. 54, 121, and 123.)

scribed in Sec. II.A. Then the spreading of objective rheological methods favored the use of dynamic oscillatory shear measurements at low strains (less than 1%) to better evaluate the dough structure (11). By application of polymer science concepts, these studies have allowed us to underline the role of protein composition and molecular features in the gluten network (133, 134) and the influence of other components, like water (135). In the last ten years, there has been growing attention to performing objective rheological measurements in conditions of strain and strain rate more similar to those encountered in breadmaking. This is also due to a better understanding of the various phenomena that occur at the different stages of breadmaking: mixing, fermentation, and baking (13). If simple shear deformation may dominate during the first stage, elongational strain becomes more significant during the two others, mainly because of gas-cell creation and growth, the dynamics of which will be partly controlled by the elongational properties of the dough matrix.

1. Elongational Tests and Results

a. Uniaxial Deformations.   Various systems were employed to measure uniaxial elongational properties. De Bruijne et al. (136) used a mercury bath extensometer in which the sample was floating, to discard gravity effects. This sample had a dumbbell shape to minimize end effects on strain. Constant strain rates ($\dot{\varepsilon} = 10^{-4}$–$10^{-2}$ s^{-1}) were obtained by modifying exponentially the speed of the Instron machine head. Although the test allowed them to discriminate doughs with various breadmaking performances, the final bread volume could not be predicted, which was attributed to the fact that the deformation was uniaxial. Morgenstern et al. (137) measured average elongation and elongation rates in the case of a dough sheet deformed by means of the crosshead of a universal testing machine. In these types of experiments, correction has to be made on the sample section area to compute the stress σ from the measured force F:

$$\sigma = \frac{F}{A} \tag{29}$$

where section A:

$$A(t) \cdot h(t) = A_0 h_0 \tag{30}$$

where A_0 and h_0 are, respectively, the initial cross-sectional area of the sample and its initial length and h is the sample length at time t.

Stress σ, deformation ε, and deformation rate $\dot{\varepsilon}$ are all functions of time, defined by:

$$\sigma(t) = \frac{F(t)}{A_0} \exp \varepsilon(t) \tag{31}$$

$$\varepsilon(t) = \ln \frac{h(t)}{h_0} \tag{32}$$

$$\dot{\varepsilon}(t) = \frac{1}{h(t)} \frac{dh(t)}{dt} \tag{33}$$

Depending on the machine used, one can impose either a constant force [$F(t) = F_0$] or a constant speed [$(dh(t)/dt = V_0$] or a constant strain rate [$\dot{\varepsilon}(t) = \dot{\varepsilon}_0$]. A transient elongational viscosity $\eta_E(t)$ can be always defined by the ratio $\sigma(t)/\dot{\varepsilon}(t)$. De Bruijne et al. (136) and Morgenstern et al. (137) found that, for flours at 45% moisture content but with differing protein content, this viscosity followed a power law ($\eta_E(t) = K\dot{\varepsilon}^{n-1}$) with a power law index n of 0.24. Strain hardening was observed, which means that the stress increased with strain more rapidly than would be predicted by the theory of linear viscoelasticity.

Schweizer and Condé-Petit (57) used a Meissner caterpillar-type elongational rheometer, which allows one to operate at constant strain rate, controlled by a video system. The measurement of the traction force directly provides the elongational viscosity as a function of time (or deformation). They observed strain-hardening behavior above a critical strain, equal to 0.3 and independent of the elongation rate. The strain-hardening index ($d \ln \sigma/d \ln \varepsilon$) varied from 3 to 7, according to the elongation rate (0.03–0.3 s^{-1}).

b. Biaxial Deformations. The most widely employed method for measuring the biaxial properties of a dough is lubricated squeezing flow, presented in Sec. II.B.1.c. In this case, for a constant compression speed V_0, the strain and strain rate are linked by the following relationship:

$$\dot{\varepsilon}_B = \frac{V_0}{2h_0} \exp 2\varepsilon_B \tag{34}$$

where the biaxial deformation is defined as:

$$\varepsilon_B = -\frac{1}{2} \ln \frac{h(t)}{h_0} \tag{35}$$

Janssen et al. (24) and Kokelaar et al. (29) have applied this technique to various flours and gluten doughs, with different compression speeds in order to separate the effects of strain and strain rate. Both types of materials presented the same characteristics, also similar to the ones described before:

For a constant total strain, elongational viscosity decreased with strain rate, according to a power law ($n = 0.11$–0.30) (Fig. 21a).

For a constant strain rate, they all exhibited strain hardening, the index of which ($d \ln \sigma/d\varepsilon_B$) varied in the range 1.3–3 (Fig. 21b).

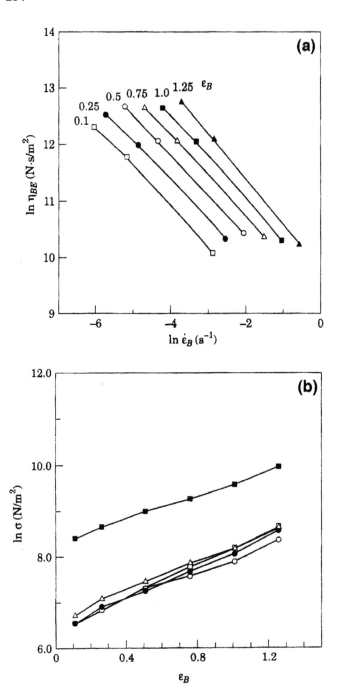

The same kind of observations were done by Launay and Buré (26), as well as by Dobraszczyk and Roberts (12), using the Alveograph test (see Sec. II.A). However, in this last work, the strain-hardening index was derived directly from the stress–strain experimental curve, i.e., not for constant strain rates, which makes difficult the comparison of the values obtained with those from other works.

2. Application to the Breadmaking Process

Kokelaar et al. (29) found that gluten doughs behaved more differently than wheat flour doughs, according to wheat origin. At first glance, this result challenges the possible use of extensional tests for predicting flour breadmaking performance. However, this discrepancy may be due to the way the samples were prepared, since various amounts of water had to be added to wheat flour to reach a 500-BU consistency in the Farinograph test. For flours, stress was found to increase when the temperature increased from 20 to 55°C, which may be due to the beginning of starch granule swelling, since the opposite trend was noticed for gluten dough. Arabinoxylan solubilization and the occurrence of enzyme activities should not be neglected either. A significant increase in the strain-hardening index was also found when ascorbic acid was added (57).

Gas retention is important for the stability of the foam created by dough fermentation. Indeed, cell-wall rupture should not occur too early before solidification. Van Vliet et al. (138) proposed using extensibility and the strain-hardening characteristics of the dough in biaxial extension as a criterion for gas retention, expressed by:

$$\left[\frac{d \ln \sigma}{d\varepsilon_B}\right]_{\dot{\varepsilon}_B=\text{cst}} + \left[\frac{d \ln \sigma}{d \ln \dot{\varepsilon}_B}\right]_{\varepsilon_B=\text{cst}} \left[\frac{d \ln \dot{\varepsilon}_B}{d\varepsilon_B}\right] > 2 \tag{36}$$

The first term of the left-hand member is the strain-hardening index (m), whereas the first part of the second term is the elongational flow index (n). Van Vliet et al. (138) suggested that the last term ($d \ln \dot{\varepsilon}_B/d\varepsilon_B$) could equal -3 during fermentation and that this value would be closer to 2 during baking. A summary of the results obtained by this group is presented in Figure 22. Regions below the value of 2 indicate poor gas retention properties of the dough, either during fermentation (for $m - 3n$) or during baking (for $m + 2n$). Except for some values

Figure 21 Examples of experimental results of wheat flour doughs in biaxial extension: (a) Viscosity as a function of elongation rate, for different values of total strain; (b) stress as a function of strain, for different values of elongation rate. (Reprinted from Ref. 29, with permission.)

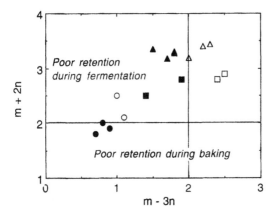

Figure 22 Summary of the gas retention criterion results (Eq. 36). Open symbols corre-spond to good baking quality, closed symbols to poor baking quality. ●, ○: flour dough (Ref. 29); ■, □: gluten dough (Ref. 29); ▲, △: flour dough (Ref. 24). (Data from Refs. 24 and 29.)

obtained at 50°C, the criterion seems able to discriminate varieties of gluten and wheat flour doughs.

Other information could be derived from results gained in extensional mea-surements. Recently, Launay and Bartolucci (139) showed that the analysis of stress relaxation curves obtained after a squeezing-flow or alveograph test could help in predicting some dimensional characteristics of biscuits, for which oven rise is not as important as for bread. If gas retention is linked to cell-wall rupture, then the strain at break should also play an important role. Opposite variations of strain at break and strain rate have been found, according to the flour quality (136). The relevance of such measurement is enhanced by the fact that, near the rupture, dough cannot be considered a continuous homogeneous medium, but rather a composite material (protein network–starch granules–water), the me-chanical behavior at rupture of which is influenced by inhomogeneities.

Several techniques are available to measure the elongational properties of dough. However, their extensive use as tools for improving the knowledge and control of the breadmaking process require progress in two directions. First, com-parison studies should be made between these different tests to solve some techni-cal points, such as the decoupling between the effects of strain and strain rates. Sample preparation is the first difficulty to overcome in this way. In some cases (biscuits, for instance), these rheological data could be directly used for modeling the dough-shaping process. But generally, and this is the second direction, the conditions of use, especially temperature, should be closer to those of the baking process, in order to provide results in actual relation to the performance of the material. This would provide useful information to improve the global modeling of heat and mass transfers during baking, which are presently limited to a global scale (140), by taking into account local phenomena, such as cell-wall rupture.

IV. CONCLUSIONS

Rheology is undoubtedly a major issue in cereal processing. The knowledge of rheological properties in connection with structural features at different scales (macromolecular structure, network and complexes, macroscopic properties) is absolutely necessary to understand the elementary mechanisms involved in the processes and thus to optimize processing conditions, equipment size, and final end-use properties. In the last 20 years, a progressive shift has been observed from empirical techniques (which are still widely used and offer interesting possibilities) to more objective measurements, partially through input from synthetic polymer science. This shift is all the more difficult because cereal products are very complex materials for which classical approaches are sometimes unsuitable and consequently that often need the development of new methodologies.

Another important issue is the rise of numerical modeling and the use of continuum mechanics to compute the flow parameters in processing operations. Extrusion cooking has already been approached by such methods, but other processes may also be characterized: bread baking (140), pasta extrusion (141), mixing (142), dough sheeting (143), and dough extrusion (144). For these applications, it is obvious that the rheological properties of the products one wants to study have to be perfectly known.

Although less investigated than biopolymer solutions, the viscous shear behavior of molten cereals and doughs is now beginning to be well understood. Viscoelastic properties are sometimes more difficult to measure, mainly at high temperatures and for large deformations. The characterization of elongational properties, which play a major role in some cereal-forming processes, remains a real challenge for the future.

REFERENCES

1. RK Schofield, GW Scott-Blair. The relationship between viscosity, elasticity and plastic strength of soft materials as illustrated by some mechanical properties of flour doughs. Proc R Soc 141:72707–72719, 1933.
2. PJ Frazier, CS Fitchett, PW Russell Eggitt. Laboratory measurement of dough development. In: H Faridi, ed. Rheology of Wheat Products. St. Paul, MN: AACC, 1990, pp 151–175.
3. R Andersson, M Hamaainen, P Aman. Predictive modelling of the bread-making performance and dough properties of wheat. J Cereal Sci 20:129–138, 1994.

4. AH Bloksma. Rheology of breadmaking. 8th International Cereal and Bread Congress, Lausanne, Switzerland, 1988.

5. VF Rasper. Dough rheology at large deformations in simple tensile mode. Cereal Chem 52:24–41, 1975.

6. JA Menjivar, C Kivett. Rheological properties of dough in shear deformation and their relation to gas holding capacity. 71st AACC Annual Meeting, Toronto, 1986.

7. JA Menjivar. Fundamental aspects of dough rheology. In: H Faridi, JM Faubion, eds. Dough Rheology and Baked Product Texture. New York: Van Nostrand, 1990, pp 1–28.

8. JI Amemiya, JA Menjivar. Comparison of small and large deformation measurements to characterize the rheology of wheat flour doughs. J Food Eng 16:91–108, 1992.

9. CA Stear. Handbook of Breadmaking Technology. Amsterdam: Elsevier, 1990.

10. JE Bernardin, DD Kasarda. Hydrated protein fibrils from wheat endosperm. Cereal Chem 50:529–536, 1973.

11. JM Faubion, RC Hoseney. The viscoelastic properties of wheat flour doughs. In: H Faridi, JM Faubion, eds. Dough Rheology and Baked Product Texture. New York: Van Nostrand, 1990, pp 29–67.

12. BJ Dobraszczyk, CA Roberts. Strain hardening and dough gas cell-wall failure in biaxial extension. J Cereal Sci 20:265–274, 1994.

13. AH Bloksma. Rheology of the breadmaking process. Cereal Foods World 35:228–236, 1990.

14. MC Olewnik, K Kulp. The effect of mixing time and ingredient variation on farinograms of cookie doughs. Cereal Chem 61:532–637, 1984.

15. JE Dexter, RR Matsuo. Relationship between durum wheat protein properties and pasta dough rheology and spaghetti cooking quality. J Agric Food Chem 28:899–902, 1980.

16. JW Dick, JS Quick. A modified screening test for rapid estimation of gluten strength in early-generation durum wheat breeding lines. Cereal Chem 60:315–318, 1983.

17. BL d'Appolonia, WH Kunerth. The Farinograph Handbook. St Paul, MN: AACC, 1969.

18. CW Brabender. Physical dough testing: past, present and future. Cereal Sci Today 10:291, 1965.

19. CO Swanson, EB Working. Testing the quality of flour by the recording dough mixer. Cereal Chem 10:1, 1933.

20. KF Finney, MDA Shogren. A ten-gram mixograph for determining and predicting functional properties of wheat flours. Baker's Dig 46:32, 1972.

21. R Spies. Application of rheology in the bread industry. In: H Faridi, JM Faubion, eds. Dough Rheology and Baked Product Texture. New York: Van Nostrand, 1990, pp 331–361.

22. S Zounis, KJ Quail. Predicting test bakery requirements from laboratory mixing tests. J Cereal Sci 25:185–196, 1997.

23. S Nagao. The Do-corder and its application in dough rheology. Cereal Foods World 31:231–240, 1986.

24. AM Janssen, T van Vliet, JM Vereijken. Fundamental and empirical rheological

behaviour of wheat flour doughs and comparison with breadmaking performance. J Cereal Sci 23:43–54, 1996.

25. AH Bloksma. A calculation of the shape of the alveograms of some rheological model substances. Cereal Chem 34:126–136, 1957.

26. B Launay, J Buré. Use of the Chopin Alvéographe as a rheological tool. Cereal Chem 54:1152–1158, 1977.

27. VF Rasper, KM Hardy, GR Fulcher. Constant water content vs. constant consistency techniques in alveography of soft wheat flours. In: H Faridi, ed. Rheology of Wheat Products. St Paul, MN: AACC, 1985, pp 51–73.

28. J Chen, BL d'Appolonia. Alveograph studies on hard red spring wheat flour. Cereal Foods World 30:433–443, 1985.

29. JJ Kokelaar, T Van Vliet, A Prins. Strain hardening properties and extensibility of flour and gluten doughs in relation to breadmaking performance. J Cereal Sci 24: 199–214, 1996.

30. AR Miller. The use of a penetrometer to measure the consistency of short doughs. In: H Faridi, ed. Rheology of Wheat Products. St Paul, MN: AACC, 1985, pp 117–132.

31. JF Tarrault. Actualisation d'un test de cuisson biscuitier. Bull Liaison CTUC 23–28, 1994.

32. DH Hahn. Applications of rheology in the pasta industry. In: H Faridi, JM Faubion, eds. Dough Rheology and Baked Product Texture. New York: Van Nostrand, 1990, pp 385–404.

33. J Chen, RC Hoseney. Development of an objective method for dough stickiness. Lebensm-Wiss Technol 28:467–473, 1995.

34. D Khelifi, G Branlard. The effects of HMW and LMW subunits of glutenin and of gliadins on the technological quality of progeny from four crosses between poor breadmaking quality and strong wheat cultivars. J Cereal Sci 16:195–209, 1992.

35. R Banguir, IL Batey, E MacKenzi, F MacRitchie. Dependence of extensograph parameters on wheat protein composition measured by SE-HPLC. J Cereal Sci 25: 237–241, 1997.

36. P d'Haene, GG Fuller, J Mewis. Shear thickening effect in concentrated colloidal dispersions. In: P. Moldenaers, R. Keunings, eds. Theoretical and Applied Rheology. Amsterdam: Elsevier, 1992, pp 595–597.

37. B Rabinowitsch. Über die Viskosität und Elastizität von Solen. Z Physik Chem a145:1–26, 1929.

38. EB Bagley. End correction in the capillary flow of polyethylene. J Appl Phys 28: 624–627, 1957.

39. FN Cogswell. Polymer Melt Rheology. New York: Wiley, 1981.

40. HM Laun, NA Schuch. Transient elongational viscosities and drawability of polymer melts. J Rheol 33:119–175, 1989.

41. CD Han. Measurements of the rheological properties of polymer melts with slit rheometer. I: Homopolymer systems. J Appl Polym Sci 15:2567–2570, 1971.

42. HM Laun. Polymer melt rheology with a slit die. Rheol Acta 22:171–185, 1983.

43. P Mourniac, JF Agassant, B Vergnes. Determination of wall slip velocity in the flow of rubber compounds. Rheol Acta 31:565–574, 1992.

44. CD Han, W Philipoff, M Charles. Rheological implication of the exit pressure and

die swell in steady capillary flow of polymer melts. II: The primary normal stress difference and the effects of L/D ratio on elastic properties. Trans Soc Rheol 14: 393–404, 1970.

45. CD Han. Slit rheometry. In: AA Collyer, DW Clegg, eds. Rheological Measurements. London: Elsevier, 1988, pp 25–48.

46. CD Han. Rheology in Polymer Processing. New York: Academic Press, 1976.

47. K Igashitani, WE Pritchard. A kinematic calculation of intrinsic errors in pressure measurements made with holes. Trans Soc Rheol 16:687–696, 1972.

48. DG Baird. A possible method of determining normal stress differences from hole pressure error data. Trans Soc Rheol 19:147–151, 1975.

49. AS Lodge. Normal stress differences from hole pressure measurements. In: AA Collyer, DW Clegg, eds. Rheological Measurements. London: Elsevier, 1988, pp 345–382.

50. DS Malkus, WG Pritchard, H Yao. The hole pressure effect and viscometry. Rheol Acta 31:521–534, 1992.

51. M Padmanabhan, M Bhattacharya. Flow behaviour and exit pressures of corn meal under high shear and temperature extrusion conditions using a slit die. J Rheol 35: 315–343, 1991.

52. M Bhattacharya, M Padmanabhan. On-line rheological measurements of food dough during extrusion-cooking. In: JL Kokini, CT Ho, MV Karwe, eds. Food Extrusion, Science and Technology. New York: Marcel Dekker, 1992, pp 213–231.

53. A Senouci, AC Smith. An experimental study of food melt rheology. II: End pressure effects. Rheol Acta 27:649–655, 1988.

54. G Della Valle, B Vergnes, P Colonna, A Patria. Relations between rheological properties of molten starches and their expansion by extrusion. J Food Eng 31: 277–296, 1997.

55. K Seethamraju, M Battacharya. Effect of ingredients on the rheological properties of extruded corn meal. J Rheol 38:1029–1044, 1994.

56. J Meissner, J Hostettler. A new elongational rheometer for polymer melts and other highly viscoelastic liquids. Rheol Acta 33:1–21, 1994.

57. T. Schweizer, B Condé-Petit. Bread dough elongation. Proceedings of 1st International Symposium on Food Rheology and Structure, Zurich, 1997, pp 391–394.

58. SH Chatraei, CW Macosko, HH Winter. Lubricated squeezing flow: a new biaxial extensional rheometer. J Rheol 25:433–443, 1981.

59. H Huang, JL Kokini. Measurement of biaxial extensional viscosity of wheat flour doughs. J Rheol 37:879–891, 1993.

60. EB Bagley. Constitutive models for doughs. In: JL Kokini, CT Ho, MV Karwe, eds. Food Extrusion, Science and Technology. New York: Marcel Dekker, 1992, pp 203–212.

61. DM Binding. An approximate analysis for contraction and converging flows. J Non-Newt Fluid Mech 27:173–189, 1988.

62. AG Gibson. Converging dies. In: AA Collyer, DW Clegg, eds. Rheological Measurements. London: Elsevier, 1988, pp 25–48.

63. FN Cogswell. Converging flow of polymer melts in extrusion dies. Polym Eng Sci 12:64–73, 1972.

64. C Carrot, J Guillet, P Revenu, A Arsac. Experimental validation of non-linear network models. In: JM Piau, JF Agassant, eds. Rheology for Polymer Melt Processing. Amsterdam: Elsevier, 1996, pp 141–198.
65. A Senouci, GDE Siodlak, AC Smith. Extensional rheology in food processing. In: H Giesekus, ed. Progress and Trends in Rheology. Vol. 2. Darmstadt, Germany: Steinkopff, 1988, pp 434–437.
66. G Marin. Oscillatory rheometry. In: AA Collyer, DW Clegg, eds. Rheological Measurements. London: Elsevier, 1988, pp 297–343.
67. JD Ferry. Viscoelastic Properties of Polymers. 3rd ed. New York: Wiley, 1980.
68. B Vergnes, G Della Valle, J Tayeb. A specific slit die rheometer for extruded starchy products. Design, validation and application to maize starch. Rheol Acta, 32:465–476, 1993.
69. G Della Valle, A Buléon, PJ Carreau, PA Lavoie, B Vergnes. Relationships between structure and viscoelastic behavior of plasticized starch. J Rheol 42:507–525, 1998.
70. A Redl, MH Morel, J Bonicel, S Guilbert, B Vergnes. Rheological properties of gluten plasticized with glycerol: dependence on temperature, glycerol content, and mixing conditions. Rheol Acta 38:311–320, 1999.
71. WP Cox, EM Merz. Correlation of dynamic and steady viscosities. J Polym Sci 28:619–622, 1958.
72. S Berland, B Launay. Shear softening and thixotropic properties of wheat flour doughs in dynamic testing at high shear strain. Rheol Acta 34:622–625, 1995.
73. N Phan-Thien, M Safari-Ardi, A Morales-Patino. Oscillatory and simple shear flows of a flour–water dough: a constitutive model. Rheol Acta 36:38–48, 1997.
74. AA Collyer, DW Clegg. Rheological Measurements. London: Elsevier, 1988.
75. CW Macosko. Rheology: Principles, Measurements and Applications. New York: VCH, 1994.
76. JM Dealy, KF Wissbrun. Melt Rheology and Its Role in Plastics Processing. New York: Van Nostrand, 1990.
77. B Vergnes, JP Villemaire. Rheological behavior of low moisture molten maize starch. Rheol Acta 26:570–576, 1987.
78. B Vergnes, JP Villemaire, P Colonna, J Tayeb. Interrelationships between thermomechanical treatment and macromolecular degradation of maize starch in a novel rheometer with preshearing. J Cereal Sci 5:189–202, 1987.
79. SI Fletcher, TJ McMaster, P Richmond, AC Smith. Rheology and extrusion of maize grits. Chem Eng Commun 32:239–262, 1985.
80. A Senouci, AC Smith. An experimental study of food melt rheology. I: Shear viscosity using a slit die viscometer and capillary rheometer. Rheol Acta 27:546–554, 1988.
81. TJ McMaster, A Senouci, AC Smith. Measurement of rheological and ultrasonic properties of food and synthetic polymer melts. Rheol Acta 26:308–315, 1987.
82. RE Altomare, M Anelich, R Rakos. An experimental investigation of the rheology of rice flour dough with an extruder-coupled slit die rheometer. In: JL Kokini, CT Ho, MV Karwe, eds. Food Extrusion, Science and Technology. New York: Marcel Dekker, 1992, pp 233–254.
83. R Parker, AL Ollett, RA Lai-Fook, AC Smith. The rheology of food melts and its

application to extrusion processing. In: RE Carter, ed. Rheology of Food, Pharmaceutical and Biological Materials. London: Elsevier, 1990, pp 57–74.

84. G Della Valle, P Colonna, A Patria, B Vergnes. Influence of amylose content on the viscous behavior of low hydrated molten starches. J Rheol 40:347–362, 1996.

85. G Della Valle, Y Boché, P Colonna, B Vergnes. The extrusion behavior of potato starch. Carbohyd Polym 28:255–264, 1995.

86. LS Lai, JL Kokini. The effect of extrusion operating conditions on the on-line apparent viscosity of 98% amylopectin (Amioca) and 70% amylose (Hylon 7) corn starches during extrusion. J Rheol 34:1245–1266, 1990.

87. KL Mackey, RY Ofoli. Rheology of low to intermediate moisture whole wheat flour doughs. Cereal Chem 67:221–226, 1990.

88. P Colonna, J Tayeb, C Mercier. Extrusion-cooking of starch and starchy products. In: C Mercier, P Linko, JM Harper, eds. Extrusion-Cooking. St Paul, MN: AACC, 1989, pp 247–319.

89. E Charun, JL Dournaux, AS Contamine, B Vergnes. Rheological characterization of biscuit doughs. Proceedings of 1st International Symposium on Food Rheology and Structure, Zurich, 1997, pp 193–197.

90. A Baltsavias, A Jurgens, T van Vliet. Rheological properties of short doughs at small deformation. J Cereal Sci 26:289–300, 1997.

91. SJ Dus, JL Kokini. Prediction of the non-linear viscoelastic properties of a hard wheat flour dough using the Bird-Carreau constitutive model, J Rheol 34:1069–1084, 1990.

92. L Lindhal, AC Eliasson. A comparison of some rheological properties of durum and wheat flour doughs. Cereal Chem 69:30–34, 1992.

93. W Burchard, SB Ross-Murphy. Physical Networks—Polymers and Gels. London: Elsevier, 1990.

94. SB Ross-Murphy. Structure–property relationships in food biopolymer gels and solutions. J Rheol 39:1451, 1995.

95. AS Contamine, J Abecassis, MH Morel, B Vergnes, A Verel. Effect of mixing conditions on the quality of doughs and biscuits. Cereal Chem 72:516–522, 1995.

96. JL Dournaux, B Vergnes, AS Contamine. Influence of the mixing conditions on the rheological behavior of a biscuit dough. In: I Emri, R Cvelbar, eds. Progress and Trends in Rheology V. Darmstadt, Germany: Steinkopff, 1998, pp 187–188.

97. A Buléon, P Colonna, V Planchot, S Ball. Starch granules: structure and biosynthesis. Int J Biol Macromol 23:85–112, 1998.

98. C Mercier, P Linko, JM Harper. Extrusion-Cooking. St Paul, MN: AACC, 1989.

99. JL Kokini, CT Ho, MV Karwe. Food Extrusion, Science and Technology. New York: Marcel Dekker, 1992.

100. D Lourdin, G Della Valle, P Colonna. Influence of amylose content on starch films and foams. Carbohyd Polym 27:261–270, 1995.

101. R Chinnaswammy, MA Hanna. Macromolecular and functional properties of native and extrusion-cooked corn starch. Cereal Chem 67:490–499, 1990.

102. JL Willett, BK Jasberg, CL Swanson. Rheology of thermoplastic starch: effects of temperature, moisture content and additives on melt viscosity. Polym Eng Sci 35:202–210, 1995.

103. WW Graessley. Viscoelasticity and flow in polymer melts and concentrated solu-

tions. In: Physical Properties of Polymers. Washington, DC: ACS Books, 1993, pp 97–145.

104. JL Willett, MM Millard, BK Jasberg. Extrusion of waxy maize starch: melt rheology and molecular weight degradation of amylopectin. Polymer 38:5983–5989, 1997.

105. RG Morgan, JF Steffe, RY Ofoli. A generalized viscosity model for extrusion of protein doughs. J Food Proc Eng 11:55–78, 1989.

106. JR Mitchell, JAG Areas, S Rasul. Modifications chimiques et texturation des protéines à faibles teneurs en eau. In: P Colonna, G Della Valle, eds. La Cuisson-Extrusion. Paris: Lavoisier Tec & Doc, 1994, pp 85–108.

107. RL Tomas, JC Oliveira, KL McCarthy. Rheological modelling of enzymatic extrusion of rice starch. J Food Eng 32:167–177, 1997.

108. L Levine, S Symes, J Weimer. A simulation of the effect of formula and feed rate variations on the transient behavior of starved extrusion screws. Biotechnol Progress 3:221–230, 1987.

109. IO Mohamed, RY Ofoli. Prediction of temperature profiles in twin screw extruders. J Food Eng 12:145–164, 1990.

110. KLB Chang, GW Halek. Analysis of shear and thermal history during co-rotating twin-screw extrusion. J Food Sci 56:518–531, 1991.

111. J Tayeb, B Vergnes, G Della Valle. A basic model for a twin screw extruder. J Food Sci 54:1047–1056, 1989.

112. G Della Valle, C Barrès, J Plewa, J Tayeb, B Vergnes. Computer simulation of starchy products transformation by twin screw extrusion. J Food Eng 19:1–31, 1993.

113. C Barrès, B Vergnes, J Tayeb, G Della Valle. Transformation of wheat flour by extrusion-cooking: influence of screw configuration and operating conditions. Cereal Chem 67:427–433, 1990.

114. B Vergnes, G Della Valle, L Delamare. A global 1D model for polymer flows in corotating twin screw extruders. Polym Eng Sci 38:1781–1792, 1998.

115. L Delamare, B Vergnes. Computation of the morphological changes of a polymer blend along a twin screw extruder. Polym Eng Sci 36:1685–1693, 1996.

116. F Berzin, B Vergnes. Transesterification of ethylene acetate copolymer in a twin screw extruder. Intern Polym Proc 13:13–22, 1998.

117. R Parker, AL Ollett, AC Smith. Starch melt rheology: measurements, modelling and applications to extrusion processing. In: P Zeuthen, JC Cheftel, C Eriksson, TR Gormley, P Linko, K Paulus, eds. Processing and Quality of Foods. London: Elsevier, pp 1290–1295, 1990.

118. RE Altomare, P Ghossi. An analysis of residence time distribution patterns in a twin screw cooking extruder. Biotechnol Progress 2:157–163, 1986.

119. B Vergnes, C Barrès, J Tayeb. Computation of residence time and energy distribution in the reverse screw element of a twin-screw cooker-extruder. J Food Eng 16: 215–237, 1992.

120. JL Kokini, CN Chang, LS Lai. The role of rheological properties on extrudate expansion. In: JL Kokini, CT Ho, MV Karwe, eds. Food Extrusion, Science and Technology. New York: Marcel Dekker, 1992, pp 631–652.

121. J Fan, JR Mitchell, JMV Blanshard. A computer simulation of the dynamics of

bubble growth and shrinkage during extrudate expansion. J Food Eng 23:337–356, 1994.

122. J Fan, JR Mitchell, JMV Blanshard. The effect of sugars on the extrusion of maize grits: I. The role of the glass transition in determining product density and shape. Int J Food Sci Technol 31:55–65, 1996.

123. B Launay. Expansion des matériaux amylacés en sortie de filière. In: P Colonna, G Della Valle, eds. La Cuisson-Extrusion. Paris: Lavoisier Tec & Doc, 1994, pp 165–202.

124. L Alvarez-Martinez, KP Kondury, JM Harper. A general model for expansion of extruded products. J Food Sci 53:609–615, 1988.

125. KH Park. Elucidation of the extrusion puffing process. PhD dissertation, University of Illinois, Urbana-Champaign, 1976.

126. B Launay, JM Lisch. Twin screw extrusion cooking of starches: flow behavior of starch pastes, expansion and mechanical properties of extrudates. J Food Eng 2: 259–280, 1983.

127. JM Ferdinand, RA Lai-Fook, AL Ollett, AC Smith, SA Clark. Structure formation by carbon dioxide injection in extrusion cooking. J Food Eng 11:209–224, 1990.

128. H Kumagai, T Yano. Critical bubble radius for expansion in extrusion cooking. J Food Eng 20:325–338, 1993.

129. M Amon, CD Denson. A study of the dynamics of the growth of closely spaced spherical bubbles. Polym Eng Sci 24:1026–1034, 1984.

130. RA Neff, CW Macosko. Simultaneous measurement of viscoelastic changes and cell opening during processing of flexible polyurethane foam. Rheol Acta 35:656–666, 1996.

131. RCE Guy, AW Horne. Extrusion and co-extrusion of cereals. In: JMV Blanshard, JR Mitchell, eds. Food Structure—Its Creation and Evaluation. London: Butterworths, 1988, pp 331–349.

132. S Pasquet, A Arhaliass. Etude en ligne des propriétés rhéologiques, de l'expansion et de la contraction des extrudés lors de la cuisson-extrusion. Cahiers Rhéol 15: 168–177, 1997.

133. M Cornec, Y Popineau, J Lefebvre. Characterization of gluten subfractions by SE-HPLC and dynamic rheological analysis in shear. J Cereal Sci 19:131–139, 1994.

134. AA Tsiami, A Bot, WGM Agterof. Rheological properties of glutenin subfractions in relation to their molecular weight. J Cereal Sci 26:15–29, 1997.

135. P Masi, S Cavella, M Sepe. Characterization of dynamic viscoelastic behavior of wheat flour dough at different moisture contents. Cereal Chem 75:429–433, 1998.

136. DW De Bruijne, J de Loof, A Van Eulem. The rheological properties of bread dough and their relation to baking. In: RE Carter, ed. Rheology of Food, Pharmaceutical and Biological Materials. London: Elsevier, 1990, pp 269–283.

137. MP Morgenstern, MP Newberry, SE Holst. Extensional properties of dough sheets. Cereal Chem 73:479–482, 1996.

138. T Van Vliet, AM Janssen, AH Bloksma, P Walstra. Strain hardening of dough as a requirement for gas retention. J Texture Studies 23:439–460, 1992.

139. B Launay, JC Bartolucci. Charactérisation du comportement viscoélastique des pâtes de farine en extension biaxiale. Applications technologiques. Cahiers Rhéol 15: 594–607, 1997.

140. SS Sablani, M Marcotte, OD Baik, F Castaigne. Modeling of simultaneous heat and water transport in the baking process. Lebensm-Wiss u-Technol 31:201–209, 1998.

141. D Le Roux, B Vergnes, M Chaurand, J Abecassis. A thermomechanical approach to pasta extrusion. J Food Eng 26:351–368, 1995.

142. PA Tanguy. A computer simulation perspective on the processing of rheologically complex materials. Proceedings of 1st International Symposium on Food Rheology and Structure, Zurich, 1997, pp 91–94.

143. L Levine, BA Drew. Rheological and engineering aspects of the sheeting and laminating of dough. In: H Faridi, JM Faubion, eds. Dough Rheology and Baked Product Texture. New York: Van Nostrand, 1990, pp 513–557.

144. L Levine, E Boehmer. The fluid mechanics of cookie dough extruders. J Food Proc Eng 15:169–186, 1992.

8

Stress and Breakage in Formed Cereal Products Induced by Drying, Tempering, and Cooling

Betsy Willis
Southern Methodist University, Dallas, Texas, U.S.A.

Martin Okos
Purdue University, West Lafayette, Indiana, U.S.A.

I. INTRODUCTION

The quality of value-added cereal products is judged by many criteria, including flavor, aroma, color, texture, and overall appearance. These quality attributes are directly influenced by raw materials and processing conditions. One of the most challenging quality characteristics to address is breakage. Drying, tempering, and cooling affect the moisture and temperature gradients in a product and thus have the potential to cause breakage.

Optimal drying, tempering, and cooling conditions depend on the material behavior as reflected in its material properties. Temperature gradients in cereal products equilibrate rapidly, so the main cause of stress development and breakage is moisture gradients. Process conditions are optimized to maximize product quality and minimize moisture gradients. For example, pasta's dense structure is achieved by shrinkage during moisture removal under high temperature and relative humidity conditions. Optimization of the drying conditions requires knowledge of the effects of variations in raw materials and processing (e.g., mixing, extrusion, sheeting) on raw product structure. The behavior of the raw product is reflected in the material properties, which influence mass transfer and stress development and necessary change with temperature and moisture. To prevent breakage, a product should be dried, tempered, and cooled under conditions that minimize moisture gradients.

Drying, tempering, and cooling conditions are optimized either by plant tests or by numerically simulating the process using mechanistic models. Plant tests are lengthy and costly due to the time required to see the result of a process change and due to the loss of salable product produced during the tests. Additionally, processing conditions are often tested by a trial-and-error method, which is extremely inefficient if multiple parameters are being changed. Numerical simulations greatly reduce the cost and time required for process optimization. Multiple processing conditions can be changed and the results of these changes observed quickly without any lost product. Thus, numerical simulations based on mechanistic models are a more efficient method of process optimization than plant tests.

Drying, tempering, and cooling are numerically simulated using mechanistic mass transfer and stress development models that require material property inputs as functions of temperature and moisture content. Information gained from such models includes drying rates, drying curves, moisture profiles, shrinkage, and stresses in multiple directions. The models require several material properties, including isotherms, glass transition temperature, storage modulus, diffusion coefficient, Young's modulus, failure stress, and Poisson's ratio, as functions of temperature and moisture content. The temperature and moisture dependence of the material properties varies among raw products made with different raw materials and processes (e.g., extrusion versus sheeting). Therefore, the effect of raw material variations on final product quality is observed by testing and simulating drying, tempering, and cooling of various final products. The effects of variations in processing parameters and raw materials on final product quality are observed rapidly using numerical simulations.

Besides food, breakage occurs across a variety of materials, including concrete, polymers, clay and soils, thin coatings, and grains during drying and storage. Most often, this breakage is due to variations in material properties within the material caused by moisture gradients. As moisture or solvent is removed, the material shrinks and may transition from a pliable, rubbery state to a brittle, glassy state. This transition, known as glass transition, inhibits shrinkage, resulting in stress development and potential failure. Product formulation or processing conditions can be adjusted to avoid the undesirable transition to the glassy state.

Reducing breakage requires understanding the relationship among raw materials, processing conditions, and final product quality. These relationships can be determined experimentally, but exhaustive testing of all raw materials and processing parameters is impossible due to the large cost and amount of time involved. Therefore, effective mechanistic models of drying, tempering, and cooling processes need to be developed to aid in the understanding of these relationships and to speed process optimization.

II. EFFECTS OF RAW MATERIALS ON FINAL PRODUCT QUALITY

Raw materials and the subsequent changes they go through during processing affect the glass transition of the product, which in turn affects the material properties, transport phenomena, and stress development. Understanding the influence of raw material properties and how the product changes during processing is important for stress prediction.

The main ingredients of cereal products are flour and water, and the attributes of the flour influence final product quality. One physical attribute of importance is particle size. A small particle-size distribution is necessary for even hydration and the formation of a strong structure. For example, pasta is made from semolina, and the particle size should be within 200–300 μm, with less than 10% outside of this range. Coarse particles, greater than 500 μm, do not hydrate evenly, and excess fines result in thermal stress (1). Another important attribute of the flour is the macromolecular composition. For example, protein content is a determining factor in selecting flour for pasta versus cakes. The relative amounts of other macromolecular components, such as carbohydrates, lipids, and ash, also affect final product quality. Both physical and chemical attributes affect final product quality.

Variations in raw materials are due to environmental conditions and varietal differences. Environmental conditions affect protein content, rheological properties, and final product quality. The protein content of wheat is closely linked to the environment (2). Frost adversely affects wet gluten yields and mixograph results (3). Additionally, sprouting in wheat increases checking of pasta (4). Genetic differences affect rheological properties of gluten and final product quality. Medium to strong glutens have high gluten indices, SDS sedimentation, and gluten viscoelasticity, which are all linked to genotype (2). The differences in macromolecular components of different varieties of durum wheat affect the cooking quality of pasta (5, 6). The relationship between raw material differences and final product quality is complex due to both the environmental and genetic effects.

The composition of the macromolecular components that make up cereal grains varies with growing season and genetics. For example, genetics control the protein quality and pigment content of durum wheat, whereas seasonal conditions affect the protein content, ash content, moisture content, and falling number of durum wheat (7). The major macromolecular components of cereal grains include carbohydrates, protein, and lipids; and protein is most responsible for final product structure and quality. The structure and amount of these components affect ingredient interactions, material properties, and final product quality.

A. Carbohydrates

Carbohydrates found in cereal grains include simple sugars, such as glucose, and starch, in the forms of amylose and amylopectin. Simple sugars participate in nonenzymatic browning during baking to provide cereal products with a golden appearance. Starch granules must be disrupted before the amylose and amylopectin can interact with each other and other molecules. Granules are disrupted naturally during gelatinization or mechanically during milling. During gelatinization, granules swell in the presence of water with applied heat. At some point, the granules become so swollen that they break and the starch is released. Gelatinization during baking sets the structure (8). Starch also plays an important role in staling. During staling, molecules realign to form a crystalline-like structure that is perceived as a tough crumb, or staleness. Starch can also provide structure, as observed in pasta-like products made from nongluten flours. Mestres et al. (9) produced pasta from maize by pregelatinizing the starch to form a support structure of starch rather than proteins. Carbohydrates play important roles in the appearance, structure, and texture of cereal products.

Starch granules damaged during milling affect final product quality. Damaged starch granules are more susceptible to attack by enzymes. Enzymes break the starch into smaller molecules that may act as plasticizers, thus lowering the glass transition temperature of the product. Decreasing the glass transition temperature is beneficial during drying because it lowers the temperature required for the product to remain rubbery. However, for cooked cereal products, such as pasta, the small molecules leave the product during cooking, resulting in increased cooking losses and decreased cooked product quality. Also, damaged starch granules absorb more water, 100% of their weight as opposed to 30% for undamaged starch (10). Increased water absorption leads to a weaker loaf and sticky crumb (11). To achieve high product quality, damaged starch granules must be minimized.

B. Lipids

Lipids are a minor component in cereal grains, but they participate in the formation of the structure and texture of cereal products. Lipids act as a lubricant, stabilize air bubbles, provide texture, and provide flavor. Oxidation may adversely affect flavor and aroma but may increase dough resistance (12). During extrusion, lipids undergo chemical changes and/or complexation with starch and proteins (13). In addition, free lipids bind during processing, especially during high-temperature drying. Total lipid content varies among wheat varieties and is less in durum than in other wheat flours (14). However, lipids decrease stickiness in pasta because the monoglycerides complex with amylose (13). Increased free lipid content in pasta decreases the breaking strength of pasta made from hard

Table 1   Effect of Lipid Content on Dry-Noodle
Breaking Strength

Wheat	Percent free lipids	Breaking strength (g/mm^2)
Hard red winter	0	3587
	1.08	3397
	1.62	1918
Soft white	0	1957
	0.94	1462
	1.41	940

Source: Ref. 15.

red winter wheat and soft white wheat (15), as shown in Table 1. Lipids provide several other important functionalities in cereal products, including the following (16):

Modify gluten structure during mixing
Catalyze the oxidation of sulfhydral groups to disulfide groups
Catalyze the polymerization of protein via lipid peroxidation
Prevent interaction between starch granules during gelatinization
Support gluten
Retard water transport from the protein to the starch
Retard starch gelatinization
Act as an antistaling agent

Although a minor component, lipids are important contributors to the structure and texture of cereal products.

C. Protein

Protein is the major macromolecular component contributing to the structure and texture of cereal products. Current theories on the relationship between final product quality and protein center on the presence of high-molecular-weight proteins.

Cereal proteins possess the unique ability to form a matrix that contributes to the structure and texture of the product. The most common example is the gluten matrix formed by wheat proteins present in products such as bread, biscuits, pasta, cookies, and cakes. Other cereal proteins, such as zein from corn and kafirin from sorghum, also form a matrix with the application of shear and heat. The ability of cereal proteins to form a matrix allows for a variety of food products to be produced.

Wheat protein is divided into five fractions based on solubility: insoluble residue, albumins, globulins, gliadin, and glutenin. Gliadin and glutenin are the major proteins that compose the gluten matrix. Gliadin has a lower molecular weight than glutenin, and glutenin is a high-molecular-weight protein composed of gliadin and other protein fractions. Shewry et al. (17) classified gliadins and glutenins as one group, called *prolamins*, which can be subdivided into sulfur-rich, sulfur-poor, and high-molecular-weight proteins. Gliadins and glutenins possess several unique characteristics important for dough structure:

> Low charge density, which allows for strong associations via hydrogen bonding, hydrophobic interactions, and disulfide bonds
> High proline content, resulting in periodic disruptions or sharp turns in the protein polymer
> Insoluble at their pI of 6–9
> Soluble at low and high pH
> Precipitate at low salt concentrations

Thus, prolamins are more extended than coiled, which allows them to form strong aggregations in the pH range and salt concentrations of most cereal products. The unique characteristics of gliadins and glutenins result in the formation of a gluten matrix.

The gluten matrix is formed with the input of mechanical work, such as kneading, because the sulfhydral groups near the ends of the prolamin chains interact. The sulfhydral–disulfide interchange forms a viscoelastic protein matrix. Elongation or relaxation of the chains strengthens the matrix; otherwise the structure would be too rigid for phenomena such as bread rise (1). Mechanical work is necessary to form the strong, viscoelastic gluten matrix that supports wheat cereal products.

The viscoelastic nature of gluten is attributed to the unique characteristics of prolamins. Low-molecular-weight subunits contribute dough viscosity, and high-molecular-weight subunits contribute dough elasticity (18). Low-molecular-weight prolamins, such as gliadins, have sulfhydral groups at the ends of their chains that interact with other low-molecular-weight subunits (19, 20). Additionally, shear and heat expose previously buried hydrophobic regions, resulting in hydrophobic interactions among low-molecular-weight subunits (19, 20). High-molecular-weight subunits that contribute elasticity to dough have been associated with good breadmaking quality. The number of high-molecular-weight subunits and the amino acid sequence are important. MacRitchie (21) adds two more parameters for high gluten elasticity: (a) the fraction of subunits above a certain molecular weight and (b) the molecular weight distribution of this fraction. Barro et al. (22) found that genetically altering wheat to increase the number of high-molecular-weight subunits resulted in increased mixing characteristics. Since both low-molecular-weight and high-molecular-weight subunits contribute to the viscoelasticity of gluten, understanding the impact of these fractions on final product quality is necessary for product improvement.

Proteins from other cereal, such as corn and sorghum, are contained in protein bodies that must be disrupted before the proteins can interact with each other and with other raw materials. Application of shear and heat to the protein bodies frees the proteins and aids in the formation of a protein network. Batterman-Azcona and Hamaker (23) found that cooking corn caused disulfide-bound protein polymers to form, but further processing, either flaking or extrusion, was necessary to disrupt the protein bodies. Flaking slightly disrupted the protein bodies, and extrusion completely disrupted the protein bodies. Their results showed that both heat and high shear are necessary to release corn proteins contained in bodies and to form a protein network. Chandrashekar and Desikachar (24) improved the hydration of sorghum by adding organic acids and by grinding the flour to smaller particle sizes. Improved hydration plasticizes the proteins, allowing them to form a matrix. Bergman et al. (25) incorporated cowpea flour into pasta made from soft wheat. They found that breakage decreased and color and cooking quality increased when the product was dried with high temperatures. These results were attributed to the increased protein content of pasta supplemented with cowpea flour and the superior ability of these proteins to form a matrix. By simply altering processing conditions, cereal products can be made from underutilized grains.

D. Summary of the Effects of Raw Material Variation on Final Product Quality

Both physical and chemical attributes of the raw materials affect final product quality. Physical attributes, such as particle size, are important even for hydration and the formation of a uniform product. Macromolecular components vary with variety and growing season, making consistent product quality a challenge. The main macromolecular components of cereals include carbohydrates, lipids, and proteins. Proteins are most responsible for the support structure of cereal products. Understanding the effects of raw material variations on material property characteristics is necessary to reduce breakage.

III. EFFECT OF PROCESSING ON FINAL PRODUCT QUALITY

The basic unit operations in cereal product production include mixing, forming, and baking or drying. During mixing, the raw ingredients are combined and mechanical energy begins dough development. Kneading and forming continues dough development. Forming shapes the dough into the desired product. Drying or baking removes moisture, sets the structure, increases shelf stability, and enhances product appearance through browning reactions. By varying processing parameters, a variety of products are produced.

A. Mixing

During mixing, dough development begins as raw materials are combined and macromolecules are hydrated. Icard-Verniere and Feillet (26) measured specific mechanical energy during mixing and found that mixing has three stages, differing in torque. First, flour particles are hydrated, which is marked by a low, constant torque. Hydration increases the mobility of the macromolecular chains, which enables the proteins to polymerize. Second, dough development occurs as sulfhydral groups react to form disulfide bonds. This stage is marked by sharply increasing torque. Last, the torque remains constant at a maximum value, indicating fully developed dough. Overmixing the dough results in breakdown of the gluten matrix. However, optimal mixing produces dough with even hydration and a strong gluten structure.

B. Forming

After mixing, the dough is formed into the desired shape for the final product. Loaves of bread may be formed or pieces may be cut from a large sheet. To form a large sheet of dough, the dough is passed through sets of rollers that gradually reduce the thickness. The gradual thickness reduction further develops the gluten structure without tearing the dough. Important sheeting parameters include reduction ratio and roller speed. Oh et al. (27) found that reduction ratio and roll speed affect the surface of cooked pasta. Cooked-noodle surface was firmer for slower roll speeds and for greater reduction ratios, implying that high pressure results in a smooth surface. Once the large dough sheet is formed, it can be cut into the desired shape for the final product. Alternatively, pieces may be cut from the large dough mass and formed into the final product's shape, such as loaves of bread. Forming takes the developed dough and shapes it into the desired form for the final product.

C. Extrusion

Extrusion combines mixing, kneading, and forming into one unit operation. An extruder consists of a screw within a barrel that conveys the material to the outlet, where it is pushed through a die. Raw materials are metered into the extruder, where they are mixed and conveyed by the screw. Throughout the mixing section, the gap between the screw and barrel remains constant. To develop the protein matrix, compression is applied to the material by decreasing the gap between the screw and barrel or by configuring the screw to hold the dough in the kneading section. The dough is then conveyed to the die. Dough is pushed through the die to shape the final product. Teflon dies are often used due to their ability to form products with minimal surface irregularities and a smooth protein coat. Surface irregularities result in weak spots and are often the site of crack initiation. Worn dies produce pieces of uneven thickness resulting in uneven drying, the development of moisture gradients, and the potential for cracking (9). The extrusion process is an efficient unit operation to mix, knead, and shape a product and is commonly used in the breakfast cereal and pasta industries.

Shear, heat, and air incorporation are important parameters to control during extrusion. Shear is applied to the material to promote dough formation as the screw rotates. Excessive shear breaks down the dough and produces excessive viscous heating. Often heat is applied to aid in the aggregation of proteins by the formation of disulfide bonds (29). When calculating the amount of heat to add, one must account for both the external heat applied and heat due to friction from the screw shear. Overheating the dough results in overworked dough and the gluten denatures (28, 30). An additional concern with extrusion is air incorporation. Air produces weak spots in the extrudate that may result in breakage, so a vacuum is used to remove air (1, 28, 30, 31). Extrusion is an efficient method for the production of a variety of food products, but care must be taken to control shear, heat, and air incorporation.

D. Drying, Tempering, and Cooling

Drying, tempering, and cooling set the final product structure and texture. Drying or baking serves to remove moisture from the product, and the processing conditions affect final product quality. Tempering equilibrates moisture gradients to reduce stresses. Cooling brings the product to room temperature from the processing temperature. Drying, tempering, and cooling conditions must be optimized to produce the desired final product quality.

The texture of a product may be light and airy or dense, depending upon the drying or baking conditions. Puffed products are produced by subjecting the dense pellets to high heat under atmospheric conditions and then quickly lowering the pressure to below atmospheric conditions. Moisture flashes off the product, causing it to expand. Alternatively, dense products may be formed by drying the material under high temperature and relative humidity conditions. The material shrinks as a result of moisture loss. Dexter et al. (32) found that high-temperature drying resulted in a stronger dry pasta that when cooked had improved characteristics. Fujio and Lim (33) found that heat-treated gluten was an amorphous bio-polymer, thus exhibiting glass transition behavior as observed with DSC. The state transition affects diffusion and deformation. Product deformation due to moisture removal continues until a point where the material transitions into the glassy state. Once the structure becomes immobile, moisture removal may continue, but the final product texture and integrity may be compromised. Another condition detrimental to final product quality is the formation of moisture gradients, since they can result in breakage as the gradients equilibrate. External drying conditions, such as relative humidity and temperature, can be adjusted to prevent transition to the glassy state and to prevent moisture gradients from forming.

Drying conditions affect the final product quality through changes in the protein structure of the cereal product. High-temperature drying increases product output and causes proteins to aggregate to form an insoluble network (34). Increased pasta strength and cooking quality have been associated with higher drying temperature (32). Resmini and Pagani (35) explained that a protein network is formed without starch gelatinization during pasta drying, which traps the starch and prevents it from leaving the product during cooking. This resulted in increased cooked quality. Optimal drying or baking conditions enhance product texture and set the structure through positive effects on the protein matrix.

Various researchers have studied the effect of temperature and moisture content on the failure stress of pasta at all stages of drying. Harsh predrying, such as overdrying, reduces the strength of the pasta from 3,000 to 1,500 psi, as shown by Earle (36). Akiyama and Hayakawa (37) studied the temperature dependence of failure stress. They found that tensile fracture stress was independent of temperature, and the average failure stress for commercial pasta at 8.4% moisture content (dry basis) in tension was 3.3 MPa. Liu et al. (38) found that tensile failure stress decreased for increasing moisture content. Failure stresses were very low for samples with high moisture content (greater than 25% wb), and increased twofold for every 5% moisture reduction and tenfold for every 15% moisture reduction. Failure stress of pasta is affected by drying conditions and the moisture content of the pasta.

Dry products are often tempered and cooled before packaging. During tempering, a product is held at constant temperature and relative humidity conditions to allow temperature and moisture gradients to equilibrate. Tempering is sometimes used in the pasta-drying process in conjunction with convective drying periods. The dryer consists of alternating convection and tempering sections. Regular equilibration of moisture gradients reduces the stresses and breakage of the final product. During the cooling process, the temperature decreases from the drying or baking temperature to room temperature. Temperature gradients equilibrate rapidly and are not a major source of stress development and breakage (39). Tempering and cooling serve to decrease moisture and temperature gradients that could result in breakage.

E. Summary of the Effects of Processing Conditions on Final Product Quality

A variety of cereal products can be produced simply by varying processing conditions. The basic unit operations include mixing, forming, and drying. Each unit operation must be optimized for each product to maintain high final product quality. Processing conditions affect the structure of the cereal product and in turn affect the product's material properties. The material properties govern the drying kinetics and stress development. Understanding the effects of processing on material behavior is necessary to optimize each process to reduce breakage.

IV. GLASS TRANSITION AND MATERIAL PROPERTIES

The glass transition phenomenon can be used to describe cereal product behavior during processing. Glass transition governs an amorphous material's behavior and can be explained using the theory of free volume. Many polymers and macromolecules in foods, such as proteins, form a cross-linked structure with branches. The cross-linking and branches prevent the formation of a well-ordered crystalline structure. The resulting structure is termed *amorphous*. Amorphous structures can be liquid, semisolid, or brittle.

Free volume, the volume not occupied by the amorphous chains, is affected by temperature and moisture content. The free volume is a minimum at the glass transition temperature. At and below the glass transition temperature, the amorphous material is glassy, where deformation and diffusion are restricted. Increasing the temperature and/or moisture content increases the free volume. Thus, the amorphous chains have more volume in which to move, so shrinkage or expansion can occur. Additionally, more volume between chains is available for diffusion of small molecules, such as water. Material properties, such as viscosity and diffusion, change by several orders of magnitude at the glass transition. Therefore, the state of a material greatly affects the material properties that influence mass transfer and stress development.

The point of glass transition varies for different raw materials and is represented by a glass transition temperature as a function of moisture content. Above the glass transition temperature curve, the material is rubbery and below the curve glassy. Glass transition temperatures for different proteins vary with season and variety (40), and glass transition temperature depends on molecular weight (41). Kokini et al. (42) developed the state diagrams for gliadin, zein, and glutenin. Gliadin had an anhydrous glass transition temperature of 121.5°C, which was lower than that of zein, 139°C, and glutenin, 145°C, and was attributed to its lower molecular weight (42). The variation in glass transition temperature among raw materials is due to structural and molecular weight differences.

Glass transition affects the properties, such as viscosity, diffusion, and stiffness, of a material, that are observed in processing phenomena, such as collapse, stickiness, reaction rates, and crispness (43–46). Below the glass transition temperature, movement is localized; but above the glass transition temperature, segmental motion occurs. Material properties provide a link between the state of a material and processing phenomena.

Viscosity (and other material properties such as the storage modulus) reflects the ability of a material to flow, which is observed in collapse and stickiness. In the glassy state, the viscosity or storage modulus is high, approximately 10^{12} Pa $\cdot$ s. The viscosity decreases by several orders of magnitude at the glass transition temperature, so above the glass transition in the rubbery state, the viscosity is low. The lower viscosity allows the material to deform or flow. For example, while drying at temperatures above the glass transition temperature, pasta shrinks as moisture is removed to produce a dense product. Stickiness is an important phenomena in particle processing. Above the glass transition temperature, the particle surface becomes plasticized, which leads to interparticle bonding (46). Caking is an issue in the handling of particles such as sugar. Higher-molecular-weight materials, such as maltodextrins with high glass transition temperatures, are added to the low-molecular-weight particles to reduce caking (45). Controlling the viscosity of a material controls the collapse and stickiness.

Mass transfer and reaction rates are affected by diffusion. Diffusion depends on the solubility and diffusivity of the diffusing species. The diffusing species and other species present must be compatible for diffusion to occur. In addition, a space for the diffusing species is required, which is related to the free volume. The theory of free volume is based on the following (47):

1. The probability that a molecule will obtain sufficient energy to overcome attractive forces
2. The probability that a fluctuation in density will result in a hole of sufficient size to accommodate the diffusing species

The formation of a hole is related to the viscosity of the solid material. In the glassy state, molecular movement is inhibited and the free volume is a minimum. But in the rubbery state, free volume increases and more movement occurs. During pasta drying, Ollivier (48) observed changes in the diffusion coefficient of water corresponding to changes in pasta behavior. Increasing the temperature decreased the moisture content at which the transition occurred. Although not recognized at the time, the pasta went through glass transition during drying, which affected the diffusion coefficient and viscosity. Reaction rates are also affected by glass transition via moisture content, temperature, and diffusion (45). Browning reactions, enzymatic activity, and nutrient loss reactions decrease with temperatures below the glass transition temperature (46). The diffusion coefficient reflects the mobility of a diffusing species, which is revealed in mass transfer kinetics and reaction rates.

Young's modulus, or stiffness, changes with glass transition and is observed in crispness. Stiffness decreases with plasticization, which affects stress development and texture. During drying, few stresses and cracks develop if the material remains rubbery, when stiffness is low. Crisp products, such as biscuits, retain their characteristic texture in the glassy state, when stiffness is high. Wollney and Peleg (49) studied the effect of plasticization on stiffness of Zwiebacks and cheese balls. Stiffness versus water activity curves for both products remained constant over a low water-activity range, then dropped sharply. The abrupt change in stiffness was indicative of glass transition. Nikolaidis and Labuza (50) studied the bending modulus of crackers as a function of moisture content. The bending modulus decreased at the glass transition, as the crackers became rubbery. Comparing the glass transition temperature curve to the isotherm revealed a critical water activity of 0.65 at room temperature (23°C) at a moisture content of 10.7% (wb), below which the cracker was glassy and above which the cracker was rubbery. The stiffness of a material affects the stress and texture of a material.

The relationship between relative humidity and equilibrium moisture content is given by isotherms. Relative humidity is a critical processing parameter in drying and product storage. High relative humidities plasticize the material, resulting in collapse or loss of crispness. Thus, material behavior is dictated by the equilibrium moisture content at the processing relative humidity and temperature conditions. The state of a material is based on the relationship between the processing temperature and glass transition temperature of the product at the equilibrium moisture content. From a plot containing two curves, glass transition temperature versus water activity and moisture content versus water activity at the processing temperature, the critical water activity and corresponding moisture content that depress the glass transition temperature to the storage or processing temperature are determined (46). Isotherms are a valuable tool for determining the behavior of a material under a given set of processing conditions.

The glass transition temperature varies among materials, so their behavior under a given set of processing conditions is different. For example, drying at a temperature of 130°C allows zein to remain rubbery to its anhydrous state, but gliadin becomes glassy as it approaches zero moisture content. Generally, higher-molecular-weight materials have higher glass transition temperatures. Increasing the moisture content of a material depresses the glass transition temperature. Drying is a highly dynamic process, since the material constantly changes as moisture is removed and proteins aggregate to form larger molecules. Both of these changes result in an increasing glass transition temperature. Knowledge of the glass transition temperature as a function of moisture content for a given material is important for setting optimal drying conditions. If a dense product is desired, the drying temperature should remain above the glass transition temperature throughout drying to allow the product to collapse. If a recipe change is made and another raw material is used, optimal processing conditions change since the new material's glass transition temperature is different than the original raw material's glass transition temperature. Therefore, processing must be optimized for each product to attain optimal final product quality.

State diagrams relating processing temperature to product glass transition temperature and moisture content can be used to understand the effects of processing on final product quality. This can be demonstrated by looking at state diagrams for two very different products—pasta and extrusion-puffed cereal.

During pasta drying, collapse is desired to form a dense structure; Figure 1 shows the relationship between ideal drying conditions and the glass transition of pasta. Initially, the pasta enters the dryer at a high moisture content and is quickly heated. The drying temperature must be greater than the glass transition temperature so the structure can collapse. Near the outlet of the dryer or in the cooling section, the temperature is less than the glass transition temperature, so the pasta becomes glassy. This sets the structure, and the resulting product is a dense, hard piece of pasta. If the pasta transitions to the glassy region early in the drying process, all mobility is greatly decreased, often by multiple orders of magnitude. The matrix becomes rigid, with a drying front moving from the outer edge of the pasta toward the center. A shell forms, which decreases the moisture mobility and "locks" it into the pasta. The residual moisture in the form of moisture gradients causes the pasta to crack as the moisture equilibrates (34). Moisture equilibration can occur during later stages of production or once the pasta is packaged and on the store shelf. Therefore, basing the drying protocol on the glass transition curve is vital to the production of uncracked pasta.

Processing conditions of extruded puffed cereals depend on the glass transi-

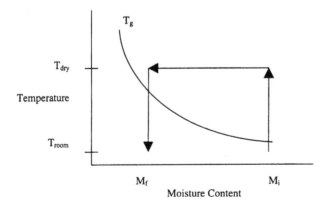

Figure 1   Phase-state diagram for pasta drying.

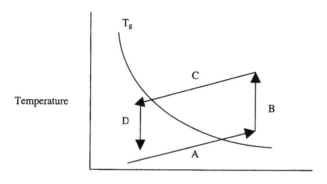

Figure 2   Phase-state diagram for cereal: A—increasing moisture content and temperature of flour to form a melt; B—further increase in the temperature of melt; C—exit through the die to a lower temperature and pressure, resulting in moisture leaving rapidly; D—cool to room temperature.

tion of the material, as shown in Figure 2. In the case of puffed extruded cereal, flour is metered into the extruder at a low moisture content and temperature. The moisture content and temperature are increased with water addition and heating to a temperature greater than the glass transition temperature. This forms a melt that can easily be worked into a dough and shaped. Pressure is also increased to greater than atmospheric. Upon exit through the die into lower temperature and pressure conditions, moisture is flashed off, forcing the product to expand. The puffed product is rapidly cooled to a temperature less than the glass transition temperature to preserve the structure. The resulting product is a light, airy, crunchy cereal. As demonstrated by these two cases, the state of a material under a given set of processing conditions greatly affects the final product texture.

Glass transition governs cereal product behavior during processing. Processing phenomena, such as collapse, stickiness, and diffusion, depend on the state of the material. The state of the material, rubbery or glassy, is related to the free volume. Moisture content, temperature, and molecular weight influence the free volume. Understanding material behavior as related to free volume is necessary for process optimization.

V.   NUMERICAL MODELS: MASS TRANSFER, STRESS DEVELOPMENT, AND MATERIAL PROPERTIES

The most efficient method for process optimization is the use of mechanistic models. Drying, tempering, and cooling can be optimized with regard to kinetics and product failure through the use of mass transfer and stress development models, which require the input of material properties.

A. Mass Transfer Models

According to the polymer literature, mass transfer can be classified according to the governing mechanisms into three basic categories: zero order, case I, and case II (51). *Zero-order transport*, termed *constant-rate drying*, refers to the case where the moisture content varies linearly with time, such as during the initial stages of drying, when the rate of moisture removal at the surface is equal to the rate of moisture transport to the surface. Zero-order transport is generally observed only during the initial stages of drying. *Case I diffusion* is characterized by a falling rate of moisture transport and is governed by Fick's law and Luikov's model. Luikov's model accounts for Soret and Dufour effects. *Case II diffusion* describes the case where material relaxation is important in moisture transport. Case II diffusion has been the subject of much research in the polymer field (52–54) and has application to moisture transport in foods. Achanta et al. (55) developed a mechanistic model, the thermomechanical model, for mass transfer in biopolymers during drying based on Case II diffusion. Since constant-rate drying (Case 0) occurs for a very short time during drying, models for falling-rate drying (Cases I and II) are of major interest for predicting mass transfer in cereal products.

One particular challenge in predicting mass transfer in cereal products, such as pasta, is the anomalous profiles observed with nuclear magnetic resonance (NMR). These profiles are fairly flat in the center of the pasta and sharply decrease near the edge of the pasta (56). Case I models predict parabolic profiles, which are nonexistent or observed late in the drying process. Case II models are capable of predicting the non-Fickian profiles due to the inclusion of a relaxation term. Following is a discussion of two popular Case I mass transfer models, Fick's law and Luikov's model, and a recently developed model of Case II diffusion, the thermomechanical model.

1. Case I: Fick's Law

Fick's law relates the moisture flux to the varying moisture content with distance via an effective diffusion coefficient. The effective diffusion coefficient lumps together vapor diffusion (due to Knudsen, Stefan, mutual diffusion, evaporation/condensation, and Poiseuille flow) and liquid diffusion (due to capillary flow, liquid flow, and surface diffusion). The model fails to reveal the importance of vapor and liquid diffusion at various stages during drying and produces parabolic moisture profiles. Therefore, Fick's law gives only a general idea of the moisture transport.

2. Case I: Luikov's Model

Luikov's model, also known as *irreversible thermodynamics*, includes the Soret and Dufour effects in addition to the moisture flux due to a concentration gradient. The irreversible thermodynamic approach requires energy, vapor, and liquid transport equations to model coupled heat and mass transfer in a capillary porous media. Solution of the partial differential equations provides drying profiles and curves for unsaturated flow.

Assumptions of the irreversible thermodynamic model include:

Media is an isotropic continuum.
Shrinkage and the effect of total pressure are neglected.
Quantities of different tensorial character do not interact (Curie's principle).
Local equilibrium exists.

Luikov's model has been used to predict mass transfer during pasta drying by Cummings (57) and Andrieu et al. (58).

Results obtained by Cummings revealed the minor influence of the Soret and Dufour effects and the presence of a moving drying front. Liquid flux dominated at high moisture contents, >20% dry basis, and vapor flux was favored at lower moisture contents, <10% dry basis. Increasing the temperature increased the moisture flux in both the liquid and vapor forms. Cummings' work showed the negligible influence of the Soret and Dufour effects and the importance of both liquid and vapor flux.

Andrieu et al. (58) also used the irreversible thermodynamic approach (Luikov's model) and made the following additional assumptions:

Vapor phase is negligible.
Drying conditions are isothermal.
Heat and mass transfer occur in only the radial direction.
Desorption enthalpy is negligible.
Total pressure is constant.
Thermomigration is negligible.
Evaporation occurs near the external surface.

Results showed parabolic moisture gradients. Moisture transport occurred in the liquid phase (adsorbed phase), and heat transfer was much more rapid than moisture transport, validating the assumption of isothermal drying conditions.

Application of Luikov's model to predict mass transfer during pasta drying revealed the negligible influence of Soret and Dufour effects, parabolic moisture profiles, the importance of liquid and vapor transport, and the negligible effect of heat transfer. Although the model offers insight into mass transfer during pasta drying, it fails to predict the non-Fickian moisture profiles observed experimentally.

3. Case II

The principle of glass transition influences mass transfer in Case II diffusion because time is required for the material to relax and for diffusion to occur. At the beginning of drying, the entire material is rubbery. As drying proceeds, a drying front moves inward—behind which the material is glassy and ahead of which the material is rubbery. At the drying front, the abrupt change from glassy to rubbery results in large changes in material properties and the potential for stress development.

Achanta et al. (55) developed a thermomechanical model using the hybrid mixture theory approach and Case II diffusion to predict the moisture profiles in a shrinking gel. This model has application to viscoelastic, isotropic, homogeneous, axisymmetric food systems. The model separates the viscous effects and diffusive effects on moisture transport into two terms. A modified Darcy's law, containing two terms (viscous and diffusive), is used to describe the moisture flux. This flux is used in the overall mass balance to predict drying rates and profiles. The thermomechanical model accounts for the influence of both diffusion and relaxation on mass transfer in a system experiencing glass transition.

The following assumptions were made in the model development:

There are no solutes in the water.
Resistivity tensor is isotropic.
Gravitational effects are negligible.
Heat transfer is rapid, so the drying process is assumed to be isothermal.
The surface is in equilibrium with the surroundings.
Flow is saturated.

The magnitude of the Deborah number, a ratio of the relaxation time to diffusion time, reveals the controlling mass transfer mechanism, and the mass transfer Biot number, the ratio of external mass transfer to internal mass transfer, determines the rate of surface drying. At low Deborah numbers, relaxation time is small compared to diffusion time, so relaxation is negligible and mass transfer follows Fickian kinetics. At high Deborah numbers, relaxation time is large compared to diffusion time, and relaxation controls the mass transfer, resulting in non-Fickian moisture profiles. At low mass transfer Biot numbers, external mass transfer limits drying; at high mass transfer Biot numbers, internal mass transfer limits drying. The magnitudes of the Deborah and mass transfer Biot numbers indicate the controlling mechanisms for mass transfer in deforming cereal products.

Achanta et al. (55) proposed the Shell number to describe surface drying. The Shell number is the ratio of the mass transfer Biot number to the Deborah number. The magnitude of the Shell number determines which mechanism, relaxation or diffusion, governs surface drying.

Achanta et al. (55) used the thermomechanical model to predict drying curves, drying rates, shrinkage, and moisture profiles during drying of biopolymers. Simulated drying curves agreed well with experimental data. Additionally, moisture profiles exhibited the non-Fickian moisture profiles observed for pasta with NMR. This indicated the potential applicability of the model to predicting mass transfer during cereal product drying.

4. Summary of Mass Transfer Models for Foods

The transport models presented here provide three examples of moisture transport modeling. Although Fick's Law is simple, it fails to account for phenomena such as vapor and liquid flow, various types of flow, and the importance of viscous relaxation during drying. The irreversible thermodynamics model accounts for both liquid and vapor flow and couples the mass and heat transfer. However, the model fails to predict the non-Fickian moisture profiles observed with NMR during pasta drying. Additionally, simulations reveal the negligible contribution of the Soret and Dufour effects in pasta drying. Case II diffusion, the thermomechanical model, is able to predict the flat profiles observed experimentally and the formation of a shell. The Deborah, mass transfer Biot, and Shell numbers provide useful information concerning material behavior. The thermomechanical model appears to have great potential for predicting mass transfer during cereal product drying, such as pasta.

B. Stress Development Models

Stresses during drying, tempering, and cooling are attributed to thermal and moisture gradients in the material. In foods, moisture transport is most responsible for stress development. A drying front is observed as moisture is removed first from the surface and then from interior points. During tempering and cooling, equilibration of moisture gradients results in stress. Therefore, both mass transfer and stress development models are needed to predict failure. Stress development models proposed to predict stresses in cereal products include pseudoelastic, viscoelastic, plastic, and elastic-plastic. The applicability of each stress model depends on product behavior.

1. Pseudoelastic Stress Model

The pseudoelastic stress model is applicable to brittle materials. Nakamura (59) used a pseudoelastic model with a bilayer, the crust and the crumb, to model stress development in a cracker during cooling via normal, forced, cold, and gradual practices. The crackers were assumed to be brittle, so the pseudoelastic model was most applicable. Results predicted failure for all cooling regimes due to moisture gradients. The crumb lost moisture, while the crust gained moisture, as found also by Kim (60). The largest stresses developed in the crust and resulted in failure for all cooling regimes. Experimental results confirmed checking for normal, forced, cold, and gradual cooling.

The pseudoelastic model with the thermomechanical mass transfer model was used by Achanta et al. (55) to predict stress development in shrinking biopolymer slabs during drying. The magnitude of the mass transfer Biot number greatly impacted the magnitude of the maximum stress. High mass transfer Biot numbers, on the order of 1500, resulted in crust formation and large stress buildup. The point of maximum stress occurred at the beginning of drying where the crust met the interior. Low mass transfer Biot numbers, on the order of 15, indicated less rapid moisture removal at the surface and resulted in less stress buildup. The maximum stress for drying at a low mass transfer Biot number was 250% less than the previous case involving a shell formation. No comparison was made to experimental results, but the stress development predictions illustrate the influence of crust formation on stress development.

2. Viscoelastic Stress Model

The viscoelastic stress model accounts for the time required for deformation and stress relaxation to occur. Several researchers have modeled heat and mass transfer with Luikov's model and viscoelastic stresses in foods. Cummings (57) and Litchfield (61) modeled pasta drying. Itaya et al. (62) modeled three-dimensional heat and mass transfer and stress development in starch/sucrose bars, and Akiyama et al. (63) studied crack formation and propagation in viscoelastic foods. Kim (60) modeled checking in crackers during cooling.

Cummings predicted failure to occur when the shear or normal stresses were greater than the calculated failure stress, as calculated and modeled from experimental results. The maximum stress occurred at the radial point where the largest moisture gradient occurred. Experimental results showed two types of cracks developing during drying: tangential surface cracks and surface cracks propagating inward. Table 2 compares the experimental and predicted times at

Table 2 Times at Which Experimental and Predicted Tangential Cracks Appeared

Temperature (°C)	Experimental time (hours)	Predicted time (hours)
53.1	3.5	4.4
66.0	2.33	3.6
79.4	2.25	2.8

Source: Ref. 57.

which the tangential surface cracks occurred for each drying temperature. The model overpredicted the time at which these cracks appeared, since the model did not predict cracks due to surface irregularities and weak spots. The second type of crack, surface cracks that propagated inward, was attributed to axial and shear failure. The actual time at which failure occurred was difficult to determine experimentally since the crack started on the surface but did not propagate immediately. The viscoelastic model overpredicted the time of failure as compared to experimental results.

Litchfield (61) modeled the pasta-drying process as consisting of a five-stage dryer. Stress increased due to moisture gradients; during cooling, stress increased as moisture and temperature gradients equilibrated.

Itaya et al. (62) modeled three-dimensional heat and mass transfer and stress development in starch/sucrose bars. Equations were solved using finite element analysis, and simulation results were compared to experimental results. Larger moisture gradients corresponded to larger stresses, and large predicted stresses agreed reasonably well with observed crack formation.

Akiyama et al. (63) studied crack formation and propagation in viscoelastic foods. A critical tensile stress criterion was used to predict failure, and equations were solved using finite element analysis. Simulated drying and heating curves agreed with experimental results. Additionally, simulated crack sizes and times agreed with experimental results.

Kim (60) used the viscoelastic model to predict checking in crackers for forced, ambient, and tempering cooling practices. Temperature and moisture profiles were determined using Fick's law for diffusion in two directions. The crackers were modeled as homogeneous, isotropic, viscoelastic cylinders. Results showed that moisture gradients caused stress development as moisture migrated from the crumb to the crust and as moisture was adsorbed on the crust from the air. Stresses on the crust were compressive, while stresses in the crumb were tensile. Failure was predicted for only the forced convection condition, but experimental results showed checking for all cooling processes.

3. Plastic Stress Model

Materials that deform experience plastic stresses. Plastic stresses in pasta were modeled by Andrieu et al. (58), but the viscoelasticity of pasta was ignored. Andrieu et al. (58) modeled pasta drying using Luikov's model and the method of successive elastic solutions for the plastic stresses. At the beginning of drying, the axial stress was tensile at the surface of the pasta and compressive in the core. These stresses inverted, such that at the end of drying, the surface was under compression and the core was under tension. No comparison of predicted stresses to experimentally determined failure stresses was made. Most deformation in cereal products, such as dough, requires time for the deformation to occur. Therefore, caution must be taken when applying a non-time-dependent plastic stress model to such materials.

4. Elastic-Plastic Stress Model

The elastic-plastic stress model describes stress in both deforming and non-deforming states. An elastic-plastic stress model has been tested on food and a sensitivity study conducted. Tsukada et al. (64) studied mass transfer and stress development in starch bars using Luikov's model and an elastic-plastic stress model. Elastoplasticity was assumed, since these materials were more susceptible to cracking than viscoelastic materials. Experimentally observed cracks qualitatively corresponded to stress distributions predicted using the simulations. Liu et al. (38) conducted a sensitivity analysis for stress development during drying using Luikov's model and the elastic-plastic stress model. Air humidity and temperature had the greatest effect on time at which cracks formed. These results incorporate the importance of plastic deformation in stress development, but neglect the phase transition, glass transition, that occurs during drying.

5. Summary of Stress Development Models for Foods

Several models have been proposed to predict stress development in foods during drying, cooling, and tempering, and model selection depends on the food system of interest. The pseudoelastic and plastic models predict stresses in the undeforming and deforming states, respectively, but neglect viscoelasticity. Viscoelastic stress models predict the time-dependent behavior of cereal products, but neglect plastic stress. The elastic-plastic stress model accounts for the material behavior in both the glassy and rubbery states. The elastic-plastic model can be modified to account for viscoelasticity by summing stresses over time. Knowledge of the mechanisms contributing to stress development in a food system allows the researcher to select the most applicable model.

C. Mass Transfer and Stress Development in Other Materials

Much of the work on mass transfer and stress development in foods neglects shrinkage and the role of glass transition, with the exception of the thermomechanical model (55). Studies on mass transfer and stress development in other areas offer insight into modeling mass transfer and stress development in a material undergoing glass transition. Stress development during drying of thin films has been experimentally studied and numerically modeled. The ceramics literature addressed changes in material properties during drying. The ideas from these areas can be applied to mass transfer and stress development in foods during drying.

Thin coatings are particularly susceptible to cracking during drying, due to constrained shrinkage. Croll (65) postulated that internal strain developed from a difference between the volume of the solvent present at solidification and the volume of the solvent at the conclusion of drying:

$$\sigma = \frac{E}{1 - \nu} \frac{\phi_s - \phi_r}{3} \tag{1}$$

where σ is stress, E is Young's modulus, ν is Poisson's ratio, ϕ_s is the volume fraction of solvent at solidification, and ϕ_r is the volume fraction of solvent in the dry coating. Solidification was thought to occur at the glass transition. Croll experimentally measured stress in polystyrene and poly(isobutyl methacrylate) thin coatings, with toluene as the solvent, by cantilever deflection. Table 3 compares the experimentally determined stresses and predicted stresses. Although the predicted stresses were a little low, the model provided a basic mechanism for stress development in solid materials.

Sato (66) and Perera and Eynade (67) recognized the importance of glass transition in coatings on internal stress due to volumetric shrinkage. Croll (68) studied the effect of solvent concentration, evaporation rate, coating thickness, and cross-linking kinetics on strain in epoxy coatings. The solidification point was taken as the time at which a cross-linked network was formed. Strain increased with increasing film thickness, and strain decreased with increasing evaporation rate and solvent concentration above 50%. These studies on coatings provide direction for modeling stress development in materials that undergo a phase transition during drying.

Stress development during drying of tape-cast ceramic layers was linked to material properties by Lewis et al. (69). Stress was higher in nonplasticized materials than plasticized materials by an order of magnitude. The stress was due to constrained volume shrinkage associated with solvent loss. With drying, viscosity increased so the evaporation rate became greater than the shrinkage, which led to the formation of a drying front. Therefore, the material did not deform to accommodate the volume lost in solvent evaporation. The observed

Table 3 Experimental and Predicted Stresses in Thin Coatings

Material	Experimental stress	Predicted stress
Polystyrene	14.3 MPa	10.8 ± 1.7 MPa
Poly(isobutyl methacrylate)	4.5 MPa	3.8 ± 0.9 MPa

Source: Ref. 65.

changes in viscosity explained the behavior of the material under a given set of processing conditions and the final product quality.

Mass transfer and stress modeling in food materials neglect the plastic deformation and glass transition that occur during drying but do account for the viscoelastic nature of foods. The materials literature offers some insight as to the driving forces behind stress development due to shrinkage and how to model it. By combining the ideas from the food and materials literature, mass transfer and stress development during drying of a food material can be modeled.

D. Glassy Strain

The idea of a glassy strain is proposed to model stress development in a material that contains glassy and rubbery regions. When drying a rubbery material, all points remain deformable as the moisture content decreases to equilibrium, and the volume of water lost is equal to the change in volume of the solid as the material shrinks. When drying a glassy material, no deformation occurs as moisture is lost. Most commonly, a material is neither entirely rubbery nor entirely glassy during drying; rather, a continuum from rubbery in the interior to glassy at the exterior exists. Therefore, defining strain, which depends on deformation, is difficult. The idea of glassy strain has been developed to predict stresses in systems containing both rubbery and glassy regions (70).

Stress models for the drying of thin films provides the basic idea for glassy stress. Croll (65) defines strain in a solid coating as the difference between the volume fraction of the solvent at the point of solidification and the volume fraction of the solvent in the dry film:

$$\varepsilon = \phi_s - \phi_r \tag{2}$$

The strain can be understood as the change in volume, or any dimension, between the solidified and dried states. Croll's definition of strain can be used to calculate the overall stress in the dry film, but it cannot be used to calculate the stress during finite time steps during drying.

Using the same idea of strain due to lost moisture, Croll's equation is modified to predict the strain at finite time steps. If a material is completely rubbery, material deformation is equal to the volume of moisture lost during a given time step, so strain in the radial direction is given by:

$$\varepsilon_{ru} = \frac{r_{ru,n+1} - r_{ru,n}}{r_{ru,n}} \tag{3}$$

where ε_{ru} is the rubbery strain in the radial direction and $r_{ru,n}$ is the radial position in the rubbery state at time step "n." In the glassy state, the material does not deform as moisture is lost. However, using the idea of strain due to moisture

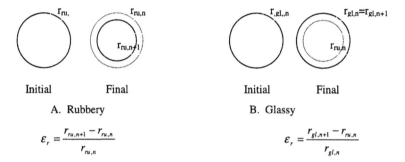

$$\varepsilon_r = \frac{r_{ru,n+1} - r_{ru,n}}{r_{ru,n}}$$

$$\varepsilon_r = \frac{r_{gl,n+1} - r_{ru,n}}{r_{gl,n}}$$

Figure 3 Glassy strain for the glassy and rubbery states. The solid lines represent the actual radial position of the material. A: Shows a differential volume element of unit thickness in the rubbery state. The material deformation is equal to the moisture lost as the material deforms from the dashed line to the solid line in the "Final" state. B: Shows a differential volume element of unit thickness in the glassy state. The material does not deform with moisture loss. In the "Final" state, the solid line is the actual position of the undeformed material, and the dashed line is the deformed position the material would take if it could deform.

loss, strain in the glassy state is defined as the difference between the glassy radial point and the rubbery radial point after moisture loss during a given time step. The rubbery radial point is the deformation that would occur if the material deforms to compensate for the moisture loss. The glassy strain is defined as follows:

$$\varepsilon_{gl} = \frac{r_{gl,n+1} - r_{ru,n}}{r_{gl,n}} \tag{4}$$

where ε_{gl} is the glassy strain and $r_{gl,n}$ is the radial position in the glassy state at time step "n." Figure 3 depicts the concept of glassy strain and compares strain in the rubbery and glassy states.

The idea of glassy strain, based on deformation due to moisture loss, is used to predict the strain in both the rubbery and glassy states. This new definition of strain enables the elastic-plastic stress model to predict stress in both the rubbery and glassy states.

E. Material Properties

Several material properties are required as inputs to the mechanistic models to perform multiple calculations. The thermomechanical model requires the following material properties: isotherms, glass transition temperature, storage modulus, diffusion coefficient, and Young's modulus. The material properties required for the stress–strain model include Young's modulus, Poisson's ratio, and failure stress. Material properties are temperature and moisture dependent, and Young's modulus and failure stress are also time dependent. The material properties can be experimentally measured and data fit to correlations that predict the material behavior in both the glassy and rubbery states.

1. Isotherms

Isotherms relate the relative humidity of the surrounding air, also called the water activity of the sample, to the equilibrium moisture content of the sample. This relationship has important implications in shelf life, texture, drying, and food formulation. Many models exist, and their applicability is determined by the goodness of fit with experimental data. One of the most common methods of determining the relationship between relative humidity and equilibrium moisture content is to measure the moisture content of samples equilibrated over saturated salt solutions. The data are then fit to one of the isotherm models. Table 4 lists some isotherm equations. Iglesias and Chirife (83) and Wolf et al. (84) have published reviews on food isotherm data and constants.

2. Glass Transition Temperature

For a given food product, the glass transition temperature is not a single temperature, but a range of temperatures due to the various macromolecular components in the food and the range of molecular weights of the macromolecular components. Glass transition is a second-order phase transition marked by discontinuities in material properties. Two common methods used to determine the glass transition temperature are differential scanning calorimetry (DSC) and dynamic mechanical thermal analysis (DMTA). Other methods include nuclear magnetic resonance (NMR), electron spin resonance (ESR) spectroscopy, and fluorescence recovery after photobleaching (FRAP) (46).

    The glass transition temperature is a function of moisture content and varies exponentially. The Arrhenius model describes the moisture dependence of the glass transition temperature:

$$T_g(M) = T_{g0} \exp(-k_1 M) \tag{5}$$

The constants T_{g0} and k_1 are constants fit to experimental data.

3. Storage Modulus

The storage modulus is a measure of the ability of a material to deform. In the glassy region, the storage modulus is high, on the order of 10^{12} Pa, and drops sharply at the glass transition by multiple orders of magnitude. In the rubbery

Table 4 Isotherm Equations

Isotherm name	Equation	Comments
BET (71)	$\dfrac{M}{M_0} = \dfrac{Ca_w}{(1 - a_w)[1 + (C - 1)a_w]}$	Applicable in a_w range of 0.05–0.45
Harkins–Jura (72)	$\ln(a_w) = B - \dfrac{A}{M^2}$	
Bradley (73)	$-\ln(a_w) = AB^M$	Based on assuming sorptive surface is polar Derived by the polarization theory
Smith (74)	$M = A - B[\ln(1 - a_w)]$	Applicable for high a_w of biopolymers and foods
Oswin (75)	$a_w = \dfrac{(M/C)^{1/n}}{1 + (M/C)^{1/n}}$	
Henderson (76)	$a_w = 1 - \exp(-CTM^n)$	Popular for agricultural products
Chung–Pfost (77)	$a_w = \exp\left[\dfrac{-C_1}{RT} \exp(-C_2M)\right]$	
Chen (78)	$a_w = \exp[K + A \exp(-BM)]$	
Alam–Shove (79)	$M = A + Ba_w + Ca_w^2 + Da_w^3$	
GAB (80, 80)	$\dfrac{M}{X_m} = \dfrac{CKa_w}{(1 - Ka_w)(1 - Ka_w + CKa_w)}$	Multilayer model Used for many food products
Iglesias and Chirife (82)	$\ln[M + (M^2 + M_{0.5})^{0.5}] = Aa_w + B$	

region, the storage modulus continues to decrease, but at a slower rate. These drastically different behaviors between the transition and rubbery regions necessitate the use of multiple models to predict the storage modulus over a wide temperature and moisture content range.

One of the classical models for the temperature dependence of a material property is the Arrhenius model:

$$E'(T) = E'_g \exp\left[\frac{E_a}{R_g}\left(\frac{1}{T} - \frac{1}{T_s}\right)\right] \tag{6}$$

where E' is the storage modulus, E'_g is the storage modulus in the glassy state, E_a is activation energy, R_g is the gas constant, T is temperature, and T_s is a reference temperature. The activation energy is determined from the slope of the plot of $\ln(E'/E'_g)$ versus $(1/T)$. The Arrhenius model is applicable above the glass transition, and application depends on the linearity of $\ln(E'/E'_g)$ versus $(1/T)$.

Polymer scientists developed the Williams–Landel–Ferry (WLF) model [Eq. (7)] based on the principle of free volume to predict material property dependence on temperature in the rubbery region (85):

$$\log\left(\frac{E'(T)}{E'_g}\right) = \frac{-C_1(T - T_g)}{C_2 + T - T_g} \tag{7}$$

where T_g is the glass transition temperature. The constants C_1 and C_2 are often taken as 17.44 and 51.6, respectively. The applicability of these standard coefficients is questioned for food products and discussed by Peleg (86). For a given product, the constants can be determined from experimental data. The model is modified to account for the effects of moisture content by incorporating the Arrhenius equation for the glass transition temperature. The WLF model is applicable at temperatures well above the glass transition temperature, $T_g + 30$ or $50°C$.

Peleg (86–90) suggests Fermi's model for modeling material properties around the glass transition. Equation (8) shows Fermi's model for a constant moisture content, and Eq. (9) shows Fermi's model for a constant temperature. The constants "a_T" and "a_M" reflect the steepness of the curve.

$$\frac{E'(T)}{E'_g} = \frac{1}{1 + \exp\left[(T - T_g)/a_T\right]} \tag{8}$$

$$\frac{E'(M)}{E'_g} = \frac{1}{1 + \exp\left[(M - M_g)/a_M\right]} \tag{9}$$

Figure 4 shows a normalized plot of any material property versus temperature of moisture using Fermi's model. Fermi's model has the downward concavity characteristic of material properties near glass transition.

Fermi's model can be modified to account for both temperature and moisture dependence. Using the constant-moisture equation [Eq. (8)], "T_g" and "a_T" are modified to account for moisture dependence as shown in Eqs. (10) and (11):

$$T_g = T_{g0} \exp(-k_1 M) \tag{10}$$

$$a_T(M) = a_0 \exp(-k_2 M) \tag{11}$$

In Eq. (10), T_{g0} is the glass transition temperature of bone-dry material. The constant k_1 is found by fitting the model for T_g versus moisture data. The constants k_2 and a_0 in Eq. (11) are found by first finding the "a_T" values for $E'(T)/E'_g$ over

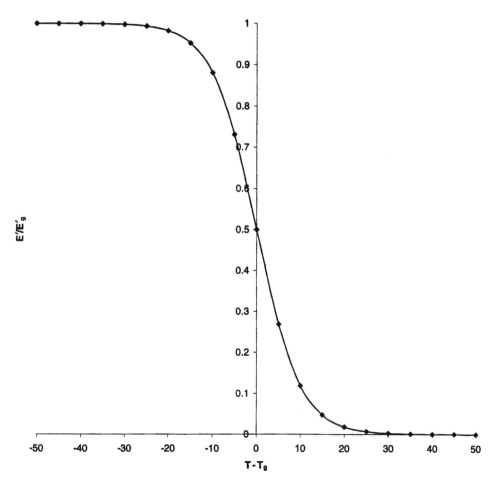

Figure 4 Normalized Fermi's model.

a range of moisture contents using Eq. (8). The a_T and moisture content data are then used to fit Eq. (11) and determine k_2 and a_0. Substituting Eqs. (10) and (11) into the original Fermi's model results in:

$$\frac{E'(T, M)}{E'_g} = \frac{1}{1 + \exp \left[(T - T_{g0} \exp (-k_1 M))/a_0 \exp (-k_2 M)\right]} \tag{12}$$

Even though drying is an isothermal process, using an equation with temperature dependence allows the model to be used over a range of processing temperatures.

To predict the storage modulus over a wide temperature and moisture content range, either a two-part model or a further modified Fermi's model may be used. The WLF and Arrhenius models have application above the glass transition temperature when the material is rubbery, and Fermi's model has application at the glass transition. Application of the Arrhenius model depends on the linearity of $\ln(E'/E'_g)$ versus $1/T$, and application of the WLF model depends on the linearity of $\log(E'/E'_g)$ versus T. For a two-part model, one can apply the Arrhenius or WLF model above the glass transition and Fermi's model at the glass transition. Alternatively, Fermi's model can be further modified to predict the storage modulus in the rubbery region with the addition of a constant:

$$\frac{E'(T, M)}{E'_g} = \frac{E'_r}{E'_g} + \frac{1}{1 + \exp\left[(T - T_{g0}\exp(-k_1 M))/a_0 \exp(-k_2 M)\right]} \tag{13}$$

Application of a two-part model or the modified Fermi's model depends upon the fit of the experimental data to the models.

4. Diffusion Coefficient

Many models of the diffusion coefficient as a function of temperature and moisture content have been proposed, and Waananen (39) provides an excellent review of diffusion coefficient models. While many of the models are empirically based, the theory of free volume from the polymer literature is theoretical in nature. The theory of free volume is used to describe the temperature and moisture dependence of the diffusion coefficient in both the rubbery and glassy states. A molecule migrates by jumping into free volume holes formed by natural thermal fluctuations. In the glassy state, the free volume of a material is a minimum. As the temperature or moisture content increases, the material becomes rubbery, and the free volume increases. The free volume in a material comes from the interstitial spaces and hole formation.

Three conditions must be met to apply the free volume theory:

1. It is valid at temperatures from T_g to $T_g + 100$.
2. The weight fraction of polymer must be greater than 0.2.
3. Relaxation time is less than diffusion time.

For most food-drying applications, conditions 1 and 2 are easily met. Condition 3 is satisfied by considering the magnitude of the Deborah number and what mechanism controls mass transfer during drying. For low Deborah numbers, diffusion time is greater than relaxation time, and condition 3 is satisfied. This type of mass transfer is Fickian. For high Deborah numbers, relaxation controls transport and diffusion is negligible. Therefore, the diffusion term in the thermomechanical model is less important than the viscous relaxation term.

Two versions of the free volume theory for diffusion have been developed: Fujita theory and Vrentas–Duda theory (91). The Fujita theory is valid only in the rubbery region, but the Vrentas–Duda theory is valid for both the glassy and rubbery regions. The Fujita theory is a special case of the Vrentas–Duda theory. Although both theories are essentially equivalent in correlating diffusivity data, the Vrentas–Duda theory is semipredictive and all parameters are given a physical interpretation. Most of the parameters can be estimated from pure component data, and only a few diffusivity data points are necessary to determine the key parameters (91). Additionally, the model can predict diffusivity over a wide temperature and concentration range. The Vrentas–Duda model is given in Eq. (14):

$$D_1 = D_0 \exp\left(\frac{-E_a}{R_g T}\right) \exp\left[\frac{-\gamma(\omega_1 \hat{V}_1^* + \omega_2 \xi \hat{V}_2^*)}{\omega_1 \hat{V}_{FH1} + \omega_2 \hat{V}_{FH2}}\right] \tag{14}$$

where D_1 is the self-diffusion coefficient, D_0 is the pre-exponential factor, E_a is the activation energy per mole, R_g is the gas constant, and T is temperature. The overlap factor, γ, is introduced because the same free volume may be available to more than one molecule. The weight fractions of the solvent and polymer are ω_1 and ω_2, respectively. The critical hole free volumes of the solvent and polymer, $\hat{V}_1^*$ and $\hat{V}_2^*$, are taken as the specific volume at 0 K. ξ is found by fitting the model to experimental data. The specific hole free volume of the solvent at a temperature T is $\hat{V}_{FH1}$, and the specific hole free volume of the pure polymer at a temperature T is $\hat{V}_{FH2}$, which depends on the state of the polymer, rubbery or glassy.

The following assumptions are made in utilizing the free volume theory for diffusion (47):

The partial specific volumes of the solvent and polymer are independent of composition. No volume change occurs during mixing.

An average thermal expansion coefficient is valid over the temperature range of interest.

The overlap factor, γ, is independent of solvent concentration.

The activation energy per mole, E_a, is independent of solvent concentration.

The Vrentas–Duda free volume theory for diffusion provides a mechanistic model in which where each parameter has physical significance, and the model reflects the material's behavior under a given set of conditions.

The mutual diffusion coefficient is usually measured gravinometrically during drying, and the self-diffusion coefficient is measured using permeability or NMR studies. The mutual diffusion coefficient is a measure of water movement over a large distance scale, several millimeters. The self-diffusion coefficient is a measure of the localized motion, Brownian motion, of the water molecules. In the limit of zero moisture content, the mutual and self-diffusion coefficients are the same.

5. Young's (Elastic) Modulus

Viscoelastic materials require time to deform; therefore, the time dependence, in addition to the temperature and moisture content dependencies, is included in the model for Young's modulus. The generalized Maxwell model is used to describe the time, moisture content, and temperature dependence of Young's modulus:

$$E(t, T, M) = \frac{T_0}{T} \left[\sum E_i \exp\left(\frac{-ta_T(T)a_M(M)}{\tau_i}\right) + E_e \right] \tag{15}$$

where E is Young's modulus, t is time, T is temperature, M is moisture content, T_0 is a reference temperature, E_i are constants, a_T and a_M are shift factors, τ_i are relaxation times, and E_e is the equilibrium value of Young's modulus. The generalized Maxwell model is based on the idea that the behavior of viscoelastic materials' can be visualized as an elastic spring and viscous dashpot.

Many materials, including pasta, are thermorheologically simple, which means that data taken at multiple temperature, moisture contents, and strain rates superimpose onto a master curve. The slope of the master curve, E_r, is the reduced elastic modulus, which follows the generalized Maxwell model and reflects temperature, moisture content, and time dependence. The model is fit to the data according to the following steps (57, 59, 60).

1.   Plot $\ln(\sigma M_0/\dot{\varepsilon}M)$ versus $\ln(\varepsilon/\dot{\varepsilon})$, where M_0 is a reference moisture content. The constant moisture curves are shifted to the reference moisture content by taking the ratio of Young's modulus at a given moisture content to Young's modulus at the reference moisture content:

$$a_M = \frac{E(T, M)}{E(T, M_0)} \tag{16}$$

2.   Plot $\ln(\sigma M_0/\dot{\varepsilon}Ma_M)$ versus $\ln(\varepsilon/\dot{\varepsilon}a_M)$, which results in one curve for all moisture contents.
3.   The temperature shift is similar to the moisture content shift:

$$a_T = \frac{E(T, M)}{E(T_0, M_0)} \tag{17}$$

4.   Plot $\ln(\sigma_{red})$ versus $\ln(\varepsilon_{red})$. Reduced stress, σ_{red}, and reduced strain, ε_{red}, are defined in Eqs. (18) and (19), respectively:

$$\sigma_{red} = \frac{\sigma M_0 T_0}{\dot{\varepsilon} M T a_M a_T} \tag{18}$$

$$\varepsilon_{red} = \frac{\varepsilon}{\dot{\varepsilon} \tau a_M a_T} \tag{19}$$

The reduced Young's modulus is:

$$E_r(t) = \frac{\sigma_{red}}{\varepsilon_{red}} \frac{d \ln \sigma_{red}}{d \ln \varepsilon_{red}} \tag{20}$$

The reduced Young's modulus is fit to the Maxwell model.

6. Failure Criterion

Failure occurs when the stress is larger than the maximum sustainable stress. Several failure criteria have been proposed, as follows (57, 60):

> *Maximum normal stress theory*: Failure occurs when the maximum normal stress exceeds the ultimate strength:
>
> $$\sigma_{max}(r, t) \geq \sigma_f \tag{21}$$
>
> where σ_{max} is the maximum normal stress, r is radial position, t is time, and σ_f is the ultimate stress. The ultimate stress is measured experimentally either in tension or compression.
>
> *Maximum shearing stress theory*: Failure occurs when the maximum shear stress at any point exceeds the ultimate shear stress of the material [Eq. (22)]. This theory is commonly used for ductile materials.
>
> $$\tau_{max}(r, t) \geq \tau_f = \frac{\sigma_f}{2} \tag{22}$$
>
> where τ_{max} is the maximum shear stress and τ_f is the ultimate shear stress.
>
> *Maximum strain theory*: Failure occurs when the maximum principle strain reaches the ultimate uniaxial tension strain. Equation (23) gives this criterion for a biaxial stress state:
>
> $$\sigma_1 - \frac{\sigma_2}{2} = \sigma_f \tag{23}$$
>
> *Huber–Hencky–von Mises theory (maximum-distortion-energy density theory)*: The strain energy that produces a volume change is responsible for failure by yielding. The distortion energy density is:
>
> $$\bar{u}_d = \frac{1 + v}{6E} [(\sigma_1 - \sigma_2)^2 + (\sigma_2 - \sigma_3)^2 + (\sigma_3 - \sigma_1)^2] \tag{24}$$

where $\bar{u}_d$ is the distortion energy density, v is Poisson's ratio, and E is the elastic modulus. Failure occurs when the maximum-distortion-energy density reaches a critical value, as determined by experiments, as shown in Eq. (25).

$$[(\sigma_1 - \sigma_2)^2 + (\sigma_2 - \sigma_3)^2 + (\sigma_3 - \sigma_1)^2] \geq \frac{\sigma_f^2}{2} \tag{25}$$

Maximum strain energy theory: Failure occurs when the maximum strain energy reaches the ultimate strain energy for uniaxial tension. The failure stress is:

$$(\sigma_1^2 - \sigma_1\sigma_2 + \sigma_2^2)^{0.5} = \sigma_f \tag{26}$$

For an amorphous material undergoing glass transition, tensile failure occurs in the glassy state when the material is brittle (37, 38, 92). The maximum normal stress theory is used as the failure criterion (57, 60). Since the material is viscoelastic and thermorheologically simple, the failure stress is temperature, moisture content, and time dependent (57). Therefore, a generalized Maxwell model for a master curve is used for the failure stress:

$$\sigma_f(t, T, M) = \frac{TM}{T_0 M_0}(\sigma_e + \Sigma\, \sigma_i \exp[t/a_M a_T \tau_i]) \tag{27}$$

where σ_f is failure stress and σ_e is the equilibrium value of failure stress. The failure stress is the maximum sustainable stress in a three-point bending test. A master curve reflects temperature, moisture content, and strain rate dependence by using moisture and temperature shifts.

7. Poisson's Ratio

Poisson's ratio is the ratio of the transverse strain to the axial strain when a uniformly distributed load is applied axially. Assuming an isotropic material, Poisson's ratio is constant. Poisson's ratio has been measured for a variety of materials, listed in Table 5. A value of 0.30 is assumed for pasta based on the values listed.

Table 5 Poisson's Ratio for Some Materials

Material	Poisson's ratio
Apple	0.21 to 0.34
Corn—horny endosperm (10% moisture)	0.32
Glass	0.24
Polyethylene	0.33

Source: Ref. 93.

8. Summary of Material Property Models

Polymer systems, whether biological or chemical, exhibit complex behavior as they undergo glass transition and due to their viscoelasticity. Successful prediction of these material properties as functions of temperature and moisture around and above the glass transition provides accurate information to be used in the mass transfer and stress development models. The free volume theory is a mechanistic basis for the prediction of the elastic modulus and diffusion coefficient around the glass transition. The viscoelastic behavior of polymer systems is accounted for by Maxwell's model. Experimental data on the food system of interest can be fit to the material property models to provide accurate descriptions of the material behavior.

F. Summary of Numerical Models

Selection of mass transfer and stress development models depends on the system of interest. Common mass transfer models include the Arrhenius model, Luikov's model, and the thermomechanical model. Only the thermomechanical model is capable of predicting mass transfer in deforming systems. Stress development models include pseudoelastic, viscoelastic, plastic, and elastic-plastic. The elastic-plastic is most applicable to materials undergoing glass transition when strain is defined using the theory of glassy strain. To maximize the accuracy of mass transfer and stress development predictions in cereal products, mechanistic models that account for glass transition during drying need to be developed with material property correlations based on principles of material behavior.

VI. CONCLUSIONS

Extruded cereal products are subjected to temperature and moisture gradients during drying, tempering, and cooling that can lead to cracking. Research has shown that temperature gradients equilibrate quickly relative to moisture gradients, so moisture gradients are responsible for stress development. The ability of a product to resist cracking depends on the quality of the raw ingredients, processing conditions, and storage conditions.

Various researchers have developed models to predict stress development and failure in pasta. Temperature and moisture profiles are predicted with transport equations, and the profiles are used in the stress prediction model. Only the thermomechanical mass transfer model predicts flat, ''non-Fickian'' profiles, and shrinkage. The elastic-plastic stress model predicts stresses in both the rubbery and glassy states. However, the definition of strain must be modified to predict glassy strains according to the glassy strain theory. Material properties as functions of temperature and moisture are required in the mass transfer and stress models. Failure is determined by comparing the predicted stress to an experimentally determined stress value causing failure. The models allow the researcher to simulate various drying and storage conditions and the effects on failure of the product.

In reviewing the previous work to predict product quality based on cracking, two research needs come to mind. The first involves the role of the raw materials and processing prior to drying. Both raw materials and processing conditions affect the structure of the raw, undried, product. Variations in raw materials and processing conditions result in variations in material properties, which influence mass transfer and stress development. Therefore, the link between raw product characteristics and material property variation around the glass transition needs to be determined. The second area of research involves the prediction of crack propagation. Thus far the models used to predict stress and failure in extrudates consider stresses during drying, tempering, and cooling. However, cracking can occur after the product is packaged. The manufacturer has less control over the quality of the product after it leaves the factory and is in distribution. Therefore, models need to be developed to predict cracking during storage and crack propagation. Previous researchers have provided insightful information that can be applied to furthering the knowledge of stress cracking in extruded cereal products.

NOMENCLATURE

A	constant in isotherm equations
a_o	constant for Fermi's model
a_M	moisture shift factor
a_T	temperature shift factor
a_w	water activity
B	constant in isotherm equations
C_1, C_2	constants
C	constant in isotherm equations
D_1	self-diffusion coefficient
D_0	pre-exponential term in free volume of diffusion
D	constant in isotherms
E	Young's modulus
$E_1 \ldots E_n$	Young's modulus constants
E'	storage modulus
E'_g	glassy storage modulus
E'_r	rubbery storage modulus
E_a	activation energy

E_e	equilibrium value of Young's modulus
E_r	reduced Young's modulus
k_1, k_2	constants
M	moisture content
M_0	reference moisture content
n	exponential constant in isotherm equations
r	radial distance
r_{ru}	rubbery radius
r_{gl}	glassy radius
R_g	gas constant
t	time
T	temperature
T_g	glass transition temperature
T_{g0}	pre-exponential term in Arrhenius equation for T_g
T_0	reference temperature
T_s	reference temperature
$\bar{u}_d$	distortion energy density
$\hat{V}_1^*$	specific critical hole free volume of the solvent
$\hat{V}_2^*$	specific critical hole free volume of the polymer
$\hat{V}_{FH1}$	solvent specific hole free volume
$\hat{V}_{FH2}$	polymer specific hole free volume
X_m	monolayer moisture content

Greek Symbols

γ	overlap factor
ε	strain
ε_{gl}	glassy strain
ε_r	radial strain
ε_{ru}	rubbery strain
ε_{red}	reduced strain
$\dot{\varepsilon}$	strain rate
ϕ_s	volume fraction of solvent at solidification
ϕ_r	volume fraction of solvent in dry coating
σ	stress
$\sigma_1, \sigma_2, \sigma_3$	stress in various directions
σ_e	equilibrium stress
σ_f	failure stress
σ_i	failure stress constants
σ_{max}	maximum stress
σ_{red}	reduced stress
τ	relaxation time

τ_{max}	maximum shear stress
τ_f	shear failure stress
ν	Poisson's ratio
ξ	constant in free volume theory for diffusivity
ω_1	solvent volume fraction
ω_2	polymer volume fraction

ACKNOWLEDGMENTS

This material is based upon work supported by the National Science Foundation under Grant No. BES 9510066. Any opinions, findings, and conclusions or recommendations expressed in this material are those of the authors and do not necessarily reflect the views of the National Science Foundation (NSF). Additionally, funds were provided by USDA-NRI Grant 97-35503-4409 and the USDA National Needs Fellowship 95-384-20-2206.

REFERENCES

1. C. Antognelli. The Manufacture and Applications of Pasta as a Food and as a Food Ingredient: A Review. J. Food Tech. 15:125–145, 1980.
2. N.P. Ames, J.M. Clarke, B.A. Marchylo, J.E. Dexter, S.M. Woods. Effect of Environment and Genotype on Durum Wheat Gluten Strength and Pasta Viscoelasticity. Cereal Chem. 76(4):582–586, 1999.
3. J.E. Dexter, B.A. Marchylo, V.J. Mellish. Effects of Frost Damage and Immaturity on the Quality of Durum Wheat. Cereal Chem. 71(5):494–501, 1994.
4. J.E. Dexter, R.R. Matsuo, J.E. Kruger. The Spaghetti-Making Quality of Commercial Durum Wheat Samples with Variable α-Amylase Activity. Cereal Chem. 67(5): 405–412, 1990.
5. N.H. Oh, P.A. Seib, A.B. Ward, C.W. Deyoe. Noodles VI. Functional Properties of Wheat Flour Components in Oriental Dry Noodles. Cereal Foods World 30(2):176–178, 1985.
6. M.I.P. Kovacs, N.K. Howes, D. Leisle, J. Zawistowski. Effect of Two Different Low Molecular Weight Glutenin Subunits on Durum Wheat Pasta Quality Parameters. Cereal Chem. 72(1):85–87, 1995.
7. G. Fabriani, C. Lintas. Durum Chemistry and Technology. St. Paul, MN: American Association of Cereal Chemists, 1988.
8. A. Eliasson. Lipid–Carbohydrate Interactions. In: R.J. Hamer and R.C. Hoseney, eds. Interactions: The Keys to Cereal Chemistry. St. Paul, MN: American Association of Cereal Chemists, 1998, pp 47–79.
9. P.C. Mestres, M.C. Alexandre, F. Matencio. Comparison of Various Processes for Making Maize Pasta. J. Cereal Sci. 17:277–290, 1993.

10. W Bushuk. Interactions in Wheat Doughs. In: R.J. Hamer and R.C. Hoseney, eds. Interactions: The Keys to Cereal Chemistry. St. Paul, MN: American Association of Cereal Chemists, 1998, pp 1–16.

11. R.L. Whistler, J.N. BeMiller. Carbohydrate Chemistry for Food Scientists. St. Paul, MN: American Association of Cereal Chemists, 1997.

12. P.S. Given. Influence of Fat and Oil Physiochemical Properties on Cookie and Cracker Manufacture. In: H. Faridi, ed. The Science of Cookie and Cracker Production. New York: Chapman & Hall, 1994, pp 163–201.

13. B.J. Donnelly. Pasta: Raw Materials and Processing. In: K.J. Lorenz and K. Kulp, eds. Handbook of Cereal Science and Technology. New York: Marcel Dekker, 1991, pp 763–792.

14. B. Laignelet. Lipids in Pasta and Pasta Processing. In: P.J. Barnes, ed. Lipids in Cereal Technology. New York: Academic Press, 1983, pp 269–286.

15. K.L. Rho, O.K. Chung, P.A. Seib. Noodles VIII. The Effect of Wheat Flour Lipids, Gluten, and Several Starches and Surfactants on the Quality of Oriental Dry Noodles. Cereal Chem. 66(4):276–282, 1989.

16. Y. Pomeranz. Composition and Functionality of Wheat Flour Components. In: Y. Pomeranz, ed. Wheat Chemistry and Technology. St. Paul, MN: American Association of Cereal Chemists, 1988, pp 219–370.

17. P.R. Shewry, A.S. Tatham, J. Forde, M. Kreis, B.J. Miflin. The Classification and Nomenclature of Wheat Gluten Proteins—A Reassessment. J. Cereal Sci. 4(2):97–106, 1986.

18. P.R. Shewry, A.S. Tatham, F. Barro, P. Barcelo, P. Lazzeri. Biotechnology of Breakmaking: Unraveling and Manipulating the Multi-Protein Gluten Complex. Bio/Technology 13:1185–1190, 1995.

19. K. Kobrehel, R. Alary. The Role of a Low Molecular Weight Glutenin Fraction in the Cooking Quality of Durum Wheat Pasta. J. Sci. Food Agric. 47(4):487–500, 1989.

20. P. Feillet, O. Ait-Mouh, K. Kobrehel, J. Autran. The Role of Low Molecular Weight Glutenin Proteins in the Determination of Cooking Quality of Pasta Products: An Overview. Cereal Chem. 66(1):26–30, 1989.

21. F. MacRitchie. Wheat Proteins: Characterization and Role in Flour Functionality. Cereal Foods World 44(4):188–193, 1999.

22. F. Barro, L. Rooke, F. Bekes, P. Gras, A.S. Tatham, R. Fido, P.A. Lazzeri, P.R. Shewry, P. Barcelo. Transformation of Wheat with High Molecular Weight Subunit Genes Results in Improved Functional Properties. Nature Biotechnology 15:1295–1299, 1997.

23. S. Batterman-Azcona, B.R. Hamaker. Changes Occurring in Protein Body Structure and α-Zein During Cornflake Processing. Cereal Chem. 75(2):217–221, 1998.

24. A. Chandrashekar, H.S.R. Desikachar. Studies on the Hydration of Starches, Flour, and Semolina from Different Cereal Grains. J. Food Sci. Technol. 21:12–14, 1983.

25. C.J. Bergman, D.G. Gualberto, C.W. Weber. Development of a High-Temperature-Dried Soft Wheat Pasta Supplemented with Cowpea (*Vigna unguiculata* (L.) Walp). Cooking Quality, Color, and Sensory Evaluation. Cereal Chem. 71(6):523–527, 1994.

26. C. Icard-Verniere, P. Feillet. Effects of Mixing Conditions on Pasta Dough Development and Biochemical Changes. Cereal Chem. 76(4):558–565, 1999.

27. N.H. Oh, P.A. Seib, D.S. Chung. Noodles. III. Effects of Processing Variables on Quality Characteristics of Dry Noodles. Cereal Chem. 62(6):437–440, 1985.

28. S.A. Matz. Pasta Technology In: S. Matz, ed. The Chemistry and Technology of Cereal as Food and Feed. 2nd ed. New York: Van Nostrand Reinhold, 1991, pp 451–496.

29. M. Li, T. Lee. Relationship of the Extrusion Temperature and the Solubility and Disulfide Bond Distribution of Wheat Proteins. J. Agric. Food Chem. 45:2711–2717, 1997.

30. D.H. Hahn. Application of Rheology in the Pasta Industry. In: H. Faridi and J.M. Faubion, eds. Dough Rheology and Baked Product Texture. New York: Van Nostrand Reinhold, 1990, pp 385–404.

31. O.J. Banasik. Pasta Processing. Cereal Foods World 26(4):166–169, 1981.

32. J.E. Dexter, R.R. Matsuo, B.C. Morgan. High Temperature Drying: Effect on Spaghetti Properties. J. Food Sci. 46:1741–1746, 1981.

33. Y. Fujio, J. Lim. Correlation Between the Glass Transition Point and Color Change of Heat-Treated Gluten. Cereal Chem. 66(4):268–270, 1989.

34. M.F. Jeanjean, R. Damidaux, P Feillet. Effect of Heat Treatment on Protein Solubility and Viscoelastic Properties of Wheat Gluten. Cereal Chem. 57(5):325–331, 1980.

35. P. Resmini, M.A. Pagani. Ultrastructure Studies of Pasta. A Review. Food Microstruct. 2:1–12, 98, 1983.

36. P.L. Earle. Studies in the Drying of Macaroni: Factors Affecting Checking. Ph.D. dissertation, University of Minnesota, St. Paul, MN, 1948.

37. T. Akiyama, K. Hayakawa. Tensile Fracture Stress of a Pasta Product at Temperatures from 293 to 343 K. Lebensm.-Wiss u. Technol. 27:93–94, 1994.

38. H. Liu, J. Qi, K. Hayakawa. Rheological Properties Including Tensile Fracture Stress of Semolina Extrudates Influenced by Moisture Content. J. of Food Sci. 62(4):813–815, 1997.

39. K.M. Waananen. Analysis of Mass Transfer Mechanisms During Drying of Extruded Pasta. Ph.D. dissertation, Purdue University, West Lafayette, IN, 1989.

40. L. Slade, H. Levine. Selected Aspects of Glass Transition Phenomena in Baked Goods. 1997 IFT Basic Symposium—Phase/State Transitions in Food: Chemical, Structural, and Rheological Changes, Orlando, FL, June 13, 1997.

41. Y. Roos, M. Karel. Water and Molecular Weight Effects on Glass Transition in Amorphous Carbohydrates and Carbohydrate Solutions. J. Food Sci. 56(6):1676–1681, 1991.

42. J.L. Kokini, A.M. Cocero, H. Madeka, E. de Graaf. The Development of State Diagrams for Cereal Proteins. Trends Food Sci. Technol. 5:281–288, 1994.

43. T.R. Noel, S.G. Ring, M.A. Whittam. Glass Transition in Low-Moisture Foods. Trends Food Sci. Technol. September:62–67, 1990.

44. Y. Roos. Characterization of Food Polymers Using State Diagrams. J. Food Eng. 24(3):339–360, 1995.

45. Y. Roos. Glass Transition-Related Physiochemical Changes in Foods. Food Technol. October:97–102, 1995.

46. Y.H. Roos, M. Karel, J.L. Kokini. Glass Transitions in Low Moisture and Frozen Foods: Effects on Shelf Life and Quality. Food Technol. November:95–108, 1996.

47. J.S. Vrentas, C.M. Vrentas. Energy Effects for Solvent Self-Diffusion in Polymer-Solvent Systems. Macromolecules 26:1277–1281, 1993.

48. J.L. Ollivier. Pasta Drying at Very High Temperature: A Fact. In: C. Mercier and C. Cantarelli, eds. Pasta and Extrusion Cooked Foods. London: Elsevier Applied Science, 1985, pp 90–97.

49. M. Wollney, M. Peleg. A Model of Moisture-Induced Plasticization of Crunchy Snacks Based on Fermi's Distribution Function. J. Sci. Food Agric. 64:467–473, 1994.

50. A. Nikolaidis, T.P. Labuza. Glass Transition State Diagram of a Baked Cracker and Its Relationship to Gluten. J. Food Sci. 61(4):803–806, 1996.

51. A. Kumar, R.K. Gupta. Fundamentals of Polymers. New York: McGraw-Hill, 1998.

52. N.L. Thomas, A.H. Windle. A Theory of Case II Diffusion. Polymer 23:529–542, 1982.

53. R.W. Cox, D.S. Cohen. A Mathematical Model for Stress-Driven Diffusion in Polymers. J. Polym. Sci., Part B: Polymer Physics 27:589–602, 1989.

54. I. Hopkinson, R.A.L. Jones, S. Black, D.M. Lane, P.J. McDonald. Fickian and Case II Diffusion of Water into Amylose: A Stray Field NMR Study. Carbohydr. Polym. 37:39–47, 1998.

55. S. Achanta, T. Nakamura, M.R. Okos. Stress Development in Shrinking Slabs During Drying. In: D.S. Reid, ed. The Properties of Water in Foods, ISOPOW 6, 1997, pp 253–271.

56. B.P. Hills, J. Godward, K.M. Wright. Fast Radial NMR Microimaging Studies of Pasta Drying. J. of Food Eng. 33:321–335, 1997.

57. D.A. Cummings. Modeling of Stress Development During Drying of Extruded Durum Semolina. MS thesis, Purdue University, 1981.

58. J. Andrieu, M. Boivin, A. Stamatopoulos. Heat and Mass Transfer Modeling During Pasta Drying. Application to Crack Formation Risk Prediction. In: S. Bruin, ed. Preconcentration and Drying of Food Materials. Amsterdam: Elsevier, 1988, pp 183–192.

59. T. Nakamura. Stress Analysis of Chemically Leavened Biscuits: Application of Pseudoelastic, Bi-Layer Model. Ph.D. dissertation, Purdue University, West Lafayette, IN, 1995.

60. M.H. Kim. Analysis of Stress and Failure for Food Products During a Simultaneous Heat and Mass Transfer Process. PhD dissertation, Purdue University, West Lafayette, IN, 1994.

61. J.B. Litchfield. Prediction of Stress and Breakage Caused by the Drying, Tempering, and Cooling of Yellow-Dent Corn Kernels and of Extruded Durum Semolina. MS thesis, Purdue University, West Lafayette, IN, 1984.

62. Y. Itaya, T. Kobayashi, K. Hayakawa. Three-Dimensional Heat and Moisture Transfer with Viscoelastic Strain-Stress Formation in Composite Food during Drying. Int. J. Heat and Mass Transfer 38(7):1173–1185, 1995.

63. T. Akiyama, H. Liu, K. Hayakawa. Hygrostress-Multicrack Formation and Propagation in Cylindrical Viscoelastic Food Undergoing Heat and Moisture Transfer Processes. Int. J. Heat and Mass Transfer 40(7):1601–1609, 1997.

64. T. Tsukada, N. Sakai, K. Hayakawa. Computerized Model for Strain–Stress Analysis of Food Undergoing Simultaneous Heat and Mass Transfer. J. Food Sci. 56(5): 1438–1445, 1991.

65. S.G. Croll. The Origin of Residual Internal Stress in Solvent-Cast Thermoplastic Coatings. J. Appl. Polym. Sci. 23:847–858, 1979.

66. K. Sato. The Internal Stress of Coatings Films. Prog. Org. Coat. 8:143–160, 1980.

67. D.Y. Perera, D.V. Eynde. Internal Stress in Pigmented Thermoplastic Coatings. J. Coat. Technol. 53(678):40–45, 1981.

68. S.G. Croll. Residual Strain Due to Solvent Loss from a Crosslinked Coating. J. Coat. Technol. 53(672):85–92, 1981.

69. J.A. Lewis, K.A. Blackman, A.L. Ogden, J.A. Payne, L.F. Francis. Rheological Property and Stress Development during Drying of Tape-Cast Ceramic Layers. J. Am. Ceram. Soc. 79(12):3225–3234, 1996.

70. O.H. Campanella, M.R. Okos, B.F. Willis. Effects of Glass Transition on Stress Development during Drying of a Shrinking Food System. In: G.V. Barbosa-Canovas and S.P. Lombardo, eds. Proceedings of the 6th Conference on Food Engineering, 1999.

71. S. Bruanuer, P.H. Emmet, E. Teller. Adsorption of Gases in Multimolecular Layers. J. Am. Chem. Soc. 60:309–319, 1938.

72. W.D. Harkins, G. Jura. A Vapor Adsorption Method for the Determination of the Area of a Solid Without the Assumption of a Molecular Area and the Areas Occupied by Nitrogen and other Molecules on the Surface of a Solid. J. Am. Chem. Soc. 66: 1366, 1944.

73. R.S. Bradley. Polymolecular Adsorbed Film. Part I. The Adsorption of Argon on Salt Crystals at Low Temperature, and the Determination of Surface Fields. J. Chem. Soc. Pp. 1467, 1936.

74. S.E. Smith. The Sorption of Water Vapor by High Polymers. J. Am. Chem. Soc. 69:646, 1947.

75. C.R. Oswin. The Kinetics of Package Life. III. The Isotherm. J. Chem. Ind. (London) 65:419–423, 1946.

76. S.M. Henderson. A Basic Concept of Equilibrium Moisture. Agri. Eng. 33:29–32, 1952.

77. D.S. Chung, H.B. Pfost. Adsorption and Desorption of Water Vapor by Cereal Grains and Their Products. Part II. Development of the General Isotherm Equations. Trans. ASAE. 10(4):552–555, 1967.

78. C.S. Chen. Equilibrium Moisture Curves for Biological Material. Trans. ASAE. 14: 924, 1971.

79. A. Alam, G.C. Shove. Hygroscopicity and Thermal Properties of Soybeans. Trans. ASAE. 16:707, 1973.

80. C. Van den Berg. Description of Water Activity of Foods for Engineering Purposes by Means of the GAB Model of Sorption. In: B.M. McKenna, ed. Engineering and Food, Vol. 1: Engineering Sciences in the Food Industry. London: Elsevier Applied Science, 1984.

81. H. Bizot. Using the ''GAB'' Model to Construct Sorption Isotherms. In: R. Jowitt, F. Escher, B. Hallstrom, H.F.Th. Meffert, W.E.L. Speiss, G. Vos, eds. Physical Properties of Foods. London: Applied Science Publishers, 1983.

82. H.A. Iglesias, J. Chirife. An Empirical Equation for Fitting Water Sorption Isotherms of Fruits and Relater Products. Can. Inst. Food Sci. Tech. J. 11(1):12–15, 1978.
83. H.A. Iglesias, J. Chirife. Handbook of Food Isotherms. New York: Academic Press, 1982.
84. W. Wolf, W.E.L. Spiess, G. Jung. Sorption Isotherms and Water Activity of Food Materials. Science and Technology, 1985.
85. A. Eisenberg. The Glassy State and the Glass Transition. In: J.E. Mark, A. Eisenberg, W.W. Graessley, L. Mandelkern, J.L. Koenig, eds. Physical Properties of Polymers. Washington, DC: American Chemical Society, 1984, pp 55–95.
86. M. Peleg. On Modeling Changes in Food and Biosolids at and Around Their Glass Transition Temperature. Crit. Rev. Food Sci. Nutr. 36(1&2):49–67, 1996.
87. M. Peleg. On the Use of the WLF Model Polymers and Foods. Crit. Rev. Food Sci. Nutr. 32(1):59–66, 1992.
88. M. Peleg. Mapping the Stiffness–Temperature–Moisture Relationship of Solid Biopolymers at and Around Their Glass Transition. Rheol. Acta 32: 575–580, 1993.
89. M. Peleg. A Model of Mechanical Changes in Biomaterials at and Around Their Glass Transition. Biotechnol. Prog. 10:385–388, 1994.
90. M. Peleg. Mathematical Characterization and Graphical Presentation of the Stiffness–Temperature–Moisture Relationship of Gliadin. Biotechno. Prog. 10:652–654, 1994.
91. J.L. Duda. Theoretical Aspects of Molecular Mobility. In: Y. Ross, R.B. Leslie, P.J. Lilliford, eds. Water Management in the Design and Distribution of Quality Foods. Lancaster, PA: Technomic, 1999, pp 237–254.
92. L. Piazza, A. Schiraldi. Correlation between Fracture of Semi-Sweet Hard Biscuits and Dough Viscoelastic Properties. J. Texture Stud. 28: 523–541, 1997.
93. N.N. Mohsenin. Physical Properties of Plant and Animal Materials. Volume I: Structure, Physical Characteristics, and Mechanical Properties. New York: Gordon and Breach Science, 1970.

9
Textural Characterization of Extruded Materials and Influence of Common Additives

Andrew C. Smith
Institute of Food Research, Norwich, U.K.

I. INTRODUCTION

Cereals are complex foods, and in order to understand their textural characteristics it is important to study their mechanical properties as they vary with composition and structure. The combination of the protein and starch polymers and the addition of other constituents in the process, such as water and sugar, alters the glass transition (T_g) of the system (1, 2). The onset of molecular mobility with increasing temperature corresponds to a change from the glassy to the rubbery state (3), which has implications for the mechanical behavior of biopolymers. Studies of mechanical properties and T_g in wheat starch and amylopectin, gluten, and their mixtures as a function of water content have been reviewed by Slade and Levine (3). The stiffness decreases more or less sharply at the glass transition (4). At high strains the brittle–ductile transition is associated with T_g. Ward (5) defines brittle and ductile behavior from the stress–strain curve (Fig. 1). Brittle fracture is typified by failure at the maximum stress (curve A), whereas ductile fracture shows a load maximum before failure (curve B). The two other graphs depict cold drawing (curve C) and rubberlike behavior (curve D), progressively above T_g. The distinction between brittle and ductile failure is also shown in differences in the energy dissipated in fracture and also the appearance of the fracture surface (5, 6). These comments were, however, made with reference to

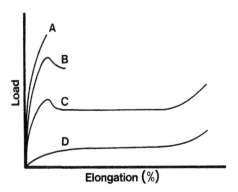

Figure 1   Load–elongation relationships for polymers. A, brittle fracture; B, ductile failure; C, cold drawing; D, rubberlike behavior. (From Ref. 5 with permission.)

isotropic polymers. The next stage of complexity invokes orientation in the polymers and the presence of other phases, making the class of materials known as composites. The situation with foods is that they are usually composites comprising more than one polymer, with other low-molecular-weight materials and inert fillers present, including air. Mechanical properties and structure are seen as the principal building blocks of physical texture. This is borne out by the classification of texture terms by Jowitt (7), including those that relate to the behavior of materials under stress or strain and those relating to structure.

It is usually assumed that there is a relationship between sensory texture and instrumental measurements. The Scott–Blair approach divides instrumental methods of texture measurement into three types: empirical, imitative, and fundamental (8, 9). *Empirical* tests have been developed from practical experience and are often marked out as arbitrary, poorly defined, lacking an absolute standard, and effective for a limited number of foods (10). *Imitative* tests are often seen as a subset of empirical tests that subject the food to a process that partially mimics the consumer. Empirical tests cannot easily be expressed in fundamental terms and are dependent on test geometry, friction, and sample size (11). *Fundamental* tests are more rigorously defined, usually in engineering units, whereas empirical tests are often more successful than their fundamental counterparts. Many fundamental tests use low stresses, which do not cause the material to break or fail, and also use rectilinear motion, whereas the movement of the teeth is along an arc and much faster than speeds in the universal test machine. Physical tests often produce single values, whereas consumers may change rates and manipulate the food during mastication.

II. MECHANICAL MEASUREMENTS

A. Mechanical Properties

Many tests for cereal products are based on those that are well developed in metallurgy and adapted in polymer science. They are often based on a universal testing machine, and in some cases there is an overlap with instrumental tests. Mechanical property tests are marked by the end result of absolute properties, in engineering units, principally strength, stiffness, and toughness. The mechanical properties of viscoelastic materials depend on temperature, strain rate, and the presence of plasticisers. Low-strain behavior is described by linear and nonlinear elasticity, which do not involve failure of the sample. In addition to stress–strain measurements, thermomechanical analysis or mechanical spectroscopy allows measurement of real (storage, E') and imaginary (loss, E'') moduli from oscillatory experiments in tension or flexure. The ratio E''/E', defined as tan δ, shows a maximum at T_g. Polymers can also show other lower-temperature transitions, with minor peaks in tan δ and small changes in modulus (3–5).

Large-strain behavior may be complex and involve necking, plastic yield, or fracture. Strength and stiffness can be measured in compression, tension, and shear. Flexure is also used, although it is not a pure stress experiment (Table 1) (12). Ward (5) described the Bauschinger effect, whereby compressive and tensile strengths are unequal in polymers, although they are equal for solid foams (13). Orientation of polymers can affect properties. For example, drawing amorphous or semicrystalline polymers causes molecular alignment such that the modulus depends on the draw ratio and the testing direction, although the effects are less for amorphous polymers, as would be expected (5). Orientation in solid foams has been analyzed by Huber and Gibson (14), who considered identical but elongated cells and found that the mechanical properties were anisotropic, being enhanced in the direction of elongation. Young's modulus was greater by a factor of 8 and the yield or breaking strength greater by a factor of 2.6 for cells that were twice as long as their square cross section.

Fracture mechanics is a subject in its own right (15, 16). Linear elastic fracture mechanics based on Griffith theory is at the heart of quantitative fracture. Griffith postulated that the energy to create a new surface is balanced by a decrease in elastically stored energy, which is not distributed uniformly in the specimen but is concentrated at flaws. This is the origin of the strength of materials being below their theoretical values. Fracture occurs from the growth of cracks that originate in flaws in the specimen. The fracture stress of a plate containing a small elliptical crack and tested orthogonally to the axis of the crack is inversely proportional to the square root of the crack length. The absolute mechanical properties associated with fracture mechanics are the critical stress intensity factor, K_{1c}, for loading normal to a crack, which is called Mode 1 fracture. In terms of

Table 1 Instrumental Measures of Texture in Extrudates

Test	Measurement $f^a(\ )$	Typical properties	References
Compression	Force, displacement f (time)	Modulus, strength, failure strain	5, 12, 17–25, 27, 46, 63, 65, 86
Tension	Force, displacement f (time)	Modulus, strength, failure strain	5, 12, 17, 20, 24, 26, 66
Flexure	Force, displacement f (time)	Modulus, strength, failure strain	12, 19, 20, 27–30, 64, 68, 93, 96
Charpy pendulum	Potential-energy loss	Toughness, strength	5, 12, 31, 32
Instrumented compression	Displacement f (time)	Strength	50
Stress relaxation	Stress f (time)	Relaxation modulus	5, 22
Creep	Displacement f (time)	Creep compliance	5
Warner–Bratzler cell	Force, displacement f (time)	Shear force or "strength," work	8, 26, 29, 38–40, 93–95, 98
Kramer shear press cell	Force, displacement f (time)	Breaking/shear force or "strength," sometimes relative to mass	8, 38, 41–43, 90, 99
Shortometer	Force, displacement	Force or "toughness"	38, 44
Cutting blade	Force, displacement f (time)	Force or "strength," work	45, 47
Penetration/puncture	Force, displacement f (time)	Strength	21, 38, 48, 64, 91
Texture profile analysis	Force f (displacement)	See Figure 2	38, 43, 49, 61, 91

^{a} f, function of.

available energy the material property is G_c, the critical strain energy release rate. The two properties are related through Young's modulus and the Poisson ratio.

Compression techniques are popular because of the ease of performing the test (17). The initiation of the failure may be inside the test piece and advance in a complex manner. Antila et al. (18) used a piston of area 100 mm^2 at a speed of 10 mm s^{-1} to break specimens from flat-bread extrusion and expressed their data in units of energy per unit area. Launay and Lisch (19) used 30-mm-length extruded starch samples tested at 10 mm min^{-1} to calculate the elastic modulus. Hutchinson et al. used a compression test axially (20) (15mm long) and radially (21) on cylindrical maize samples and determined stiffness and strength using a crosshead speed of 5mm min^{-1}. Halek et al. (22) tested 10-mm-long cylindrical maize extrudates at a crosshead speed of 25 mm min^{-1} and calculated strength and stiffness. Mohamed (23) also used a compression test at a speed of 50 mm min^{-1} on cylindrical maize and maize starch extrudates and measured the maximum stress before fracture, which was termed the hardness (in newtons). Fontanet et al. (24) tested 30 × 15 × 7.5-mm strips excised from crispbreads parallel and perpendicular to the extrusion direction in compression at 20 mm min^{-1}. They characterized their data by fracture stress, Young's modulus, and deformation at fracture (strain). Attenburrow et al. (25) tested 10 × 10 × 5-mm specimens of sponge cake in compression at a crosshead speed of 10 mm min^{-1} and calculated failure stress and strain.

Tensile testing is not as common as compression for some foods because of the difficulties of gripping the sample. However, the commencement of failure is often easily observed on the surface of the sample (17). Hutchinson et al. (20) and Shah et al. (26) used a tensile test with specimens of length 50 mm and at speeds of 5 and 50 mm min^{-1}, respectively. The former tested maize extrudates and reported the tensile modulus and strength, whereas the latter examined maize–fish mixtures and gave tensile strengths in newtons. Fontanet et al. (24) tested 30 × 15 × 7.5-mm strips excised from crispbreads parallel and perpendicular to the extrusion direction and used a crosshead speed of 10 mm min^{-1}. They characterized their data by fracture stress, Young's modulus, and deformation at rupture (strain).

Bending is a combination of compression and tension with shear. As with compression, this type of test is easy to perform. The three-point bend test has been used for crackers and biscuits (27, 28) and also extrudates. Faubion and Hoseney (29) recorded the breaking force (in kilograms) for starch and flour extrudates. Launay and Lisch (19) used a three-point bend test for starch extrudates and calculated the apparent elastic modulus and rupture strength. Hutchinson et al. (20) also used a three-point bend test at a crosshead speed of 5 mm min^{-1} to obtain the flexural modulus and strength of 150-mm-long cylindrical maize extrudates according to standard formulae (12). Van Hecke et al. (30)

tested extruded flat breads in three-point bending and calculated the modulus and strength.

Van Zuilichem et al. (31) recommended the Charpy test from their study of extrusion cooked maize. The pendulum starting height gives it a known potential energy that is used in breaking the sample on the down swing, the remainder appearing as the potential energy on its upswing. The energy used to break the sample was divided by the cross-sectional area of the cylindrical sample to give a "strength" in units of kilograms per centimeter. Kirby and Smith (32) also used a Charpy test (Table 1), but with extrudates that were notched. The energy to break decreased for notches greater than the pore size but notches were ineffective flaws when smaller than the pore size. The critical strain energy release rate, G_c, was calculated from the energy to break notched specimens using the approach of Plati and Williams (33). Samples were cut from extrudates obtained using a rectangular die, and cylindrical extrudates were also notched. In the latter case a variation of the method due to Plati and Williams was necessary to accommodate circular-cross-section specimens using foamed plastics as standards. The Izod test is a similar impact test, where the sample is gripped at one end as a cantilever and impacted (12).

Reports of stress-relaxation and creep experiments are relatively few. Halek et al. (22) performed stress-relaxation experiments on 10-mm-long maize extrudates in triplicate, for which the decay of the force was recorded after sample compression at a speed of 4 mm s^{-1} to 10%. The sample was then held and the force relaxation monitored for 10 min. The relaxation modulus was calculated following the approach of Shama and Sherman (34) at different times up to 10 min and the relaxation time obtained by fitting an exponential decay law.

B. Acoustic Emission

Vickers (35, 36) has reviewed the evaluation of crispness and the hypothesis that auditory sensations are involved in the perception of crispness. One means of assessing sensory attributes such as crispness is to measure the sounds produced during compression of foods. This may occur with instrumentally or manually deformed samples or by holding a microphone against the outer ear (35). Andersson et al. (27) manually crushed crispbreads with a cylindrical plunger and recorded the sounds. The signal was played back through a frequency analyzer, and the data were presented as amplitude–time curves at different frequencies. Acoustic emission has also been used to characterize extrudates (37).

C. Instrumental Tests

A number of general texture measuring devices has been used for testing instrumental texture of extrudates. Full details are given in texture publications, e.g.,

Refs. 8 and 38. The rate of compression between the teeth is of the order of 20 mm s^{-1} (38), although many instrumental and mechanical tests use test speeds much lower than this.

The Warner–Bratzler instrument (Table 1) was originally employed for testing meat in the form of cylinders cut using a cork borer. The cylinder is positioned through a triangular hole in a blade and cut by two metal anvils that force the cylinder into the V of the triangle. The anvils were set to move at 230 mm min^{-1} in the original test cell rather than as adapted for a universal testing machine. Among the studies that have reported this technique are Faubion and Hoseney (29), who measured the shear strength (in kilograms) of starch and flour cylindrical extrudates, and Owusu-Ansah et al. (39), who observed the breaking strength of extrusion cooked corn starch. They also observed that the force–deformation pattern was related to microstructure, in that highly porous extrudates showed more force peaks, which were related to the number and distribution of the pores. Falcone and Phillips (40) extruded cowpea and sorghum mixtures and measured force at failure, stress at failure, and total energy dissipated in comparison with those for commercial expanded snacks. The anvils were set to move at 50 mm min^{-1}. Shah et al. (26) also used it with the same crosshead speed to study maize–fish mixtures and expressed the results as shear strength in newtons. They extruded maize grits with freeze-dried cod at levels of 1–11% in a Brabender laboratory extruder. The tensile strength and Warner–Bratzler shear strength decreased or remained approximately constant with increasing fish content, depending on extrusion temperature. No particular correspondence between the results from the two tests was evident.

The Kramer shear press uses a sample box with a base comprising bars spaced equally. The crosshead carries a multiblade device that is driven into the box such that the blades move through the slats between the bars (Table 1). Material in the box is deformed and moves upward around the blades and downward around the bars. Bhattacharya and Hanna (41) used the shear press to calculate the shear strength (in megapascals) as a function of extrusion moisture content and temperature for corn with different levels of amylose. Matthey and Hanna (42) used a twin shear cell with a 50-mm length of extrudate placed across the width of the cell perpendicular to the plates. They used a crosshead speed of 10 mm min^{-1} and calculated the shear strength from the average force and the average circular cross-sectional area of the extrudate. Martinez-Serna and Villota (43) used a shear compression cell and measured the maximum force to break 25.4-mm extrudate pieces filling the cell using a crosshead speed of 50 mm min^{-1}.

Maga and Cohen (44) used a Chatillon shortometer on 60-mm samples of extruded potato flake and reported the results as a "toughness," in terms of mass to break the extrudate. A shortometer is a bending or snapping test for bars or sheets using a three-point bend geometry, as described in detail by Bourne (38).

Mercier and Feillet (45) used a cutting test in which a blade was driven through a cylindrical extrudate using a crosshead speed of 47 mm min^{-1}. The lower platen of the test apparatus was slotted so that the blade could fully penetrate the sample. The maximum force, expressed in kilograms, was termed the breaking strength. In experiments on extruded maize, this breaking strength decreased with increasing extrusion temperature.

Variations of standard mechanical tests have also been used, for example, the force required to compress an expanded half product specimen weighing 0.45 g to 60% between parallel plates using a crosshead speed of 5 mm min^{-1} (46). Sauvageot and Blond (47) used a blade descending at 0.5 mm s^{-1} and compressing a small sample located in a groove. They measured the initial slope of the force–deformation curve, the displacement to the first force peak, and the work up to the first force peak.

Puncture or penetration techniques have been well described (38), analyzed in terms of shear and compression contributions to the force. The use of a pin to probe extrusion cooked maize cylinders has also been reported (21, 48). Hayter and Smith (48) measured the first peak force and the average of the peaks over 10 mm as a function of bulk density for cylinders tested axially. They calculated the failure stress from the force and the cross-sectional area of the pin and the modulus from the stress and strain, where strain was defined as the crosshead movement relative to the average pore diameter. Hutchinson et al. (21) measured the stiffness and strength of maize extrudate cylinders tested radially and by pin indentation into the circumference. They compared the stress to break the extrudate outer skin using a pin indentation test, where the local compressive strength was interpreted as the force divided by the cross-sectional area of the pin and the local shear strength was the force divided by the area sheared (the circumference of the pin multiplied by the wall thickness). The third interpretation of the test was that the skin covering one cell undergoes flexure relative to the walls of adjacent cells and the flexural strength is obtained from a three-point bend test equation, where the span equals the width. One interesting feature of the puncture or pin indentation test is that the local shear rate, defined as the crosshead movement speed divided by the original sample length, is higher (typically by an order of magnitude) than for bulk specimen deformation, since to calculate the strain the original dimension is more appropriately the pore size than the sample size.

Texture profile analysis has also been used, based on work at General Foods, and was adapted for the universal testing machine (38, 49). It comprises two successive compressive movements, and instrumental forces of fracturability (brittleness) and hardness are given directly from the force–time plot (Fig. 2). Other characteristics may be deduced from the test: springiness (elasticity), cohesiveness (ratio of Area 2 to Area 1), gumminess (hardness × cohesiveness), and chewiness (gumminess × springiness). Halek et al. (22) used this approach for

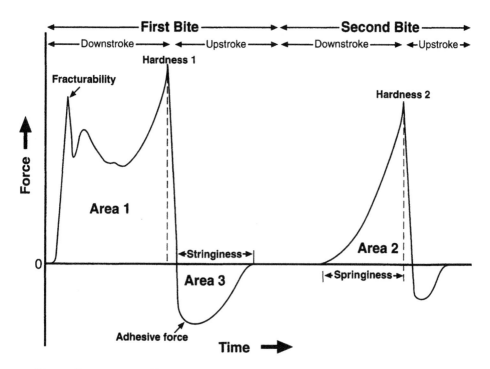

Figure 2 Texture profile analysis curve. (From Ref. 49 with kind permission from the Institute of Food Technologists.)

10-mm-long cylindrical samples compressed axially. Martinez-Serna and Villota (43) tested rehydrated 35-mm-long extrudates using a two-bite compression cycle, which reduced the sample to a quarter of its diameter. Cohesiveness and elasticity were then calculated.

Testing of multiple specimens may also be carried out, and a compression test may be carried out on stacked crispbreads, for example, Ref. 27. The force–deformation curve was recorded and the work determined for compression to 85% of the original height.

Instrumental tests at higher shear rates are often based on pendula and are justified by the higher rates than are often possible using universal testing machines. One such technique was based on a portable pendulum for testing impact damage of potato tubers (50) (Table 1). It monitors the angular displacement of the pendulum and can be used to record the displacement of the sample over which the potential energy of the hammer is dissipated. The energy dissipated can be corrected for the cases of rebound from the sample. This test is actually a rapid compression test using energy measurement rather than force. Hayter et al. (51) divided the dissipated energy by the penetration into the sample and the cross-sectional area of the hammer to give a strength (in pascals).

III. SOLID FOAMS

Extrusion cooking often produces expanded solid foams. While these structures are complex in their pore size distributions, the understanding of the relationship between mechanical properties and structure is important for understanding sensory texture. The publication of a comprehensive treatment of the mechanical properties of idealized cellular structures by Ashby (13) paved the way for examining food foams in some detail. This treatment originally considered the mechanical properties of idealized closed- and open-cell foams, which deform elastically, plastically, and by fracture. Gibson and Ashby (52) have extended their work to include anisotropic pores, fluid-filled pores, and pore size and shape. The original treatment considered three-dimensional cubic cells whose relative density was expressed in terms of the wall thickness and length. The deformation of the foams is understood from a two-dimensional treatment of hexagonal cells. Linear elastic deformation was derived by considering the cell walls to be bending beams (Fig. 3a), nonlinear behavior by treating the cell walls as Euler struts, and plastic yielding by the creation of plastic hinges at the cell wall intersections. Compressive failure was understood from the bending of the cell wall (Fig. 3b) and tensile fracture by considering the stress concentration resulting from a crack and the bending moment on cell walls ahead of the crack (Fig. 3c). In all cases the mechanical property of the foam normalized by the wall value scales as a power of the normalized density:

$$\frac{\sigma}{\sigma_w} = C \left(\frac{\rho}{\rho_w} \right)^n \tag{1}$$

where σ, σ_w are the mechanical properties of the foam and matrix, respectively, and ρ, ρ_w are the densities of the foam and matrix, respectively. The value of n depends on whether the foam cells are open or closed and which mechanical property is involved.

Gibson and Ashby provided evidence that the response of foamed polymers, ceramics, and metals obeyed Eq. (1). Despite the simplifying assumptions of the treatment, it provides an elegant framework for the study of extruded foams (53). An immediate deviation from the model is apparent from the structure of real extrudates, which do not contain identical isotropic pores. In the case of foamed materials σ_w in Eq. (1) was identified with the bulk value. The analogous use of the mechanical property of an unfoamed food extrudate would be unrepresentative of a foam wall, not least because of the different processing conditions

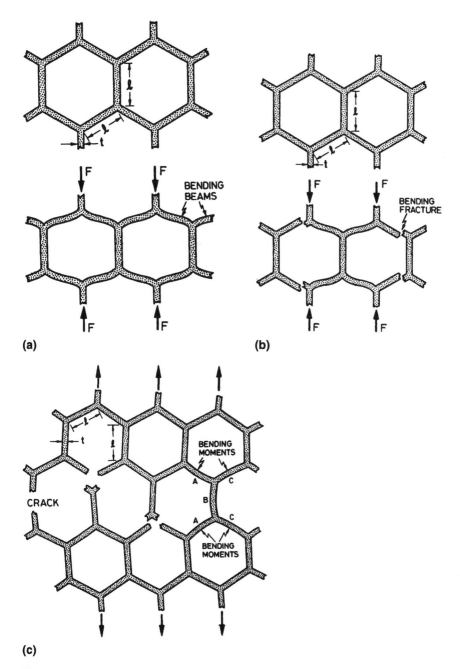

Figure 3 Derivation of the relationship between mechanical properties and cell dimensions: (a) linear elasticity; (b) crushing; (c) tensile fracture. (From Ref. 13 with kind permission from ASM International.)

required in the two cases. A further, more general reservation about using the bulk value results from orientation of the material in the cell walls, which is known to affect the mechanical properties of plastics, as described earlier.

Initial studies used an approximation of Eq. (1) for the case of constant wall properties, which reduces to:

$$\sigma \propto \rho^n \tag{2}$$

which had earlier been used to describe the mechanical properties of foamed plastics (54). Other approaches have come to similar results based on considering solids containing spherical holes (55), which reduced to an equation of the form:

$$G = G_0 \left[1 - A \frac{\rho}{\rho_w} + B \left(\frac{\rho}{\rho_w} \right)^2 \right] \tag{3}$$

where G, G_0 are the shear moduli of the composite and matrix, respectively, and A and B are constants.

A typical relationship between compressive strength and density (20) is shown in Figure 4. The data are seen to fit Eq. (2), and in general the modulus and strength of starchy foams obey this law in each of the testing modes, as shown in Table 2. In all cases the value of n was greater for the modulus than the strength, in agreement with the predictions. However, the values of n were generally more consistent with the open-cell predictions of the model. Although Ashby argued that closed-cell foams often behave like open-cell foams due to concentration of material in the cell edges, the experimental evidence points to extrudates being clearly of the closed-cell type.

The more recent treatments of Gibson and Ashby (52) consider the different contributions to the deformation of closed-cell foams. The simple power law [Eq. (1)] applies for face bending only. In the more detailed treatment it is recognized that the volume fraction of material in the cell edges, δ, contributes as for open-celled foams. The contribution to the modulus, E, is therefore $\delta^2(\rho/\rho_w)^2$ (as in the simple treatment). Bending of faces contributes $(1 - \delta)^3(\rho/\rho_w)^3$ and stretching faces $(1 - \delta)(\rho/\rho_w)$. Hence:

$$\frac{E}{E_w} = C_1 \delta^2 \left(\frac{\rho}{\rho_w} \right)^2 + C_2(1 - \delta) \left(\frac{\rho}{\rho_w} \right) + C_3(1 - \delta)^3 \left(\frac{\rho}{\rho_w} \right)^3 \tag{4}$$

where E_w is the modulus of the walls. Similarly for strength, σ:

$$\frac{\sigma}{\sigma_w} = C_4 \delta^{3/2} \left(\frac{\rho}{\rho_w} \right)^{3/2} + C_5(1 - \delta) \left(\frac{\rho}{\rho_w} \right) + C_6(1 - \delta)^2 \left(\frac{\rho}{\rho_w} \right)^2 \tag{5}$$

where σ_w is the strength of the walls. This refinement of the model actually fits the experimentally observed power laws better than the simple closed-cell equa-

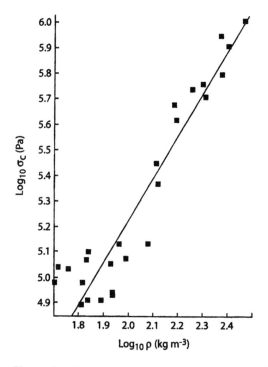

Figure 4 Compressive strength of extruded maize foams as a function of bulk density. (From Ref. 20 with kind permission from Kluwer Academic Publishers.)

Table 2 Power Law Values in Eq. 2

Stress state	Power law index
Modulus	
Tension	1.7
Compression	2.3
Flexure	2.0
Theory (13)	2 (open cells), 3 (closed cells)
Strength	
Tension	1.5
Compression	1.6
Flexure	2.0
Theory (13)	1.5 (open cells), 2 (closed cells)

Source: Ref. 20 with kind permission from Kluwer Academic Publishers.

tion (Table 2). It also predicts a transition in the curve when δ changes, as occurred when the cell shape changes. This emphasizes that in addition to density and expansion, foamed solids may be characterized by cell shape and size (56). With increasing density, the cells generally change from spherical to polyhedral (57). Polyhedral cells appear to be a consequence of bubbles growing together and reaching an equilibrium structure, with some drawing of material from the cell faces to the cell edges before growth stops. Spherical cells tend to represent a rapidly set structure of growing bubbles. The volume fraction of material in the cell edges, δ, is given by:

$$\delta = \frac{t_e^2}{t_e^2 + Z_f t_f l / \langle n \rangle}$$

where l is the cell size, t_e is the cell edge thickness, t_f is the cell face thickness, $\langle n \rangle$ is the mean number of edges per face, and Z_f is the face connectivity. δ is typically ~0.5 for polyhedral cells and ~0.02 for spherical cells (58).

The anisotropy of extrudate properties has been investigated by local mechanical testing, in which a pin of 0.58-mm diameter was driven into cylindrical foams (21, 48). The force to successively break the foam walls was recorded, although conversion to strength units required assumptions to be made on the type of local deformation of the cell wall. At low densities, the surface of an extrudate was some 100 times stronger than the interior, although at higher densities similar values were recorded (Fig. 5). Not surprisingly, the surface layer showed only a slight dependence on density, although the interior strength increased strongly with density more closely in agreement with Ashby's closed-cell prediction. The principal contribution in the local test would be expected to be face bending, hence the higher power law, in agreement with prediction. The pin indentation test is also able to give one-dimensional structural information, since a characteristic force spectrum is obtained (Fig. 6). The separation of the force peaks has been compared with the distance between pore walls, derived from diameters drawn across scanning electron micrographs of cylindrical extrudate cross sections. This shows that the pin test provides some overestimation of the pore size because some spurious force events occurred, although this could be improved with signal filtering (48). Owusu-Ansah et al. (39) had earlier indicated extrudate porosity from the force–displacement signal of a modified Warner–Bratzler test, albeit carried out on a bulk sample.

Experiments using a Charpy pendulum (Table 1) allowed calculation of the critical strain energy release rate G_c (32). Ashby (13) gives the value of the stress intensity factor, K_{1c}, for tensile fracture of foams:

$$K_{1c} \propto (\pi l)^{1/2} \sigma_w \left(\frac{\rho}{\rho_w} \right)^{3/2} \tag{6}$$

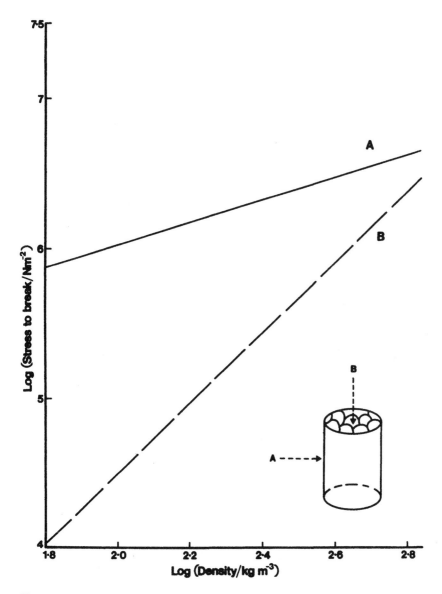

Figure 5 Local strength of extruded maize cylinders measured by a pin test axially (— — —) and radially (———). (From Ref. 53. Reproduced with kind permission of the Royal Society of Chemistry.)

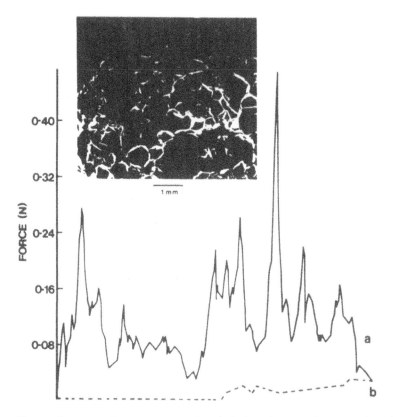

Figure 6 Force–distance spectrum by pin indentation: (a) pin penetration, (b) pin removal. (From Ref. 48 with kind permission from Kluwer Academic Publishers.)

The strain energy release rate may be derived from K_{1c} and E (5):

$$G_c \propto \frac{(K_{1c})^2}{E} \propto \frac{\sigma_w^2 l \rho}{E_w \rho_w} \tag{7}$$

For constant wall properties Eq. (7) reduces to:

$$G_c \propto \rho l \tag{8}$$

This highlights the problem of defining a single pore size for an extrudate. Experimentally G_c increases with density, as also found for polyurethane foams (59).

Most applications of the Gibson and Ashby models have concerned extrudates (60). In addition to the work in this laboratory, other groups have considered the applicability of the Gibson and Ashby models. Halek and Chang (61) used the hardness from texture profile analysis of maize extrudates, which increased as relative density to a power of 1.4. Van Hecke et al. (30) found power laws of 1.6 and 3.1 for wheat flour mixtures extrusion cooked into flat breads, depending on composition. Another early test of the models employed a compressive test on well-defined specimens of sponge cakes by Attenburrow et al. (25). Interestingly, this work did comprise open-cell foams rather than the closed cells of the extrudates, and they similarly tested out the scaling laws [Eq. (2)], assuming the matrix properties were constant. They found general agreement with theory, although in some cases the strength was more sensitive to density than theory predicted. Amemiya and Menjivar (62) applied these models to cookies and crackers. Keetels et al. (63) tested the relationship for starch breads, which have mainly open cells, using pore-wall mechanical properties extrapolated from more dilute starch gels. They examined the proportionality constants in Eq. (1) and monitored their change with storage time. Shogren et al. (64) produced baked foams based on starch batters and found a power law of 1.28 for bending modulus and relative density, whereas breaking stress in a penetration test scaled with relative density to a lower power of 1.12. Barrett and Peleg (65) compressed cubes excised from corn meal extrudates and found that breaking stress correlated positively with bulk density for transverse and longitudinal samples. Nussinovitch et al. (66) found no evidence for a relationship between tensile force and density for different commercial breads. It should be pointed out that a number of authors have related mechanical properties to porosity. Earlier, Mizukoshi (67) used Eq. (3) and found good agreement with the model for cakes of different density, produced by altering the mixer vacuum.

One way of causing a variation in wall properties is to vary their water content, as would occur in postextrusion conditioning. Equation (1) may be rewritten in terms of volumes:

$$\frac{\sigma}{\sigma_w} = C \left(\frac{V_w}{V}\right)^n \tag{9}$$

A variation in moisture content will change the volume ratio, depending on swelling of the foam structure. It will also profoundly affect σ_w (see later). The foam mechanical properties would be expected to fall dramatically with increasing moisture content. This is what is observed in practice (68). Figure 7 shows the stiffness of extruded wheat starch as a function of density, obtained by varying extrusion conditions. Increasing or decreasing the moisture content of these foams also causes a density increase or decrease with the mechanical behavior typical of pasta. These data show that postextrusion processing is as important as the extrusion step itself in determining mechanical properties. Figure 7 may be seen as a physical texture "map" showing how to choose extrusion and drying conditions to get particular structure (density) and mechanical properties (69).

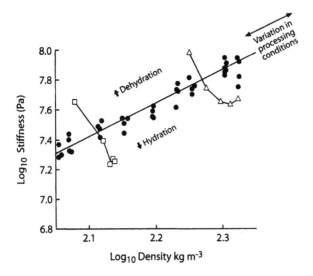

Figure 7 Flexural modulus, E_F, as a function of density, ρ, for extruded wheat starch: samples obtained by varying extrusion conditions (●), samples obtained by subsequent drying or hydration steps (□, △). (From Ref. 53. Reproduced with kind permission of the Royal Society of Chemistry.)

IV. MATRIX PROPERTIES

Studies have been carried out to define the base material properties without the complication of the gross structure, using specimens of simple geometry obtained by hot pressing or extrusion or a combination thereof. Mechanical properties as a function of water content have been studied in wheat starch, amylopectin, and gluten and mixtures of these components with themselves and with sugars, polyols, and lipids (70–80). With the addition of glucose to starch, the fall in modulus occurred at a lower moisture content corresponding to a depression of the glass transition of starch due to monomer plasticization (Fig. 8). The magnitude of the fall in modulus and its value at low moisture fell with increasing glucose content. The modulus difference between glassy and rubbery states is less than for a purely amorphous polymer, and the rubbery modulus is indicative of partially crystalline material (79). The increase in moisture was also accompanied by a change in the failure mode, with a general trend from fracture through tearing to yielding (Fig. 9). This statistical nature of the failure mode is also emphasized in the results of Nicholls et al. (72). For the starch–glycerol samples at 6.5% water content, the fractured pieces fit back perfectly and the fracture faces are mirror-like. At a water content of 9.9%, fractured samples do not fit back together due

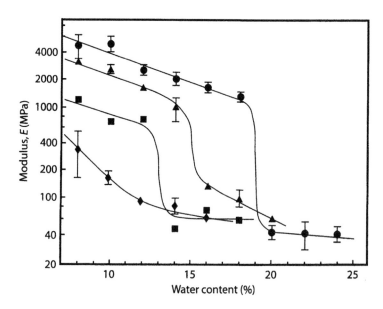

Figure 8 Flexural modulus as a function of water content for wheat starch containing different levels of glucose: ● 0%, ▲ 8.4%, ■ 17.4%, ◆ 67.8%. (From Ref. 79 with kind permission from Kluwer Academic Publishers.)

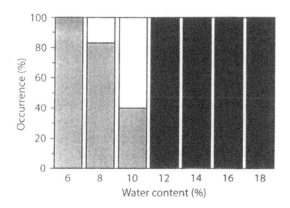

Figure 9 Modes of failure of starch–glycerol (20% nonaqueous components wt/wt) mixtures: ▓ fracture, ☐ tear, ■ bend. (From Ref. 80 with kind permission from Kluwer Academic Publishers.)

to permanent deformation where the fracture initiated. The surfaces are seen to be coarse textured in scanning electron micrographs (79, 80).

Even more complex systems are processed cereals and their mixtures with sugars and other components, which show that the same concepts can be applied (81–85). Breakfast cereal flake materials, comprising a control formulation and examples where one or more of the components had been subtracted, were milled and then molded as bars and conditioned to different water contents. A hot-press technique was used to reconstitute the ground flakes as bar-shaped specimens to remove the geometry and structure effects and allow comparison of the matrix properties. Dynamic mechanical thermal analysis was used to compare the response with that for a simpler one and for two-component materials (4, 71, 74–78), which were also in the form of hot-pressed bars. In general the decrease in modulus is less well defined and the tan δ peak height changes less with water content. The tan δ peak height decreased with increasing water content for wheat and cereal formulations without added sugar and the converse for systems with added fructose. Sucrose additions cause a greater fall in modulus at the glass transition (Fig. 10a), as observed for simpler systems by Kalichevsky et al. (74, 78), and a sub-T_g peak can be discerned (Fig. 10b).

Barrett et al. (86) pressed corn–sucrose extrudates produced at two extrusion moistures that had then been conditioned to 17% moisture. They were tested using dynamic mechanical spectrometry and revealed a one- to two-decade reduction in modulus. The temperature of the tan δ peak decreased with addition of sucrose but also depended on process history through extrusion moisture content.

The combination of wheat and other constituents in the process, principally water and sugar, alters the glass transition and the mechanical properties of the matrix (81–83). At any particular water content from 10 to 23% (w.w.b.), the bending modulus, E', was lower in the presence of sugars, with the difference decreasing to zero as the water content decreased to 10% (w.w.b) (Fig. 11).

An estimate of the critical energy release rate, G_c, was made, based on the energy to break, W^*, for notched samples of different notch sizes, as described by Plati and Williams (33):

$$W^* = G_c BD\phi \tag{10}$$

where B is the sample width, D is the sample depth, and ϕ is a calibration factor that depends on the ratios a/D and $2L/D$, where a is the crack depth and $2L$ is the span. Values of ϕ were calculated by Plati and Williams for this geometry and were used in the calculations reported here. Data from Charpy tests conform to this relationship (Fig. 12).

An estimate of the fracture toughness, G_c, indicates that this was highest for the wheat samples and lowest for the fructose-containing samples at a water content of 7–9%. The value of G_c increased dramatically from 100 to 10,000 J m^{-2}, corresponding to a decrease in the modulus from 1000 to 10 MPa with

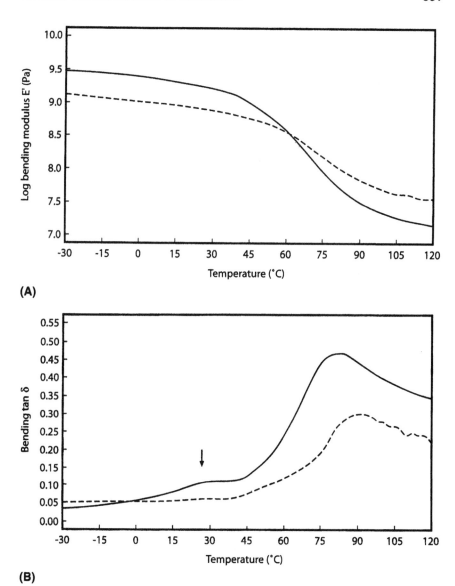

Figure 10 DMTA (A) Log E' and (B) tan δ for pressed wheat (– – –), wheat and sucrose (———) samples [ratio wheat:sucrose 6:1] for water contents of 9–10% as a function of temperature. Sub-T_g peak arrowed. (From Ref. 82 with kind permission from Akademiai Kiado RT.)

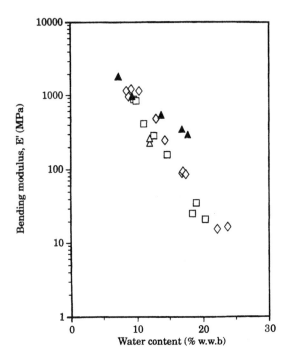

Figure 11 Bending modulus, E', of pressed wheatflake material as a function of water content: control ▲, control and sucrose □, control and fructose ◊, control and maltose △ [ratio wheat:sugar 6:1]. (From Ref. 82 with kind permission from Akademiai Kiado RT.)

increasing water content as samples approach their glass transition. The present study shows that although both samples were glassy and brittle at ambient temperature, the addition of sucrose made the samples easier to break or less tough. It should be emphasized that these values of G_c are not absolute, since they depend on particle size (84). In the context of sample breakage, Kalichevsky et al. (74, 78) reported that the addition of fructose (at the 1:2 level) to gluten or amylopectin reduced the water content below which their samples were brittle in three point-bend tests. Kalichevsky and Blanshard (77) also reported the maximum force to break or yield was much greater for amylopectin alone than in the presence of sugars.

In related studies on flattened wheatflakes, a probe deflection technique was used (87) that showed a decrease of Young's modulus with increasing water content and also enabled calculation of failure stress and strain. A summary of the stiffness and failure data on flattened flakes and pressed blocks as a function of water content is given in Figure 13, which shows the toughness increasing

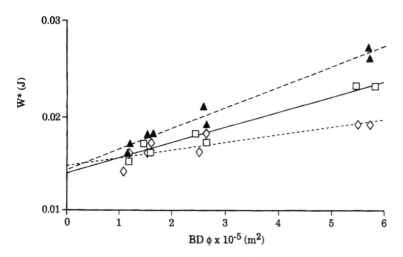

Figure 12 Charpy test data of energy to break, W^*, for the pressed wheatflake material [ratio wheat:sucrose 6:1] as a function of $BD\phi$ at water contents of control ▲, control and sucrose □, control and fructose ◇ at water content of 7–9% (w.w.b.), where B is the sample width, D is the sample depth, and ϕ is a calibration factor. (From Ref. 82 with kind permission from Akademiai Kiado RT.)

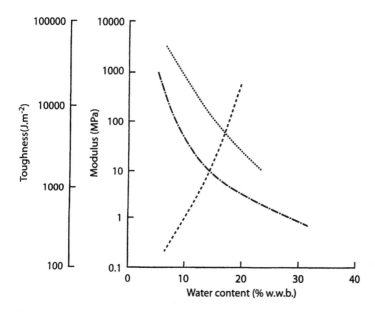

Figure 13 Mechanical properties of wheatflakes --- toughness of pressed milled flake; - - - - - - modulus of flattened flakes; ·········· modulus of pressed milled flake.

and the modulus decreasing with increasing water content. The modulus for the flattened flake is consistently lower than that for the pressed blocks. This follows from the lower bulk density of the former and the proportionality between mechanical properties and density for solid foams discussed earlier. Notably, recent data on the impact properties of starch–synthetic polymer composites showed an increase in the impact energy with increasing water content (88).

V. EFFECT OF VARIABLES

Many studies of extrudates have altered protein sugar, salt, and fat levels, usually relative to a starch control. Expansion or solubility properties are usually measured; mechanical or textural properties are less commonly assessed.

A. Effect of Water Content or Relative Humidity

Here, process water content can be very different from actual sample water content. In the studies of Faubion and Hoseney (29) using the Warner–Bratzler method, the shear strength fell by a factor of 3 for starch and by a factor of 2 for flour with increasing extrusion moisture content from 17 to 24%. Owusu-Ansah et al. (39) observed the breaking strength of extrusion cooked corn starch to decrease with increasing extrusion moisture content. Bhattacharya and Hanna (41), using the Kramer shear press, found the shear strength of extrusion cooked waxy and nonwaxy corn that had been dried to 1–2% moisture to decrease with increasing extrusion moisture content from 15 to 45%. Martinez-Serna and Villota (43) found that "breaking force" showed a maximum or minimum in the extrusion moisture range 17–25%, depending on the manipulation of whey protein isolates (WPI) blended with corn starch. Brittleness increased with increasing extrusion moisture content in all cases of WPI modification. Cohesiveness of rehydrated extrudates decreased with increasing extrusion moisture content, except for acidic and alkaline WPI, which showed the opposite trend.

Hutchinson et al. (68) found that the flexural modulus and strength of wheat starch and wheat flour extrudates decreased with increasing density, caused by increasing the sample water content (Fig. 7). Attenburrow et al. (25) found that compressive modulus and strength decreased with increasing RH in the range 33–75% for sponge cakes. Seymour and Hamann (89) used a shear cell for single or multiple sample testing. The force per unit mass and the work to failure increased with increasing water activity, a_w, from 0 to 0.65, indicating greater toughness. Experiments by Zabik et al. (90) using a Kramer shear press, a texturometer cell, and a breaking test were carried out on cookies conditioned to different RH. Breaking strength was the most sensitive mechanical measure, showing a decrease with increasing RH from 52 to 79%. Katz and Labuza (91) used breaking and puncture tests and texture profile analysis to test snacks equilibrated to a_w from 0 to 0.85. The initial slope of force–deformation curves (kilograms per millimeter), and the compression work and cohesiveness from texture profile analysis decreased with increasing a_w. Similarly, Sauvageot and Blond (47) found that the initial slope decreased with increasing a_w, particularly above $a_w = 0.50$. Loh and Mannell (92) used the General Foods Texturometer with cereal flakes and milk contained in a cup and compressed with a plunger. They showed a decrease in texture retention, defined as the peak force relative to that for a dry sample, with time in milk.

Studies by Attenburrow et al. (73) have monitored the changing acoustic emission amplitude of crushing starch and shown a marked reduction in intensity with increasing moisture content. Nicholls et al. (72) reported the decrease of acoustic events with increasing relative humidity when crushing extruded granules of gluten wheat starch and waxy maize starch (Fig. 14). Seymour and Hamann (89) also found a general decrease in the acoustic emission parameters, sound pressure and sound intensity, with increasing water activity.

B. Protein

The effect of protein enrichment was investigated in studies on extruded crispbreads by Antila et al. (18). Increasing the wheat content increased the breaking force more than did protein enrichment. Faubion and Hoseney (93) found that the Warner–Bratzler shear strength and the flexural strength of wheat starch de-

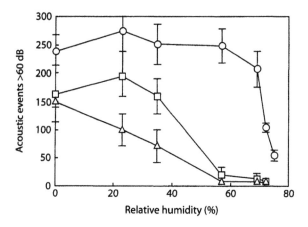

Figure 14 Acoustic emission produced when crushing extruded granules as a function of conditioning relative humidity: ○ gluten, △ wheat starch, □ waxy maize starch. (From Ref. 72 with kind permission from Academic Press.)

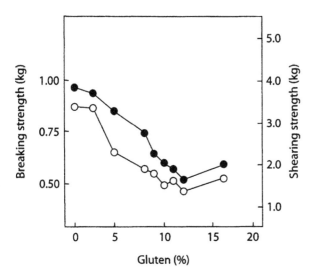

Figure 15 Shearing and breaking strengths of extruded wheat starch as a function of gluten addition: ● shearing strength, ○ breaking strength. (From Ref. 93 with kind permission from American Association of Cereal Chemists.)

creased as gluten was added, up to 15% (Fig. 15). Interestingly, soya protein isolate had the reverse effect up to the 10% level. Mohamed (23) found that adding soya protein up to 25% to corn grits decreased the compressive strength. Matthey and Hanna (42) studied whey protein concentrate (WPC) addition to corn starch of different amylose content as an extrusion cooking feedstock. The shear strength increased with increasing amylose content, but no significant effect of WPC was observed. They commented that WPC could not be added in excess of 20% because of decreased expansion and low sensory scores. Martinez-Serna and Villota (43) used native and modified whey protein isolates (WPI) blended with cornstarch before extrusion cooking. They considered that the breaking force was influenced by two factors, the level of expansion and the strength of the cell walls, which is consistent with variations on the foam model described earlier [Eq. (1)]. Whey proteins decreased the expansion relative to cornstarch alone. They found the "brittleness" to be least for the alkaline WPI—corn starch mixtures increasing for the esterified and acidic WPI and highest for the native WPI with starch. "Cohesiveness" was least for the native WPI relative to the modified WPI mixtures. They concluded that disulfide bonds accentuated toughness and inelasticity in extrudates while predominance of hydrophobic interactions made extrudates more brittle and less cohesive.

C. Sugars

A number of authors has considered the effect of sucrose on extrudate expansion and reported a decrease in expansion or an increase in bulk density. Mohamed (23) found that added sugar to 10% caused a decrease in compressive strength. Hsieh et al. (94) added sucrose to corn meal up to the 8% level and observed that the Warner–Bratzler breaking strength and bulk density decreased with increasing sucrose. However, Jin et al. (95) added sucrose, salt, and soy fiber to corn meal and found the Warner–Bratzler breaking strength, defined as the maximum shear force relative to the sample cross-sectional area, increased up to a level of 12% of sucrose. Moore et al. (96) measured the modulus and strength from a flexural test on sucrose–wheat flour mixtures and found that as the sucrose increased from 0 to 16%, the modulus increased while the strength remained little changed. Barrett et al. (86) considered the effect of sucrose on the structure and mechanical properties of corn extrudates. They tested extrudate discs in compression. The compressive strength varied with extrusion moisture and also postextrusion-conditioned moisture content (Fig. 16). They suggested that dry extrudates and those equilibrated to 12% moisture failed by brittle fracture and plastic yielding, respectively. This change was superimposed on the changes with respect to porosity. At low extrusion moisture, bulk density varied little with sucrose content and the compressive strength was almost constant when the samples were dry, but equilibration to the higher moisture caused the matrix properties to dominate. At a higher extrusion moisture, the bulk density increased with increasing sucrose content; and when dry the compressive strength also increased. On equilibrating the moisture content to 12%, the matrix properties compensated for the structural changes and the strength remained almost constant.

D. Oils and Emulsifiers

Pan et al. (97) reported the effect of soybean oil added to extruded rice increased sectional expansion and shear force (grams). Ryu et al. (98) studied emulsifier addition to wheat flour with and without sucrose and shortening powder. Sucrose esters generally increased the shear strength (kilograms) obtained by the Warner–Bratzler test, whereas glyceryl monostearate decrease it. In the presence of the sucrose and shortening powder, the breaking strength fell with increasing concentration of emulsifiers (Fig. 17). Faubion and Hoseney (93) extracted lipids from wheat flours and added lipids to wheat starch. In both types of experiment the presence of lipids decreased the Warner–Bratzler shear strength and the flexural strength. Mohamed (23) found that adding corn oil up to 3% to corn grits did not affect the compressive breaking force.

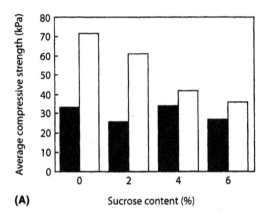

(A)

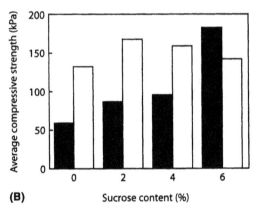

(B)

Figure 16 Effect of sucrose on average compressive stress at extrusion moistures of: (A) 15%, (B) 20%—■ dry, □ equilibrated to 12% moisture. (From Ref. 86 with kind permission from Elsevier Science.)

E. Starch Composition

Launay and Lisch (19) made measurements of mechanical properties—stiffness in compression and flexural strength—of extrudates from mixtures of high amylose and waxy maize starches. They found a complex interaction between amylose content and extrusion temperature. Chinnaswamy and Hanna (99) considered the effect of amylose content at 1% addition of urea, sodium bicarbonate, and sodium chloride. Apart from urea inclusion, which showed a maximum with amylose content addition the shear strength increased with increasing amylose content (Fig. 18).

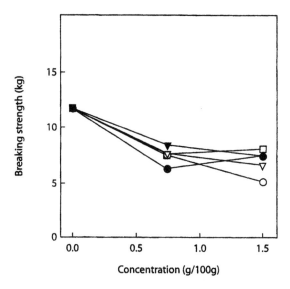

Figure 17 The effects of emulsifiers on breaking strength of wheat flour extrudates containing 6% sucrose and 2.5% shortening powder as a function of emulsifier addition: □ ▽ ▼ sucrose esters, ● sodium stearoyl-2-lactylate, ○ glyceryl monostearate. (From Ref. 98 with kind permission from Academic Press.)

F. Salt

Chinnaswamy and Hanna (99) studied the shear strength using a Kramer shear press and considered the effect of additives to corn starch that would promote expansion. They added urea, sodium chloride, and sodium bicarbonate, the last to produce CO_2 by decomposition. They found that the shear strength was inversely related to expansion ratio, varying in a nonlinear way with concentration up to 4%. Only urea caused an increase relative to no addition. Mohamed (23) found a decrease in the compressive strength of extruded corn grits on adding salt to the 2.5% level. Hsieh et al. (94) found a decrease in Warner–Bratzler breaking strength and bulk density with increasing salt level up to 3% (Fig. 19).

G. Fiber

The extrusion process is often used to incorporate dietary fiber, particularly bran, into expanded snack foods and crispbreads (18, 100). Bjorck et al. (101) reported an increase of dietary fiber and a redistribution from soluble to insoluble components in extruded material. Guy and Horne (102) found that particles of bran

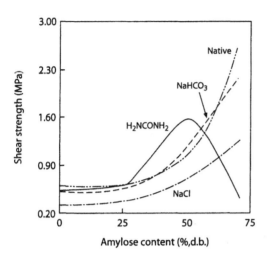

Figure 18 Shear strength of extruded corn starch mixtures as a function of amylose content — ··· — ··· —. Second component added at 1% relative to dry starch: – · – · NaCl, ------ sodium bicarbonate, _____ urea. (From Ref. 99 with kind permission from Wiley–VCH.)

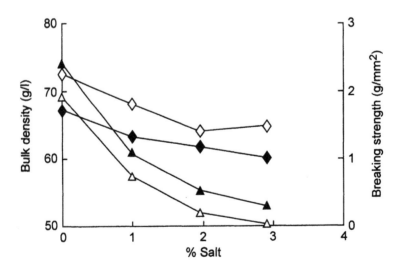

Figure 19 Effect of salt addition on breaking strength (◆, ◊) and bulk density (▲, △) of corn meal extrudates at screw speeds (rpm) of: 250, open symbols and 200, closed symbols. (From Ref. 94 with kind permission from the Institute of Food Technologists.)

caused premature rupture of the expanding gas cells. Studies of the effect of fiber have emphasized its deleterious effect on expansion, as in addition of wheat bran, oat fiber, and sugar beet fiber (103, 104) to corn meal. Artz et al. (105) found that shear force per gram of sample using a shear cell was not statistically significantly related to barrel temperature, screw speed, or fiber content for corn fiber added to corn starch. The shear strength was expressed as force to shear 1 g of sample. Moore et al. (96) studied bran–wheat flour mixtures and found that as the fiber content increased from 0 to 16%, the modulus increased while the strength remained little changed. Jin et al. (95) added soy fiber, salt, and sucrose to corn meal and found the Warner–Bratzler breaking strength increased with fiber addition up to a level of 40%.

VI. SENSORY TEXTURE

Peleg (11) points out that sensory terms can be used interchangeably and cites *crunchy, crisp*, and *brittle* as suffering some overlap, as do *firm, tough*, and *hard*. This is particularly true in the case of low-moisture cereal products considered here. Vickers (36) has reviewed the evaluation of crispness and the hypothesis that auditory sensations are involved in the perception of crispness (106). She concluded that the number of emitted sounds per unit biting distance and the loudness of the sounds changed with perceived crispness. One means of assessing sensory attributes such as crispness of baked foods is to measure the sounds produced during compression of foods. Studies have indicated that vibratory stimuli can lead to the distinguishing of crisp and crunchy foods (35). Although crispness and crunchiness were closely related sensations, crisper sounds were higher in pitch and louder than the crunchier sounds (107). Loudness, crunchiness, and crispness were judged to be very closely related. A good correlation between sensory crispness and acoustic emission has been reported (73, 89, 92).

Some examples will be given of studies that relate sensory attributes to structure, mechanical properties, and instrumental texture of starch-based foods; other studies related to starchy products are reviewed elsewhere (60). Sherman and Deghaidy (28) identified sensory brittleness with the first bite and with the maximum force in fracture, while crispness was associated with the later stage of mastication and the slope of the force–deformation curve from a flexural test. Attenburrow et al. (25) conducted sensory testing on sponge pieces that were crushed between the panelists' molars. Perceived hardness was scored against a control. A good correlation was found between sensory hardness and both stiffness and strength. Antila et al. (18) measured breaking force and made a sensory evaluation of a ''quality number,'' which included texture. This was fitted to a quadratic equation to describe the dependence on rye content, feed rate, barrel temperature, and extrusion moisture. Guraya and Toledo (46) considered expanded products that had been produced from drum-dried or extruded half-products by fluid-bed drying or frying. Drum-dried collets required higher forces to achieve 60% compression than the extruded collets, whether fried or dried. Fried specimens required higher forces for both half-products. The force exhibited a high negative linear correlation with sensory textural attributes of crunchiness, hardness, and density. They concluded that the compressive force alone as defined was not a good indicator of sensory texture. Barrett et al. (108) considered sensory attributes of crispness, hardness, and crunchiness, which correlated linearly with bulk density, although "perceived density" did not correlate with bulk density. Vickers and Christensen (109) studied a range of foods, including crackers and biscuits. They found that peak force in a three-point bend test indicated tactile "firmness" and that Young's modulus was correlated with sensory "crispness" Sauvageot and Blond (47) found a good correlation between sensory crispness and the deformation to the first fracture for breakfast cereals. Loh and Mannell (92) linked instrumental and sensory measures of breakfast cereal texture. They compared the effect of atmospheric-induced moisture content on cereals with large-scale increases on contact with milk and termed these processes "plasticiza-

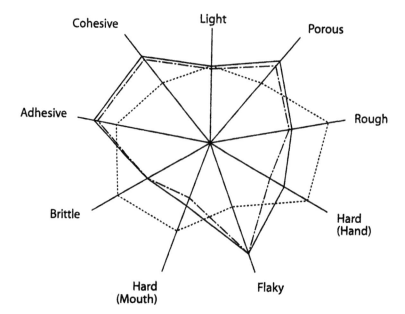

Figure 20 Polar coordinate graph of adjusted means of panel scores for nine sensory attributes. Extrusion conditions—fibre (%)/monoacylglycerol (%)/feed rate (kg h^{-1})/ screw speed (rpm): ——— 15/0.375/36.25/350, ·········· 15/1.125/36.25/350, – · – · – 10/0/42.5/300. (From Ref. 112 with kind permission from Food and Nutrition Press, Inc.)

tion,'' corresponding to a slight increase in shear force, increased rubberiness, and decreased brittleness, and ''solubilization'' accompanied by increased sogginess and decreased crispness.

Brennan and Jowitt (9) found that their instrumental results correlated well with sensory crispness and hardness. Sensory tests revealed that textural crispness was more sensitive but less reproducible than hardness to changes in moisture content. Seymour and Hamann (89) found that sensory crispness and crunchiness of low-moisture foods decreased with increasing water activity, whereas hardness tended to increase. Katz and Labuza (91) found that sensory crispness intensity and sensory acceptability decreased with increasing water activity. Sauvageot and Blond (47) found that sensory crispness of breakfast cereals fell rapidly at water activities greater than 0.50.

The mechanical properties of extrudates and their component matrices have been related to their glass transitions (see earlier sections). Kaletunc and Breslauer (110) reported sensory crispness and denseness, which were found to increase and decrease, respectively, with increasing glass transition temperature. Maga and Cohen (44) ranked acceptability of extrusion cooked potato flakes on a scale from 1 to 7 and found that the most acceptable product used the highest extrusion temperature, the highest compression ratio, and the smallest die.

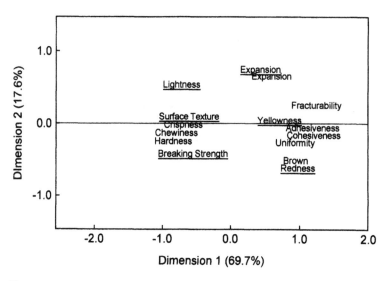

Figure 21 Procrustes consensus for sensory attributes and physical properties for texture of extruded potato. *Sensory attributes* and **physical properties**, the first letter represents the position. (From Ref. 113 with kind permission from Food and Nutrition Press, Inc.)

Hu et al. (111) considered the effects of additives on the sensory attributes of corn meal extrudates. They showed that soy fiber (Fig. 20) and emulsifier significantly affected most of the sensory attributes studied. Increased fiber content increased both sensory hardness and porosity but decreased brittleness (compare physical measurements earlier). Faller and Heymann (112) used generalized Procrustes analysis for extrusion cooked expanded potato products and showed that breaking strength was most similar to sensory hardness among the sensory texture attributes. Breaking strength was highly positively correlated with chewiness and crispness but negatively correlated with sensory fracturability. Low-feed-moisture (16%) extrudates were classed as more easily fractured, whereas those of a moisture content of 18% were harder, chewier, and crisper. Increasing oil content from 0 to 4% increased expansion. Figure 21 shows their Procrustes attribute diagram. Bramesco and Setser (113) considered moisture sorption in relation to mouth dryness, cohesiveness of cereal product types, and adhesion to teeth. They emphasized the importance of salivary flow and the time at which the product is evaluated relative to time in the mouth.

VII. CONCLUSIONS

The glass transition applied to food materials enables a fundamental framework to be established between mobility at the molecular level and mechanical properties, first with a single polymer as influenced by water and ultimately with mixed polymers and small molecules where miscibility and distribution of small molecules are an issue (3). Cereal products are usually composites, and their mechanical properties depend on matrix properties and structure. In particular, extrudates are often solid foams, which can be understood from the matrix properties and porosity, to a first approximation using models. A more detailed modeling treatment reveals the importance of cell size and shape and disposition of the matrix material at the cell edges. Similarly for the matrix properties, the unfoamed material will not always represent the material in the foam walls, since orientation and process history can affect the mechanical properties of cereals. In bridging from mechanical properties and structure to texture, the ideal is to have fundamental tests with predictive power or to analyze empirical tests rigorously, in effect making them fundamental (10). Finally, texture is more than physical properties and structure; its perception can also be affected by chemical variables as well as psychological and cultural factors (11).

ACKNOWLEDGMENTS

The author acknowledges funding from the BBSRC Competitive Strategic Grant.

REFERENCES

1. H Levine and L Slade. Influences of the glassy and rubbery states on the thermal, mechanical and structural properties of doughs and baked products. In: H Faridi and JM Faubion, eds. Dough Rheology and Baked Product Texture. New York: Van Nostrand Reinhold, 1990, pp 157–330.
2. H Levine and L Slade. The glassy state in applications for the food industry, with an emphasis on cookie and cracker production. In: JMV Blanshard and PJ Lillford, eds. The Glassy State in Foods. Nottingham; UK: University Press, 1993, pp 333–373.
3. L Slade and H Levine. The glassy state phenomenon in food molecules. In: JMV Blanshard and PJ Lillford, eds. The Glassy State in Foods. Nottingham, UK: University Press, 1993, pp 35–101.
4. MT Kalichevsky, JMV Blanshard, and RDL Marsh. Applications of mechanical spectroscopy to the study of glassy biopolymers and related systems. In: JMV Blanshard and PJ Lillford, eds. The Glassy State in Foods. Nottingham, UK: University Press, 1993, pp 133–156.
5. IM Ward. Mechanical Properties of Solid Polymers. New York: Wiley Interscience, 1983.
6. R Parker and AC Smith. The mechanical properties of starchy food materials at large strains and their ductile–brittle transitions. In: JMV Blanshard and PJ Lillford, eds. The Glassy State in Foods. Nottingham, UK: University Press, 1993, pp 519–522.
7. R Jowitt. The terminology of food texture. J Text Studies 5:351–358, 1974.
8. AS Szczesniak. Objective measurements of food texture. J Food Sci 28:410–420, 1963.
9. JG Brennan and R Jowitt. Some factors affecting the objective study of food texture. In: GG Birch, JG Brennan, and KJ Parker, eds. Sensory properties of foods. London: Applied Science, 1977, pp 227–245.
10. MC Bourne. Converting from empirical to rheological tests on foods—it's a matter of time. Cereal Foods World 39:37–39, 1994.
11. M Peleg. The semantics of rheology and texture. Food Technol 37(11):54–61, 1983.
12. RP Brown. Handbook of Plastics Test Methods. London: George Godwin, 1981.
13. MF Ashby. The mechanical properties of cellular solids. Acta Metall. 14A:1755–1769, 1983.
14. AT Huber and LJ Gibson. Anisotropy of foams. J Mater Sci 23:3031–3040, 1988.
15. JG Williams. Fracture mechanics of polymers. Chichester, UK: Ellis Horwood, 1984.
16. AG Atkins and Y-W Mai. Elastic and Plastic Fracture. Chichester, UK: Ellis Horwood, 1985.
17. H Luyten, T Van Vliet, and P Walstra. Comparison of various methods to evaluate fracture phenomena in food materials. J Text Studies 23:245–266, 1992.
18. J Antila, K Seiler, W Seibel, and P Linko. Production of flat breads by extrusion cooking using different wheat/rye ratios, protein enrichment and grains with poor baking ability. J Food Eng 2:189–210, 1983.

19. B Launay and JM Lisch. Twin screw extrusion cooking of starches: flow behavior of starch pastes, expansion and mechanical properties of extrudates. J Food Eng 2:259–280, 1983.

20. RJ Hutchinson, GDE Siodlak, and AC Smith. Influence of processing variables on the mechanical properties of extruded maize. J Mater Sci 22:3956–3962, 1987.

21. RJ Hutchinson, I Simms, and AC Smith. Anisotropy in the mechanical properties of extrusion cooked maize foams. J Mater Sci Lett 7:666–668, 1988.

22. GW Halek, SW Paik, and KLB Chang. The effect of moisture content on mechanical properties and texture profile parameters of corn meal extrudates. J Text Studies 20:43–55, 1989.

23. S Mohamed. Factors affecting extrusion characteristics of expanded starch-based products. J Food Proc Pres 14:437–452, 1990.

24. I Fontanet, S Davidou, C Dacremont, and M Le Meste. Effect of water on the mechanical behavior of extruded flat bread. J Cereal Sci 25:303–311, 1997.

25. GE Attenburrow, LJ Taylor, RM Goodband, and PJ Lillford. Structure, mechanics and texture of a food sponge. J Cereal Sci 9:61–70, 1989.

26. AJ Shah, GE Warburton, AEJ Morris, AJ Rosenthal, and J Lamb. Expansion, color and textural properties of extrusion-cooked fish/corn grit mixtures. In: RE Carter, ed. Rheology of Food, Pharmaceutical and Biological Materials with General Rheology. London: Elsevier, 1990, pp 27–42.

27. Y Andersson, B Drake, A Granquist, L Halldin, B Johansson, RM Pangborn, and C Akesson. Fracture force, hardness and brittleness in crisp bread, with a generalized regression analysis approach to instrumental–sensory comparisons. J Text Studies 4:119–144, 1973.

28. P Sherman and FS Deghaidy. Force-deformation conditions associated with the evaluation of brittleness and crispness in selected foods. J Text Studies 9:437–459, 1978.

29. JM Faubion and RC Hoseney. High-temperature short-time extrusion cooking of wheat starch and flour. I. Effect of moisture and flour type on extrudate properties. Cereal Chem 59:529–533, 1982.

30. E Van Hecke, K Allaf, and JM Bouvier. Texture and structure of crispy-puffed food products I: Mechanical properties in bending. J Text Studies 26:11–25, 1995.

31. DJ Van Zuilichem, G Lamers, and W Stolp. Influence of process variables on the quality of extruded maize. Proceedings of Sixth European Symposium on Engineering and Food Quality, Cambridge, UK, 1975, pp 380–406.

32. AR Kirby and AC Smith. The impact properties of extruded food foams. J Mater Sci 23:2251–2254, 1988.

33. E Plati and JG Williams. The determination of the fracture parameters for polymers in impact. Poly Eng Sci 15:470–477, 1975.

34. F Shama and P Sherman. Stress relaxation during force-compression studies with the Instron Universal Testing Machine and its implications. J Text Studies 4:353–362, 1973.

35. ZM Vickers. The relationships of pitch, loudness and eating technique to judgments of the crispness and crunchiness of food sounds. J Text Studies 16:85–95, 1985.

36. ZM Vickers. Evaluation of crispness. In: JMV Blanshard and JR Mitchell, eds. Food Structure: Its Creation and Evaluation. London: Butterworths, 1988, pp 433–448.

37. A-L Hayter, AR Kirby, and AC Smith. The physical properties of extruded food foams. In: HG Ang, ed. Trends in Food Processing. Vol. 1. Singapore: Institute of Food Science and Technology, 1989, pp 265–274.

38. MC Bourne. Food Texture and Viscosity: Concept and Measurement. London: Academic Press, 1982.

39. J Owusu-Ansah, FR van de Voort, and DW Stanley. Textural and microstructural changes in corn starch as a function of extrusion variables. Canad Inst Food Sci Tech J 17:65–70, 1984.

40. RG Falcone and RD Phillips. Effects of feed composition, feed moisture, and barrel temperature on the physical and rheological properties of snack-like products prepared from cowpea and sorghum flours by extrusion. J Food Sci 53:1464–1469, 1988.

41. M Bhattacharya and MA Hanna. Textural properties of extrusion-cooked corn starch. Lebensm wiss u Technol 20:195–201, 1987.

42. FP Matthey and MA Hanna. Physical and functional properties of twin-screw extruded whey protein concentrate–corn starch blends. Lebensm wiss u Technol 30: 359–366, 1997.

43. MD Martinez-Serna and R Villota. Reactivity, functionality and extrusion performance of native and chemically modified whey proteins. In: JL Kokini, C-T Ho, and MV Karwe, eds. Food Extrusion Science and Technology. New York: Marcel Dekker, 1992, pp 387–414.

44. JA Maga and MR Cohen. Effect of extrusion parameters on certain sensory, physical and nutritional properties of potato flakes. Lebensm wiss u Technol 11:195–197, 1978.

45. C Mercier and P Feillet. Modification of carbohydrate components by extrusion-cooking of cereal products. Cereal Chem 52:283–297, 1975.

46. HS Guraya and RT Toledo. Microstructural characteristics and compression resistance as indices of sensory texture in a crunchy snack product. J Text Studies 27: 687–701, 1996.

47. F Sauvageot and G Blond. Effect of water activity on crispness of breakfast cereals. J Text Studies 22:423–442, 1991.

48. A-L Hayter and AC Smith. The mechanical properties of extruded food foams. J Mater Sci 23:736–743, 1988.

49. MC Bourne. Texture profile analysis. Food Technol 32:62–67, 1978.

50. A-L Hayter, EHA Prescott, and AC Smith. Application of the IFR portable pendulum for the assessment of the mechanical properties of solid foams. Polym Test 7:27–38, 1987.

51. A-L Hayter, AC Smith, and P Richmond. The physical properties of extruded food foams. J Mater Sci 21:3729–3736, 1986.

52. LJ Gibson and MF Ashby. Cellular Solids—Structure and Properties. London: Pergamon Press, 1996.

53. AC Smith. Solid Foams. In: RD Bee, P Richmond, and J Mingins, eds. Food Colloids. London: Royal Society of Chemistry, 1989, pp 56–73.

54. E Baer. Engineering Design for Plastics. New York: Chapman and Hall, 1964.

55. JK MacKenzie. The elastic constants of a solid containing spherical holes. Proc Phys Soc Lond B63:2–11, 1950.

56. AC Smith. Studies on the physical structure of starch-based materials in the extru-

sion cooking process. In: JL Kokini, C-T Ho, and MV Karwe, eds. Food Extrusion Science and Technology. New York: Marcel Dekker, 1992, pp 573–618.

57. SC Warburton, AM Donald, and AC Smith. The deformation of brittle starch foams. J Mater Sci 25:4001–4007, 1990.
58. SC Warburton, AM Donald, and AC Smith. Structure and mechanical properties of brittle starch foams. J Mater Sci 27:1468–1474, 1992.
59. A McIntyre and GE Anderton. Fracture behavior of rigid polyurethane foam in impact. Europ J Cell Plast July:153–158, 1978.
60. AC Smith, Starch-based foods. In: AJ Rosenthal, ed. Food Texture Perception and Measurement. New York: Aspen, 1999.
61. GW Halek and KLB Chang. Effect of extrusion operation variables on functionality of extrudates. In: JL Kokini, C-T Ho, and MV Karwe, eds. Food Extrusion Science and Technology. New York: Marcel Dekker, 1992, pp 677–691.
62. J Amemiya and JA Menjivar. Mechanical properties of cereal based food cellular systems. Abstract of 77th Annual Meeting of American Association of Cereal Chemists, Minneapolis, 1992.
63. CJAM Keetels, T Van Vliet, and P Walstra. Relationship between the sponge structure of starch bread and its mechanical properties. J Cereal Sci 24:27–31, 1996.
64. RL Shogren, JW Lawton, WM Doane, and KF Tiefenbacher. Structure and morphology of baked starch foams. Polymer 39:6649–6655, 1998.
65. AH Barrett and M Peleg. Extrudate cell structure–texture relationships. J Food Sci 57:1253–1257, 1992.
66. A Nussinovitch, I Roy, and M Peleg. Testing bread slices in tension mode. Cereal Chem 67:101–103, 1990.
67. M Mizukoshi. Model studies of cake baking. V. Cake shrinkage and shear modulus of cake batter during baking. Cereal Chem 62:242–246, 1985.
68. RJ Hutchinson, SA Mantle, and AC Smith. The effect of moisture content on the mechanical properties of extruded foams. J Mater Sci 24:3249–3253, 1989.
69. AC Smith. Brittle textures in processed foods. In: JFV Vincent and PJ Lillford, eds. Feeding and the texture of food. Cambridge, UK: University Press, 1991, pp 185–210.
70. G Attenburrow, DJ Barnes, AP Davies, and SJ Ingman. Rheological properties of wheat gluten. J Cereal Sci 12:1–14, 1990.
71. MT Kalichevsky, EM Jaroszkiewicz, S Ablett, JMV Blanshard, and PJ Lillford. The glass transition of amylopectin measured by DSC, DMTA and NMR. Carbohydr Polym 18:77–88, 1992.
72. RJ Nicholls, IAM Appelqvist, AP Davies, SJ Ingman, and PJ Lillford. Glass transitions and the fracture behavior of gluten and starches within the glassy state. J Cereal Sci 21:25–36, 1995.
73. GE Attenburrow, AP Davies, RM Goodband, and SJ Ingman. The fracture behavior of starch and gluten in the glassy state. J Cereal Sci 16:1–12, 1992.
74. MT Kalichevsky, EM Jaroszkiewicz, and JMV Blanshard. Glass transition of gluten. 1: Gluten and gluten–sugar mixtures. Int J Biol Macromol 14:257–266, 1992.
75. MT Kalichevsky and JMV Blanshard. A study of the effect of water on the glass transition of 1:1 mixtures of amylopectin, casein and gluten using DSC and DMTA. Carbohydr Polym 19:271–278, 1992.
76. MT Kalichevsky, EM Jaroszkiewicz, and JMV Blanshard. Glass transition of glu-

ten. 2: The effect of lipids and emulsifiers. Int J Biol Macromol 14:267–273, 1992.

77. MT Kalichevsky and JMV Blanshard. The effect of fructose and water on the glass transition of amylopectin. Carbohydr Polym 20:107–113, 1993.

78. MT Kalichevsky, EM Jaroszkiewicz, and JMV Blanshard. A study of the glass transition of amylopectin–sugar mixtures. Polymer 34:346–358, 1993.

79. A-L Ollett, R Parker, and AC Smith. Deformation and fracture behavior of wheat starch plasticized with water and glucose. J Mater Sci 26:1351–1356, 1991.

80. AR Kirby, SA Clark, R Parker, and AC Smith. The deformation and fracture behavior of wheat starch plasticized with water and polyols. J Mater Sci 28:5937–5942, 1993.

81. DMR Georget and AC Smith. The mechanical properties of wheatflake components. Carbohydr Polym 24:305–311, 1995.

82. DMR Georget and AC Smith. The mechanical properties of processed cereals. J Therm Analy 47:1377–1389, 1996.

83. DMR Georget and AC Smith. A study of the effect of water content on the mechanical texture of breakfast wheat flakes. In. GR Fenwick, C Hedley, RL Richards, and S Khokhar, eds. Agri-Food Quality. London: Royal Society of Chemistry, 1996, pp 196–199.

84. DMR Georget, PA Gunning, ML Parker, and AC Smith. The mechanical properties of pressed processed wheat material. J Mater Sci 31:3065–3071, 1996.

85. DMR Georget and AC Smith. Mechanical properties of wheat flakes and their components. In: PJ Frazier, AM Donald, and P Richmond, eds. Starch: Structure and Function. London: Royal Society of Chemistry, 1997, pp 96–104.

86. A Barrett, G Kaletunc, S Rosenburg, and K Breslauer. Effect of sucrose on the structure, mechanical strength and thermal properties of corn extrudates. Carbohydr Polym 26:261–269, 1995.

87. DMR Georget, R Parker, and AC Smith. An assessment of the pin deformation test for measurement of mechanical properties of breakfast cereal flakes. J Text Studies 26:161–174, 1995.

88. BK Jasberg, CL Swanson, RL Shogren, and WM Doane. Effect of moisture on injection molded starch–EA–HDPE composites. J Polym Mater 9:163–170, 1992.

89. S Seymour and D Hamann. Crispness and crunchiness of selected low moisture foods. J Text Studies 19:79–95, 1988.

90. ME Zabik, SG Fierke, and DK Bristol. Humidity effects on textural characteristics of sugar-snap cookies. Cereal Chem 6:29–33, 1979.

91. E Katz and TP Labuza. Effect of water activity on the sensory crispness and mechanical deformation of snack food products. J Food Sci 46:403–409, 1981.

92. J Loh and W Mannell. Application of rheology in the breakfast cereal industry. In: H Faridi and JM Faubion, eds. Dough Rheology and Baked Product Texture. New York: Van Nostrand Reinhold, 1990, pp 405–420.

93. JM Faubion and RC Hoseney. High-temperature short-time extrusion cooking of wheat starch and flour. II. Effect of protein and lipid on extrudate properties. Cereal Chem 59:533–537, 1982.

94. F Hsieh, IC Peng, and HE Huff. Effects of salt, sugar and screw speed on processing and product variables of corn meal extruded with a twin-screw extruder. J Food Sci 55:224–227, 1990.

95. Z Jin, F Hsieh, and HE Huff. Effects of soy fiber, salt, sugar and screw speed on physical properties and microstructure of corn meal extrudate. J Cereal Sci 22:185–194, 1995.

96. D Moore, A Sanei, E Van Hecke, and JM Bouvier. Effect of ingredients on physical structural properties of extrudates. J Food Sci 55:1383–1387, 1402, 1990.

97. BS Pan, M-S Kong, and H-H Chen. Twin-screw extrusion for expanded rice products: processing parameters and formulations of extrudate properties. In: JL Kokini, C-T Ho, and MV Karwe, eds. Food Extrusion Science and Technology. New York: Marcel Dekker, 1992, pp 693–709.

98. GH Ryu, PE Neumann, and CE Walker. Effects of emulsifiers on physical properties of wheat flour extrudates with and without sucrose and shortening. Lebensm wiss u Technol 27:425–431, 1994.

99. R Chinnaswamy and MA Hanna. Expansion, color and shear strength properties of corn starches extrusion-cooked with urea and salts. Starch 40:186–190, 1988.

100. Y Andersson, B Hedlund, L Jonsson, and S Svensson. Extrusion cooking of a high-fiber cereal product with crispbread character. Cereal Chem 58:370–374, 1981.

101. I Bjorck, M Nyman, and N-G Asp. Extrusion cooking and dietary fiber: effects on dietary fiber content and on degradation in the rat intestinal tract. Cereal Chem 61:174–179, 1984.

102. RCE Guy and AW Horne. Extrusion and co-extrusion of cereals. In: JMV Blanshard and JR Mitchell, eds. Food Structure: Its Creation and Evaluation. London: Butterworths, 1988, pp 331–349.

103. S Lue, F Hsieh, IC Peng, and HE Huff. Expansion of corn extrudates containing dietary fiber—a microstructure study. Lebensm wiss u technol 23:165–173, 1990.

104. S Lue, F Hsieh, and HE Huff. Extrusion cooking of corn meal and sugar-beet fiber—effects on expansion properties, starch gelatinization, and dietary fiber content. Cereal Chem 68:227–234, 1991.

105. WE Artz, CC Warren, and R Villota. Twin screw extrusion modifications of a corn fiber and corn starch extruded blend. J Food Sci 55:746–750, 1990.

106. ZM Vickers and MC Bourne. A psychoacoustical theory of crispness. J Food Sci 41:1158–1164, 1976.

107. ZM Vickers. Crispness and crunchiness: a difference of pitch? J Text Studies 15:157–163, 1984.

108. AH Barrett, AV Cardello, LL Lesher, and IA Taub. Cellularity, mechanical failure and textural perception of corn meal extrudates. J Text Studies 25:77–95, 1994.

109. ZM Vickers and CM Christensen. Relationships between sensory crispness and other sensory and instrumental parameters. J Text Studies 11:291–307, 1980.

110. G Kaletunc and K Breslauer. Glass transitions of extrudates: relationship with processing-induced fragmentation and end-product attributes. Cereal Chem 70:548–552, 1993.

111. L Hu, HE Huff, H Heymann, and F Hsieh. Effects of emulsifier and soy fiber addition on sensory properties of corn meal extrudate. J Food Qual 19:57–77, 1996.

112. JY Faller and H Heymann. Sensory and physical properties of extruded potato puffs. J Sens Studies 11:227–245, 1996.

113. NP Bramesco and CS Setser. Application of sensory texture profiling to baked products: some considerations for evaluation, definition of parameters and reference products. J Text Studies 21:235–251, 1990.

10

Utilization of Rheological Properties in Product and Process Development

Victor T. Huang
General Mills, Inc., Minneapolis, Minnesota, U.S.A.

Gönül Kaletunç
The Ohio State University, Columbus, Ohio, U.S.A.

I. INTRODUCTION

The previous four chapters covered the characterization of mechanical properties of cereal flours, prior to, during, and after processing. In this chapter, we will discuss the application of rheological information in product and process development. Specifically, we want to discuss the following three areas: the rheological comparison between synthetic and food polymers; the utilization of basic knowledge of rheological properties to analyze the stability of materials before, during, and after processing; the influence of rheological properties on the selection of appropriate processing methods.

II. RHEOLOGICAL COMPARISON BETWEEN SYNTHETIC AND FOOD POLYMERS

It was not until the early 1980s that food scientists, led by Drs. Harry Levine and Louise Slade, realized that synthetic polymer principles are applicable to food systems (1–8). Various thermal analysis techniques have demonstrated the similarity between them (9–11). Figure 1 shows the five regions of viscoelasticity of a synthetic polystyrene: AB glassy region, BC glass transition region, CD

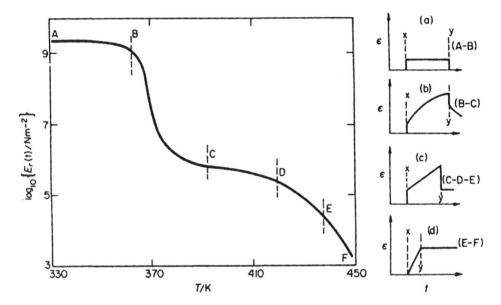

Figure 1   Five regions of viscoelasticity, illustrated by using a polystyrene sample. Also shown are the strain (ε)–time curves for stress applied at x and removed at y: (a) glassy region, (b) leathery state; (c) rubbery state; and (d) viscous state. (From Ref. 12.)

rubbery plateau region, DE rubbery flow curves for stress applied at x and removed at y. In the glassy state (AB), the material behaves like an elastic solid. In the viscous state (EF), it is a liquid, while it behaves like a viscoelastic material in the temperature ranges from B to E. In The modulus–temperature curve is very sensitive to many structural factors, such as molecular weight, degree of cross-linking, percentage crystallinity, copolymerization, plasticization, and phase separation (13).

The glass transition temperature (T_g) is a function of product composition, molecular weight of the continuous structural matrix, degree of branching, degree of cross-linking, crystallinity, and degree of plasticization. For un-cross-linked molecules, the drop in modulus is about three decades near T_g. The magnitude of drop in modulus in the glass transition region decreases as the degree of cross-linking or molecular entanglement increases, which is the case for low-moisture gluten samples. As shown in Figure 2, for a gluten sample equilibrated in 65% relative humidity (RH), the drop in modulus is less than two decades (13). The degree of viscosity drop at a constant ($T - T_g$) has been used as an index of "fragility." Unfortunately, organic glass usually is more fragile than inorganic glass. Thus, there is minimal opportunity for applying this concept in stabilizing food systems. A high-crystallinity sample has lower modulus drop at T_g due to

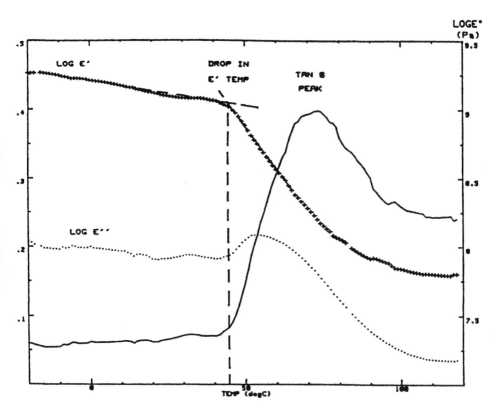

Figure 2 Typical DMTA plot for gluten (RH = 65%), showing tan δ, log loss modulus (E″), and log elastic modulus (E′) as a function of temperature. (From Ref. 13.)

the reduced amount of amorphous region, and is directionally higher in T_g. Plasticizers decrease both T_g and the rubbery modulus of a PVC-diethylhexyl succinate system (Fig. 3), while Figure 4 shows that water is a powerful plasticizer for gluten (13). Crispy breakfast cereal has modulus of around 10^9 Pa. The modulus for glucose glass is 8×10^9 Pa, while it is 8×10^8 Pa for glucose/sucrose glass at 2–3% moisture, regardless of their ratios (15).

As shown in Figure 1, the modulus in the rubbery plateau region is relatively constant, and its temperature range (CD) is a function of molecular weight and the number of entanglements per molecule. Thus, as shown in Figures 2 and 5, a gluten sample at 65% relative humidity, due to its high molecular weight and degree of entanglement, has very long rubbery plateau, while sorbitol has almost no visible one (15). A slice of microwave-heated bread has the modulus of 10^6–10^7 Pa at room temperature, which is in between the glass transition and rubbery plateau regions as typified by its leathery-rubbery texture.

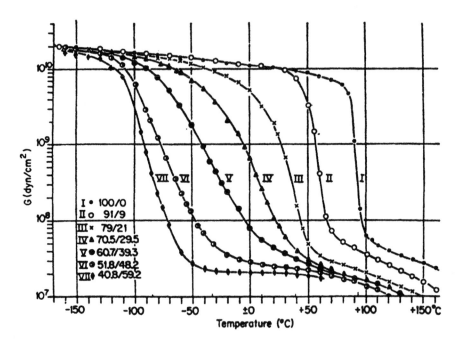

Figure 3 Dynamic shear modulus of polyvinyl chloride plasticized with various amounts of diethylhexyl succinate plasticizer. (From Ref. 14.)

In contrast to sorbitol, which goes into rubbery and viscous flow right after the glass transition region, the gluten DMTA curve lacks the rubbery flow and viscous flow regions. This is probably due to the relatively high degree of cross-linking and/or entanglement. Molasses, honey, and batters are examples of food systems in the viscous flow region at room temperature.

A blend of incompatible polymers will phase-separate and show more than one T_g, as indicated in Figure 6 for sodium caseinate–water and fructose–water systems (16). From all these observations, we can conclude that food systems can indeed be viewed as synthetic polymer systems.

Air cells or pockets help soften some products, such as breakfast cereals, that would otherwise be too hard. Gas cells inside some products can be from many sources: entrapped air during mixing, yeast leavening, chemical leavening, injected inert gases, carbon dioxide, ethanol. Those gas cells will eventually have the same gaseous composition as that of the packaged environment. One area that differs from the synthetic polymer system is the presence of fat in food matrices. In bakery products, fats and emulsifiers do not affect T_g, but they do decrease the rubbery modulus (17). That is why they are sometimes called tenderizers instead of plasticizers. The tenderizing effect at serving temperature is a

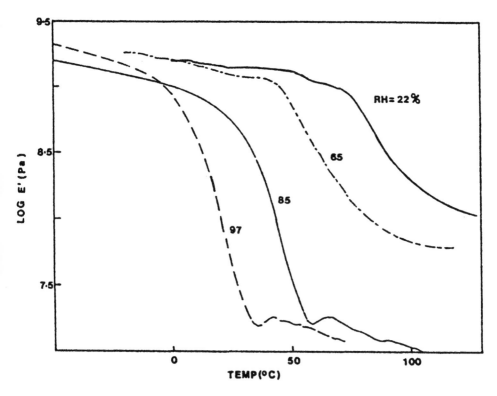

Figure 4 DMTA plot for gluten samples stored under different RH values. (From Ref. 13.)

function of the solid fat index, fat content, and fat crystalline form. Due to their lubricating effect, fats and emulsifiers also enhance the perceived moistness of bakery products. Moistness also results partly because the true moisture content of the nonfat portion is higher than the apparent moisture content, which is based upon the total system. No wonder that fat-free bakery products usually taste dry and not as tender.

III. EFFECT OF RHEOLOGICAL PROPERTIES ON THE STABILITY OF MATERIAL BEFORE, DURING, AND AFTER PROCESSING

In addition to having high hedonic quality during consumption, any successful product in the marketplace needs to be stable throughout distribution. During manufacturing, ingredients and in-process intermediates should not have caking,

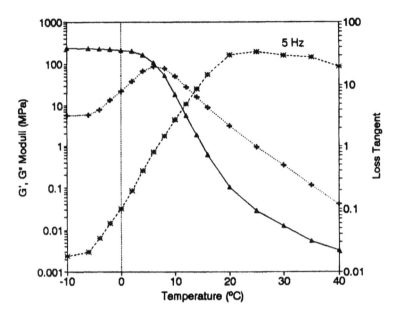

Figure 5   Annular shear test for amorphous sorbitol: (δ) storage modulus G', (+) loss modulus G'', and (*) loss tangent. (From Ref. 15.)

shrinkage, or other physical instability problems. In new-product development, significant effort is on stabilizing the ingredient during storage, the processing intermediates during manufacturing, and the finished products during storage, consumer preparation, and consumption. Directly and indirectly, rheological properties affect physical, chemical, and microbiological stabilities. In this chapter, we will focus more on the physical stability aspect.

How do rheological properties affect stability? Well, in the previous section we discussed the effect of physical state on rheological properties. Thus, it is physical state that determines stability. A state diagram shows T_g as a function of water content and solubility as a function of temperature. It also shows information on various physical changes that may occur due to the metastable state of amorphous food solids and their approach toward equilibrium (18). Roos (19) summarized the methods of applying state diagrams in food processing and product development. LeMeste et al. (20) reviewed the relationship between physical states and the quality of cereal-based foods. A product in the glassy state during storage will be more stable, while products stored at temperatures above the T_g curve will undergo physical changes at a rate according to William–Landel– Ferry (WLF) kinetics (4–8). The Gordon–Taylor equation can relate the effect of composition on T_g (19, 21). Thus, by combining the WLF and Gordon–Taylor

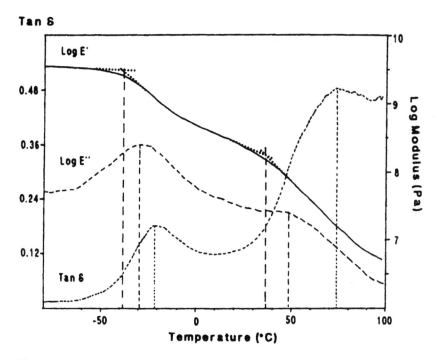

Figure 6 DMTA plot of sodium caseinate and fructose at a ratio of 2:1 stored at 75% RH (16% water). (From Ref. 16.)

equations, one can express rheological properties, such as viscosity, of food materials as a function of temperature and moisture content. However, the coefficients in those equations are not yet readily available, even for common food systems. Some of the stability problems encountered in product development will be discussed here. These include caking during storage and spray-drying, structural collapse during processing, and loss of crispness during storage.

A. Caking During Storage and Processing

As described in Chapter 7, caking, or the loss of free flow, is one of the most often encountered stability problems associated with spray-dried or freeze-dried amorphous ingredients. Ingredient caking affects plant operation efficiency. Caking of wheat flour usually is of no concern if it is stored properly. However, under humid conditions, e.g., at 90% relative humidity, mold growth or infestation might occur even at room temperature (22).

Caking is a mobility-related phenomenon associated with the physical state of the continuous structural matrix on the state diagram. When either storage temperature is too high (e.g., above T_g) or moisture content is too high (e.g., higher than W_g), a powder ingredient might loss its free-flowing property. Usually, caking problem is worse for lower-molecular-weight amorphous carbohydrate powders with relatively low T_g values, such as 42DE corn syrup solids or brown sugar when stored at high temperature or under humid conditions. When an ingredient absorbs enough moisture to depress T_g below storage temperature, caking may occur. In the extreme case, such as brown sugar, some dissolved sucrose will recrystallize and the matrix will harden.

Using a sample of sucrose/fructose glass at 7:1 ratio, Roos (23) has shown that the sticky point is about 20–25°C above the onset temperature of the increase in heat capacity, which is close to the end point of glass transition as measured by DSC at a heating rate of 5°C/min. This corresponds to the critical viscosity of 10^{10} mPa $\cdot$ s for caking (24). The critical moisture content is the moisture content at which the ingredient or product has critical viscosity, which is only slightly higher than W_g. When viscosity is too low due to high temperature or moisture content, the particle surfaces tend to cement together through liquid bridging, resulting in caking (25). If the critical moisture content is known, then from the moisture absorption isotherm, one can find the critical relative humidity above which caking will occur during storage at room temperature.

Mannheim (26) suggested the following formulation and processing methods for minimizing the caking problem during storage: (a) drying to low moisture content, (b) treating powders at low-humidity atmospheres and packaging in high-barrier packages, (c) storing at low temperatures, (d) in-package desiccation, (e) agglomeration, (f) adding anticaking agents. This confirms again the importance of temperature and moisture content on controlling caking. Other solutions include modifying the design of the hopper and agitator for conveying or using agglomerated ingredients (27–28).

Caking during spray-drying is the same phenomenon, except it occurs at higher temperature during a shorter time. From the T_g curve and the temperature profile inside the spray-dryer, one can determine the maximally allowable moisture content beyond which stickiness will occur. In terms of formulation, by adding 10-15DE maltodextrin to orange juice prior to spray-drying, one can reduce caking during processing and storage (29). In Sec. IV.C, we will cover dryer design modification to reduce caking.

B. Structural Collapse During Processing

Phase transitions in foods have a strong impact on food behavior during processing as well. As expected, understanding rheological properties of in-process intermediates can facilitate product formulation and process design for achieving optimum food quality and stability. During manufacturing, structural collapse can occur if the product is not properly formulated or processed. Collapse in amorphous material is usually related to the glass transition (4–8). Levi and Karel (30) show that volume shrinkage occurs in freeze-dried carbohydrates above their T_g. The rate of shrinkage is a strong function of $(T - T_g)$ and can be modeled using the WLF equation, if the coefficients C1 and C2 are known. The effect of moisture is due to T_g depression and the resultant increase in $(T - T_g)$ at constant temperature. Structural shrinkage during extrusion and cake baking can be predicted and manipulated through rheological control.

1. Shrinkage of Extrudates During Cooling

The role of rheological properties on extrudate expansion has been extensively reviewed (31). In the glass transition region, the elastic modulus can range from 10^6 to 10^9 Pa, with viscosity ranges of 10^{11}–10^{15} mPa · s, which is highly dependent on temperature, moisture content, and measurement techniques. The estimated elastic modulus for collapse is about 10^5–10^6 Pa, with viscosity ranges of 10^8–10^{11} mPa · s (32). The rheological properties of extrudates will dictate the die design and extruding temperature. The extrudate temperature will in turn affect the final product moisture content and the degree of shrinkage during cooling. Breakfast cereal extrudates with high sugar content will shrink during the initial stage of cooling due to the plasticization effect of sugar on the continuous polymeric matrix, resulting in lower elastic modulus (G') when the extrudate is still hot. To reduce shrinkage, in addition to formulating a higher T_g matrix, one can modify the exit temperature and cooling rate so that the product does not stay too long in the rubbery state or warmer, i.e., in the range of $(T_g + 30°C)$ to exit temperature (33). Thus, rheological measurements in the proper temperature range can help product scientists determine the critical temperatures above which product expands during heating and shrinks during cooling.

2. Collapse of Cake During Cooling

In high-ratio cakes, due to an excessively high amount of sucrose and a relatively lower amount of water, the effectiveness of the aqueous phase in batter in plasticizing the starch polymer is significantly reduced. Thus, during baking, not enough starch was gelatinized to provide structural strength for converting the closed foam in batter into an open foam, and the cake collapses upon cooling. Chlorinated flour is usually used to minimize this problem, probably by providing higher G' during the later stage of baking (34). Dea (35), reported that a mechanical spectrometer can be used to measure rheological properties during the baking of cake batter. The technique of mechanical spectroscopy is uniquely suited to the rheological characterization of a material with a dynamic temperature profile during baking. It typically shows that batters that produce a collapsed cake after cooling usually do not have a high enough G' at temperatures higher than 80°C.

C. Loss of Crispness During Storage or Consumption

As mentioned in Chapter 9, crispness is a desirable textural attribute of some cereal-based products. However, during storage or consumer preparation, moisture pickup from the air (cotton candy), from other food components (ice cream sandwich), or from milk (breakfast cereal) will significantly decrease the crispness score and acceptability. Most of the crisp materials are in glassy state with modulus of 10^9–10^{10} Pa, e.g., gliadin glass with 1–3 × 10^9 Pa (36), glucose glass with 8 × 10^9 Pa (37), wheat grain glass with 2 × 10^9 Pa (38), and sodium caseinate glass with 1.8 × 10^9 Pa (39). However, as the product absorbs enough moisture, T_g is depressed to below serving temperature, and modulus drops dramatically (40).

The Fermi distribution function describes a normalized distribution of Y ranging from 0 to 1 as a function of temperature with midpoint at T_c and the spread factor "a" that defines the shape of the curve (41). It is useful in describing the sensitivity of normalized Y values around T_c. Peleg adapted it to describe the temperature, water activity, and/or moisture content dependency of Young's modulus or crispness score (42). For example, the following equation describes Young's modulus as a function of moisture content:

$$Y(m) = \frac{Y_g}{1 + \exp[(m - m_c)/a]} \tag{1}$$

where $Y(m)$ is Young's modulus at moisture content m, Y_g is Young's modulus in the glassy state, a is the spread factor, and m_c is the critical moisture content, i.e., the moisture content at which Y is half of Y_g.

It is important to note that the critical temperature, critical water activity, and critical moisture content are material dependent (43). For extruded bread, a significant loss of crispness at room temperature was observed between 9 and 10% moisture. The brittle–ductile transition, defined as the temperature or moisture content above which the product texture changes from brittle to ductile, happens between 9 and 13.7% while W_g is 15% (44). This means the loss in crispness happens at a temperature below T_g.

"Moisture toughening" means the modulus or stiffness increases with moisture content at moisture content below m_c. This happens in extruded bread (44) and cheese puffs (45), and a linear term of moisture needs to be added to the numerator of the original equation, leading to the following equation:

$$Y(m) = \frac{Y_g + bm}{1 + \exp[(m - m_c)/a]} \tag{2}$$

where b is the slope factor at a moisture lower than m_c.

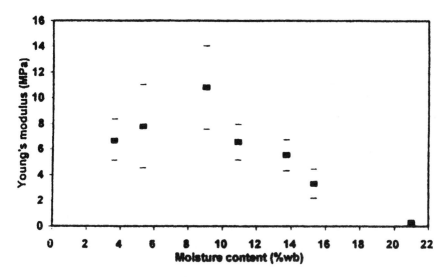

Figure 7   Effect of water on Young's modulus of extruded bread. (From Ref. 44.)

Figure 7 shows that, in extruded bread, the apparent Young's modulus vs. moisture curve has a peak. The modulis at 4% and 11% moisture are both 6.5 MPa, but the modulus is 11 MPa at 9%, then it drops to 5 MPa at 13.7% and 3 MPa at 15.3%, respectively. A significant loss of crispness already occurs between 9% and 10%. However, if we use 11 MPa at 9% for the modulus before the brittle–ductile transition, then the critical moisture content, the moisture content at which the modulus is decreased to 5.5 MPa, would be between 11% and 13.7% moisture. Does this mean that a slight decrease in modulus can significantly decrease crispness or that modulus alone cannot explain the complex sensory attribute of crispness? Vickers (46) reported that sensory crispness contains both hardness and fracturability and has a strong correlation with the peak force and Young's modulus. Roos et al. (43) showed that as moisture content increases, the initial slope of the force–deformation curve determined by a snap test decreases at a rate in between crispness scores measured by a bite task panel and a bite-and-chew task panel, with the crispness measured by the bite-and-chew task panel decreasing the fastest.

How can one apply all this rheological information to product development? It seems that products with T_g or the brittle–ductile transition well above serving temperature will stay crisp after moisture pickup; i.e., the product modulus will remain high even at relatively high moisture content. How do we keep a breakfast cereal crisp as long as possible in a bowl of milk? Well, this is one of the new areas for potential technological innovation in the food industry. One way to increase the bowl life of cereal is to reduce the bulk density of the cereal so that it will float in the milk.

IV. EFFECT OF RHEOLOGICAL PROPERTIES ON THE SELECTION OF PROCESSING METHODS

Rheology is key to various aspects of food process engineering, such as fluid dynamics, heat transfer, and reaction kinetics. However, the level of sophistication of process engineering in the food industry is still at an early stage. Thus, rheological properties have been used more regularly for product formulation than for process development and equipment design. Rheological properties have not yet been widely applied in food processing, except in some areas, such as high-temperature extrusion. The good news is that product scientists and process engineers are starting to realize that rheological data are useful in process design, especially during scale-up.

The rheological properties of food ingredients, in-process intermediates, and finished products are highly dependent on the physical state of the structural matrix. The physical states range from the glassy state, the glass–rubber transition region, the rubbery plateau region, the rubbery flow region, and the viscous flow region. Depending upon the combination of temperature and/or moisture content, the viscosity could be anywhere from 10^{15} mPa · s for glassy solid materials to as low as 10^2 mPa · s for very thin liquids. During the glass transition range alone, the rheological properties can change as much as 1000 times. The rheological properties of the states listed earlier affect not only physical and chemical stability during storage and final textural quality during consumption, but also the selection of processing techniques, equipment design, and operating conditions. The physical states of the in-process intermediates and products, in turn, are a function of the formulation, the temperature, and the associated changes during processing. In the following, we will use several examples to demonstrate the effect of rheological properties on the selection, design, and operation of some processing equipment.

A. Conveying Fragile, Glassy Ingredients

Bread crumbs of up to 10 mm in length at around 3–4% moisture are commonly used along with batter to coat the surface of meat products, such as fish or chicken, to provide a crispy surface texture. Bread crumbs are an example of glassy ingredients with a modulus of 10^9-10^{10} Pa, which are brittle and easily fractured during conveying (12). In a breading applicator, excess bread crumbs are recycled over and over again, resulting in crumb sizes as small as 2–3 mm in length, which is not as acceptable to consumers. Thus, the ideal conveyor should apply a minimal amount of shear on the ingredient. This is important during scale-up since flow rate is much higher in a plant production line. By minimizing the number of recycling passes and/or by using crumbs of slightly higher moisture content, e.g., above 8–9% moisture, one can directionally minimize breakage during processing.

B. Pasta-Drying Process

Dry pasta is in the glassy state with a Young's modulus of 3×10^9 Pa (47). It is brittle and cracks easily even under small stresses, resulting in an unacceptable product. *Checking* is the formation of numerous hairline cracks. Checking can appear shortly after drying, but sometimes it appears after packaging. Checking, which is caused by differential contraction during moisture removal, is worse when the drying rate at the surface is too fast (34). Thus, moisture distribution and the associated stress profiles during drying need to be optimized through the drying method and rate and through product geometry selection. Decareau and Peterson (48) reported that microwave-assisted drying has the advantage of reduced checking through moisture profile modification during drying. There is some evidence that low-protein semolina, high-alpha-amylase-activity flour, and storage in an unstable relative humidity environment can induce checking (49).

C. Drying Sticky Ingredients

Stickiness is a major issue during spray-drying of products rich in low-molecular-weight carbohydrates, due to the excessively low viscosity at processing temperatures higher than the sticky point. As shown in Figure 8, the critical viscosity for stickiness is around $10^9 - 10^{11}$ mPa · s (23). Solutions include formulating the product to be higher in T_g and modifying dryer design. Cereal-derived ingredients, such as starch and low-DE maltodextrins, due to their relatively high T_g are commonly used as drying aids for minimizing stickiness during drying. Dryer design modification includes using relatively low air temperatures and cooling of the dryer walls so that the time the product stays above its T_g is minimized (50). Freeze-drying is being used for pharmaceutical products. However, the 30–50 times higher processing cost associated with long drying times and low throughput will not make most food products commercially feasible (51).

D. Dough Processing

1. Dough Sheeting Process

Cheeses and bread doughs at room temperature are in the rubbery flow region with a modulus of 10^4 to 5×10^5 Pa (52–53). Levine and Drew (54) modeled

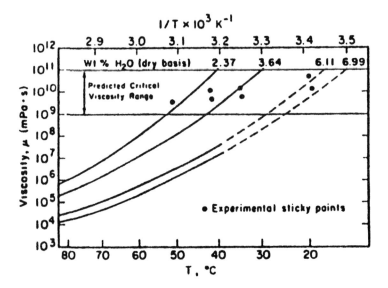

Figure 8 Viscosities of a sugar solution at the sticky point. (From Ref. 24.)

the effect of dough rheology on sheeting and laminating process designs. The amount of work input during sheeting affects the rheological properties of the sheeted dough and the specific volume of the resulting baked product. For a dough to be optimally sheeted at high speed, it needs to be extensible enough. For doughs with low extensibility, frequent relaxation is needed and the processing speed cannot be too fast, to avoid tearing. However, doughs with high resistance toward extension have better gas-holding ability during proofing and baking (55). Thus, dough rheology needs to be optimized by adjusting the flour-to-water ratio and the mixing energy input, while the sheeting speed needs to be adjusted according to dough rheological properties. Spies reviewed the application of rheology in the bread industry (56).

2. Cookie Dough Depositing Process

A soft and sticky cookie dough with G' about 10^4 Pa is usually extruded through an orifice and cut to size by a reciprocating wire. However, a rotary molding process is usually used for firmer cookie dough with a G' value of about 10^5 Pa (57), since high-G' dough is not as sticky (58). In designing a depositing process for cookie dough, one can either optimize its rheological properties for using the existing equipments or take the more costly approach of designing a new process. The rheological properties are affected by processing temperature, flour type, flour-to-water ratio, and thickener concentration (57).

E. Batter Processing

Batter and honey are examples of food materials in the viscous flow region at room temperature. Batters are mostly pseudoplastic fluids having G' around 10^2–10^3 Pa and viscosities in the range of 10^4–10^5 mPa · s. Due to the nature of pseudoplasticity, batter rheology is a strong function of shear. The yield value of batters after pumping and depositing is critical, because it affects the degree of batter spread during depositing and the degree of stickiness to the grids after baking (59). Thus, while designing a batter processing plant, one needs to minimize the total amount of shear during mixing, pumping, and depositing. This will dictate the location of the batter mixing tank relative in height and distance to all the depositing heads. The amount of recycled batter also needs to be reduced. To further minimize shear-induced thinning of batter, one needs to design the diameter of the depositor opening according to the target batter weight, flow rate, and batter viscosity.

V. CONCLUSIONS

Recent literature has shown that food systems can be treated in ways similar to synthetic polymer systems, with water being the most important plasticizer and air, fat, and emulsifier being the key tenderizers. Rheological properties, being a function of physical states, affect the stability of ingredients, in-process intermediates, and finished products. Rheological properties of food systems also determine the selection, design, and operation of food-processing equipment. If the coefficients for products are known, then using state diagrams and WLF or Fermi equations one can roughly estimate the rheological properties as functions of product composition and temperature. More research is needed to derive various coefficients of WLF, Fermi, and Gordon–Taylor equations for modeling the effect of temperature and composition on rheological properties. Thus, product developers eventually can have more quantitative rheological information for process design and product stabilization.

REFERENCES

1. L Slade, H Levine. Recent advances in starch retrogradation. In: SS Stivala, V Crescenzi, ICM Dea, eds. Industrial Polysaccharides. New York: Gordon & Breach, 1986, pp 387–430.
2. L Slade, H Levine, JW Finley. Protein–water interactions: water as a plasticizer of gluten and other protein polymers. In: RD Phillips, JW Finley, eds. Protein Quality and the Effects of Processing. New York: Marcel Dekker, 1988, pp 9–124.
3. H Levine, L Slade. Influence of the glassy and rubbery states on the thermal, mechan-

ical, and structural properties of doughs and baked products. In: H Faridi, JMF Faubion, eds. Dough Rheology and Baked Product Texture. New York: Van Nostrand Reinhold, pp 157–330, 1989.

4. L Slade, H Levine. Beyond water activity: recent advances based on an alternative approach to the assessment of food quality and safety. CRC Crit Revs Food Sci Nutr 30(2–3):115–360, 1991.

5. H Levine, L Slade. Water Relationships in Foods. New York: Plenum Press, 1991.

6. H Levine, L Slade. Glass transitions in foods. In: Schwartzberg and RW Hartel, eds. Physical Chemistry of Foods. New York: Marcel Dekker, pp 83–221, 1992.

7. L Slade, H Levine, J Ievolella, M Wang. The glassy state phenomenon in applications for the food industry: application of the food polymer science approach to structure–function relationships of sucrose in cookie and cracker systems. J Sci Food Agric. 63:133–176, 1993.

8. L Slade, H Levine. Glass transitions and water–food structure interactions. In: JE Kinsella, ed. Advances in Food and Nutrition Research. San diego: Academic Press, Vol. 38, pp 103–269, 1994.

9. L Slade, H Levine. Thermal analysis of starch and gelatin. DMA of frozen aqueous solutions. Proceedings 13th Annual Conference, North American Thermal Analysis Society, Sept. 23–26, Philadelphia, 1984.

10. TJ Maurice, L Slade, C Page, C Sirett. Polysaccharide–water interaction: thermal behavior of rice starch. In: D Simatos, JL Multon, eds. Properties of Water in Foods. Dordrecht, the Netherlands: Nijhoff, 1985, pp 211–227.

11. H Levine, L Slade. Recent advances in applications of thermal analysis to foods. J Thermal Analysis 47(5):1175–1616, 1996.

12. MG Cowie. Polymer: Chemistry and Physics of Modern Materials. New York: Chapman & Hall, 1991, p 248.

13. LE Nielsen. Mechanical Properties of Polymers and Composites. Vol. I. New York: Marcel Dekker, 1974.

14. MT Kalichevsky, EM Jaroskiewicz, JMV Blanshard. Glass transition of gluten 1: Gluten and gluten–sugar mixtures. J Biol Macromol. 14:257–266, 1992.

15. G Blond. Mechanical properties of frozen model solutions. J. Food Engineering 22(1–4):253–269, 1994.

16. MT Kalichevsky, JMV Blanshard, PF Tokarczuk. Effect of water content and sugars on the glass transition of casein and sodium caseinate. Int J Food Sci Tech 28:139–151, 1993.

17. MT Kalichevesky, EM Jaroskiewicz, JMV Blanshard. Glass transition of gluten. 2: The effect of lipids and emulsifiers. J Biol Macromol 14:257–266, 1992.

18. JL Kokini. State diagrams help predict rheology of cereal proteins. Food Technology 49(10):74–82, 1995.

19. Y Roos. Glass transition–related physicochemical changes in foods. Food Technology 49(10):97–102, 1995.

20. M LeMeste, G Roudaut, A Rolee. The physical state and quality of cereal-based foods. In: F Ortega-Rodriquez, GV Barbosa-Canovas, eds. Food Engineering 2000. New York: Chapman & Hall, 1997, pp 97–113.

21. Y Roos, M Karel. Applying state diagrams to food processing and development. Food Technology 45(12):66–68, 1991.

22. P Vijay, BPN Singh, N Maharaj. Equilibrium moisture content of some flours. J Food Sci Technology—India 19(4):153–158, 1982.
23. Y Roos, M Karel. Plasticizing effect of water on thermal behavior and crystallization of amorphous food models. J Food Sci 56:38–43, 1991.
24. GE Downton, JL Flores-Luna, J King. Mechanism of stickiness in hygroscopic, amorphous powders. Ind Eng Chem Fundam 21:447–451, 1982.
25. M Peleg. Physical characteristics of food powders. In: M Peleg, EB Bagley, eds. Physical Properties of Foods. Westport, CT: AVI, 1983, pp 293–323.
26. CH Mannheim, JX Liu, SG Gilbert. Control of water in foods during storage. J Food Engineering 22(1–4):509–532, 1994.
27. L Mancini. Handle with care. Food Engineering 65(11):85–86, 88, 1993.
28. SW Crispe. Device for metering and applying a substance, in particular a fat/flour. West German patent application 1473211, 1970.
29. AS Gupta. Spray drying of orange juice. US patent 4112130, 1978.
30. G Levi, M Karel. Volumetric shrinkage (collapse) in freeze-dried carbohydrates above their glass transition temperature. Food Res. Int 28(2):145–151, 1995.
31. JL Kokini, CN Chang, LS Lai. The role of rheological properties on extrudate expansion. In: JL Kokini, CT Ho, MK Karwe, eds. Food extrusion science and technology. New York: Marcel Dekker, 1992, pp 631–652.
32. CJ King. Application of freeze-drying in food products. In: SA Goldblith, L Rey, WW Rothmayer, eds. Freeze-Drying and Advanced Food Technology. London: Academic Press, 1975, pp 333–343.
33. J Fan, JR Mitchell, JMV Bkanshard. A computer simulation of the dynamics of bubble growth and shrinage during extrudate expansion. J Food Engr 23:337–356, 1994.
34. RC Hoseney. Principles of Cereal Science and Technology. St. Paul, MN: AACC, 1993.
35. ICM Dea, RK Richardson, SB Ross-Murphy. Characterization of rheological changes during the processing of food materials. In: GP Phillips, DJ Wedlock, PA Williams, eds. Gums and Stabilizers for the Food Industry. II. Applications of Hydrocolloids. New York: Pergamon Press, 1984, pp 357–366.
36. EM deGraaf, H Madeka, AM Cocero, J Kokini. Determination of the effect of moisture on glass gliadin transition using mechanical spectrometry and differential scanning calorimetry. Biotechnol Prog 9:210–213, 1993.
37. PB McNulty, DG Flynn. Force-deformation and texture profile behavior of aqueous sugar glass. J Texture Study 8:417, 1977.
38. JL Multon, H Bizot, JL Dublier, J Lefebvre, DC Abbott. Effect of water activity and sorption hysteresis on rheological behavior of wheat kernels. In: LB Rockland, GF Stewart, eds. Water Activity: Influence on Food Quality. New York: Academic Press, 1981.
39. MT Kalichevski, JMV Blanshard, PF Tokarczuk. Effect of water content and sugars on the glass transition of casein and sodium caseinate. J Food Sci Technol 28:139, 1993.
40. G Attenborrow, AP Davies. The mechanical properties of cereal-based foods in and around the glassy state. In: JM Blanshard, PJ Lillford, eds. The Glassy State in Foods. Loughborough, England: Nottingham University Press, 1983, pp 317–331.
41. GT Seaborg, WD Loveland. The Elements Beyond Uranium. New York: Wiley, 1990.

42. M Peleg. A model of mechanical changes in biomaterials at and around their glass transition. Biotechnology Progress 10:385–388, 1994.

43. YH Roos, K Romininen, K Jouppila, H Tuorila. Glass transition and water plasticization effect on crispness of a snack food extrudate. Int J Food Properties 1(2):163–180, 1999.

44. M LeMeste, G Roudaut, S Davidou. Thermomechanical properties of glassy cereal foods. J Thermal Analysis 47, 1361–1375, 1996.

45. M Peleg. Mechanical properties of dry brittle cereal. In: D Reid, ed. The Properties of Water in Foods. New York: Blackie Academic & Professional, 1998, pp 233–252.

46. ZM Vickers. Crispness and crunchiness—textural attributes with auditory components. In: HR Moskowitz, ed. Food Texture: Instrumental and Sensory Measurement. New York: Marcel Dekker, 1987, pp 145–166.

47. MJ Lewis. Physical properties of foods and food processing systems. Chichester, UK: Ellis Horwood, 1987.

48. RV Decareau, RA Peterson. Microwave processing and engineering. Chichester, England: VCH, 1986, p 24.

49. JE Dexter, RR Matsuo, JE Kruger. The spaghetti-making quality of commercial durum wheat samples with variable alpha-amylase activity. Cereal Chemistry 67(5): 405–412, 1990.

50. YH Roos. Phase Transitions in Foods. New York: Academic Press, 1995.

51. SA Desorby, FM Netto, TP labuza. Comparison of spray-drying, drum-drying, and freeze-drying for beta-carotene encapsulation and preservation. J Food Sci 62(6): 1158–1162, 1997.

52. MT Tunick, EL Malin, JJ Shieh, PW Smith, VH Holsinger. Detection of mislabeled butter and cheese by differential scanning calorimetry and rheology. American Laboratory 30(1):28–32, 1998.

53. JM Faubion, PC Dreese, KC Diehl. Dynamic rheological testing of wheat flour doughs. In: H Faridi, ed. Rheology of Wheat Products. St. Paul, MN: AACC, 1985, pp 91–116.

54. L Levine, B Drew. Rheological and engineering aspects of the sheeting and laminating of doughs. In: H Faridi, JMF Faubion, eds. Dough Rheology and Baked Product Texture. New York: Van Nostrand Reinhold, 1989, pp 157–330.

55. I Hlynka. Rheological properties of dough and their significance in the breadmaking process. Bakers' Digest 44(2):40–41, 44–46, 57, 1970.

56. R Spies. Application of rheology in the bread industry. In: H Faridi, JM Faubion, eds. Dough Rheology and Baked Product Texture. New York: Van Nostrand Reinhold, 1989, pp 343–362.

57. JA Menjivar, H Faridi. Rheological properties of cookie and cracker doughs. In: H Faridi, ed. The Science of Cookie and Cracker Production. New York: Chapman & Hall, 1994, pp 283–322.

58. CA Dahlquist. Creep. In: D Satas, ed. Handbook of Pressure-Sensitive Adhesive Technologies. New York: VN Reinhold, pp 97–114, 1989.

59. VT Huang, JB Lindamood, PMT Hansen. Ice cream cone baking: 1. Dependence of baking performance on flour and batter viscosity. Food Hydrocolloids 2(6):451–466, 1988.

11

Characterization of Macrostructures in Extruded Products

Ann H. Barrett
U.S. Army Natick Soldier Center, Natick, Massachusetts, U.S.A.

I. INTRODUCTION

High-temperature, short-time extrusion is used to create expanded foods that have a porous, open structure (Fig. 1). Many snack products and ready-to-eat cereals, which share a characteristic ''crunchy'' texture, are formed by this process. Typically during extrusion, flash-off of superheated water contained in the extrusion ''melt'' occurs as the product passes through the die and encounters ambient-pressure conditions. This sudden vaporization serves to expand the product into a network of cells. Expansion occurs in both radial and axial directions—and to different degrees, depending on the viscoelastic properties of the melt (1, 2). Vaporization of moisture from and cooling of the extrudate serve to bring the product from a molten to a rubbery state; further drying is usually used to produce the brittle, fracturable texture typical of these products.

However, the degree of expansion achieved during extrusion, as well as the exact product structure, is quite variable and depends on both the process parameters employed and the physical properties of the extrusion formula. There exist myriad potential extruded structures. Since expansion occurs due to the growth of multiple individual air cells, it is important to understand that extruded products need to be characterized in terms of the number, size, and distribution of these cells, in addition to general bulk properties such as density. Consider, for example, that a given volume and weight (hence, bulk density) of a porous material can be divided into an infinite number of arrangements of contiguous cells. At one extreme, the material could comprise a hollow structure containing a single cavity; progressive subdivision of the sample into ''compartments''—

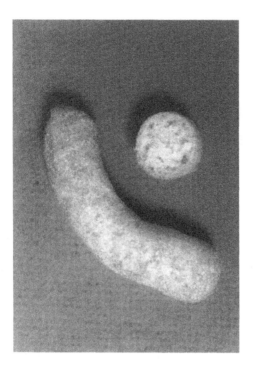

Figure 1 Representative commercially extruded product and cut radial section. Cross section shows cell structure and high porosity. (Photographs compliments of Sarah Underhill, U.S. Army Natick Soldier Center, Natick, MA.)

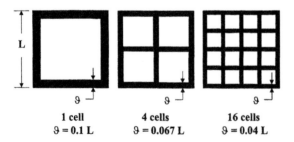

Figure 2 Schematic two-dimensional representation of increasing division of a porous structure (of outer dimension "L") and constant total and percent pore (64%) area. Within each illustration, pores are of equal size and uniformly spaced so that cell wall thickness (ϑ) is invariant throughout the structure. Demonstrates cell wall thinning with increasing number of pores. (Graphic art compliments of Don Pickard, U.S. Army Natick Soldier Center, Natick, MA.)

two, three, four void spaces, and so on—will involve the construction of an ever-increasing number of "walls" (Fig. 2). As the structure becomes increasingly finely divided, these cell walls become increasingly thinner, because of their greater number, and shorter, because of the decreasing size of individual cells. Adding to the array of possible structures is, of course, the fact the cell sizes need not be, and generally are not, uniform. Rigorous description of extrudate structure thus necessitates both careful measurement of cell sizes and statistical description of their distribution.

II. PRACTICAL IMPORTANCE OF CELL STRUCTURE

Cellularity defines our perception of extruded products from visual, auditory, and tactile perspectives. Increasing pore number and the subsequent thinning of cell wall supports will create the visual perception of a more delicate, "frothier" structure, which may be lighter in appearance due to the relatively greater translucence of cell walls. The number—and strength—of cell walls being ruptured during mastication gives rise to the number—and intensity—of audible fractures. Correspondingly, the frequency and magnitude of these fractures, which are sensorially perceived as abrupt changes in the stress level necessary for chewing, constitute a critical property of so-called "crunchy" products.

This characteristic crunchy texture of extrudates arises from the incremental structural failure that occurs in these products as a result of deformation. Such samples, when undergoing deformation—i.e., either instrumentally or by chewing—do not fail by one primary rupture. Because of their cellularity, which implies a multitude of interconnected structural units, the specimens progressively fracture, and fail by a series of breakages occurring in subunits of the network. Fracturability is a hallmark characteristic of extrudate texture, but precise fracture patterns vary widely among different products as a consequence of variation in physical structure.

III. ASPECTS OF STRUCTURE AND MEASUREMENT TECHNIQUES

A. Bulk Characteristics

The "bulk"—or gross structural—properties of extrudates can be characterized by single-parameter measurements of: bulk density (D_B),

$$D_B = \frac{M}{V_E} \tag{1}$$

where M = sample mass and V_E = expanded sample volume; percent pore volume ($\%V_p$),

$$\%V_p = 100\left(1 - \frac{D_B}{D_S}\right) \tag{2}$$

where D_s = material solid density; and expansion ratio (E),

$$E = \frac{V_E}{V_S} \tag{3}$$

where V_s = sample solid volume. Solid density is a property of the extrudate wall material,

$$D_S = \frac{M}{V_S} \tag{4}$$

and can be considered characteristic of a completely unexpanded (or completely compressed) product with $V_p = 0$ and $E = 1$.

Since expansion occurs both radially and longitudinally, overall volumetric expansion is the product of the two directional expansion indices (1, 2). Radial expansion (E_R) is equal to the proportional increase in sample radial area (perpendicular to the direction of extrusion) during puffing,

$$E_R = \frac{A_S}{A_D} \tag{5}$$

where A_s = sample cross-sectional area and A_D = die cross-sectional area. Longitudinal expansion (E_L) is the proportional increase in sample length (axial to the direction of extrusion) during puffing, and has been defined as the ratio of the velocity of the extruded product stream to that of the extrusion melt in the die (1). If V_E and D_S can be measured accurately, E_L can be calculated by

$$E_L = \frac{E}{E_R} \tag{6}$$

$$= \frac{V_E/V_S}{E_R} = \frac{V_E/(M/D_S)}{E_R} \tag{7}$$

$$= \frac{V_E D_S}{M E_R} \tag{8}$$

Accurate determination of sample volume requires appropriate estimation of geometric structure; for example, product extruded through a "slit" die of moderate length/width ratio assumes an oval- or ellipse-shaped cross section during expansion. Replicated dimensional measurements are necessary since radial measurements can vary depending on the size and location of individual, locally protruding, cells near the extrudate surface.

Solid volume for porosity and true density measurements can be determined by stereopyncnometry (3). This technique is based upon displacement of, and increased pressure in, an ideal gas within a chamber of known volume due to the presence of the sample. Stereopyncnometry is best performed on ground or crushed specimens in order to eliminate errors due to the presence of completely closed cells, which exclude the interpenetrating vapor and thus falsely increase volume determinations. However, analysis of specimen volume (V) both before and after crushing will indicate the proportion of the structure that consists of closed cells ($\%V_C$),

$$\%V_C = 100 \left(\frac{V_{\text{uncrushed sample}} - V_{\text{crushed sample}}}{V_{\text{uncrushed sample}}} \right) \tag{9}$$

A rough approximation of solid density can also be obtained based upon the true density values of the formula constituents—that for starch, for example.

Highly expanded extrudates have a high proportion of cell walls disrupted (i.e., containing pores or cracks), as evidenced by their utilization as "infusible" structures, in which the pore volume is infiltrated with a liquefied suspension, using vacuum as a driving force; such a technique was tested as a means of developing calorically dense, lipid-infused ration components for the military (4). Extrusion is, after all, usually a small-scale "explosive" process involving instantaneous vaporization of water. Localized strain, and strain rates, can exceed the extensibility of the material, leading to the development of ruptures during expansion. Microscopy of highly expanded extrudates has illustrated their permeability and also the visible stretching and thinning of cell walls that give rise to the development of between-cell pores (5).

The general intactness of wall supports can influence mechanical behavior. According to Gibson and Ashby (6), the collapse stress of brittle cellular solids is a function of $[(1 - \phi)D_B/D_S]$, where $(1 - \phi)$ is the fraction of material in the cell wall faces (as opposed to the cell wall edges, or junctions). Permeability can also be correlated with other physical attributes, since a relatively more expanded, or stretched, extrudate is likely to be a relatively more open "foam" structure— that is, to have a higher occurrence of cell wall failures. A dependence of permeability on expansion was demonstrated by Barrett and Ross (5), who determined a negative relationship between extrudate infusibility (i.e., receptivity to infiltration) and bulk density.

B. Cellularity

While microscopy can provide a general assessment of the size and configuration of cells, and also of the thickness and relative intactness of cell walls, quantification of cell size distributions by an image analysis technique is usually necessary for a complete and accurate description of structure. Different image analysis systems and procedures applied to extrudates, or similarly porous products, have been described (5, 7–13). Many of these techniques were used to evaluate cut cross sections of products (i.e., in which the interior cell structure was exposed). Most involved analysis of the configuration of and/or the relative area occupied by cell walls (9, 11) or determination of cell sizes (5, 7). Other approaches have focused on exterior features, such as relative roughness of the extrudate surface (10).

In its simplest form, image analysis involves digitization of an image and assignment to each pixel a value corresponding to its position on a "gray level" (white-to-black) intensity scale. Image "thresholding" constructs a binary image that includes only those pixels with gray levels between preselected values; these pixels are assigned a gray level of black, while every other pixel is assigned a gray level of white. The resulting image contains object silhouettes that can be counted, measured, and evaluated by statistical and geometric analysis.

The key to successful image processing is construction of a binary image that truly represents, and identifies salient characteristics of, the sample. Good contrast between structural features of interest and "background" is necessary. Objects to be evaluated should therefore be significantly darker, or lighter, than the rest of the image. For extrudates, cell size analysis can be conducted by evaluating a two-dimensional "planar" section of the sample. If it is possible to obtain very thin sections (which is generally difficult for dry products), specimens can be placed against a contrasting background. Alternatively, inking of cut surfaces serves to highlight cell structure in one plane (Fig. 3a); features deeper in the interior of the specimen can be eliminated through thresholding and construction of a binary image (Fig. 3b).

For cell size determination, it is necessary to obtain a representation of the wall structure that reveals discrete, separated cells—which typically requires further modification or enhancement of the thresholded image. One problem that arises is that binary cell structure silhouettes typically are not perfectly continuous, due either to the naturally occurring structural failures in the sample or to insufficient application of contrasting ink. Even minute gaps will significantly affect calculation of cell size distributions. For example, if the images of two actually distinct cells are connected by only one pixel due to a minute discontinuity in the projected wall structure, the program will assume that the images represent one larger cell. Automatic enhancement capability, such as dilation algorithms that iteratively add pixels to the image, are typical features of image analysis programs. Additionally, manual editing—correction of slight discontinuities in the structure using a mouse—while labor intensive, can considerably improve the quality of the projected image.

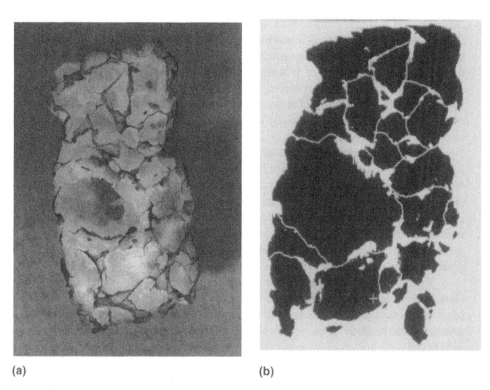

Figure 3 Image analysis procedure. (a) Inked sample cross section, in which cell structure in one plane is highlighted. (b) Binary image of sample, from which cell area size distribution can be calculated. (Photographs compliments of Sarah Underhill, U.S. Army Natick Soldier Center, Natick, MA.)

The selection and use of specific thresholding and enhancement procedures should have as a rationale the construction of a binary image that recreates as closely as possible the actual cross-sectional structure of the extruded sample. This final, processed image is subsequently analyzed, resulting in numerical data pertaining to each discrete object it encompasses. Data (size and geometric information) for individual cells can be extracted by specifying these objects using the cursor. However, for distribution analysis, it is advisable to establish a (very low) cutoff value below which cells are excluded, since extremely small measured objects, of the order of magnitude of a few pixels, are more likely to be artifacts of lighting or preparation than representative of actual cells.

Data for several (randomly distributed) sections of a sample should be combined into a cumulative distribution. Extrudates are inherently heterogeneous in structure, and a single sample cross section is unlikely to accurately represent the product. It may also be of interest to measure a sample cut in different orientations (i.e., cross-sectional vs. longitudinal), in order to determine anisotropy in the structure. Barrett and Peleg (12) reported larger mean cell size measurements for some specimens, those processed from very high-moisture formulas, when viewed in axial rather than radial orientations; such samples had limited radial expansion and, presumably, relatively higher axial velocity during extrusion due to the lowered melt viscosity.

Statistical descriptors of cellularity are, most simply, single-parameter values such as mean or median cell size; these indices are useful characteristics by which to assess process or formulation effects or to predict mechanical properties and texture. But cell "size" is in itself a complicated entity, since extrudate cells are typically irregular in shape; "diameter" therefore requires the inaccurate assumption of a circular conformation. Cell area size and perimeter are preferable measurements, since both can be obtained directly according to the number of pixels actually in, or along the border of, an image.

It is also important to consider the distribution of cell sizes—which is typically quite wide for extrudates; standard deviations of cell area distributions are often of the same order of magnitude as the mean (unpublished data). Another often-reported property of extrudate cell size distributions is their tendency to be significantly right-skewed—i.e., to have a relatively greater number of small rather than large cells (12, 13) (Fig. 4). This preponderance of small cells may reflect coalescence during expansion—several small cells subsumed into one large cell—or nonuniform distribution of moisture within the melt.

Cell size distributions have been described by mathematical functions appropriate for right-skewed data, particularly the log normal distribution,

$$f(Z) = \frac{1}{[\sigma_z(2\pi)^{1/2}] \exp[(Z - \mathbf{Z})^2/2\sigma_z^2]} \tag{10}$$

(where $Z = \ln(X)$, $X = $ cell area size, and $\mathbf{Z}$ and σ_z are, respectively, the mean and standard deviation of Z), and cumulative Rosen-Rammler distribution,

$$Gd = \exp\left[-\left(\frac{X}{C}\right)\right]^S \tag{11}$$

(where Gd is the fraction of cells with cross-sectional area greater than X, and C and S are constants) (12). Regression coefficients for fits to these functions, for a range of sample structures, were found to average 0.80 and 0.98 for the log normal and Rosen Rammler models, respectively. Given that measured cell sizes are so distinctly skewed toward smaller cells and that microporous structures within cell walls have been reported (14), a possible fractal nature of extrudate cell structure (i.e., self-similar at different length scales) has been postulated (15).

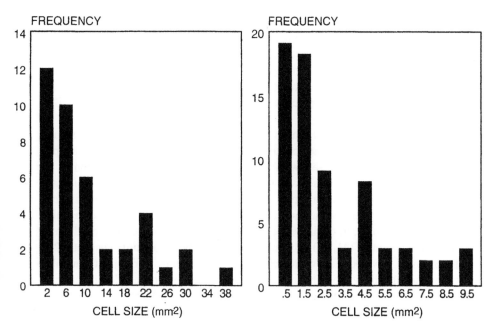

Figure 4 Representative distributions of extrudate cell area sizes, showing relative preponderance of smaller cells. (From Ref. 13, p. 84. With permission.)

IV. CONTROL AND ALTERATION OF STRUCTURE

Many process and formulation parameters will affect extrudate structure. Chief among these are variables that change the rheology of the extrusion melt or ingredients that serve as nucleating agents for expansion. Moisture content has a profound effect on expansion and cellularity. Higher-water-content formulations have often been reported to produce relatively less expanded, higher-density products (12–13, 16–18). Increased moisture reduces viscosity, thus facilitating sample shrinkage or collapse after expansion. Furthermore, the negative pressure produced by condensation of water vapor in high-moisture systems can serve as a driving force for postextrusion shrinkage (19). Correspondingly, higher-moisture products generally have relatively smaller cells (12, 13, 16) with relatively thicker cell walls. However, a few studies have also shown reduced expansion at extremely low-moisture contents (5, 20), indicating the existence of an optimal moisture level for puffing; water vapor is, after all, the carrier for expansion.

Hence, in products in which structure is adjusted by moisture content, cell size and density are typically negatively related. Product expansion or shrinkage is due to the expansion or shrinkage of individual cells. However, other factors contributing to structure, particularly in high-moisture, low-viscosity melts, are cell coalescence during expansion and the presence of "blow holes" due to localized steam pockets. Both can serve to produce a population of extremely large cells that exist among small or shrunken cells. Increased skewness in extrudate cell area size distributions—as assessed by the ratio of mean cell area/median cell area—with increasing extrusion moisture was observed by Barrett and Peleg (12).

Additives or processes that influence melt viscosity in a manner similar to that of water can have analogous effects on product structure. For example, incorporation of sucrose—a low-molecular-weight plasticizer and a common ingredient in prepared cereals—was found to simultaneously reduce cell size and increase bulk density in corn meal extrudates (16). However, this negative cell size–bulk density dependence no longer holds when structure is adjusted by other means, for example, through the incorporation of additives. Certain constituents, sodium bicarbonate (21, 22), silicon dioxide (22), and tricalcium phosphate (21), for example, have been shown to act as nucleating agents, increasing the number of individual cells without the overwhelming changes in product density effected by moisture. Other compounds reactively influence complexation/molecular interaction: cysteine for example, which inhibits polymerization during extrusion, has been shown to produce relatively finer-textured, smaller-cell-size products (23).

That process and formulation parameters so significantly affect extrudate physical characteristics allows for the "tailoring" of products to accommodate specific desired functional properties.

V. FUNCTIONAL PROPERTIES ARISING FROM STRUCTURE

A. Mechanical Strength

Relationships between the mechanical properties and the physical structure of extrudates have been reported. Obviously, a porous structure becomes stronger with increased bulk density: More mass per unit volume of material provides more resistance to deformation. Additionally, increased mechanical strength—determined by both breaking strength during compression and average compressive resistance—of extrudates with decreasing cell size has been reported (13, 21). Such a relationship between cell structure and deformability arises due to the increased number and decreased length of cell wall supports as cell size diminishes. Consider the earlier discussion of the theoretically infinite number of arrangements of cell sizes and cell wall thicknesses for a given volume and weight of a porous material: Relatively finely divided samples, with a relatively greater number of (smaller) cells, have many more cell wall supports than do coarsely divided, large-celled products. Cell walls in highly divided samples of a given mass are on average thinner, making individual structural units more prone to failure (by decreasing bending moment of inertia and resistance to buckling). However, this tendency is to some extent offset by the fact that cell walls are also much shorter (increasing bending moment of inertia) and significantly counterbalanced by the vastly greater number of cell wall supports and intercellular connections. A linear dependence of average compressive stress developed during compression (S, in kilopascals) on a function of mean cell area size (A, in square millimeters), a negative contributor, and bulk density (D, in grams per cubic centimeter), a positive contributor, in corn meal-based extrudates,

$$S = 53.1 + 634D - 5.79A \tag{12}$$

($r^2 = .91$) was reported by Barrett et al. (13).

B. Fracturability

The characteristic texture and "crunchiness" of extruded products arises from incremental fracturing of the structure brought about through progressive failure of individual cell wall components. Since extrudates are composed of a network of brittle cell wall supports surrounding void spaces, samples fail gradually during strain (effected either instrumentally or by mastication); subunits of the structure fracture sequentially, which only partially and temporarily relieves developed stress.

Such incremental failure gives rise to the distinctive "jagged" appearance of stress–strain relationships of brittle extrudates (Fig. 5a). According to Gibson and Ashby (6), compression of cellular solids yields a three-part stress–strain function that includes a linear elastic region at low levels of deformation, a generally nonrising region at intermediate levels of deformation (indicating either plastic or brittle collapse of cell wall supports), and a region of increasing stress at extremely high levels of strain (indicating densification of the structure). Plastic cellular materials have smooth stress–strain relationships; brittle cellular materials have oscillations due to fracturing superimposed on the same overall three-part shape. It has been demonstrated repeatedly that extrudates conform to this deformation behavior and that the stress–strain function changes from a brittle type to a plastic type (Fig. 5b) as equilibrium relative humidity is increased (21, 24–26).

Efforts to quantify fracturability and to use this characteristic as an intrinsic textural property of extrudates have been undertaken in a number of studies.

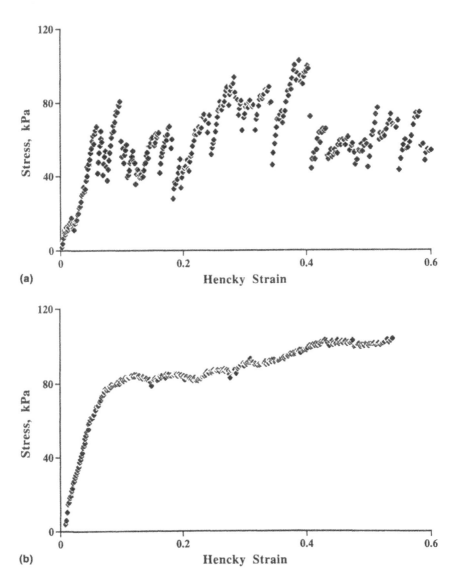

Figure 5 Representative stress–strain relationships for a porous, corn-based extrudate undergoing uniaxial compression: (a) dried product; (b) product equilibrated at 75% RH.

Barrett et al. (13, 24, 25), Rhode et al. (26), and Peleg and Normand (27) analyzed jagged stress–strain functions by fractal and Fourier analysis. Fractal analysis was accomplished through use of a "Blanket" algorithm,

$$X_{e+1}(I) = \max\{X_e(I) + 1, \max[X_e(I - 1), X_e(I + 1)]\} \tag{13}$$

(where X is height, or level, of the digitized image at position I and iteration e), which iteratively smoothed the function (Fig. 6); fractal dimension is equal to 1 minus the slope of the relationship, log[(blanket area)/$2e$] vs. log($2e$). Highly irregular stress–strain relationships, due to either a high frequency of fracturing or large magnitude fractures, require many iterations of "blanketing" in order to be smoothed—hence, a large variation of blanket area with iteration and a relative high fractal dimension (close to 2, the theoretical maximum for images). A fast Fourier transform was used to determine power spectra of stress–strain relationships, from which mean power magnitudes in specific frequency ranges—representing the relative intensity of the frequencies—were used as indices of fracturability. Barrett et al. (25) additionally fitted the distribution of the intensity

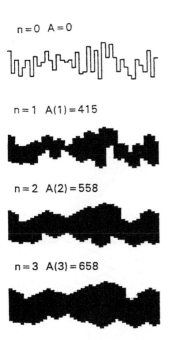

Figure 6 Schematic of the blanket algorithm, demonstrating progressive smoothing of a jagged relationship. "A" is the blanket area after iteration "n." (From Ref. 27, p. 303. With permission.)

of fractures (Δx, or reductions in stress) occurring during the compression of extrudates to an exponential function,

$$Y = \text{frequency} = C \exp(-b\Delta x) \tag{14}$$

in which both the fitted exponent and the coefficient of this relationship were descriptors of fracture behavior. A characteristic of the fracture intensity distributions was their tendency to be skewed toward lower-intensity fractures—i.e., a preponderance of smaller rather than greater reductions in developed stress.

Quantified fracturability, like compressive strength, was demonstrated to be structurally determined—and similarly related to cellularity—by Barrett et al. (13). Both the fractal dimensions (FD) and the mean power spectrum magnitudes [within selected frequency ranges (P)] derived from stress–strain functions of corn-based extrudates varied according to a linear function of mean cell size in square millimeters (A) and bulk density in grams per cubic centimeter (D) (Fig. 7). Equations for these relationships were

$$FD = 1.37 - 0.0112A + 1.93D \tag{15}$$

($r^2 = .94$) and

$$\ln(P_{3.75-5.63 \text{ mm-1}}) = -2.24 - 0.123A + 23.9D \tag{16}$$

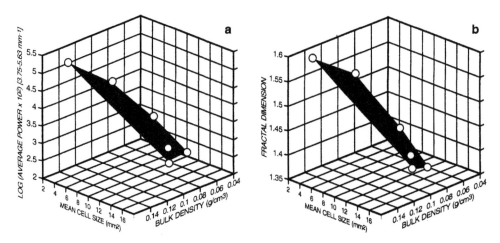

Figure 7 Dependence of power spectrum parameters (a) and fractal dimension (b) (derived from stress–strain functions) on extrudate mean cell size and density. Shows linear relationship between fracturability and structural properties. (From Ref. 13, p. 88. With permission.)

(r^2 = .90). Mean magnitudes of all frequency ranges in the power spectrum were closely correlated with structure. Additionally, in subsequent investigations both fracturability indices were found to be strongly correlated with fitted parameters of Eq. (14) (25), indicating that all three methods are consistent in providing a reliable measurement of relative fracturability.

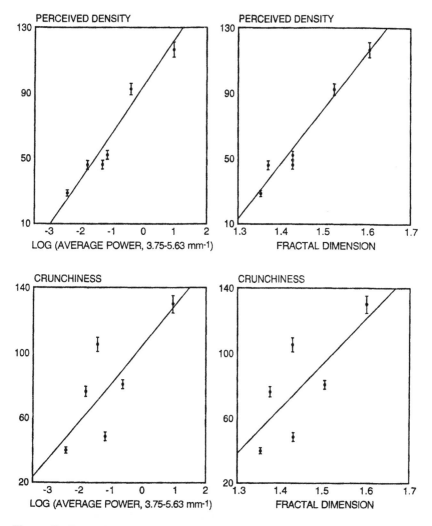

Figure 8 Dependence of sensory crunchiness and sensory density on power spectrum parameters and fractal dimension. (From Ref. 13, p. 89. With permission.)

The physical basis for the dependence of fracturability on structure is analogous to that for strength–cellularity relationships. A relatively dense structure with thick cell walls can withstand relatively greater stress prior to failure and can be expected to fail with fractures that are individually of higher intensity. A small-celled product, with a relatively greater number of cell wall supports, might expectedly fail with a higher frequency of fracturing—although for such very interconnected structures, many fractures may extend throughout large subunits of the structure. Fracturability, and sensory attributes such as crunchiness that depend on fracture behavior, can thus be tailored by adjusting and optimizing the two physical properties.

In fact, subjective attributes have been shown to depend on fracture behavior, as demonstrated by positive relationships between pertinent sensory properties—such as those between perceived crunchiness or perceived denseness and either fractal dimension or mean magnitude of the power spectrum (13) (Fig. 8). Such correlations between sensory attributes and mechanical properties, and those between failure behavior and structure, can be used as predictive tools for designing extruded foods with specific, desired characteristics.

VI. OTHER POROUS-STRUCTURED FOODS

The relationships between mechanical or textural properties and structure observed for extrudates are potentially applicable to other brittle-porous products, such as baked flat breads, crackers, and popcorn. Furthermore, many baked goods, while plastic, are cellular and therefore subject to similar relationships between mechanical strength and structure or between mechanical properties and texture. Such correspondence between structure, failure properties, and functionality provides a convenient means of "tailoring" foods to possess desired attributes and acceptance.

ACKNOWLEDGMENT

This work was supported by the U.S. Army SBCCOM, Natick Soldier Center, and was conducted as part of ration research and development efforts.

REFERENCES

1. Alvarez-Martinez, L., Kondury, K.P., and Harper, J.M. 1988. A general model for expansion of extruded products. J. Food Science 53(2):609–615.
2. Launey, B., and Lisch, J.M. 1983. Twin-screw extrusion cooking of starches: flow

behavior of starch pastes, expansion and mechanical properties of extrudates. J. Food Engineering 2:259–280.

3. Hicsasmaz, Z., and Clayton, J.T. 1992. Characterization of the pore structure of starch-based food materials. Food Structure 11:115–132.

4. Barrett, A.H., Ross, E.W., and Taub, I.A. 1990. Simulation of the vacuum infusion process using idealized components: effects of pore size and suspension concentration. 1990. J. Food Science 55(4):989–993.

5. Barrett, A.H., and Ross, E.W. 1990. Correlation of extrudate infusibility with bulk properties using image analysis. J. Food Science 55(5):1378–1382.

6. Gibson, L., and Ashby, M.F. 1988. Cellular Solids. New York: Pergamon Press.

7. Moore, D., Sanei, A., Van Hecke, E., and Bouvier, J.M. 1990. Effect of ingredients on physical/structural properties of extrudates. J. Food Science 55(5):1383–1387, 1402.

8. Russ, J.C., Stewart, W.D., and Russ, J.C. 1988. The measurement of macroscopic images. Food Technology, February:94–102.

9. Gao, X., and Tan, J. 1996. Analysis of expanded food texture by image processing. J. Food Process Engineering 19:425–456.

10. Tan, J., Gao, X., and Hseih, F. 1994. Extrudate characterization by image processing. J. Food Science 59(6):1247–1250.

11. Smolarz, A., Van Hecke, E., and Bouvier, J.M. 1989. Computerized image analysis and texture of extruded biscuits. J. Texture Studies 20:223–234.

12. Barrett, A.H., and Peleg, M. 1992. Cell size distributions of puffed corn extrudates. J. Food Science 57(1):146–149, 154.

13. Barrett, A.H., Cardello, A.V., Lesher, L.L., and Taub, I.A. 1994. Cellularity, mechanical failure, and textural perception of corn meal extrudates. J. Texture Studies 25:77–95.

14. Cohen, S., Voyle, C., Harniman, R., Rufner, R., Barrett, A.H., and Hintlian, C. 1989. Microstructural evaluation of porous nutritional sustainment module extrudates. U.S. Army Natick RD&E Center Technical Report, TR-89-034.

15. Barrett, A.H., and Peleg, M. 1995. Applications of fractal analysis to food structure. Food Science and Technology 28:553–563.

16. Barrett, A.H., Kaletunc, G., Rosenberg, S., and Breslauer, K. 1995. Effect of sucrose on the structure, mechanical strength and thermal properties of corn extrudates. Carbohydrate Polymer 26:261–269.

17. Kim, C.H., and Maga, J.A. 1993. Influence of starch type, starch/protein composition and extrusion parameters on resulting extrudate expansion. In: G. Charalambous, ed. Food Flavors, Ingredients and Composition. New York: Elsevier Science, pp 957–964.

18. Faubion, J.M., and Hoseney, R.C. 1982. High-temperature short-time extrusion cooking of wheat starch and flour. I. Effect of moisture and flour type on extrudate properties. Cereal Chemistry 59(6):529–537.

19. Fan, J., Mitchell, J.R., and Blanshard, J.M.V. 1994. A computer simulation of the dynamics of bubble growth and shrinkage during extrudate expansion. J. Food Engineering 23:337–356.

20. Chinnaswamy, R., and Hanna, M.A. 1988. Optimum extrusion-cooking conditions for maximum expansion of corn starch. J. Food Science 53(3):834–836, 840.

21. Barrett, A.H., and Peleg, M. 1992. Extrudate cell structure–texture relationships. J. Food Science 57(5):1253–1257.
22. Lai, C.S., Guetzlaff, J., and Hoseney, R.C. 1989. Role of sodium bicarbonate and trapped air in extrusion. Cereal Chemistry 66(2):69–73.
23. Koh, B.K., Karwe, M.V., and Schiach, K.M. 1996. Effects of cysteine on free radical production and protein modification in extruded wheat flour. Cereal Chemistry 73(1):115–122.
24. Barrett, A.H. Normand, M.D., Peleg, M., and Ross, E.W. 1992. Characterization of the jagged stress–strain relationships of puffed extrudates using the fast Fourier transform and fractal analysis. J. Food Science 57(1):227–232, 235.
25. Barrett, A.H., Rosenberg, S., and Ross, E.W. 1994. Fracture intensity distributions during compression of puffed corn meal extrudates: method for quantifying fracturability. J. Food Science 59(3):617–620.
26. Rhode, F., Normand, M.D., and Peleg, M. 1993. Effect of equilibrium relative humidity on the mechanical signatures of brittle food materials. Biotechnology Progress 9:497–501.
27. Peleg, M., and Normand, M.D. 1993. Determination of the fractal dimension of the irregular, compressive stress-strain relationships of brittle, crumbly particulates. Particle and Particle Systems Characterization, 10:301–307.

12
Understanding Microstructural Changes in Biopolymers Using Light and Electron Microscopy

Karin Autio and Marjatta Salmenkallio-Marttila
VTT Biotechnology, Espoo, Finland

I. INTRODUCTION

Most foods are derived from raw materials that have a well-organized microstructure, e.g., cereal products from grains. Processing, including malting, milling, dough mixing, and heating, produce great microstructural changes in proteins, cell wall components, and starch. These changes can have a large effect on the quality of the end product. The microstructure determines the appearance, texture, taste perception, and stability of the final product. A variety of microscopic techniques is available for studying the different chemical components of grains and cereal products (1). Bright-field and fluorescence microscopic methods are frequently used because they allow selective staining of different chemical components. These staining systems can also be used in confocal scanning laser microscopy (CSLM). One of the main advantages of this technique is the minimal degree of sample preparation that is required (2, 3). And CSLM is well suited for high-fat doughs, which are difficult to prepare for conventional microscopy. Compared with light microscopy (LM), the resolution is considerably improved with electron microscopy. Scanning electron microscopy is used to examine surfaces and transmission electron microscopy the internal structure of food (1).

Although the morphologies of cereal grains share many similar features, differences exist, especially in chemical composition and the distribution of components. Wheat is rich in protein and has very thin walls, especially in the outer endosperm. Rye is rich in cell walls, and the cell walls are thick throughout the kernel. Barley may also have thick cell walls throughout the kernel, but there are great variations in thickness between varieties. In barley, wheat, and rye, the size distribution of starch granules, A- and B-type, vary from one sample to an other. In oat, cell walls can be very thick in the subaleurone layer (4). This concentration of cell wall material can be of benefit for milling of fiber-rich oat bran. Oat bran has a higher protein content than the brans of other cereals. The starch granules of oat grain are smaller than in other cereals and are clustered into a greater unit.

Germination and malting induce great changes in the microstructure of cell walls, proteins, and starch granules. The most important quality requirement of malted barley is rapid and even modification of the grain. Although in Western countries high viscosities of cereal cell wall components are largely beneficial in bread baking and in human nutrition, in brewing these same properties may decrease the rate of wort separation and beer filtration.

During milling the outer layers of the grain and the embryo must be separated from the starchy endosperm to produce a high yield of wheat flour (5). Grain properties affect the milling quality of different cultivars. Flour yield, amount of starch damage, losses of storage proteins, and efficiency of embryo removal during milling can vary from one wheat to another. In baking, the major purpose of wheat dough mixing is to blend the flour and water into a homogeneous mixture, to develop the gluten matrix. The major microstructural changes that take place during the heating of wheat dough are starch gelatinization, protein cross-linking, melting of fat crystals, and sometimes fragmentation of cell walls. Although most wheat breads are made from refined flour rather than from whole grain flour, the raw material of traditional rye breads is whole grain flour. In rye dough, the cell wall components make a greater contribution to structure formation than do the proteins.

Pasta products have gained popularity in many countries. Both the raw material, typically durum semolina, durum flour, and hard wheat flour, and the pasta process affect the quality of dry and cooked pasta. To a great extent, the microstructure of fresh, dry, and cooked pasta determines the textural properties, such as stickiness. Protein quality and quantity are important in the production of high-quality pasta products. However, additional research is needed on the role of starch, the major constituent of flour (6).

Oat products have captured the attention of the food industry because the soluble β-glucans abundant in oat bran have been shown to have cholesterol-lowering effects in rats and humans (7). Several commercial products have come to the market: oat bran, oat flour, oat flakes, rolled oats, oat chips, oat biscuits, oat breads, etc. The nutritional value of the dietary fiber components of the other cereals is also receiving great attention.

Table 1 Chemical Composition of Cereal Grains (% Dry Weight)

	Wheat	Rye	Oat	Barley
Fat	2.1–3.8	2.0–3.5	3.1–11.6	0.9–4.6
Protein	9–17	8–12	11–15	8–13
Starch	60–73	50–63	39–55	53–67
Pentosan	5.5–7.8	8.0–10.0	3.2–12.6	4.0–11.0
β-Glucan	0.5–3.8	1.0–3.5	2.2–5.4	3.0–10.6

II. MICROSTRUCTURE OF CEREAL GRAINS

The microstructure of wheat, barley, oat, and rye is shown in Figures 1–4 (see insert). The sections were stained with fuchsin acid and Calcofluor White M2R New and examined under a fluoresence microscope. The primary cell walls appear blue, protein brown to red, starch black, and outer lignified layers yellow. Information about the chemical composition of the grains is summarized in Tables 1 and 2.

A. Wheat

This chapter reviews the microstructure of mature grain only. The review of Evers and Bechtel (5) is recommended for a more detailed discussion, including the microstructure of developing grain. The microstructure of transversely sectioned mature wheat grain is shown in Figure 1 (see insert). The four morphologically different tissues of the grain are: the layers of pericarp and seed coat, embryo, aleurone, and starchy endosperm. The embryo is located on the dorsal side of the grain. The pericarp—the lignified, dead tissue—is composed of an outer epidermis, hypodermis, parenchyma, intermediate cells, cross cells, and tube cells

Table 2 Structure, Size, and Composition of Cereal Starches

Cereal	Size (μm)		Amylose content (%)	Lipid content (%)
	A-type	B-type		
Wheat	15–25	4–6	28–29	0.8–0.9
Rye	23–40	<10	22–26	1.0–1.6
Barley	10–25	3–5	25–30	0.7–1.3
Oat[a]	2–15		18–19	2.1–2.5

[a] Unlike wheat, rye, and barley, oat grains contain only one type of starch granules.

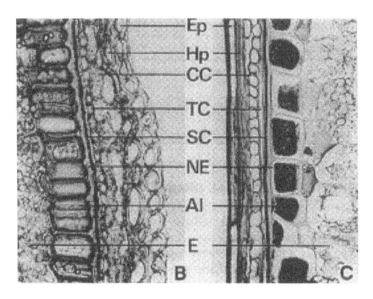

Figure 5 Photomicrographs of outer layers of the wheat grain sectioned transversely
(B) and longitudinally (C). Ep = epidermis, Hp = hypodermis, CC = cross cell, TC =
tube cell, SC = seed coat, NE = nucellar epidermis, Al = aleurone layer, E = starchy
endosperm (×200). (Reprinted from Ref. 5.)

(Fig. 5). The aleurone cells form the outermost layer of the endosperm and represent live tissue at maturity. Pericarp, seed coat, and aleurone layer are the main
parts of the bran fraction produced during milling. Endosperm texture influences
milling and is an important criterion for the end use of various wheat types (8).
In wheat grain, the cell walls of outer starchy endosperm are much thinner than
those of inner endosperm. Usually the cell walls of aleurone layer are very thick.
    The structural characteristics of the grain play an important role in the
sprouting resistance of wheat (9) and the susceptibility to insects and fungi and
in further processing, e.g., milling (5), malting, or hydrothermal treatment of
grains, and in baking, where whole wheat grains or bran are used. The seed coat
of wheat is the outermost layer of the seed and together with the pigment strand
completely covers the seed. The seed coat is fused to the pericarp. The outer
epidermis is composed of long, narrow cells (10), and it covers the whole grain
surface except for the point of attachment to the rachilla. It does not prevent the
penetration of water into the grain (11). Hypodermis lying below epidermis
forms, together with the epidermis, the outer pericarp. Thin-walled parenchyma
cells lie below the hypodermis. The epidermis and hypodermis are attached to

parenchyma only at the ventral crease (12); elsewhere they form a loose layer called the outer pericarp. Fungal mycelia are often found between the outer and inner pericarp (10). Intermediate cells are located below the thin-walled parenchyma cells at the brush and germ ends of the grain, but they do not form a complete layer. Cross cells and tube cells play an important role in the developing caryopsis, where they function as photosynthetic cells. In the mature grain, a layer of empty epidermal cells called *nucellus* is located between the endosperm and the seed coat. An amorphous layer, *lysate*, was shown to exist between the nucellar epidermis and aleurone layer (13).

The aleurone layer of wheat is typically only one cell layer thick. The aleurone cell walls are typically much thicker than the endosperm cell walls and are autofluorescent due to ferulic acid (14, 15). The aleurone cells are filled with protein bodies (16) and lipid droplets (17). Two inclusions are deposited in the protein bodies: phytate and niacin (18).

The main contents of endosperm cells are starch and protein. The protein exists as a continuous matrix rather than in protein bodies. Peripheral cells have the lowest starch content and the highest protein content (19). Typical features of soft wheats are a relatively loose interaction between starch and protein, and endosperm characterized by interruptions with air spaces (20, 21). Starch is the major component of wheat endosperm, composing 64–74% (14% moisture basis) of milled endosperm (22).

B. Barley

Malting barley is one of the most studied plants, and the structure and physiology of the germinating grain have been especially well characterized (23–26). The ungerminated grain consists of an embryo and endosperm enclosed by a layer composed of the fruit coat, pericarp, together with the seed coat, testa. More than 80% of the grain volume is endosperm, the store of food reserves for the embryo. A fibrous husk surrounds the grain. The husk of barley is, unlike that of wheat, strongly attached to the grain and is not detached from it during threshing. The husks of malting barley play an important role in wort filtration. The microstructure of the barley grain varies, depending on genotype, environment, and growing conditions (26–31). There are, e.g., naked barley varieties (the husks are separated from the grains during threshing), thin-walled low-β-glucan varieties, and barley varieties producing waxy or high-amylose starch. Hull-less barley is nutritionally superior to hulled barley for swine and poultry and offers the advantage that it can be directly milled and sieved to obtain meal for food use (32).

As in other cereals, the outer layers of the barley grains are characterized by having relatively high concentrations of phenolic compounds, such as lignin, lignans, flavonoids, and ferulic and coumaric acids (33, 34). The chemical composition and structural features of these cell layers have a major influence on the permeability of the grain to water during germination and on the high levels of insoluble dietary fiber of the grain. This is also important in malting, where fast and even germination is required, and the latter is an important aspect in the food use of cereal brans.

The embryo is located at the basal end of the grain. The cells of the embryo have thin walls and contain a large nucleus. They also contain protein bodies and fat globules called *spherosomes*. The germ is composed of two functionally different parts: the embryo proper and the scutellum, which is a shieldlike structure appressed to the endosperm. At germination the scutellum becomes a digesting and adsorbing organ, which transfers the stored nutrients from the endosperm to the growing parts of the embryonic axis.

The outer layer of the endosperm is called the *aleurone* layer. In barley it is one to four cells thick and almost completely surrounds the starchy endosperm and germ (35). The aleurone cells are cubic and their cell walls are thick. The aleurone cell walls of barley or wheat are reported to contain about 65–67% arabinoxylan and about 26–29% mixed linkage $(1 \rightarrow 3)$, $(1 \rightarrow 4)$-β-D-glucan (36). The cells contain large nuclei and storage materials, including lipids, phytin, minerals, protein, and phenolic compounds. The aleurone is important both botanically and industrially. It plays a significant role both in germination and in animal nutrition. The aleurone cells are an important source of the hydrolytic enzymes that degrade the starchy endosperm during germination. Nutritionally the aleurone is a rich source of dietary fiber, minerals, vitamins, and other health-promoting compounds.

The inner part of the endosperm, the starchy endosperm, is a nonliving tissue. The cell walls of the starchy endosperm of barley contain about 20% arabinoxylan and 70% β-glucan (36). The thickness of the cell walls in the starchy endosperm may vary within the grain or between different barley cultivars (34, 37). The peripheral areas in the barley kernel have been shown to have a lower concentration of β-glucan than the central endosperm (38). The cells are full of starch granules, which are embedded in a protein matrix. The starchy endosperm is the primary source of the starch (53–67% of the grain dry weight, Table 1) that is eventually converted to fermentable sugars during mashing. There are two types of starch granules, large lenticular A-granules (15–25 μm in diameter) and smaller polygonal B-granules (<10 μm in diameter, Table 2). The small granules account for 80–90% of the total by number but only for 10–15% of the total by mass (39). In the subaleurone layer, located under the aleurone layer, cells are smaller and they contain more protein and fewer starch granules than the cells of the inner starchy endosperm. A compressed layer of empty endosperm cells, the crushed cell layer, exists adjacent to the scutellum. The contents of the cells were used in the growth of the embryo during development of the grain.

C. Oat

The greatest differences in oat in comparison to other cereal grains are the size of the grain, the distribution of chemical components, and the chemical composition. Fulcher (40) has made an extensive review of oat microstructure. As shown in Figure 3, the oat grain is longer and thinner than wheat grain. One great difference between oat grain and other cereal grains is the distribution of cell walls, proteins, and lipids within the grain. In contrast to barley, rye, and wheat, the cell walls in the subaleurone layer of some oat varieties can be very thick. This is an important structural feature allowing production of bran fractions with high β-glucan content.

The oat aleurone layer is approximately 50–150 μm thick. The cell walls of aleurone contain phenolic acid–rich cell walls that are insoluble, in contrast to those of endosperm. Ferulic acid is the major fluorescent compound in the aleurone cell walls (38). Although the concentration of mixed linked β-glucans is not high, they have a great effect on the water-binding properties of oat bran. Oat bran has a much higher water hydration capacity than barley or wheat bran (41). Inside the tough cell wall, each aleurone cell contains numerous individual protein bodies. Oat bran including the aleurone layer contains approximately half of the total grain protein (42). The oat bran contains the highest level of basic amino acids, in comparison to other cereals (43). Each aleurone grain is surrounded by lipid droplets. The total lipid content in oat grain may differ greatly (3.1–11.6%, Table 1) (44), and, as with other cereals, the aleurone layer and germ are rich in lipids. The neutral lipid fraction (primarily triglycerides) usually causes the variation in the lipid content of different varieties. Lipases present in oat products catalyze the hydrolysis of triglycerides to free fatty acids, causing an unpleasant off-flavor (45).

The microstructure of oat starchy endosperm differs from that of other cereals mainly in terms of protein, starch, and lipids. At maturity, barley, rye, and wheat endosperm proteins appear microscopically as a homogeneous matrix in which starch granules are embedded. In oat endosperm, proteins exist in spherical protein bodies (46). The size of the protein bodies can range from 0.2 to 6.0 μm in diameter. The larger bodies are located in the subaleurone region. High-protein varieties, e.g., Hinoat, contain a lot of large protein bodies in the region. As with other cereal grains, the lipid content of aleurone and germ are high. In contrast to other cereals, the endosperm may also contain a lot of lipids (47).

In oat grain, starch exists as an aggregate composed of several individual starch granules (48). The size of the aggregates ranges from 20 to 150 μm in diameter and that of individual granules from 2 to 15 μm across (Table 2). The oat starch has higher lipid content than other cereal starches, which is reflected in the high value of the transition enthalpy measured for the amylose–lipid complex (49).

The distribution of mixed linked β-glucan differs in different varieties. In some oat varieties β-glucan content is highest in the subaleurone cell walls and lowest in the inner endosperm (4). In a high-β-glucan variety, Marion, no concentration of β-glucan in the subaluerone cell walls was observed, but the β-glucan was distributed evenly throughout the grain. In contrast to other cereal grains, the mixed linked β-(1→3)(1→4)-D-glucans in oat endosperm cell walls are more soluble than β-glucans in barley or wheat. This affects the structure of oat-based products, such as porridge and bread.

D. Rye

The microstructure of rye grain is shown in Figure 4. The main morphological characteristics are similar to those of other grains. A review on the microstructure of developing and mature rye grain has been made by Simmonds and Campbell (50). The size of rye grains differs greatly between varieties. Hybrid varieties commonly have larger grains than the population varieties. The thickness of cell walls is very similar in the different parts of the grain (Fig. 4). However, the chemistry and solubility of the arabinoxylans in the endosperm cell walls differ in different parts of the grain. The precise chemistry of the fluorescing compounds in rye is not known, but rye cell walls contain ferulic acid esterified to arabinoxylans (51). The ratio of soluble to total arabinoxylans is 25% in the aleurone and 14% in the outer bran (52). Furthermore, the ratio of arabinose to xylose differs in different parts of the grain, being 0.40 in the aleurone and 0.63 in whole rye. The arabinoxylans in endosperm differ both in solubility and in the ratio of arabinose to xylose. The ratio of soluble to total arabinoxylans was 71% in the endosperm, and the ratio of arabinose to xylose was 0.75 (0.63 for whole rye).

III. MILLING

Although grains have been used whole in various ways as human food, usually they have been ground in preparation for cooking. Grinding changes the mechanical properties of the various tissues of the grain. Pericarp, seed coat, and aleurone layer are the main parts of the bran fraction produced during milling. The miller separates the bran and the embryo from the starchy endosperm to produce a high yield of flour. The toughness of the bran enables it largely to withstand the crushing and tearing, which is sufficient to detach the endosperm initially attached to the bran. The potential of flourescence imaging as a quality control tool in commercial milling for flour refinement determination by aleurone and pericarp flourescence has been established (53). Strong linear correlations of pericarp fluorescence to ash content were found in different wheat classes representing diverse biotypes and a wide range of kernel hardness.

Endosperm texture influences milling and is an important criterion for the end use of various wheat types (8). Typical features of soft wheats are a relatively loose interaction between starch and protein, and endosperm characterized by interruptions with air spaces (20, 21). In durum wheat milling, when the desired end product is course semolina flour, vitreous kernels having higher protein content usually produce higher yield than mealy kernels with lower protein content.

When wheat is milled into flour, a portion of the wheat starch becomes mechanically modified as a consequence of the grinding action of the mill rolls. Damage to starch during milling of wheat flour affects the properties of the dough and the bread baked from it. A moderate amount of damaged starch is beneficial; excessive damage is undesirable. Damage to starch granules increases susceptibility to enzymatic action. The hardness of the endosperm affects the extent of starch damage during milling. It has been shown that starch damage is largely responsible for differences in water absorption and handling properties of the dough and affects loaf volume and crumb structure (54).

Milling can also be used for the enrichment of different grain tissues, e.g., pericarp/testa, aleurone, and endosperm. The fractions obtained vary in gross chemical composition. For example, the viscosity and dietary fiber characteristics are of importance with respect to the technological and nutritional properties of cereal products (55, 56). Interest in the production of oat bran has been stimulated by the possibility of labeling foods containing sufficient amounts of β-glucan with statements indicating that consumption of oat bran products may reduce the risk of heart disease (7, 57).

Wheat, barley, and rye grains are harder in texture than oat grains and can be dry-milled to produce bran. The endosperm part of oat does not easily separate from the outer layers, and, because of the lipids, sieving problems occur (41). This means that the oat bran fraction separated from oat with hull is usually composed of pericarp, seed coat, nucellus, aleurone layer, and, to some varying extent, starchy endosperm. In the case of dehulled oat, seed coat is missing (4). In dry-milled products, the adhering material from starchy endosperm and germ is the major feature of the bran. In wet-milled material it is considerably minor. One great difference between oat grain and other cereal grains is the distribution of cell walls, proteins, and lipids within the grain. In contrast to barley, rye, and wheat, the cell walls of some oat varieties in the subaleurone layer can be very thick. This is a very important structural feature allowing production of bran fractions with high β-glucan content. The oat bran also typically contains more protein than other cereal brans (10–22%) (4, 42, 58).

IV. MICROSTRUCTURE OF MALTED GRAINS

Malting has been used for centuries for domestic use in baking, brewing, and distilling. Many cereals can be malted. Barley is the cereal most often used for this purpose, for it is the most efficient producer of enzymes during germination.

Microstructural changes in barley grain during malting have been extensively studied, and there is also some data on the effects of other processing methods on the structure of the grain (59, 60). Microscopy can also be used to help identify new and improved cultivars (61).

Quality requirements of malting barley are different from those of feeding barley. The most important quality criterion of malting barley is rapid and even germination leading to homogeneous modification of the grain. Modification means enzymatic degradation of cell wall materials β-glucan, arabinoxylan, and associated protein and of the protein matrix in which starch granules are embedded. Grain structure and cell wall thickness have a significant effect on the malting performance of barley. Water uptake during steeping and diffusion of enzymes during germination determine the rate of modification. A loosely structured endosperm and thin cell walls improve the water uptake during steeping and the diffusion of growth regulators and enzymes during germination. In addition, a loose endosperm structure may mean that the surfaces of starch granules and protein bodies are more exposed to the hydrolyzing enzymes.

Thick cell walls act as barriers in the transportation of different compounds during modification of the endosperm. β-Glucans are the main components of the cell walls. The amount of total β-glucan has been shown to correlate with the rate of modification, and special low-β-glucan mutants with thin cell walls have been developed (28, 37). β-Glucan polymers are easily observed by means of flourescence microscopy (61). The composition of cell walls is the basis for good modification, but the modification rate also greatly depends on the production of cell-wall-degrading enzymes. The high viscosity of β-glucans can also inhibit filtering during brewing.

The distribution of protein within the endosperm cells, and in particular the binding of the protein matrix with small starch granules, has been studied with electron microscopy (62, 63). A higher level of starch-associated protein is proposed to mean poor malting quality. These differences in the starch-associated proteins are also correlated with variation in the textural characteristics (hardness or softness) of the grain. Also, protein associated with the cell walls of the endosperm may affect the rate of cell wall degradation and therefore the modification of the grain.

Degradation of the endosperm tissue during malting starts at the embryo end, and the transitional zone between intact and degraded endosperm is approximately one cell layer wide (64). At the beginning, the degradation of the large starch granules affects only their outer edge. Later, corroding channels form toward the center of the granules and elicit a rapid breakdown of the central part. Further degradation proceeds from the center, giving rise to hollow shells, which finally break up into pieces. The sawtooth pattern observed in the corroding channels indicates, in agreement with earlier studies, that enzyme susceptibility varies periodically in a radical direction. At temperature below the starch gelatinization temperature there is a preferential hydrolysis of small starch granules, most of which disappear during malting; but large granules also show signs of hydrolysis when viewed in a scanning electron microscope (26, 39, 65). Small starch granules are hydrolyzed by surface erosion.

V. MICROSTRUCTURE OF BREADS

A. Wheat Breads

During development of a wheat kernel, protein bodies within the endosperm cells fuse to form the protein matrix (66, 67). In the milling process, the outer layers of the grain and the germ are separated from the starchy endosperm, the flour. In the flour particles, gluten forms a network around the starch granules. During the mixing of bread dough, wheat-endosperm-storage proteins become hydrated and interact to form gluten, an extensive and gas-retentive film. The unique viscoelastic properties are derived from the mixture of glutenin and gliadin (68): Hydrated glutenin is an elastic material; hydrated gliadin is viscous liquid. In wheat bread dough, the gluten forms the continuous phase in which starch granules, lipid, added yeast cells, and cell wall fragments are dispersed. Although the structural properties of gluten have a great effect on the final texture of bread, the other components affecting the distribution of water in the dough, such as starch and cell walls, have a direct influence on gluten, too. Figure 6 (see insert) presents the microstructure of wheat dough. Cell walls, which appear blue, exist as cell wall fragments (Fig. 6) (69).

During mixing, the flour, water, and other ingredients are blended into a homogeneous mixture to develop the gluten matrix and to incorporate air. Amend and Belitz (70) have suggested that the arrangement of protein matrix at the end of the maturation of the wheat kernel is conserved during hydration of the flour particles. The protein networks of individual flour particles stick together when the individual particles come into contact. All researchers do not agree with this hypothesis (71). Studies with transmission electron microscopy have shown that during mixing protein strands consisting of randomly arranged protein threads are transformed into a plateletlike structure (72, 73). The thickness of the platelets is less than 10 nm.

The mixing process has a great effect on the structure of dough. In an undermixed dough, starch and proteins are not evenly distributed (74). Many starch granules are grouped together, and gluten exists as bulky strands with unevenly distributed vacuoles and cellular debris (74). Some flours resist overmixing, whereas others change rapidly (75). Overmixing usually results in a sticky dough (76). Overmixing may cause damage to the gluten network (77). Mixing generally has less effect on starch granule structure or cell walls. Mechanically damaged starch can be demonstrated by Chlorazol Violet R. Damaged starch has a great effect on the distribution of water in the dough (78). Since damaged starch granules are more accessible to α-amylase digestion, starch granules with holes may be observed even before the baking stage (79). After mixing, dough contains occluded gas cells with diameters in the range of 10–100 μm. The mixing process affects the crumb structure. Uniform crumb structure can be achieved when mixing is carried out under a partial vacuum (80). Repeated sheeting also affects gas cell structure by reducing the bubble size (81).

Yeast fermentation generates carbon dioxide, and the dough expands as a result of excess pressure in the gas cells (82). The final crumb structure is dependent not only on the number and size of gas bubbles but also on the structural properties of gluten–starch matrix and probably on liquid film composed of surface-active material. Gan et al. (83) have suggested a model for dough expansion (Fig. 7). At an early state of fermentation the gas cells are embedded in starch-protein matrix. At a later stage of expansion, the matrix fails to enclose the gas cells. As a result, areas that contain only the liquid film are formed between adjacent gas cells. At the end of oven spring, the liquid film will rupture. It has been proposed that surface-active proteins and lipids are present at the gas–dough interphase.

Dough ingredients also affect the size distribution of gas cells in the dough. The use of shortenings stabilizes the gas cells by allowing bubbles to expand during baking without rupture (84). Recent studies on the role of fats in the stabilization of gas cells in cake batters (85) and bread doughs (86) have shown that fat crystals originating from shortenings move from the liquid phase to the gas–liquid interface, where they are in direct contact with the gas in the bubble. During baking, the crystals melt, allowing the bubbles to grow without rupture. Figure 8 illustrates the role of fat in the stabilization of gas bubbles. The proposed mechanism suggests that shortenings containing small crystals are more effective in the stabilization of gas cell structure than those containing large crystals.

Baking has little effect on the microstructure of gluten (87) but a great effect on starch granule structure. Bechtel (71) studied the effect of baking on starch granule structure by transmission electron microscopy (TEM) and scanning electron microscopy (SEM). He reported that gelatinized starch appeared fibrous and was highly swollen. The advantage of light microscopy is that the location of the two types of starch molecules, amylose and amylopectin, can be demonstrated after staining with iodine (88). Our own studies (69) have suggested that in normal wheat bread, amylose is not leached out from a granule to such an extent that it can be observed by the magnification of light microscopy (Fig. 9, see insert). If cell-wall-degrading enzymes are added to the dough, fragmentation of cell walls occurs during fermentation and baking (89).

B. Rye Breads

The microstructure of rye doughs and that of breads made of whole meal differ greatly from that of wheat bread. Whereas in the wheat-milling process, the bran

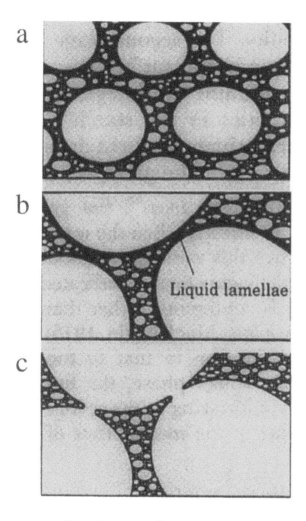

○ Starch granules

⊗ Starch-protein matrix

○ Gas cell lined with a liquid film

Figure 7 A model of dough expansion. (a) At early stages of fermentation, the expanding gas cells are embedded in a starch–protein matrix. (b) At advanced stages of fermentation and early stages of baking, the starch–protein matrix fails to enclose the gas cells, leaving a thin liquid film. (c) At the end of oven spring, the liquid film will rupture. (Reprinted from Ref. 103.)

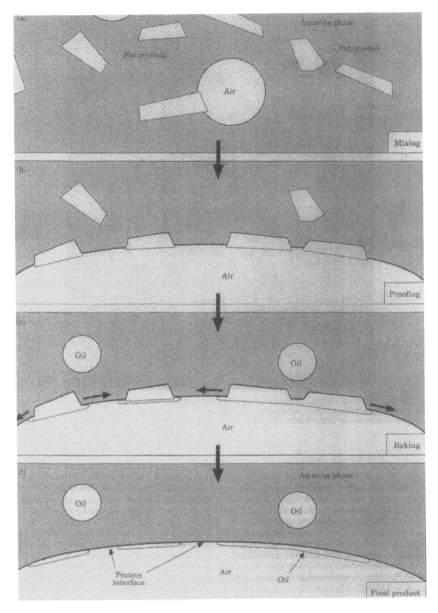

Figure 8 Schematic representation of the behavior of fat during the mixing, proofing, and baking of bread dough: (a) Dispersion of fat crystals and their occasional adsorption to the gas–liquid interface during mixing; (b) an increasing number of crystals adsorb during proofing as the surface of the bubble expands; (c) during baking, crystals melt and the fat–liquid interface is incorporated into the bubble surface; (d) in the baked bread, the fat forms a discontinuous layer on the inside surface of bubbles. (Reprinted from Ref. 86.)

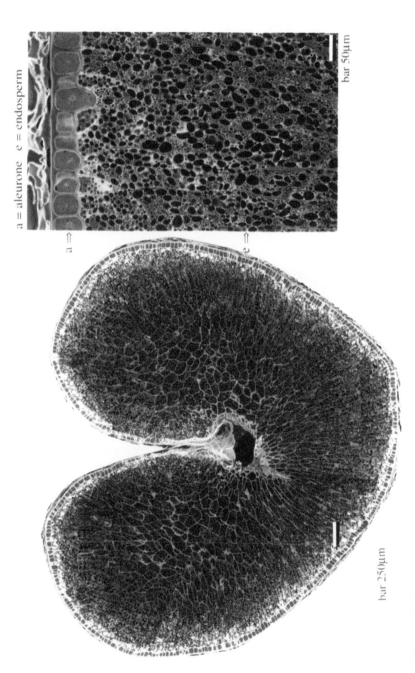

a = aleurone   e = endosperm

bar 50μm

bar 250μm

Figure 12.1 Embedded section of a wheat grain. The section was stained with fuchsin acid and Calcofluor White M2R New and examined under a fluorescence microscope. The primary cell walls appear blue, protein appears brown to red, starch appears black, and the outer lignified layers appear yellow. a = aleurone layer, e = endosperm.

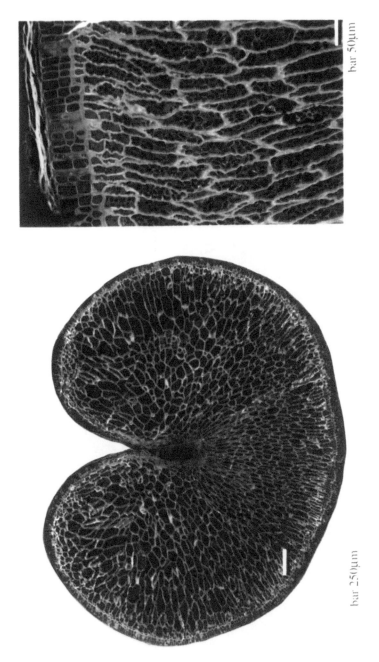

bar 50μm

bar 250μm

Figure 12.2 Embedded section of barley grain. The section was stained as in Figure 1.

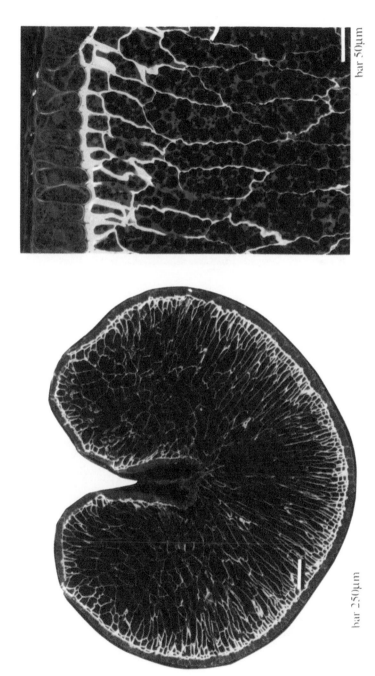

bar 50µm

bar 250µm

Figure 12.3 Embedded section of oat grain. The section was stained as in Figure 1.

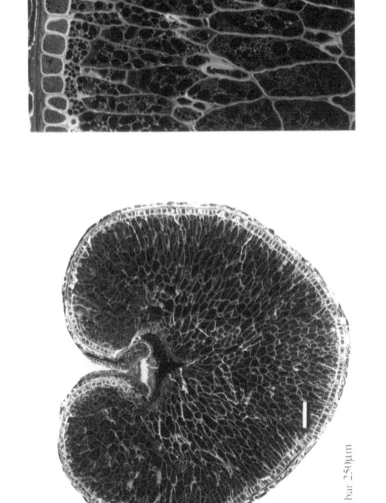

Figure 12.4 Embedded section of rye grain. The section was stained as in Figure 1.

bar 250 μm

Figure 12.6 Embedded section of wheat dough. The section was stained as in Figure 1. (Reprinted from Ref. 69.)

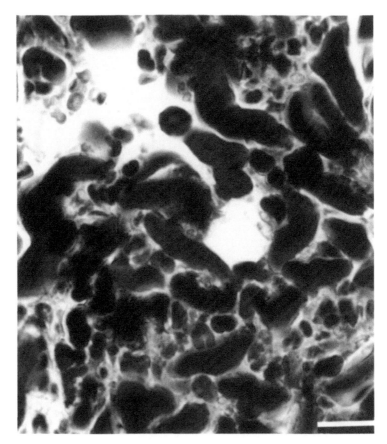

bar 250 μm

Figure 12.9 Embedded section of wheat bread. The staining is with iodine and light green. Starch granules appear violet and proteins green.

bar 250 μm

Figure 12.10 Embedded section of rye dough. The staining is as in Figure 1.

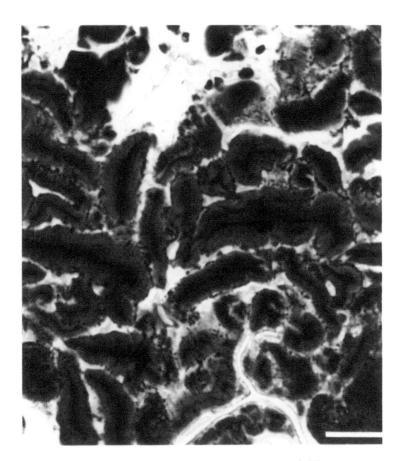

bar 250 µm

Figure 12.11 Embedded section of rye bread. The staining is as in Figure 9. Starch amylose appears blue, amylopectin brown, and proteins yellow.

and germ are separated from the starchy endosperm, in rye milling whole grains are used. Lactic acid fermentation belongs to the traditional rye-baking process and influences taste and resistance against mold. The particle size distribution of rye flours is much higher than that of wheat flours. The protein of rye does not form a gluten network, and the gas retention of rye dough is weak in comparison to wheat doughs. Fermented wheat dough is much more porous than rye dough (69). In rye dough, just after mixing, the continuous phase is composed of protein–starch matrix in which huge aleurone and endosperm particles are dispersed, as shown in Figure 10 (see insert) (90). In wheat dough, the cell walls exist as small fragments, whereas in rye dough much of the original cell wall structures in the particles are detectable (69). As the baking process proceeds, the cell walls fragment and become part of the continuous phase. Because of the higher water content (typically 70–75%) and α-amylase activity of the dough, the starch in rye bread is much more swollen than that in wheat bread. If the pH is around 5.0, much of the amylose has leached out of the granule (Fig. 11, see insert).

Starter cultures are commonly used in rye bread baking, and the acid conditions have a great effect on enzyme activities and starch gelatinization. Both the cell-wall-degrading enzymes and α-amylase affect the baking properties of rye doughs and the quality of breads. Softening of dough during fermentation has been shown to be partly dependent on the swelling or fragmentation of cell walls (91–93) induced by xylanase, whereas the softening during baking is mainly due to α-amylase.

VI. DURUM WHEAT PASTA PRODUCTS

Durum semolina, durum flour, and hard wheat flour are the most important raw materials of pasta products (94). The extrusion temperature and both the humidity and temperature of the drying process have a great effect on the microstructure and texture of the final product. In freshly extruded pasta, a continuous protein film forms the surface, and an amorphous protein fraction with embedded starch forms the inner structure (95). Too high an extrusion temperature or improper drying conditions can disrupt the continuous protein fraction in the surface and internal parts of the pasta product (96).

During cooking of pasta products, starch gelatinization and coagulation of proteins take place (96). In addition to the pasta process, the quality of pasta products is dependent on the protein quantity and the quality of the flour and the composition of cooking water (97). The microstructure of the surface and that of the inner part are very different, and the degree of starch gelatinization is highest in the surface and least in the inner part. Only 5 min after immersion in boiling water, starch granules in the surface were highly swollen (96). Probably also some amylose had leached out. The extent of disruption of protein matrix by swollen starch granules is related to the surface stickiness and cooking loss of noodles (98). Starch granules in the center part were only partly gelatinized after 13 min of cooking time. A typical third zone, between the surface and the inner part, was characterized by swollen granules embedded in a dense protein network. It has been suggested that in good-quality pasta, protein forms the continuous phase, whereas in poor-quality pasta, protein exists in discrete masses (95). Interactions between starch and the protein matrix are probably higher when starch is more swollen, and this is the case for the outer and intermediate layers (96).

VII. ROLLED OATS

Several commercial oat-based products are on the market: oat bran, oat flour, oat meal, oat flakes, rolled oats, oat chips, crushed oat flakes, steel-cut groats, oat bits, and whole oat groats (99). Rolled oats are prepared either from whole dehulled grains or from cut groats by steaming and rolling. Oat porridge is prepared by cooking rolled oats either in water or milk. Rolled oats can be further processed to breakfast cereals.

Rolled oats contain all the different tissues of oat grain: the bran, germ, and endosperm (100). Increasing flaking induces breakdown of starch granules and breakage of cell walls in the endosperm, subaleurone, and aleurone region. Processing of rolled oats had, however, the greatest effect on lipid droplets and proteins. In unprocessed oat, protein exists mainly as protein bodies, whereas in rolled oats protein has aggregated. Also, lipid droplets were aggregated in rolled oats. Cooking has a major effect on starch. Only after 3 min of cooking, birefringence of starch granules is lost, and the granule structure breaks down. The structural change of starch was most drastic in the quick and instant rolled oat samples, compared to old-fashioned rolled oat, which had thicker flakes. Cooking had less effect on the subaleurone and aleurone cell walls, whereas endosperm cell walls were greatly altered.

Conventional and microwave cooking have different effects on the structure of rolled-oat porridge (101). In conventionally cooked porridge, the starch granules were more swollen and the cell walls more degraded than in microwave-cooked porridge. Thicker oat flakes released less solubilized β-glucans and gelatinized starch than thinner oat flakes.

VIII. CONCLUSION

Light and electron microscopies are useful tools for studying the microstructure of cereals and cereal-based products. They can be used, e.g., to study the effects of germination and processing on the structure and composition of cereal grains.

Microscopy of grains can also be used to help identify new and improved cultivars. The structural characteristics of cells and tissues are important in studying physiological and structural interactions within cereal grains that influence grain quality and performance in industrial processes. The structure and composition of cell walls especially are the basis of many process applications of cereal grains (36). As an addition to chemical analysis, microscopy has helped us to understand and visualize structural changes and textural differences in cereal grains, food, and feed.

Microscopy yields visual qualitative information about the structures and changes in the samples studied, but this type of information is difficult to handle if one wants to make comparisons between a number of samples. Therefore methods to quantify the visual information have been developed (102). Computer-assisted image analysis has been used to quantity microscopic data. Quantification of the structural parameters enables the objective correlation of microstructure with other properties. The method can be used, for example, to study the efficiency of nutrient digestion, in enzyme activity research, and in the optimization of processing conditions.

Understanding of the structure and chemistry of grain components is the basis for developing industrial processes for cereals. In process and product development, knowledge of the interactions between the ingredients and the effects of processing conditions is required. The knowledge of structure–function relationships of the raw materials and products is valuable in order to achieve the quality required of successful products.

REFERENCES

1. MM Kalab, P Allan-Wojtas, SS Miller. Microscopy and other imaging techniques in food structure analysis. Trends Food Sci Technol 6:177–186, 1995.
2. JCG Blonk, H van Aalst. Confocal scanning light microscopy in food research. Food Res Int 26:297–311, 1993.
3. Y Vodovotz, E Vittadini, J Cuopland, DJ McClements, P Chinachoti. Bridging the gap: use of confocal microscopy in food research. Food Technol 50:74, 76–82, 1996.
4. RG Fulcher, SS Miller. Structure of oat bran and distribution of dietary fiber components. In: PJ Wood, ed. Oat Bran. St. Paul: American Association of Cereal Chemists, 1993, pp 1–24.
5. AD Evers, DB Bechtel. Microscopic structure of the wheat grain. In: Y Pomeranz, ed. Wheat Chemistry and Technology. St. Paul: American Association of Cereal Chemists, 1988, pp 47–95.
6. DH Hahn. Application of rheology in the pasta industry. In H Faridi, JM Faubion, eds. Dough Rheology and Baked Product Texture. New York: Van Nostrand Reinhold, 1990, pp 385–404.
7. JW Anderson, WJL Chen. Cholesterol-lowering properties of oat products. In: FH

Webster, ed. Oats: Chemistry and Technology. St. Paul: American Association of Cereal Chemists, 1986, pp 309–333.

8. JA Shellenberg. Production and utilization of wheat. In: Y Pomeranz, ed. Wheat Chemistry and Technology. 3rd ed. St. Paul: American Association Cereal Chemists, 1971, pp. 1–14.

9. G Huang, AJ Mccrate, E Varriano-Marston, GM Paulsen. Caryopsis structural and inbibitional characteristics of some hard red and white wheats. Cereal Chem 60: 161–165, 1983.

10. D Bradbury, MM MacMasters, IM Cull. Structure of the mature wheat kernel. II. Microscopic structure of pericarp, seed coat, and other coverings of the endosperm and germ of hard red winter wheat. Cereal Chem 33:342–360, 1956.

11. JJC Hinton. Resistance of the testa to entry of water into the wheat kernel. Cereal Chem 32:296–306, 1955.

12. SY Zee, TP O'Brien. A special type of tracheary element associated with "xylem discontinuity" in the floral axis of wheat. Aust J Biol Sci 23:783–791, 1970.

13. AD Evers, M Reed. Some novel observations by scanning electron microscopy of seed coat and nucellus of the mature wheat grain. Cereal Chem 65:81–85, 1988.

14. DL Wetzel, V Pussayanawin, RG Fulcher. Determination of ferulic acid in grain by HPLC and microspectrofluorometry. Proceedings of the 5th International Flavor Conference, Porto Carras, Greece, 1988, pp 409–428.

15. V Pussayanawin, DL Wetzel, RG Fulcher. Fluorescence detection and measurement of ferulic acid in wheat milling fractions by microscopy and HPLC. J Agric Food Chem 36:515–520, 1988.

16. IN Morrison, J Kuo, TP O'Brien. Histochemistry and fine structure of developing wheat aleurone cells. Planta 123:105–116, 1975.

17. DB Bechtel, Y Pomeranz. Ultrastructure of the mature ungerminated rice (*Oryza sativa*) caryopsis. The caryopsis coat and the aleurone cells. Am J Bot 64:966–973, 1977.

18. RG Fulcher. Observations on the aleurone layer with emphasis on wheat. PhD dissertation, Monash University, Melbourne, 1972.

19. AD Evers. Development of the endosperm of wheat. Ann Bot 34:547–555, 1970.

20. P Greenwell, JD Schofield. A starch granule protein associated with endosperm softness of wheat. Cereal Chem 63:379–380, 1986.

21. GM Glenn, RM Saunders. Physical and structural properties of wheat endosperm associated with grain texture. Cereal Chem 67:176–182, 1990.

22. Y Pomeranz. Chemical composition of kernel structures. In Y Pomeranz, ed. Wheat Chemistry and Technology. St. Paul: American Association Cereal Chemists, 1988, pp. 97–158.

23. Lermer. Beiträge zur Kenntniss der Gerste. Munich: Georg Holzner, 1888.

24. HT Brown, GH Morris. Researches on the germination of some of the Gramineae. J Chem Soc 57:458–528, 1890.

25. DE Briggs. Barley. London: Chapman and Hall, 1978.

26. GH Palmer. Cereals in malting and brewing. In: GH Palmer, ed. Cereal Science and Technology. Aberdeen: Aberdeen University Press, Scotland, U.K., 1989, pp 61–242.

27. S Aastrup. The effect of rain on β-glucan content in barley grains. Carlsberg Res Commun 44:381–393, 1979.

28. S Aastrup. Selection and characterization of low β-glucan mutants from barley. Carlsberg Res Commun 48:307–316, 1983.

29. RS Bhatty, AW McGregor, BG Rossnagel. Total and acid-soluble β-glucan content of hulless barley and its relationship to acid-extract viscosity. Cereal Chem 68: 221–227, 1991.

30. R Schildbach, M Burbridge. Barley varieties and their malting and brewing qualities. In: L Munck, ed. Barley Genetics VI. Copenhagen: Munksgaard International, 1992, pp 953–968.

31. M Oscarsson, T Parkkonen, K Autio, P Åman. Composition and microstructure of waxy, normal and high amylose barley samples. J Cereal Sci 26:259–264, 1997.

32. RS Bhatty. Dietary and nutritional aspects of barley in human foods. In: L Munck, ed. Barley Genetics VI. Copenhagen: Munksgaard International, 1992, pp 913–923.

33. GB Fincher. Ferulic acid in barley cell walls: a fluorescence study. J Inst Brew 82: 347–349, 1976.

34. RG Fulcher, T Deneka, SS Miller. Structure/function relationships in barley quality: analysis by microscopy and quantitative imaging. In: L Munck, ed. Barley Genetics VI. Copenhagen: Munksgaard International, 1992, pp 711–724.

35. BA Stone. Aleurone cell walls—structure and nutritional significance. In: RD Hill, L Munck, eds. New Approaches to Research on Cereal Carbohydrates. Amsterdam: Elsevier Science, 1985, pp 340–354.

36. GB Fincher, BA Stone. Cell walls and their components in cereal grain technology. In: Y Pomeranz, ed. Advances in cereal science and technology. Vol 8. St. Paul: American Association Cereal Chemists, pp 207–295, 1986.

37. S Aastrup, K Erdal, L Munck. Low β-glucan barley mutants and their malting behaviour. Proceedings of 21st Congress of European Brewery Convention, Helsinki, Finland, 1985, pp 387–393.

38. SS Miller, GR Fulcher. Distribution of (1→3),(1→4)-β-D-glucan in kernels of oats and barley using microspectrofluorometry. Cereal Chem 71:64–68, 1994.

39. AW MacGregor, GB Fincher. Carbohydrates in the barley grain. In: AW MacGregor, RS Bhatty, eds. Barley. Chemistry and technology. St. Paul: American Association Cereal Chemists, pp 73–130, 1993.

40. RG Fulcher. Morphological and chemical organization of the oat kernel. In: FH Webster, ed. Oats: Chemistry and Technology. St. Paul: American Association Cereal Chemists, 1986, pp. 47–74.

41. D Paton, MK Lenz. Processing: current practice and novel processes. In: PJ Wood, ed. Oat Bran. St. Paul: American Association Cereal Chemists, 1993, pp 25–47.

42. VL Youngs. Protein distribution in oat kernel. Cereal Chem 49:407–411, 1972.

43. CG Zarkadas, HW Hulan, FG Proudfoot. A Comparison of the amino acid composition of two commercial oat groats. Cereal Chem 59:323–327, 1982.

44. CM Brown, JC Craddock. Oil content and groat weight of entries in the world oat collection. Crop Sci 12:514–515, 1972.

45. VL Youngs. Oat lipids and lipid-related enzymes. IN: FH Webster, ed. Oats: Chemistry and Technology. St. Paul: American Association Cereal Chemists, 1986, pp 205–226.

46. RH Saigo, DM Peterson, J Holy. Development of protein bodies in oat starchy endosperm. Can J Bot 61:1206–1215, 1983.

47. VL Youngs, M Puskulcu, RR Smith. Oat lipids. I. Composition and distribution of lipid components in two oat cultivars. Cereal Chem 54:803–812, 1977.

48. DB Bechtel, Y Pomeranz. Ultrastructure and cytochemistry of mature oat (*Avena sativa* L.) endosperm. The aleurone layer and starchy endosperm. Cereal Chem 58: 61–69, 1981.

49. JL Doublier, D Paton, G Llamas. A rheological investigation of oat starch paste. Cereal Chem 64:21–16, 1987.

50. DH Simmonds, WP Campbell. Morphology and chemistry of the rye grain. In: W Bushuk, ed. Rye: Production, Chemistry and Technology. St. Paul: American Association Cereal Chemists, 1976, pp 63–110.

51. CJA Vinkx, CG Van Nieuwenhove, JA Delcour. Physicochemical and functional properties of rye nonstarch polysaccharides. III. Oxidative gelation of a fraction containing water-soluble pentosans and proteins. Cereal Chem 68:617–622, 1991.

52. LV Glitsø, KE Bach Knudsen, H Adlercreutz. Chemical composition of rye milling fractions. Proceedings of International Rye Symposium Technology and Products, Espoo, Finland, 1995, p 208.

53. SJ Symons, JE Dexter. Aleurone and pericarp fluorescence as estimators of mill stream refinement for various Canadian wheat classes. J Cereal Sci 23:73–83, 1996.

54. RM Sandstedt, H Schroeder. A photomicrographic study of mechanically damaged wheat starch. Food Technol 14:257, 1960.

55. DC Doehlert, WR Moore. Composition of oat bran and flour prepared by three different mechanisms of dry milling. Cereal Chem 74:403–406, 1997.

56. LV Glitsø, KE Bach Knudsen. Milling of whole grain rye to obtain fractions with different dietary fiber characteristics. J Cereal Sci 29:89–97, 1999.

57. FDA. Food labelling. Health claims: oats and coronary disease. Federal Register 61:296–344, 1996.

58. M Luhaloo, A-C Mårtensson, R Andersson, P Åman. Compositional analysis and viscosity measurements of commercial oat brans. J Sci Food Agric 76:142–148, 1998.

59. GH Palmer. Ultrastructure of endosperm and quality. Ferment 6:105–110, 1993.

60. JDJ Gallant, F de Monredon, B Bouchet, P Tacon, J Delort-Laval. Cytochemical study of intact and processed barley grain. Ferment 6:111–114, 1993.

61. RG Fulcher, DW Irving, A de Francisco. Fluorescence microscopy: applications in food analysis. In: L Munck, ed. Fluorescence Analysis of Foods.: Longman Scientific and Technical, 1989, pp 59–109.

62. CS Brennan, N Harris, D Smith, PR Shewry. Structural differences in the mature endosperms of good and poor malting barley cultivars. J Cereal Sci 24:171–177, 1996.

63. CS Brennan, MA Amor, N Harris, D Smith, I Cantrell, D Griggs, PR Shewry. Cultivar differences in modification patterns of protein and carbohydrate reserves during malting of barley. J Cereal Sci 26:83–93, 1997.

64. NH Gram. The ultrastructure of germinating barley seeds. II. Breakdown of starch granules and cell walls of the endosperm in three barley varieties. Carlsberg Res Commun 47:173–185, 1982.

65. AW MacGregor, RR Matsuo. Starch degradation in endosperms of barley and wheat kernels during initial stages of germination. Cereal Chem 59:210–216, 1982.
66. DB Bechtel, RL Gaines, Y Pomeranz. Protein secretion in wheat endosperm—formation of the matrix protein. Ceral Chem. 59:336–343, 1982.
67. DB Bechtel, BD Barnett. A freeze-etch, freeze-fracture study of wheat endosperm development. Am J Bot 71:18–19, 1984.
68. JS Wall. The role of wheat proteins in determining baking quality. In: DL Laidman, RG Wyn-Jones, eds. Recent Advances in the Biochemistry of Cereals. London: Academic Press, 1979, pp 275–311.
69. K Autio, T Parkkonen, M Fabritius. Observing structural differences in wheat and rye breads. Cereal Foods World 42:702–705, 1997.
70. T Amend, H-D Belitz. Microstructural studies of gluten and a hypothesis on dough formation. Food Struct 10:277–288, 1991.
71. DB Bechtel. The microstructure of wheat: its development and conversion into bread. Food Microstruct 4:125–133, 1985.
72. JC Grosskreutz. A lipoprotein model of wheat gluten structure. Cereal Chem 38: 336–349, 1961.
73. T Amend, H-D Belitz. Electron microscopic studies on protein films for wheat and other sources at the air/water interface. Z Lebensm Unters Forsch 190:217–222, 1990.
74. R Moss. A study of the microstructure of bread doughs. CSIRO Food Res Quart 32:50–56, 1972.
75. RC Hoseney, DE Rogers. The formation and properties of wheat flour dough. Crit Rev Food Sci Nutr 29:73–93, 1990.
76. WZ Chen, RC Hoseney. Development of an objective method for dough stickiness. Lebensm-Wiss Technol 28:467–473, 1990.
77. LG Evans, AM Pearson, GR Hooper. Scanning electron microscopy of flour–water doughs treated with oxidizing and reducing agents. Scanning Electron Microscopy 3:583–592, 1981.
78. FO Flint. The evaluation of food structure by light microscopy. In: JMV Blanshard, JR Mitchell, eds. Food Structure—Its Creation and Evaluation. London: Butterworths, 1988, pp 351–365.
79. TP Freeman, DR Shelton. Microstructure of wheat starch: from kernel to bread. Food Technol 45:162–168, 1991.
80. Anonymous. Controlling Structure, the Key to Quality. Food Review April/May, 33–37, 1995.
81. NL Stenvert, R Moss, G Pointing, G Worthington, E Bond. Bread production by dough rollers. Bakers Dig 53:22–27, 1980.
82. AH Bloksma. Dough structure, dough rheology and baking quality. Cereal Foods World 35:237–244, 1990.
83. Z Gan, RE Angold, MR Williams, PR Ellis, JG Vaughan, T Galliard. The microstructure and gas retention of bread dough. J Cereal Sci 12:15–24, 1990.
84. RR Baldwin, ST Titcomb, RG Johansen, WJ Keogh, D Koedding. Fat systems for continuous mix bread. Cereal Sci Today 10:452–457, 1965.
85. BE Brooker. The stabilization of air in cake batters—the role of fat. Food Struct 12:285–296, 1993.

86. BE Brooker. The role of fat in the stabilisation of gas cells in bread dough. J Cereal Sci 24:187–198, 1996.

87. U Khoo, DD Christianson, GE Inglett. Scanning and transmission microscopy of dough and bread. Bakers Digest 49:24–26, 1975.

88. M Langton, AM Hermansson. Microstructural changes in wheat dispersions during heating and cooling. Food Microstruct 8:29–39, 1989.

89. K Autio, T Laurikainen. Relationships between flour/dough microstructure and dough handling and baking properties. Trends Food Sci Technol 8:181–185, 1997.

90. T Parkkonen, H Härkönen, K Autio. The effect of baking on microstructure of rye cell walls and proteins. Cereal Chem 71:58–63, 1994.

91. K Autio, H Härkönen, T Parkkonen, T Frigård, M Siika-aho, K Poutanen, P Åman. Effects of purified endo-β-glucanase on the structural and baking characteristics of rye doughs. Lebens-Wiss Technol 29:18–27, 1997.

92. K Autio, M Fabritius, A Kinnunen. Effect of germination and water content on the microstructure and rheological properties of two rye doughs. Cereal Chem 75:10–14, 1998.

93. M Fabritius, F Gates, H Salovaara, K Autio. Structural changes in insoluble cell walls in wholemeal rye doughs. Lebensm-Wiss Technol 30:367–372, 1997.

94. J Snewing. Analyzing the texture of pasta for quality control. Cereal Foods World 42:8–12, 1997.

95. P Resmini, MA Pagani. Ultrastructure studies of pasta. Food Microstruct 2:1–12, 1983.

96. C Cunin, S Handschin, P Walther, F Escher. Structural changes of starch during cooking of durum wheat pasta. Lebensm-Wiss Technol 28:323–328, 1995.

97. E Schreurs, W Seibel, A Menger, K Pfeilsticker. Effect of cooking water on cooking quality of pasta in relation to raw material quality. Getreide, Mehl Brot 40:281–286, 1986.

98. R Moss, PJ Gore, IC Murray. The influence of ingredients and processing variables on the quality and microstructure of Hokkien, Cantonese and instant noodles. Food Microstruct 6:63–74, 1987.

99. WH Smith. Oats can really "beef up" cookies. Snack Food 62:33–35, 1973.

100. SH Yiu. Effects of processing and cooking on the structural and microchemical composition of oats. Food Microstruct 5:219–225, 1986.

101. SH Yiu, PJ Wood, J Weisz. Effects of cooking on starch and β-glucan of rolled oats. Cereal Chem 64:373–379, 1987.

102. T Parkkonen, R Heinonen, K Autio. A new method for determining the area of cell walls in rye doughs based on fluorescence microscopy and computer-assisted image analysis. Lebensm-Wiss Techol 30:743–747, 1997.

103 Z Gan, PR Ellis, JD Schofield. Gas cell stabilization and gas retention in wheat bread dough. J Cereal Sci 21:215–230, 1995.

13

NMR Characterization of Cereal and Cereal Products

Brian Hills and Alex Grant
Institute of Food Research, Norwich, U.K.

Peter Belton
University of East Anglia, Norwich, U.K.

I. INTRODUCTION

In this chapter we review the application of nuclear magnetic resonance (NMR) spectroscopy to cereal science. The NMR technique actually encompasses a wide diversity of different techniques, including relaxometry, diffusometry, imaging (MRI), NMR microscopy as well as high-resolution solid and liquid spectroscopy. Bearing in mind that these various techniques can be applied at each stage in the conversion of a cereal crop to processed food, it has been necessary, for the sake of brevity, to assume familiarity with the basic principles of NMR and MRI. These principles are described in many texts, including an eight-volume *Encyclopaedia of NMR* (1). The applications of NMR and MRI to more general classes of food are reviewed in the book *Magnetic Resonance Imaging in Food* (2) as well as in one of the authors' recent book *Magnetic Resonance Imaging in Food Science* (3).

II. MRI CHARACTERIZATION OF CEREAL CROPS IN THE FIELD

Optimizing the production of cereal crops in the field requires detailed knowledge of the effects of environmental factors such as drought, temperature, and nutrient availability on plant development as well as the possible negative effects of herbicides, pollutants, and disease. By permitting noninvasive, "in vivo" monitoring of the cereal grains as they develop on the plant under realistic environmental conditions, functional NMR imaging has become a powerful tool in cereal science. Small seedlings can be conveniently grown inside adapted NMR tubes, provided the root system is constantly bathed in nutrient and the whole plant maintained in an environment of controlled humidity, temperature, and luminescence intensity.

The technique can be illustrated by a recent comparison of the effect of osmotic stress on maize and pearl millet plants, which show very different drought tolerances (4). In this study, osmotic stress was induced in 4-week-old cultured maize and pearl millet plants by replacing normal root medium with well-aerated PEG-6000. Some NMR images of a 2.5-mm slice across the stem were acquired at regular time intervals. This was done using a multiple spin-echo imaging sequence with a recycle delay of 1.8 s, an echo time of 4.3 ms, with 48 echoes per echo train, resulting in a $48 \times 128 \times 128$ matrix of complex data. Two-dimensional maps or images of the water proton transverse relaxation time, T_2, across the stem were then calculated on a pixel-by-pixel basis, together with proton density maps, obtained by extrapolating the echo decays to zero time. Along with the NMR measurements, the time course of water uptake was measured over a 30-hour period by weighing the root medium vessel outside the imaging magnet, and photosynthetic activity was measured on the second top leaf in a fixed position using a modulated fluorometer under high (100 lux) and low (10 lux) light conditions.

The pearl millet showed a strong correlation between the proton and T_2 maps, water uptake, and osmotic stress, revealing subtle differences between different types of tissue in the stem. This is probably a result of adaptation by elastic shrinking under osmotic stress in combination with changes in membrane permeability, leading to a slower, gradual loss of water from all parts of the plant, delaying necrosis of the leaves and irrecoverable loss of photosynthetic ability, which are the results of severe osmotic stress. In contrast, the NMR parameter maps in maize hardly changed during stress periods.

Functional imaging of the development and effectiveness of the root systems of cereals is also possible. The newly formed primary root of corn has been imaged (5). By placing a vessel containing soil and roots inside the spectrometer magnet, the development of a plant root system can be followed, noninvasively, in real time (6–8). In this way the relationship between the root development and water-depletion zones set up in the soil as the root extracts water and nutrients can be studied quantitatively. The noninvasive aspect is mandatory because any disturbance of the soil by removal of a root can change the soil porosity, the spatial distribution of water and nutrient, and the distribution of microbiological populations within it. A recent application of this "underground" functional MRI has been to follow the development of pine roots using three-dimensional T_1-weighted images to distinguish the soil and root (8). The water depletion zones around the tap root, lateral roots, and fine roots were quantified for water content and volume and the effects on water uptake of symbiotic relationships, such as infection with mycorrhizal fungus, has been investigated. The time dependence of the root network volume and surface area were determined using seeding algorithms from the 3D images. Surprisingly little is known about the factors controlling the spatial distribution and development of the root network, and this approach will help clarify them.

Of course, it is not just the root and stem tissues that can be imaged. The microscopic flow of water within a ripening wheat grain has been imaged in a classic study by Jenner et al. (9). Wheat plants were raised in small pots under controlled environment conditions, and the ears selected about 14 days after anthesis, when they were approximately halfway through the grain-filling stage. The pot was inverted and a wheat ear inserted into the spectrometer probe. Illumination at ca. 100 Wm^{-1} was provided by a lamp standing 2.3 m from the plant. Water proton density, velocity, and diffusion maps were then obtained within an attached grain using combined velocity and NMR q-space microscopy. These were compared with the results for a detached grain. The analysis of flow patterns within the grain suggested that flow was associated with the unloading and/or distribution of assimilates in the vicinity of the furrow of the grain, and this involved extensive recirculation of water throughout the grain. The ability to observe flow and diffusion within a single grain, in vivo, is remarkable, but much remains to be done in exploiting this technique for understanding the relationship between intragrain flow, grain development, and environmental stresses imposed on the plant. It is disappointing that little appears to have been done in imaging intragrain flow in vivo, since the original work was reported in 1988.

III. NMR RELAXATION STUDIES OF OIL AND MOISTURE CONTENTS OF SEEDS, FLOUR, AND DOUGH

Once a cereal crop has been harvested it is often necessary to determine relevant compositional factors, such as oil and moisture content, since these will affect subsequent postharvest storage, milling, and processing conditions. For this purpose a number of simple nondestructive "benchtop" NMR methods have been developed for analyzing oil and water content of seeds, and these have been reviewed (10). The nondestructive aspect of NMR, which leaves genetic material undamaged, has also significantly contributed to the selection of grains in corn-breeding programs.

The oil content of a seed can be determined without weighing or drying from the free induction decay (FID) recorded after a single hard, 90° radiofrequency pulse. The solidlike biopolymer components of the seed give rise to a fast-decaying component in the FID, while the liquid oil (and moisture) give a slower-decaying component. Accordingly, the signal amplitude, S, at a very short time, say, 10 μs, immediately after the instrument deadtime corresponds to signal from both the solid and oil components. In contrast, the signal amplitude, L, acquired at a time, say, 70 μs, sufficiently long for the solid contribution to have decayed away, arises from liquid oil. The oil content (%, w/w) can therefore be calculated as $100L/(L + fS)$, where the correction factor f accounts for the decay of solid signal during the instrument deadtime, proton density differences in the solid and liquid phases, and residual moisture signal at 70μs.

A similar idea can be used with the standard CPMG (Carr–Purcell–Meiboom–Gill) pulse sequence used for measuring transverse relaxation times. The shortest echo time in the CPMG sequence is about 200 μs, so solid components do not contribute to the CPMG echo intensity, only the liquid oil and water, and these can be distinguished on the basis of their different transverse relaxation times. This is usually done by deconvoluting the CPMG echo decay envelope into a series of decaying exponential components, which can be assigned to water and oil, so that their relative amplitudes give the oil:water ratio. By combining this measurement with a solid:total liquid ratio measurement using the FID method, it is therefore possible to determine the moisture and oil content of cereal grains simultaneously, without weighing (11).

IV. MRI STUDIES OF THE REHYDRATION OF INTACT CEREAL GRAINS

The rehydration of cereal grains is an essential step in the manufacture of many types of extruded cereal products, from shredded wheat to rice crispies. Wheat grains are usually partially rehydrated by soaking in water, a process known as *tempering*, prior to milling. However, some wheat varieties, such as the Scout 66, Karl, and Newton varieties, are very resistant to the ingress of water during soaking after harvesting and require many weeks, or even months, of storage before they can be tempered. The reasons for this phenomenon (called the *new crop syndrome*) are being investigated by NMR microimaging. Preliminary studies (12) show that, during soaking, the water first penetrates the embryo, then is drawn into the vascular tissue in the grain, and finally enters the ventral, but not dorsal, endosperm tissue. However, the reasons for the resistance of some varieties has yet to be elucidated. In these qualitative studies the water ingress was inferred from the changes in Hahn-echo image intensity. Similar qualitative microimaging studies were used to distinguish dormant from nondormant grains. Dormancy in immature grains is a very desirable characteristic because it prevents early sprouting under moist conditions (including rain), which leads to starch damage. Microimaging indicated that water could penetrate the root cap region of the nondormant (germinatable) grains during soaking, but not the dormant kernels (12).

The effects of boiling and steaming wheat grains have been compared by Shapley and co-workers (13) using a spin-echo imaging sequence. The moisture images were obtained by compensating the spin-echo signal for transverse relaxation and by calibration of the spin density map with gravimetric measurements of water content. During boiling it was found that water penetrates evenly into the wheat grain from all points on the boundary, including the inside of the crease. The moisture profiles show a clearly defined water front ingressing into the center of the grain. A perfectly sharp water boundary would suggest Case II Fickian diffusion, which arises when the rate-limiting step is relaxation of the biopolymer (mainly starch) matrix rather than water diffusion. In contrast, steaming gives slower rehydration and a gradually increasing, spatially uniform water content throughout the grain. This suggests that, with steam, the transport through the grain surface is the rate-limiting step and not the diffusion of water within the bulk of the grain. It could be that the outer layers of endosperm, particularly those cells just below the aleurone, present the main moisture permeability barrier.

V. MRI STUDIES OF THE PROCESSING OF CEREAL PRODUCTS

A. The Baking of Doughs

Imaging the baking of doughs is an attractive proposition because it provides real-time data on the changing moisture and fat distribution as well as the development of voids in the dough. Such information can assist in process optimization (3) and also affect product quality. Unfortunately, a number of factors combine to complicate this type of imaging experiment. To begin with, the low moisture content of the final product (typically less than 10% in a biscuit) shortens the water proton transverse relaxation time, and this eventually limits the liquid-type imaging experiments to moisture contents above ca. 10–15%. The relatively high fat content of biscuits (e.g., 5–15%), which is characterized by a longer transverse relaxation time than the water protons, also contributes significant signal intensity and makes it necessary to devise ways of separating the fat and moisture contributions. A third complication is the increase in void formation and therefore of porosity during baking. This not only complicates the quantitative relationship between image intensity per voxel and the gravimetric moisture or oil content but can also introduce local susceptibility-induced field gradient artifacts, especially at high main spectrometer field strengths.

Despite these difficulties, Heil and co-workers (14) succeeded in using two-dimensional imaging and volume selective spectroscopy to follow the time course of the baking of American-style biscuits. Data were collected by removing the biscuit dough from the oven every 3 minutes, imaging the sample, and then returning it to the oven. The increase in porosity of the dough during baking was measured both from the overall increase in volume caused by the increasing number of voids (assuming the dough volume remains constant) and by calculation of the void volume from the images after defining a lower threshold-intensity limit for the dough and an upper background-intensity limit. These two methods correlated very well, with a coefficient of correlation of 0.996. The distribution of water and oil was determined with volume selective spectroscopy by calculating the ratio of water and oil peak areas in spectra taken from various regions in the sample. These were compared with parallel gravimetric and destructive measurements on another sample. As expected, the data showed that the oil content distribution is relatively constant in time, in contrast to the moisture content, which showed an overall decrease with baking time. Less expected was the observation that the moisture content remained highest and, in some cases increased, in the central regions of the dough. This may well indicate that water is being driven to the cooler central regions of the dough by evaporation and condensation.

In the American biscuit studies, the initial water content of the dough was relatively high (50% w/w), which greatly facilitated MRI measurements. In a separate study on the imaging of the baking of wheatflake biscuits, Duce et al. (15) measured the dependence of the water proton transverse relaxation times of both water and liquid lipid in the wheat biscuits and showed that, although the liquid lipid maintained a nearly constant T_2 of 45 ms at all moisture contents between 5.8 and 20.6%, the water transverse relaxation time decreased significantly from ca. 1.7 ms at 20.6% moisture to 300 µs at 5.8%. Because their shortest available echo time in the imager was 3.5 ms, no moisture signal could be detected at moisture contents below ca. 15%. However, the image intensity increased monotonically (but nonlinearly) with increasing moisture contents above 15% and could be used to image the changing moisture distribution during baking. As with the American biscuits, the results showed that 5 minutes after baking the moisture content was highest (between 19 and 20%) in the center of the biscuit. During a 2-hour equilibration period after baking, the moisture in the center transported to the rest of the biscuit and resulted in a loss of moisture signal because the final moisture content was less than 15%. This result suggests that solid imaging techniques, especially single-point imaging, are more appropriate to baking than the conventional spin-echo imaging methods. In single-point imaging, an image can be generated by acquiring the signal from a single point on the FID by repeatedly acquiring it in increasingly strong field gradients. In this way the signal can be taken from a point just a few hundred microseconds in the FID, where almost all the moisture signal is available.

B. Extrusion Cooking of Cereals

Extrusion processing is widespread throughout the food industry because it is a continuous operation, is very flexible in its range of application, and gives end products of reproducible quality. Common types of extruded cereal products include pasta, cereal snacks, pet foods, and animal feeds, and extrusion's potential in the creation of new products continues to be actively researched.

During extrusion, the raw ingredients are fed into the barrel of the extruder, where they are subjected to shear stress by rotating screws; to a pressure gradient, because the size of the die exit orifice is usually smaller than the barrel; and to heating, both from external heat sources along the barrel and from internal heat generated by chemical reactions and friction. Under these conditions a complex series of physical and chemical changes take place, such as mixing, gelatinization, denaturization, evaporation, flavor production (and loss), and rheological changes. Not surprisingly, understanding and mathematically modeling these complex and rapid changes is a major challenge in cereal processing science. Magnetic resonance imaging can greatly assist in understanding this endeavor by providing real-time, noninvasive images of the flow within the extruder and, potentially, spatial maps of other NMR parameters, such as relaxation times, chemical shifts, and diffusion coefficients, as well as images of temperature and/ or chemical changes. To date only the isothermal flow aspect has been researched, so, for the present, only the velocity field need be considered.

Magnetic resonance images of flow within a single-screw extruder were obtained by nulling the signal from a slice across the extruder in an orientation perpendicular to the barrel and taking an image of it at later times. These pioneering measurements were made by McCarthy and co-workers (16) with a special extruder made of nonmagnetic materials and operated inside the magnet of the NMR spectrometer and using a model carboxymethyl cellulose solution. The flow data for this model food system were shown to agree with the velocity field calculated assuming an incompressible, Newtonian fluid obeying stick boundary conditions (i.e., where there is no slip at the walls). The extension of these studies to more complex fluids and nonisothermal extruder conditions is to be anticipated and will aid the development of more realistic numerical models of the extrusion process.

C. The Rheology of Doughs and Batters: The MRI Rheometer

It is worth noting that the rheology of doughs and batters can be investigated with an MRI rheometer. An MRI rheometer is basically a rotational rheometer, such as a cylindrical Couette cell, placed inside an NMR spectrometer, permitting imaging of the flow within the rheometer. The simplest arrangement is probably one where the sample is placed between two concentric NMR tubes. By rotating the inner tube at a known angular velocity with a applied torque and imaging the radial velocity distribution in the sample it is possible to deduce the rheological flow curve in a model-independent way (17). The MRI rheometer has many potential advantages over conventional rheometers. By imaging the velocity field it is possible to see shear-induced phase transitions within the sample, induced, for example, by biopolymer aggregation. Combining flow imaging with measurements of relaxation times and diffusion coefficients provides additional insights into the microstructural and macromolecular origins of complex rheological behavior. Britton and Callaghan (17) have studied cornflour–water mixtures in this way and observed a shear-induced phase transition resulting from shear thickening. What was particularly surprising was the observation that the thickening process occurred at some distance from the shear stress application. The potential of this method for studying the microstructural basis of dough rheology is considerable.

D.   The Drying of Extruded Cereal Products

The drying of extruded pasta has been investigated with MRI by Hills and colleagues (18). During the manufacture of pasta, such as spaghetti, the extrusion step is followed by drying in a hot air stream so that the spaghetti can be packed and stored as the familiar bundles of straight, hard spaghetti cylinders. The drying step has to be very carefully optimized because drying at too fast a rate sets up severe moisture gradients in the pasta, which causes differential shrinkage, bending, and stress cracking. On the other hand, drying too slowly in a hot, humid atmosphere permits the growth of spoilage microorganisms. To optimize the drying conditions it is therefore necessary to understand the mechanism of pasta drying. For this purpose the drying kinetics have been investigated with fast radial microimaging. To permit unhindered shrinkage and a radially symmetric flow of dry air around the pasta, a specially designed holder is required. In this MRI study no attempt was made to simulate real industrial drying conditions in the probe. Rather, the purpose was to investigate the drying mechanism itself. Accordingly, the drying process was made essentially isothermal by using an air temperature of 26°C, which is only slightly above room temperature. The small dimensions of spaghetti also meant that moisture gradients were rapidly established, so the imaging acquisition time was minimized by exploiting the cylindrical symmetry of the sample by taking only one-dimensional projections across the pasta cylinder (after slice selection) and using an inverse Abel transform to extract the radial signal profile, $P(r, t)$. A maximum-entropy signal optimization algorithm was also used to smooth and enhance the radial profiles. In this way profiles could be acquired every few seconds with spatial resolutions of ca. 30 μm.

The MRI moisture profiles reveal the real complexity of the pasta-drying process. Shrinkage is observed, both in the images and with a traveling optical microscope with samples dried under similar conditions to those in the probe. Comparison of the MRI and optical radii suggests that a type of case hardening may be occurring in which the surface rapidly dries to form a drier shell around the wet inner core. The existence of this drier shell makes it difficult to determine the time dependence of the surface moisture content, because this depends critically on the choice of the position of the outer edge of the pasta cylinder. Not surprisingly, attempts to model these observations with a non-Fickian diffusion model were only partly successful. The moisture-dependent diffusion coefficient was assumed to have an exponential form such that

$$D(r, t) = A \exp(BS(r, t)) \tag{1}$$

where $S(r, t)$ is the local degree of saturation, defined as the ratio of the actual moisture content to that for the saturated sample. A crude attempt to model non-uniform shrinkage was made by reducing the number of cells in the numerical simulation using the measured dependence of shrinkage on moisture content and the equation expressing local conservation of mass in a cylindrical geometry. To incorporate the rapid surface drying, the surface was assumed to dry essentially instantaneously, giving a boundary condition, $W(R, t) = 0$, where R is the cylinder radius and W is the moisture content. The theoretical moisture profiles succeed in reproducing the qualitative trends in the data, such as shrinkage and profile shape, but fail to provide a good quantitative fit to the data. Indeed, the MRI results highlight the need for more sophisticated drying models where case hardening is more explicitly treated, perhaps as a rubber–glass transition.

E. The Rehydration of Extruded Cereal Products

Many dried cereal-based products, such as pasta, require rehydration by the consumer before consumption, so it is important to ensure that the rehydrated product has a satisfactory quality. A detailed MRI study of the rehydration of spaghetti has been reported (19). With spaghetti, the stickiness and cohesion after rehydration are important quality factors and the rehydration rate should obviously not be too slow. These quality considerations affect both the choice of wheat used to make the spaghetti and the extrusion temperature. Hard ''durum'' wheat, soft wheats, and mixtures of the two types can be used in the extrusion process, and the choice profoundly alters the rehydration kinetics. Magnetic resonance imaging can assist in optimizing the choice of raw material by providing detailed information about the moisture profiles within individual spaghetti strands during rehydration (19).

To do this, dry spaghetti, extruded from various types of wheat, was immersed in water at 80°C and samples removed every few minutes for radial imaging with a spin-echo sequence. Because the water content of dry pasta is very low and remains low during the early stages of rehydration, the radial images have to be compensated for signal loss due to fast relaxation. This problem was circumvented by directly imaging radial maps of the effective water proton transverse relaxation time measured by increasing the number of echoes in a CPMG preparation sequence introduced before the radial imaging pulses sequence. The T_2 maps were then converted to water content maps using a calibration curve obtained for small, uniformly rehydrated pasta samples of varying gravimetrically determined water content.

The rehydration results were modeled in an analogous way to drying by numerically solving the diffusion equation, taking account of sample expansion during rehydration as well as of the dependence of the water self-diffusion coefficient on local water content. Although the theory provided only a semiquantitative explanation of the imaging data, it showed that increasing the proportion of soft wheat in the extrusion mix shifts the diffusion kinetics away from the Case II type of Fickian diffusion kinetics, observed with hard wheat pasta, toward the profiles expected for classical Fickian diffusion.

The fact that rehydration starts with a dry food suggests the use of solid imaging techniques. This has not yet been attempted with pasta, though a STRAFI solid imaging study on the rehydration of a related model system (a starch glass) by liquid water and by water vapor has been reported by Hopkinson and Jones (50). The MRI profiles showed that rehydration by water vapor followed Fickian diffusion kinetics. In contrast, rehydration with liquid water followed Case II diffusion kinetics, at least semiquantitatively.

F. Freezing Cereal Products

Frozen cereal products, such as doughs and pastries, are gaining commercial importance, because freezing can extend shelf lives and permit long-distance transport. High-power NMR relaxometry can be used to quantify the amount of unfrozen water in a sample at subzero temperatures, and this is an important factor influencing the rate of slow chemical and enzymatic spoilage reactions. The principle is similar to that used to determine solid/liquid ratios. In fact ice has such a short transverse relaxation time compared to liquid water that it contributes no signal in most relaxation and imaging experiments. The signal intensity is therefore proportional to the amount of unfrozen water in the sample. It is also worth noting that MRI has been used to image the spatial distribution of unfrozen water (and therefore, indirectly, of ice) during the freezing of rehydrated pasta. This is not a commercially important process but was undertaken to test various theoretical models of freezing (20). The results revealed a sharp ice front moving into the pasta, and the time course could be fitted with a simple Plank freezing model.

VI. NMR CHARACTERIZATION OF THE MACROMOLECULAR CONSTITUENTS OF CEREALS

A. Starch

The structure and functionality of starch has been the subject of intensive study and has been extensively reviewed (21). Figure 1 shows what might be called the "standard" model of a native starch granule, which has been formulated to fit small-angle and wide-angle X-ray scattering data. The granule is too small to permit direct imaging of its internal structure with an NMR microscope, but NMR relaxometry and diffusometry can provide unique insights into the dynamic state of water within the granule, although data interpretation is not always straightforward. One might expect water in the amorphous region to be more mobile and have a longer transverse relaxation time than water in the semicrystalline lamellae, and this appears to be confirmed in the observed bimodal distribution of water proton transverse relaxation times in packed beds of native corn starch granules of varying water contents.

However in, Figure 2 we see that the larger and more irregular potato starch granules show hints of four proton transverse relaxation time peaks, which are well resolved at 277 K and have been assigned to four water compartments (54). The longest relaxation time peak arises from water outside the granules; the next

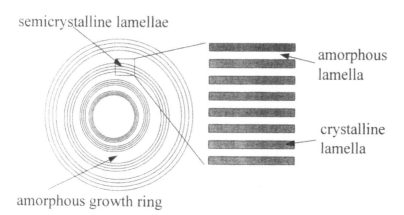

semicrystalline lamellae

amorphous lamella

crystalline lamella

amorphous growth ring

Figure 1 Schematic of the internal structure of a starch granule based on SAXS data. The whole granule is composed of stacks of semicrystalline lamellae separated by amorphous growth rings. On the right is a magnified view of one such stack, showing that it is made up of alternating crystalline and amorphous lamellae. The crystalline lamellae comprise regions of lined-up double helices formed from amylopectin branches. The amorphous lamellae are where the amylopectin branch points sit. (Based on data from Ref. 21.)

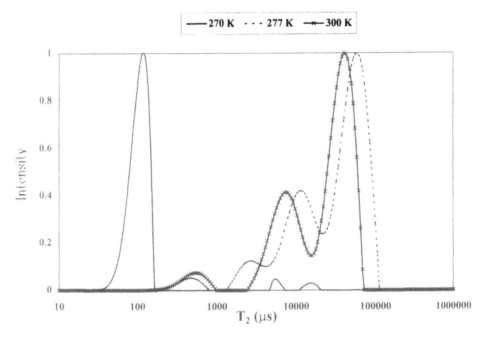

Figure 2 Distribution of water proton transverse relaxation times measured with the CPMG pulse sequence at the indicated temperatures for a water-saturated packed bed of native potato starch granules.

two shorter relaxation time peaks observed at 277 K arise from slowly exchanging water in the amorphous growth rings and semicrystalline lamellae. Finally, the shortest relaxation time peak, at about 200 µs, has been assigned to water in the hexagonal channels of B-type amylopectin crystals and possibly to mobile starch chains (54). Raising the temperature to 300 K causes the merging of the two middle peaks because of fast diffusion of water between the amorphous growth rings and the semicrystalline lamellae. Freezing at 270 K removes most of the water proton signal, because ice has a transverse relaxation time of just a few microseconds. The new peak at about 100 µs presumably corresponds to starch CH protons and also to residual signal from small amounts of nonfreezing water hydrating the starch chains. Replacing the water with D_2O permits the relaxation time distribution of the nonexchanging CH starch protons to be observed, and this reveals three transverse relaxation components, though it is difficult to assign them to particular structural components. The mobility of water and its relationship to starch granule structure can be measured directly with NMR diffusion pulse sequences, such as the standard pulsed gradient stimulated echo method in q-space microscopy (1).

Figure 3 shows the stimulated echo attenuation for a packed bed of native corn starch containing 40% water by weight, as a function of the NMR parameters $q^2\Delta$, where q is the wavevector corresponding to the pulsed gradient area, and Δ is the diffusion time. The results were found to be independent of whether q^2 or Δ were varied, and attempts to fit the data with models based on restricted diffusion or diffusion in fractal structures failed. However, a model based on two-dimensional diffusion provided a good fit, indicating, perhaps, that the semi-crystalline stacks within the granule are confining the water mobility, at least on the millisecond time scale of the NMR diffusion experiment (53). Clearly the potential of NMR relaxation and diffusion methods for following water mobility as a function of thermal and/or pressure processing of the starch is considerable and is largely unexplored.

Figure 4 shows the effect of heating a D_2O-saturated packed bed of native potato starch granules. Here all water has been replaced with D_2O so that only

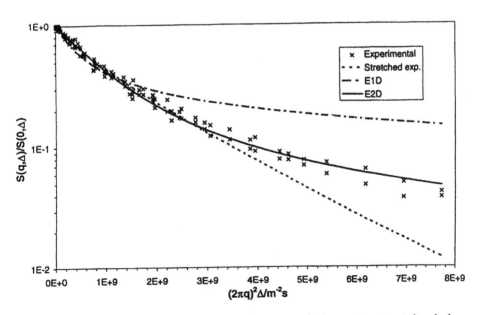

Figure 3 Dependence of stimulated echo amplitude on $q^2\Delta$ for a water-saturated packed bed of native potato starch granules. The lines show the best fits of a 1- and 2-dimensional diffusion models and for a stretched exponential. Results for three sample replicates are superimposed.

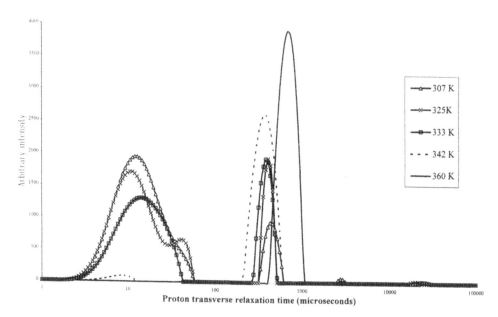

Figure 4 Distribution of transverse relaxation times for nonexchanging starch protons measured with the free induction decay at the indicated temperatures for a packed bed of native potato starch granules in D_2O.

the changing mobility of the starch chains are observed as the granules are gelatinized. The longer relaxation time peak corresponds to amylose gel, and this increases in relative intensity as the granules pass through the gelatinization temperature (see Fig. 5).

Clearly there is considerable scope for using NMR to follow the processing response of granules as well as the effects of different harvesting and irrigation conditions on the subsequent processing response of the granule. Recent (unpublished) work at the authors' laboratory shows significant differences in processing response of cassava starch granules harvested at different times and for different exposures to the rainy and dry seasons. Such effects will be important when optimizing the functional behavior of starches.

The more complicated (and realistic) three-component maize–sugar–water system has been studied by NMR relaxation methods (22). Extrusion was found to enhance both the molecular mobility of the rigid "solidlike" and the more mobile "liquidlike" components of the NMR free induction decay. There was also evidence for preferable hydration, in that, above a moisture content of ca. 15%, the sucrose apparently preferentially partitioned into the mobile water

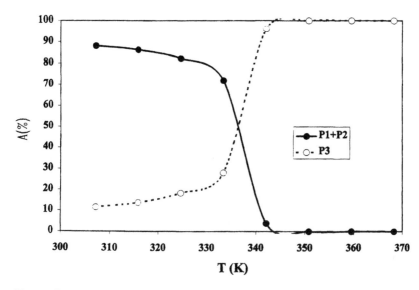

Figure 5 Temperature dependence of the relative percentage area of the longest relaxation time peak (labeled P3) in Figure 4. P1 and P2 refer to the less well-resolved short relaxation time peaks in Figure 4. Additional temperatures not plotted in Figure 4 have been included.

phase, resulting in a decreased amplitude of the rigid component and a reduced relaxation time of the mobile component. It is possible that multilayers of hydration water are formed above the 15% moisture contents, and these permit the solvation of the sucrose. Evidence for multilayer formation above ca. 15% moisture contents has also been found in recent NMR relaxation studies on gelatine gels (51).

Proton-decoupled NMR water oxygen-17 and deuterium (D_2O) relaxometry has also been used to study water mobility in sucrose–starch systems and its relationship to the gelatinization and glass transition temperatures (23). Proton-decoupled water oxygen-17 relaxometry has the advantage that it directly probes water mobility without the added complications of proton exchange between water and hydroxyl protons on the sucrose and starch. Although there was an increase in gelatinization temperature with increasing sucrose concentration it was found to be a result of reduced plasticization of the starch by water as the water preferentially associated with the sucrose. This meant that the correlation between gelatinization temperature and water mobility depended sensitively on the starch: sucrose ratio (23).

B. Cereal Proteins

Cereal proteins play a key role in determining the quality of breads, biscuits, and other processed cereal–based foods. This is particularly true of cereal storage proteins, the prolamins, which in some cereals can account for half the total protein. Understanding prolamin structure–function relationships is therefore an important scientific and technical challenge. Amino acid sequencing has shown that all cereal prolamins, with the exception of the major zeins, belong to a single protein superfamily rich in glutamine and proline (hence the name) but poor in charged amino acids and contain many repeated sequences. This composition accounts for the fact that most prolamins are soluble in alcohol–water mixtures but not in water or aqueous solutions of salts. No doubt this explains why they are deposited as protein bodies in the cell, but these have no known role apart from storage. Because most prolamins have high molecular weights and are water insoluble (24, 25), NMR relaxometry and high-resolution solid-state NMR spectroscopy are the tools of choice for studying prolamin structure–function relationships, especially the effects of hydration and thermal processing. The discussion will center on the major proteins from wheat, barley, maize, and sorgum.

1. NMR Studies of Wheat Proteins

In the baking industry, the type of wheat flour dictates to a large extent the nature of the final baked product. Breads, for example, are usually prepared from hard wheat flour, while biscuits and cakes are prepared from soft wheat flour. Moreover, it has long been recognized that the gluten network in wheat flour dough is largely responsible for the dough's unique functional properties and that some wheat flours produce better breadmaking doughs than others (26). Pragmatically, gluten can be defined as the viscoelastic mass remaining after dough is washed in water or dilute salt solution to remove starch and water-soluble components. Besides lipid and residual starch (10–15%), about 75% of dry gluten consists of prolamins, so most research has focused on this component.

The wheat prolamins are usually divided into two groups: (a) the α, β, γ, and ω gliadins, which are monomers with relative molecular weights (M_r) ranging from 30,000 to 74,000 and having no interchain disulphide bond, and (b) the glutenins, which are polymeric and comprise individual subunits with M_r values up to ~160,000 that are cross-linked by interchain disulphide bonds. For this reason they are not soluble in alcohol–water mixtures. The subunits may be classified as high-molecular-weight (HMW) or low-molecular-weight (LMW), the former ranging between 69,000 and 88,000. The LMW subunits resemble the monomeric α-type and γ-type gliadins in amino acid composition, while the HMW subunits contain less proline but are high in glycine.

A number of early NMR studies (27) attempted to relate flour quality to differences in the ^{13}C solid-state (CPMAS) spectrum of dry gluten extracted from different wheat varieties. The first attempts were complicated by the appearance of lipid signals, but a judicious choice of NMR acquisition conditions permits separation of the lipid and protein spectra. The lipid spectrum is obtained at long contact times and is characterized by sharp peaks, while the protein spectrum, which consisted of much broader peaks, was obtained with short NMR contact times and could be simulated by assuming that the signals arose from proteins whose amino acid content reflected that of the whole gluten. These early CPMAS results showed, not surprisingly, that the lipid is considerably more mobile than the protein but failed to shed light on the problem of the origin of gluten quality (28).

The low-resolution proton transverse relaxation of gluten can be deconvoluted into three components: a fast-relaxing Gaussian component arising from the more rigid protein component; an intermediate exponential component, which represents between 5 and 15% of the magnetization, and finally, a third, slowly relaxing component, which can be assigned to mobile lipid (29). The observation that the intermediate component depends on the origin of the gluten, with pasta- and breadmaking glutens showing different relaxation times, suggests that it arises from less mobile water associated with the protein matrix and possibly plasticized domains in the gluten matrix and that these are important in determining functionality. Below the glass transition temperature, the T_2 of the fast-decaying Gaussian component was found to be independent of temperature; but after passing through the rigid lattice limit temperature, T_{RLL}, it was found to increase with temperature, which clearly indicates the onset of mobility in the more rigid protein chain component. Interestingly, the decrease of T_{RLL} with increasing water content was found to be the same for gluten, soluble glutenin, and gliadin. Since comparison with DSC indicates that T_{RLL} corresponds to the onset of the glass transition, this shows that glutenin, gliadin, and gluten have similar glass transition temperatures.

Further insight is obtained by following the proton transverse relaxation as dry gluten is hydrated with D_2O. The proton transverse relaxation then becomes double exponential, with fast- and slow-decaying components. The ratio of these components changes with temperature until, at 90°C, most of the signal arises from the slowly relaxing "mobile" component, and this change is retained on cooling. This is surprising, because the heat treatment also causes the gluten to harden, presumably by formation of interchain disulphide bonds. The data suggest that the molecular domains between the cross-links within glutenin retain the more mobile, random-coil structure of the denatured material.

The proton longitudinal relaxation of dry gluten also shows double exponential behavior, but, unlike the transverse relaxation, the components are not sensitive to gluten type. However, when the interchain gluten disulphide bonds are removed, single-exponential T_1 relaxation is observed (28, 30). Data analysis suggests that the double exponential behavior results from restricted motion about disulphide linkages in a region that has dimensions of the order of 10 nm and supports the dynamic model by which more mobile domains punctuate more rigid regions created by interchain disulphide bonds.

Water NMR relaxometry based on deuterium (D_2O) and oxygen-17 ($H_2^{17}O$) have also been applied to the gluten (31). The increase in deuterium signal in gluten with increasing water content correlated remarkably well with the decrease in relative stiffness, $R(M)$, at a water content, M, in dynamic mechanical analysis (DMA). In fact,

$$\% \text{ detected signal } = 100 \, [1 - R(M)] \tag{2}$$

This was interpreted as showing that the onset of motion (plasticization) in the gluten corresponded to an increase in the deuteron relaxation time and hence of observable signal intensity. It is particularly surprising that two methods of such widely differing time scales give such close correspondence.

The interaction of the gluten proteins with the starch component is also a potentially important determinant of product quality. Li et al. (32) used rotating frame relaxation measurements to probe this interaction, though the starch actually came from maize, not wheat. The $T_{1\rho}(H)$ values for starch and gluten in a 1:1 mixture were separated by first spin-locking the proton magnetization for variable times with a fixed spin-lock field strength and then transferring the residual magnetization with a fixed contact time to ^{13}C spins so that the starch and gluten ^{13}C spectral resonances could be distinguished. Surprisingly the $T_{1\rho}$ values for the gluten and starch were found to be similar in the mixture and in the separated components, indicating limited interaction, at least at 20% moisture content. The moisture dependence of the gluten $T_{1\rho}$ values appeared to be greater than that of the starch, suggesting that the gluten has the higher water affinity, and this was supported by the observation that heating a drier mixture containing only 2% moisture resulted in a large increase in the $T_{1\rho}(H)$ values of the gluten but only slight increases in those of starch.

Dough quality can also be altered by the addition of reducing or oxidizing agents. Typical oxidants include benzoyl peroxide, azodicarbonamide, acetone peroxide, chlorine dioxide, and potassium bromate, while typical reducing agents include cysteine and sodium bisulphite. These presumably function by changing the development of the dough protein network, though their exact mode of action is unknown. Loaf volume, uniformity, texture, water-holding capacity, dough strength, and mixing tolerance of the dough can all be affected by such additives (33). To study this important phenomenon, deuterium NMR has been used to investigate the effect of two oxidants, potassium bromate and ascorbic acid, on water mobility in wheat gluten (34). The signal was simplified by making dough with D_2O, and comparison was made with the dough's thermomechanical properties. It was found that at any given hydration level over the range 10–58% D_2O, the signal intensity from the oxidant-treated samples was significantly lower than that of the untreated control, and the glass–rubber transition region extended to higher temperatures in the treated samples. This suggests that oxidation leads to a more rigid gluten fraction, extending the glass transition to a higher temperature range.

Deuterium NMR has been used to monitor the water-sorption capacity of gluten as the dry material is heated in 10°C intervals from 5 to 90°C in a sealed tube containing excess D_2O (35). The effect of lipid on the water sorption was examined by comparing whole and lipid-free gluten. The deuterium transverse relaxation was multiple exponential, with the slowest-relaxing component corresponding to bulk D_2O. The thermal treatment caused a decrease in the population of this slowly relaxing component as bulk D_2O was adsorbed into the gluten matrix. Such behavior is characteristic of hydrophilic biopolymers and contrasts with hydrophobic polymers like the protein elastin, which contracts and dispels water. The data for the lipid-free and whole gluten were very similar, which suggests that the lipid has little effect on the gluten chain dynamics. This is consistent with the observation that the proton transverse relaxation in whole and lipid-free gluten samples is also very similar. Indeed, ^{31}P NMR spectroscopy (36), in conjunction with freeze-fracture electron microscopy of wheat gluten, shows that the lipids are organized in small vesicles in which polar lipids exhibit a lamellar liquid crystalline phase. Taken together, the results suggest that wheat protein–lipid associations are not significant in determining the elasticity of gluten.

The HMW subunits of wheat glutenin are a relatively minor group of proteins, accounting for about 10% of the prolamins of wheat, but they are functionally very important. It has been shown that allelic variation (i.e., genetic variations resulting from mutation) resulting in changes in the number and properties of HMW subunits is also associated with variation in breadmaking quality and may be responsible for the elastic properties of doughs (37). Proton transverse relaxation measurements have been undertaken on the dry and D_2O-hydrated subunit and, like the gluten data, suggest the coexistence of rigid and mobile domains within the protein (38, 39). These results, in conjunction with FTIR studies, which show increases in β-sheet in HMW subunits on hydration, have given rise to a "loop and train" theory (39) that may partially explain the origins of elasticity in wheat flour doughs.

The NMR solid-state spectra of the total gliadin fraction of wheat (40, 41) and, more recently, of the ω-gliadin fraction (42) have been reported. Sidechain motions (methyl and amino group rotation, proline ring puckering) were found to act as relaxation sinks for the proton longitudinal relaxation, and this was unaffected by the glass transition, indicating that these motions persist in the glassy state. Magic-angle spinning experiments have been used to observe line narrowing in the proton and carbon cross-polarization spectra. In the proton spectra, at high hydration levels, backbone and sidechain NH groups are observed, indicating that whole segments of the protein chain are in the mobile regime. At the same hydration level, the carbon spectra are characterized by a loss of the proline Cδ signal, showing that this is motionally mobile and involved in the hydration process. A model was proposed involving the formation of mobile loops together with more rigid regions of strong interchain interaction.

2. NMR Studies of Barley Proteins

Barley is widely used as an animal feedstock and also, of course, in the brewing industry. The main barley storage protein, C-hordein, is structurally homologous to the ω-gliadins in what. This, together with the fact that it is relatively easy to purify in a homogenous form makes it an ideal model for studying ω-glidian-type proteins. The repetitive primary structure of C-hordein results in a simple solution-state NMR spectrum, which allows the assignment of the majority of the resonances to five residues (43). These residues, with the exception of the aromatic signals, are also present in the solid-state spectrum, and in both states the proline residues are found in the trans configuration, suggesting a β-turn-rich structure.

The effects of hydration on the domain dynamics in C-hordein have been investigated with solid-state NMR (44). In particular, a comparison of the ^{13}C CPMAS and DD-MAS spectra of dry and hydrated C-hordein (45) suggests a gel-type molecular structure in which the more rigid part of the system involves intermolecular hydrogen-bonded Gln sidechains as well as some hydrophobic "pockets" involving Pro and Phe residues. The liquidlike domain is characterized by considerable backbone and sidechain motion as well as rapid ring-puckering motion in Pro residues. Hydration results in swelling and disappearance of the Phe residue signals as they acquire flip-flop motion.

3. NMR Studies of Sorgum Proteins

The kaffirins, which are the main prolamins of sorghum, are characterized by a more regular structure than wheat proteins and have a high degree of hydrophobicity, which no doubt accounts for their much lower level of digestibility, both in the gut and in in vitro enzymatic studies, compared to other cereal proteins. Proton transverse relaxation measurements on D_2O-hydrated total kaffirin show that mobility increases with increasing hydration (46). However, a high percentage of the relaxation can be described by a fast-decaying Gaussion component, even at high levels of hydration. This is significantly different to the behavior of the highly hydrated HMW subunits in wheat, where the transverse relaxation was mostly exponential in nature, and suggests that, compared to the HMW subunits, kaffirins retain a high degree of structure. Interestingly, deuterium NMR measurements of water sorption by kaffirins (47) were similar to those mentioned previously for wheat gluten and suggest that the kaffirins are actually hydrophilic in nature, in contrast to the widely held belief that they are hydrophobic. Significant changes have been observed in the CPMAS spectrum of kaffirins on thermal denaturation, which points to a relationship between secondary structure and the decrease in protein digestibility of the heated kaffirin.

4. NMR Studies of Maize Proteins

Although the biochemistry of the major prolamins of maize (the zeins) has been studied extensively, there has been relatively little work on their physical properties or structure–function relationships. There are much smaller changes in the CPMAS spectrum of maize zeins on thermal denaturation compared to the kaffirins, indicating less disruption of the secondary structure, and this appears to correlate with the smaller loss of digestibility of zeins on heating. However, the relationship between secondary structure and digestibility in the cereal proteins has yet to be fully understood.

VII. NMR STUDIES OF QUALITY FACTORS IN PROCESSED CEREAL PRODUCTS

The staling of processed cereal products such as bread, biscuits, and cakes during storage is associated with the recrystallization (or "retrogradation") of starch or, more precisely, amylopectin chains, into A- or B-type crystalline polymorphs from its initial gelatinized state obtained by baking. The spatial distribution of retrogradation rates can be followed by MRI and related to the changing moisture distribution. Moisture maps (actually maps of the initial water proton magnetization, $M(0)$, for freshly baked biscuit "sweetrolls" and after 1, 3, and 5 days of storage have been reported by Ruan et al. (48). These showed that staling was associated with an apparent migration of moisture from the inside to the outside of the sweetroll. Surprisingly, the corresponding T_2 maps showed a longer initial T_2 in the drier, outer region of the sweetmeal and a shorter T_2 in the middle region, which gradually increased during storage. The reasons for these storage changes is still a matter of speculation. One possibility is that retrogradation proceeds faster in the center because of the higher initial moisture content and results in an increase of more mobile water with a longer T_2, which migrates to the outside of the sweetroll. Whatever the correct interpretation, it is clear that MRI has an important role to play in imaging the spatial changes associated with staling cereal products.

At a molecular level, retrogradation corresponds to an increased crystallization of the amylopectin into either A or B crystallites (depending on the water content and temperature), which causes the textural change associated with staling and results in decreased biopolymer proton mobility. The molecular factors controlling the retrogradation rates can therefore be studied with high-power NMR relaxometry.

In waxy maize starch extrudate consisting mainly of gelled amylopectin containing 35% water, both the relaxation rate for the solid, nonexchanging proton component of the amylopectin, obtained from a Gaussian analysis of the FID, and the single exponential water proton transverse relaxation rate obtained from the CPMG decay increase with storage time as the amylopectin crystallizes. These changes, which are shown in Figure 6, can be modeled with the Avrami equation, which has the form

$$R(t) = R_{max} - (R_{max} - R_0) \exp[-(Gt)^n] \tag{3}$$

Here G is the rate of retrogradation and the index n typically assumes values between 1 and 4. R_{max} and R_0 are the asymptotic relaxation rates at infinite and zero storage times, respectively. The resulting rates of crystallization, G, can be

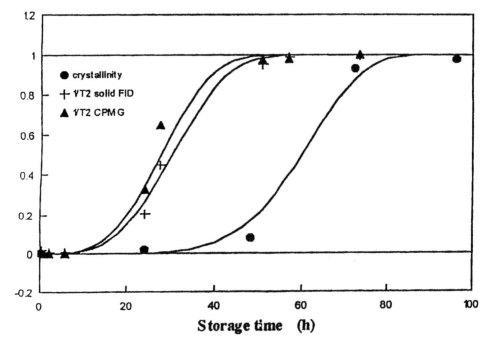

Figure 6 Effect of storage time on the CPMG water-exchangeable proton transverse relaxation rate (solid wedges); the CH proton relaxation time measured from the FID (pluses); and the crystallinity index measured by X-ray diffraction for an amylopectin extrudate containing 35% water. Only relative increases are shown, and the lines are the fits of the Avrami equation. (From Ref. 49.)

analyzed with glass–rubber transition theory. At, or below, the glass transition temperature, T_G, G is expected to be zero, because the extremely low chain mobility prevents growth of the amylopectin crystallites. The relationship between G and $(T - T_g)$ is expected to follow a modified version of the empirical Williams–Landel–Ferry (WLF) equation, which relates kinetic properties to $(T - T_g)$. Applied to retrogradation, the modified equation can be written

$$\log_{10}\left(\frac{G_{\text{ref}}}{G}\right) = \frac{-C_1(\Delta T - \Delta T_{\text{ref}})}{C_2 + (\Delta T - \Delta T_{\text{ref}})} \tag{4}$$

Here $\Delta T = (T - T_g)$ and ΔT_{ref} is $(T - T_{\text{ref}})$, where T_{ref} is an arbitrary reference temperature. The log-linear plot predicted by Eq. (4) has been confirmed by Farhat et al. (49). Because the glass transition temperature is itself a function of water content and sugar content, the rate of retrogradation, G, will also depend on the system composition, increasing with increasing water content because of the reduction in T_g. More quantitatively, the compositional dependence of the glass transition temperature can be estimated by equations such as that of ten Brinke (56) or modifications of it. The ten Brinke equation has the form

$$T_g = \frac{\Sigma_i W_i \, \Delta C_{pi} T_{gi}}{\Sigma_i W_i \, \Delta C_{pi}} \tag{15}$$

where W_i is the weight fraction of component i and ΔC_{pi} is the difference in the specific heat capacity between the liquid and glassy states at T_g. Equation (5) shows that the introduction of lower-molecular-weight species, such as water and sugars, with low intrinsic T_g values results in a lowering of the overall T_g for the mixture. This, in turn, leads to an increased rate of retrogradation via the WLF equation.

The ten Brinke equation is only a first approximation in most food systems, because it assumes that there is no selective partitioning of water between biopolymer or sugar molecules. In fact, the sorption isotherm for the amylopectin and sugar shows that this assumption is not strictly valid, with water preferably hydrating the biopolymer at low water contents and the sugar at high water contents. The foregoing analysis also assumes that the rate-limiting step in retrogradation is the growth of amylopectin crystalline regions within the biopolymer network and not the rate of crystal nucleation, which decreases up to the amylopectin melting point. If nucleation is rate limiting, the rate of retrogradation will be expected to decrease with increasing temperature, which is sometimes observed in cake products. Despite these complexities it is clear that glass–rubber transition theory offers exciting possibilities for the prediction of the effect of water content, sugar content, and temperature on the rate of staling of baked products and also clear that NMR and MRI provide the means for a detailed investigation of the space and time changes in molecular mobility associated with the staling process.

The application of ^{1}H, ^{2}H, and ^{17}O NMR to the even more complex problem of the staling of bread has been reviewed by Chinachoti (52). Both the amplitude and the relaxation rate of the rigid component in the proton free induction decay of white bread increased over a 10-day storage period, consistent with starch retrogradation and an increase in rigidity of the gluten network, though these contributions have yet to be distinguished. Somewhat surprisingly, the broadband deuterium spectrum of aging bread containing about 50% moisture showed no developing solid-state spectrum (a Pake pattern), indicative of oriented "icelike" water. This negative observation shows the high mobility of the water even when it is known that the biopolymer components are becoming more rigid. Clearly much remains to be understood about the molecular mechanisms of bread staling, and this remains an outstanding challenge for the future.

VIII. CONCLUSION

It is clear that NMR and MRI have important roles to play in optimizing all stages of cereal production, from the field to the processed food product, and that, with a few exceptions, the information emerging from NMR is of a fundamental, nonroutine nature. The fundamental nature of the information is not a disadvantage, because understanding structure–function relationships in foods over all distance scales, from the molecular to the macroscopic, is surely one of the outstanding challenges in food science. An example serves to illustrate this point.

It is now possible to genetically engineer wheat, corn, and potato to modify starch granule structure at the molecular level by altering the activities of the enzymes responsible for starch granule synthesis. Yet our understanding of starch structure–function relationships is still too primitive to predict the effect of this molecular-level engineering on the subsequent response of the granule to processing operations, such as thermal gelatinization. It is also too primitive to predict the properties of the resulting starch system, such as the rheology of the genetically modified starch gel, and it is here that the true power of NMR techniques is apparent. As we have seen, the changing dynamic state of the starch chains in the granules during processing can be monitored with techniques such as proton relaxometry and high-resolution solid-state spectroscopy (CPMAS). The changing microscopic distribution of water inside the granule can also be monitored with multinuclear relaxometry and diffusometry. Water migration during the starch processing stage can be monitored with NMR microimaging, while MRI rheology can also be used to characterize the rheological state of the starch–water system. In addition, storage changes, such as starch retrogradation, can be characterized with relaxation-weighted imaging.

Even this impressive array of NMR techniques hardly serves to scratch the surface of the large number of NMR techniques and pulse sequences that can be brought to bear on the problem of starch and starch processing. In addition it should be remembered that it is not always the NMR data themselves that are of greatest value, but rather the mathematical models of structure–function and heat and mass processing that are developed from the data and that can subsequently be used to optimize the production stages. The development and testing of these models must therefore be added to the list of outstanding future challenges in cereal science.

REFERENCES

1. D.M. Grant and R.K. Harris (eds.). Encyclopedia of NMR. New York: Wiley, 1996.
2. M.J. McCarthy. Magnetic Resonance Imaging in Foods. New York: Chapman and Hall, 1994.
3. B.P. Hills. Magnetic Resonance Imaging in Food Science. New York: Wiley, 1998.
4. L. Meulenkamp, F. Vergeldt, P.A. de Jager, D. van Dusschoten, and H. van As. Quantitative NMR microscopy, chlorophyl fluorescence and water uptake rate measurements to study water-stress tolerance in plants. Conference abstract: Towards whole plant functional imaging, Amsterdam, 1996.
5. G.P. Cofer, J.M. Brown, and G.A. Johnson. J. Magn. Reson. 83:603, 1989.
6. P.A. Bottomley, H.H. Rogers, and T.H. Foster. NMR imaging shows water distribution and transport in plant root systems in situ. Proc. Natl. Acad. Sci. USA 83:87, 1986.
7. K. Omasa, M. Onoe, and H. Yamada. NMR imaging for measuring root system and soil water content. Environmental Control Biol. 23:99, 1985.
8. J.S. MacFall, P.J. Kramer, and G.A. Johnson. New phytology 119:551, 1991.
9. C. Jenner, Y. Xia, and C. Eccles. Circulation of water within wheat grain revealed by NMR microimaging. Nature 336:399–402, 1988.
10. P.N. Gambhir. Applications of low resolution pulsed NMR to the determination of oil and moisture in oilseeds. Trends Food Sci. Technol. 3:191–196, 1992.
11. E. Brosio, F. Conti, A. DiNola, O. Scorano, and F. Balestrieri. Simultaneous determination of oil and water content in olive husk by pulsed low resolution NMR. J. Food Technol. 16:629–636, 1981.
12. D.S. Himmelsbach. Proton NMR imaging of the hydration of wheat and rice grains J. Magn. Reson. Analysis 2:163–164, 1996.
13. A.G.F. Shapley, T.M. Hyde, L.F. Gladden, and P.J. Fryer. NMR imaging of diffusion and reaction in wheat grains. J. Magn. Reson. Analysis 2:14, 1996.
14. J.R. Heil, M. Ozilgen, and M.J. McCarthy. MRI analysis of water migration and void formation in baking biscuits. In: L.G. Elmer, ed. AIChE Symposium Series, No. 297, 1993, pp 39–45.
15. S.L. Duce, S. Ablett, A.H. Darke, J. Pickles, C. Hart, and L.D. Hall. NMR imaging

and spectroscopic studies of wheat flake biscuits during baking. Cereal Chemistry 72:105–108, 1995.

16. K.L. McCarthy, R.J. Kauten, and C.K. Agemura. Application of NMR imaging to the study of velocity profiles during extrusion processing. Trends Food Sci. Technol. 3:215–219, 1992.

17. M.M. Britton and P.T. Challaghan. NMR microscopy and the non-linear rheology of food materials. Magnetic Resonance Chemistry 35:S37–S47, 1997.

18. B.P. Hills, J. Godward, and K.M. Wright. Fast radial microimaging studies of pasta drying. J. Food Engineering 33:321–335, 1997.

19. B.P. Hills, F. Babonneau, V.M. Quantin, F. Gaudet, and P.S. Belton. Radial NMR microimaging studies of the rehydration of extruded pasta. J. Food Engineering 27: 71–86, 1996.

20. B.P. Hills, J. Godward, K.M. Wright, and M. Harrison. Fast radial NMR microimaging studies of the freezing of extruded pasta. Appl. Magn. Reson. 12:529–542, 1997.

21. P.J. Frazier, P. Richmond, and A.M. Donald (eds.). Starch, Structure and Function. Cambridge: Royal Society of Chemistry, 1997.

22. L.A. Farhat, J.R. Mitchell, J.M.V. Blanshard, and W. Derbyshire. A pulsed ^{1}H NMR study of the hydration properties of extruded maize–sucrose mixtures. Carbohydrate Polymers 30:219–227, 1996.

23. P. Chinachoti, M.S. Kim-Shin, F. Mari, and L.Lo. Gelatinization of wheat starch in the presence of sucrose and sodium chloride: correlation between gelatinization temperature and water mobility as determined by oxygen-17 NMR. Cereal Chem. 68:245–248, 1991.

24. P.R. Shewry and A.S. Tatham. Recent advances in our understanding of cereal seed protein structure and functionality. Comments Agric. Food Chem. 1:71–94, 1987.

25. P.R. Shewry, J.M. Field, and A.S. Tatham. The structures of Cereal Seed Storage Proteins. In: ID Morton, ed. Cereals in a European Context. 1st Eur. Conf. on Food Sci. and Technol. Ellis Horwood Ltd, Chicester, U.K., 1987, pp 421–437.

26. F. MacRitchie. Baking quality of wheat flours. Advances Food Res. 29:201–277, 1984.

27. A.S. Tatham, P.S. Shewry, and P.S. Belton. Structural studies of cereal prolamins, including wheat gluten. In: Y Pomeranz, ed. Advances in Cereal Science and Technology. Vol. X, Ch. 1. St. Paul: AACC, 1990, pp 1–78.

28. P.S. Belton. S.L. Duce, and A.S. Tatham. Proton NMR Relaxation Studies of Dry Gluten. J. Cer. Sci. 7:113–122, 1988.

29. S. Ablett, D.J. Barnes, A.P. Davies, S.J. Ingman, and D.W. Patient. ^{13}C and pulse nuclear magnetic resonance spectroscopy of wheat proteins. J. Cer. Sci. 7:11–20, 1988.

30. P.S. Belton, S.L. Duce, I.J. Colquhoun, and A.S. Tatham. High-Power ^{13}C and ^{1}H nuclear magnetic resonance in dry gluten. Mag. Res. Chem. 26:245–251, 1988.

31. G. Cherien and P. Chinachoti. ^{2}H and ^{17}O NMR study of water in gluten in the glassy and rubbery state. Cereal Chem. 73:618–624, 1996.

32. S. Li, C. Dickinson, and P. Chinachoti. Proton relaxation of starch and gluten by solid-state nuclear magnetic resonance spectroscopy. Cer. Chem. 736:736–743, 1996.

33. K.M. Magnuson. Uses and functionality of vital wheat gluten. Cereal Foods World 30:179–181, 1985.

34. G. Cherian and P. Chinachoti. Action of oxidants on water sorption, ^{2}H nuclear magnetic resonance mobility and glass transition behavior of gluten. Cer. Chem. 743:312–317, 1997.

35. A. Grant, P.S. Belton, I.J. Colquhoun, M.L. Parker, J. Plijter, P.R. Shewry, A.S. Tatham, and N. Wellner. The effects of temperature on the sorption of water by wheat gluten studied by deuterium NMR. Cereal Chem. 76(2):219–226, 1999.

36. D. Marion, C. Le Roux, S. Akoka, C. Tellier, and D. Gallant. Lipid–protein interactions in wheat gluten. A phosphorous nuclear magnetic resonance spectroscopy and freeze-fracture electron microscopy study. J. Cer. Sci. 5:101–115, 1987.

37. P.I. Payne, K.G. Corfield, L.M. Holt, and J.A. Blackman. Correlations between the inheritance of certain high-molecular-weight subunits of glutenin and bread-making quality in progenies of six crosses of bread wheat. J. Sci. Food Agric. 32:51–60, 1981.

38. P.S. Belton, I.J. Colquhoun, J.M. Field, A. Grant, P.R. Shewry, and A.S. Tatham. ^{1}H and ^{2}H NMR relaxation studies of a high M_r subunit of wheat glutenin and comparison with elastin. J. Cer. Sci. 19:115–121, 1994.

39. P.S. Belton, I.J. Colquhoun, A. Grant, N. Wellner, J.M. Field, P.R. Shewry, and A.S. Tatham. FTIR and NMR Studies on the hydration of a high-M_r subunit of glutenin. Int. J. Biol. Macromol. 2:74–80, 1995.

40. I.C. Baianu, L.F. Johnson, and D.C. Waddel. High-resolution proton, ^{13}C and ^{15}N NMR studies of wheat proteins at high field: spectral assignments, changes in conformation with heat treatment of flinor gliadins at solution, comparison with gluten spectra. J. Sci. Food Agric. 33:373, 1982.

41. J.D. Scholfield and I.C. Baianu. Solid-state, cross-polarization magic-angle spinning carbon-13 nuclear magnetic resonance and biochemical characterization of wheat proteins. Cer. Chem. VI 59:240–245, 1982.

42. P.S. Belton, A.M. Gil, A. Grant, E. Alberti, and A.S. Tatham. Proton and carbon NMR measurements of the effects of hydration on the wheat protein ω-glaidin. Spectrochimica Acta A 54:955–966, 1998.

43. A.S. Tatham, P.R. Shewry, and P.S. Belton. ^{13}C-NMR study of C. Hordein. Biochem. J. 232:617–620, 1985.

44. P.S. Belton, A.M. Gil, and A.S. Tatham. Proton nuclear magnetic resonance lineshapes and transverse relaxation in a hydrated barley protein. J. Chem. Soc. Faraday. Trans. 90:1099–1103, 1994.

45. A.M. Gil, K. Masui, A. Naito, A.S. Tatham, P.S. Belton, and H. Saito. A ^{13}C-NMR study on the conformational and dynamic properties of a cereal seed storage protein, C-Hordein, and its model peptides. Biopolymers 41:289–300, 1997.

46. A Grant. Unpublished results.

47. NMR and FTIR Studies of Cereal Proteins. 7th European Conf. on Spectroscopy of Biol. Molecules (ECSBM) 7–12 Sept., 1997, San. Lorenzo de al Escarial, Madrid, Spain, Spectroscopy of Biological Molecules, Modern Trends, Kluwer Academic, 1997, pp 494–500.

48. R. Ruan, S. Almaer, V.T. Huang, P. Perkins, P. Chen, and R.G. Fulcher. Relationship

between firming and water mobility in starch-based food systems during storage. Cereal Chem. 73:328–332, 1996.

49. I.A. Farhat. Molecular mobility and interactions in biopolymer–sugar–water systems. Ph.D. dissertation, University of Nottingham, Sutton Bonnington Campus, 1996.

50. I. Hopkinson and R.A.L. Jones. Abstract at the British Radiofrequency Spectroscopy Group meeting at Surrey, April, 1997.

51. M.C. Vackier, B.P. Hills, and D.N. Rutledge. An NMR relaxation study of the state of water in gelatine gels. J. Magnetic Resonance 138:36–42, 1999.

52. P. Chinachoti. NMR dynamic properties of water in relation to thermal characteristics of bread. In: D.S. Reid, ed. The Properties of Water in Foods. Blackie Academic and Professional, New York: 1998, Ch 6.

53. B.P. Hills, J. Godward, C.E. Manning, J.L. Biechlin, and K.M. Wright. Microstructural characterization of starch systems by NMR relaxation and Q-space microscopy. Magnetic Resonance Imaging, Magnetic Resonance Imaging 16:557–564, 1998.

54. H.-R. Tang, J. Godward, and B. Hills. The distribution of water in native starch granules—a multinuclear NMR study. Carbohydrate Polymers 43:375–387, 2000.

55. H.R. Tang, A. Brun, and B. Hills. A proton NMR relaxation study of the gelatinisation and acid hydrolysis of native potato starch. Carbohydrate Polymers 46:7–18, 2001.

56. G. ten Brinke, Z. Karasz, and T.S. Ellis. Macromolecules 16:224, 1983.

14

Phosphorescence Spectroscopy as a Probe of the Glassy State in Amorphous Solids

Richard D. Ludescher
Rutgers University, New Brunswick, New Jersey, U.S.A.

I. INTRODUCTION

A. Physical Characterization of Foods

Biological tissues of either animal or vegetable origin and especially the foods derived from them are heterogeneous amorphous solids; these characteristics complicate any detailed description of their molecular structure and dynamics. Although biological tissues are well organized, they rarely display the regularity required for detailed structural analysis using X-ray diffraction, for example. In addition, their transformation into food often involves significant disruption and reorganization of this biological structure (compare a soufflé with an egg). Food chemists and engineers must therefore develop methods to detect and measure the many levels of structure found in foods, characterize the physical state of the biomolecules at each organizational level, and understand the role that each level of organization plays in the generation and maintenance of food quality.

Amorphous solids exhibit many of the usual physical characteristics of more easily studied crystalline solids. They are dense, elastic materials that resist flow and support their own weight. They differ from crystals, however, in lacking a regular, periodic, long-range structure. This lack of crystalline regularity complicates molecular characterization. Amorphous solids also display more complex

Research supported by the Cooperative State Research, Education, and Extension Service of the U.S. Department of Agriculture (grant #9502626) and the New Jersey Agricultural Experiment Station.

dynamic behavior than crystals, undergoing thermal transitions in which the molecules or specific groups within the molecules experience cooperative increases in the rate and/or amplitude of vibrational, rotational, or translational motion (1). Although the glass transition (at temperature T_g) seen in sugar melts and in both synthetic and biological polymers is the best-known example of a solid-state transition, one that reflects the onset temperature for translational motion in the amorphous solid (2), other, more localized transitions occur in many amorphous solids below T_g (3).

Foods are also heterogeneous in both composition and structure. They are composed of many, even hundreds to thousands, of different biological molecules. Although proteins, carbohydrates, and lipids comprise the major classes, even such seemingly homogeneous foods as butter contain many different molecules that contribute to odor and flavor. Typically, these molecules are organized into spatial domains on many different length scales. Butter, for example, is an emulsion in which water droplets, fat crystals, and air bubbles with diameters ranging from 10^{-6} to 10^{-4} m are dispersed in a continuous phase of liquid milk fat, while in cereal grains, storage proteins are organized into protein bodies with diameters of about 10^{-6} m and storage granules for starches have diameters nearly 10-fold larger.

This complexity thus presents great difficulties to food chemists and engineers for the characterization of foods on the molecular level. Such characterization requires the ability to monitor the properties of individual molecules or classes of molecules in spatially discrete locations within the food. This is a task for which luminescence spectroscopy provides a number of useful advantages (4–8). First, the technique involves the use of optical probes, specific molecules with well-characterized spectroscopic properties (4, 9–12). These probes may be either intrinsic to a specific biomolecule (tryptophan in proteins), covalently attached to a class of biomolecules (fluorescein isothiocyanate reacted with free amino groups of proteins), or physically dispersed throughout a class of biomolecules (diphenyl hexatriene in the lipid phase). The use of molecular probes allows one to study specific molecules or classes of molecules within a complex heterogeneous sample. Second, luminescence is sensitive to a number of physical and chemical properties of the matrix in which the probe is embedded (9, 11). These include, but are not limited to, polarity, pH, specific ion concentration (Ca^{2+}, Mg^{2+}), viscosity, and molecular mobility. It is thus possible to monitor these properties (and others) at specific sites within the food. Third, luminescence is routinely detectable at probe concentrations of a few parts per billion since the measurement involves detection of emitted photons against a potentially zero-level background (although in practice background luminescence is often a significant problem). Measurements of the spectroscopic properties of single molecules are not only possible but also actually routine in some laboratories. Fourth, a large number of luminescent molecules have been synthesized and characterized spectroscopically; many are commercially available (12). It is thus possible to select a chromophore with the spectroscopic and chemical properties appropriate for specific applications. Fifth, luminescence can involve either fluorescence from the excited singlet state of the chromophore (4) or phosphorescence from the excited triplet state (13–15). The spectroscopic properties of these two types of emission differ in ways that offer multiple experimental applications. The primary difference is that fluorescence occurs on the time scale of 10^{-9}–10^{-8} s, while phosphorescence occurs on time scales from about 10^{-5} to 10 s. The long lifetime of phosphorescence makes this technique sensitive to events occurring on the very long time scales characteristic of the solid state.

B. Dynamical Transitions in Amorphous Solids

Amorphous solids undergo transitions that reflect thermally activated, cooperative changes in the dynamics of the molecules (1–3). The most familiar and best-studied transition involves the change from a rigid, brittle, glassy state to a soft, flexible, rubbery state. This glass transition (with its corresponding glass transition temperature T_g) is fundamentally different from the well-known solid–liquid phase transition or the more rare solid–solid phase transitions found in some crystalline materials (triglycerides, for example). Phase transitions are order–disorder transitions that involve a change in the molecular structure of the material, i.e., either the melting of the crystal to generate an amorphous liquid or a change in the pattern of order in the crystal. A glass transition in an amorphous solid, on the other hand, is not an order–disorder transition and does not involve any change in the structure of the material (2); the material is amorphous and liquidlike (randomly oriented) in both the glassy and rubbery states. The transition, however, does involve a dramatic change in the molecular dynamics of the material. In the glassy state at $T_g - 10$ K, for example, the material may have a viscosity in excess of 10^{14} Pa s, while in the rubbery state at $T_g + 10$ K the viscosity may be less than 10^{10} Pa s (1, 16). This dramatic decrease in viscosity ($\sim 10^3$ Pa s K^{-1}) results from a cooperative change in the molecular mobility of the individual molecules. Simply put, the molecules are not capable of translational motion in the glassy state but are in the rubbery state; T_g is thus the onset temperature for translational motion (2). In sugars and small oligosaccharides, this translational motion directly reflects the motion of individual molecules; within a polymer, however, the glass transition is thought to reflect the activation of the cooperative motion of sections of the polymer backbone containing perhaps 50 atoms (1).

Nearly all materials are capable of forming amorphous glassy states (2). The primary criterion for formation of the glassy state is the frustration of crystal nucleation and growth. This can be achieved through steric constraints (common in polymers) or through an increase in viscosity in the melt that prevents crystallization within the cooling time. (Some materials, however, form glasses only under extreme conditions; formation of metallic glasses, for example, requires cooling rates in excess of 10^6 K s^{-1}.) Many biomolecules readily form amorphous solids; these include simple sugars, oligosaccharides and carbohydrates, and proteins. Synthetic polymers also readily form amorphous solids; even a simple polymer, such as polyethylene, which readily forms crystalline phases, has an appreciable fraction of amorphous-state polymer (low-density and high-density polyethylene differ in the content of amorphous and crystalline polymer).

Amorphous synthetic polymers also undergo solid–solid dynamical transitions in the glassy state corresponding to the onset of specific relaxation processes at temperatures below T_g (1, 3). These transitions reflect localized cooperative increases in the motion of specific groups or segments of the polymer. A summary of the low-temperature transitions of amorphous polystyrene (PS) and poly(methyl methacrylate) (PMMA) will illustrate this complexity in the dynamics of amorphous polymers (Table 1; Refs. 1 and 3). The structure of PS and PMMA are shown in Figure 1. The glass transition of PS is at ~373 K; three distinct dynamic transitions are seen in PS at $T < T_g$. These transitions are thought to reflect the onset of constrained oscillations of the phenyl sidechain ($T_\delta \approx 40$ K), the onset of local cooperative motions of localized (four-carbon) regions of the backbone ($T_\gamma = 130$ K), and the onset of unlimited rotational motions in the phenyl sidechain ($T_\beta = 325$ K). And PMMA also exhibits complex dynamical behavior below the glass transition ($T_g = 385$ K). These transitions are thought to reflect the onset of independent rotation of the α-methyl ($T_\gamma = 153$–173 K)

Table 1 Solid-State Transitions in Synthetic Polymers

Transition	Polystyrene		Poly(methyl methacrylate)	
	Temperature (K)	Suggested mechanism	Temperature (K)	Suggested mechanism
T_α	373	Long-range cooperative chain motions	385	Long-range cooperative chain motions
$T_{\alpha'}$			318–348	Local mode relaxation of main chain
T_β	325 ± 4	Torsional vibrations of phenyl rings	243–263	Rotation of ester group
T_γ	130	Cooperative motions of 4-carbon units	153–173	Rotation of α-methyl group
T_δ	38–48	Phenyl ring oscillations		

Source: Refs. 1, 3, 35–37.

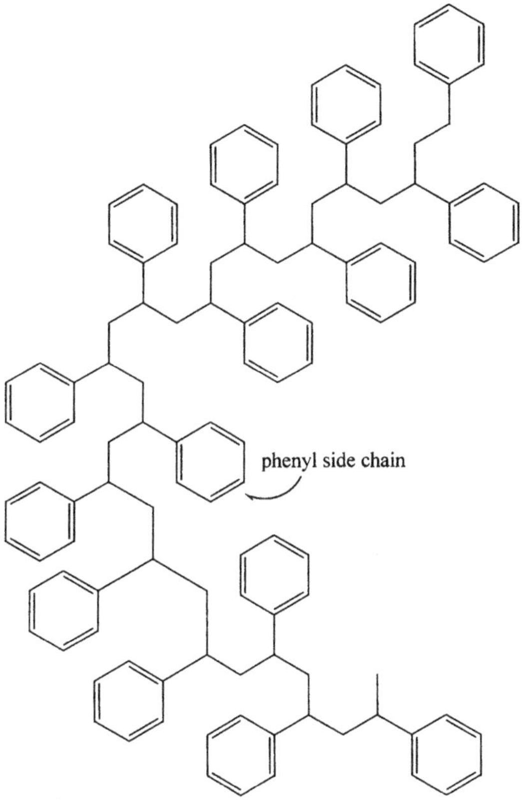

phenyl side chain

(A)

Figure 1 Chemical structures of (A) polystyrene and (B) poly(methyl methacrylate) with the sidechain groups identified that are thought to be involved in glassy-state dynamical transitions (see Table 1).

and ester groups (T_β = 243–263 K) as well as a localized cooperative relaxation of the main chain (T_α = 318–348 K). It is a general characteristic of solid-state polymer transitions that they are often difficult to detect and may only be observable in certain physical measurements (may be detectable by mechanical but not spectroscopic or thermal methods, for example).

C. Intent of This Review

This review is an introduction to the use of phosphorescence spectroscopy to monitor the molecular mobility and dynamical transitions of amorphous solids. It is not meant to be a comprehensive review of the literature of either phosphores-

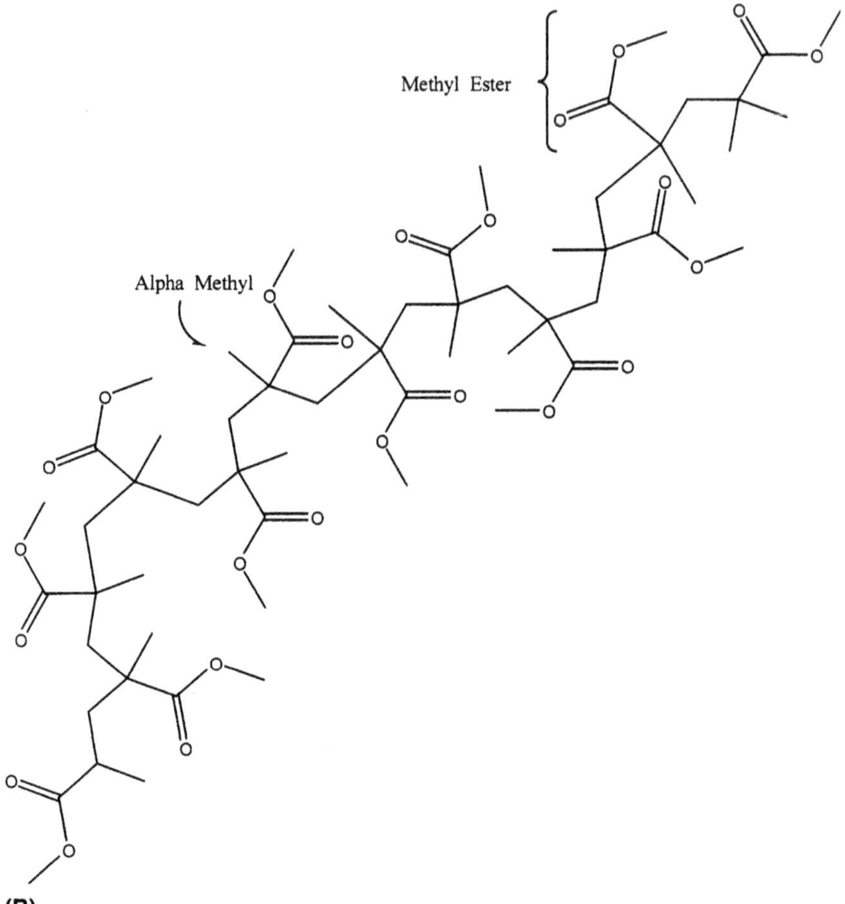

Methyl Ester

Alpha Methyl

(B)

Figure 1 Continued

cence or the measurement of the molecular mobility of amorphous solids. It is meant, however, to provide a theoretical framework for understanding how the phosphorescence emission, intensity and decay kinetics, and polarization from specific molecules are sensitive to the chemical and physical properties of the local environment, to illustrate this sensitivity with specific examples from the literature and from the author's research on sugars and proteins, and to suggest directions for future research. The success of this attempt is left for the reader to decide.

II. SPECTROSCOPIC BACKGROUND

A. Luminescence Background

All molecules absorb light; only the wavelength of the absorbed light varies. Molecules differ, however, in how this absorbed energy is dissipated; in most cases the energy is converted to heat, while in others the energy is emitted as a photon of lower energy. In order to understand how certain molecules can be used to probe molecular structure and dynamics, we must first of all analyze how the energy of a photon is absorbed and then dissipated within a molecule following excitation (for more detailed discussion of luminescence, see Refs. 4–8, 13–15, and 17–18).

The fundamental visual aid in such an analysis is an energy-level, or Jablonski, diagram (Fig. 2). In this diagram the vertical dimension represents increasing energy and the various stable states of the molecule are indicated as horizontal energy levels (the horizontal dimension does not reflect any physical property of the molecule). The unexcited molecule is said to be in the ground, or S_0, electronic state. This electronic state has many possible substates, reflecting the many possible vibrational modes of the molecule. Since there are $3N - 6$ different vibrational modes in a nonlinear molecule with N atoms, and each vibrational mode can be excited by one or more quanta of energy, there are many different vibra-

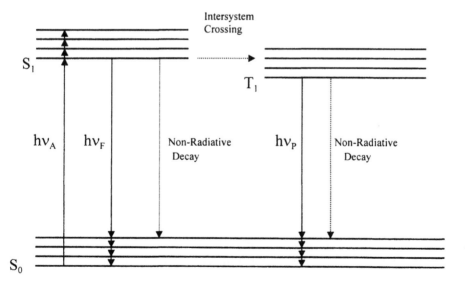

Figure 2 Jablonski energy-level diagram of the excited states of a typical luminescent molecule. See text for descriptions of the various photophysical processes illustrated.

tional states of the molecule. We can safely assume that near 20°C, electronic transitions in nearly all chromophores originate from the lowest-lying vibrational level of the ground electronic state.

Absorption of a photon results in the redistribution of electron density in the molecule; the excited molecule is in a new electronic state in which one of the electrons is in an excited (previously unoccupied) molecular orbital. This transition is represented by a vertical arrow in the Jablonski diagram; the length of the arrow is directly proportional to the energy of the absorbed photon ($h\nu_A$). Transitions to the first excited state can, however, occur at many different wavelengths because there are many different vibrational energy levels of the first excited state (S_1) and because interactions with the solvent broaden the energy of each vibrational state. The absorption band is thus broad and may display complex structure.

Following absorption of a photon, which takes approximately 10^{-15} s and is thus essentially instantaneous on the time scale of molecular vibrations and more complex chemical events, the molecule quickly ($\sim 10^{-12}$ s) relaxes to the zero vibrational level of S_1. This relaxation reflects a dissipation of the excess vibrational energy into the matrix in which the chromophore lies (either solid or liquid).

The utility of a molecule as a luminescent probe depends upon the events that occur subsequent to relaxation to the lowest vibrational level of S_1. Three processes that form the basis for understanding luminescence will be discussed here: radiative decay due to emission of a photon; nonradiative decay due to vibrational coupling to the ground state; and intersystem crossing to the lowest-lying triplet state T_1. Additional photophysical and photochemical processes (quenching due to collision with specific molecules such as oxygen, nonradiative (Forster) energy transfer, chemical reaction, etc.) will be discussed later where appropriate. We can summarize the various processes that occur during luminescence in a schematic (Fig. 3), where the arrows indicate various photophysical events (absorption, emission, nonradiative decay, quenching, etc.) and the various k_i terms indicate rate constants for the processes. These events are also represented in the Jablonski diagram.

In fluorescence the molecule emits a photon (of energy $h\nu_F$) and thus undergoes a transition directly to one of the vibrational levels of the ground state (S_0). The time scale of the fluorescence emission (the natural or intrinsic fluorescence lifetime) is typically in the range from 10^{-9} to 10^{-7} s; alternatively, emission occurs with a rate constant (k_f) in the range from 10^7 to 10^9 s^{-1} (the inverse of the lifetime). The natural lifetime of a specific electronic state is a fundamental property of the molecule. The natural lifetime should not be confused with the actual measured lifetime for decay of the excited state that reflects the sum of all possible de-excitation pathways from the excited state (as described later).

$$
\begin{array}{ccc}
 & \xrightarrow{\;k_{nr}\;} & S_0 + \text{heat} \\[2ex]
S_0 + h\nu_A \;\xrightarrow{\;I_A\;}\; S_1 & \xrightarrow{\;k_f\;} & S_0 + h\nu_F \\[2ex]
\xrightarrow{\;k_{isc}\;}\; T_1 & \xrightarrow{\;k'_{nr}\;} & S_0 + \text{heat} \\[2ex]
 & \xrightarrow{\;k_p\;} & S_0 + h\nu_P \\[2ex]
 & \xrightarrow{\;k_q[O_2]\;} & S_0 + \text{heat}
\end{array}
$$

Figure 3 Photophysical scheme for a typical luminescent molecule exhibiting both fluorescence and phosphorescence emission. The arrows indicate specific photophysical processes; the specific rate constants for each process are indicated above the arrow.

The quantum yield for fluorescence (Q_F) is the ratio of photons emitted to photons absorbed by the molecule; it is essentially the probability that an excited molecule will emit a fluorescent photon. The quantum yield for fluorescence is modulated by the rate of nonradiative decay to the ground state. Nonradiative decay is the result of coupling highly excited ground-state vibrations of the molecule to the excited electronic state; the excited electronic state (S_1) of the molecule with low vibrational energy transforms into a ground electronic state (S_0) with high vibrational energy. The process is unidirectional, because the excess vibrational energy in the ground state is rapidly dissipated due to collisions with the matrix. The rate constant for this process (k_{nr}), which is highly variable and falls in the range from about 10^6 to 10^{10} s^{-1}, is sensitive to the structure of the probe and to the chemical structure and physical state of its local environment (11, 18). The larger the rate of nonradiative decay, the lower the fluorescence quantum yield of the molecule.

In order for phosphorescence to occur, however, the excited molecule must undergo a transition from the excited singlet state S_1 to the excited triplet state T_1. The transition occurs from the lowest-lying vibrational state of S_1 to some excited vibrational state of T_1; the excited triplet-state molecule then rapidly relaxes to the zero vibrational level of T_1 (Fig. 2). This process, referred to as intersystem crossing, involves a change in the spin state of the excited electron;

it is thus a spin-forbidden process and, like other forbidden processes, occurs at a low rate (k_{isc}). The presence of heavy atoms such as bromine or iodine, however, greatly increases the rate of intersystem crossing; this is the heavy-atom effect. The larger the intersystem crossing rate, the lower the fluorescence quantum yield of the molecule. In some molecules the rate of intersystem crossing may exceed the combined rates of fluorescence and nonradiative decay; in that case the excited-state molecule undergoes intersystem crossing to populate the triplet state with a probability (triplet quantum yield Q_T) near unity.

Phosphorescence is the emission of a photon (with energy $h\nu_P$) from the triplet state T_1. This transition, which involves a transition from the lowest-lying vibrational level of the triplet state to a highly excited vibrational level of the ground singlet state, involves intersystem crossing from the triplet to the ground (S_0) state in the singlet manifold and is thus also spin forbidden. It occurs at a low rate (k_p) that can vary from about 10^6 to less than 1 s^{-1}, depending upon the molecular structure (18). As in fluorescence, nonradiative decay from the triplet to the singlet state (with rate constant k'_{nr}) competes with phosphorescence to decrease the phosphorescence quantum yield (Q_p). Unlike in the case of fluorescence, however, the presence of oxygen effectively quenches phosphorescence emission through collision with the triplet state; the rate constant for this process is the product of a rate constant and the oxygen concentration, $k_q[O_2]$.

In molecules in which the triplet state T_1 has only slightly less energy than S_1, intersystem crossing can also occur in reverse, from an excited vibrational level of T_1 to the lowest vibrational level of S_1. This process is thermally activated, since it requires that molecules be in a vibrationally excited state of T_1. The re-excited S_1 state can also emit light, but in this case the delayed fluorescence (so-called E-type fluorescence, since it was first characterized in eosin) has a lifetime equal to the lifetime of the triplet state.

B. Basic Photophysical Equations

The probability that the excited molecule in S_1 will undergo any specific process (that is, the quantum yield for that process) is given by a ratio of rate constants (7, 17, 18). The quantum yield for fluorescence (Q_F) is thus the ratio of the rate constant for emission of a photon (k_f) to the sum of the rate constants for all of the de-excitation processes that S_1 can undergo. The quantum yield for triplet-state formation (Q_T) is calculated in the same way. Within the context of the limited selection of photophysical processes discussed here, the quantum yields are expressed as follows:

$$Q_F = \frac{k_f}{k_f + k_{nr} + k_{isc}} \tag{1}$$

$$Q_T = \frac{k_{isc}}{k_f + k_{nr} + k_{isc}} \qquad (2)$$

The quantum yield for a typical fluorescent molecule varies from about 0.1 to 1 (every absorbed photon leads to an emitted photon). For those molecules that exhibit appreciable phosphorescence, the triplet quantum yield, Q_T, falls within the same range.

The quantum yield for phosphorescence (Q_p) is defined as the probability that the excited molecule in the S_1 state will emit a photon from the T_1 state. It is thus the product of the probability of triplet-state formation (Q_T) times the probability of emission from the triplet state (q_P, the quantum efficiency of phosphorescence):

$$q_P = \frac{k_p}{k_p + k'_{nr} + k_q[O_2]} \qquad (3)$$

$$Q_P = Q_T q_P$$

$$= \frac{k_{isc} k_p}{(k_f + k_{nr} + k_{isc})(k_p + k'_{nr} + k_q[O_2])} \qquad (4)$$

If other processes occur to de-excite the singlet or the triplet states, the denominator of the respective quantum yield equation is modified to include the rate constant(s) for the additional processes (collisional quenching by molecules other than oxygen or energy transfer, for example). Evaluation of these equations indicates why the presence of oxygen can effectively quench phosphorescence emission. In the presence of oxygen (which is present at about 0.3 mM concentration in aqueous solutions in equilibrium with the atmosphere), the $k_q[O_2]$ term may be several orders of magnitude larger than k_p or k'_{nr} in fluid solution.

The quantum yield is the fundamental molecular property that determines the intensity of luminescence emission. The relative magnitudes of the rate constants for the different photophysical processes thus determine the luminescence intensity of any molecule. The rate constants for emission (k_f and k_p) are determined by the structure of the molecule and, in the case of phosphorescence emission, are very sensitive to the presence of heavy atoms, increasing dramatically in the presence of bromine or iodine. The physical or chemical properties of the local environment around a chromophore modulate the luminescence intensity primarily by modulating the rate constants for nonradiative emission (k_{nr} and k'_{nr}) and, in the case of phosphorescence, by modulating the quenching rate constant and/or the oxygen concentration (11, 18).

The lifetime for fluorescence or phosphorescence is the characteristic time between absorption and emission of a photon. It is defined as the inverse of the sum of all rate constants for de-excitation of the excited state. The fluorescence

(τ_f) and phosphorescence (τ_p) lifetimes are thus given by the following equations (for the simplified photophysical scheme outlined in Fig. 3):

$$\tau_f = \frac{1}{k_f + k_{nr} + k_{isc}} \tag{5}$$

$$\tau_p = \frac{1}{k_p + k'_{nr} + k_q[O_2]} \tag{6}$$

The lifetimes are important because they determine the time course of the emission [$I(t)$] from the excited state:

$$I(t) = I(0) \exp\left(\frac{-t}{\tau}\right) \tag{7}$$

[where $I(0)$ is the intensity at time $t = 0$ and the lifetime τ is for either fluorescence or phosphorescence]. We can define the natural lifetime (τ^o) of fluorescence and phosphorescence as the inverse of the radiative rate constant for that state:

$$\tau_f^o = \frac{1}{k_f} \tag{8}$$

$$\tau_p^o = \frac{1}{k_p} \tag{9}$$

With these definitions we can redefine the quantum yields in terms of the lifetimes:

$$Q_F = \frac{\tau_f}{\tau_f^o} \tag{10}$$

$$Q_P = \frac{Q_T \tau_p}{\tau_p^o} \tag{11}$$

The fluorescence quantum yield, which determines the intensity of the fluorescence emission, is thus directly proportional to the fluorescence lifetime. The phosphorescence quantum yield, although directly proportional to the phosphorescence lifetime, is also a function of the triplet quantum yield; increases in Q_p often occur concomitantly with decreases in τ_p because any increase in intersystem crossing rate will affect both k_{isc} (to increase Q_T) and k_p (to decrease τ_p).

C. Sensitivity of Luminescence to Environment

Luminescence emission is sensitive to the chemical structure, physical state, and molecular mobility of the matrix in which the molecules are embedded (10, 11,

18). This sensitivity makes luminescence an excellent technique to monitor the structure and dynamics of biomolecules. Three aspects of luminescence emission can provide specific molecular information about the local environment around the probe: the energy (wavelength) distribution, the intensity, and the polarization of the emission (19).

1. Emission Energy (Wavelength)

The energy (wavelength) distribution of the phosphorescence emission reflects several factors: the intrinsic energy difference between the S_0 and T_1 states of a chromophore, the nature of the manifold of vibrational states available to the ground-state molecule, and the extent of any dipolar (or, more generally, dielectric) interactions that can occur between the chromophore and the amorphous matrix in which it is embedded. The effect of these dipolar interactions is mediated by the rate of matrix dipolar relaxation around the newly excited singlet and triplet states; this sensitivity can be exploited to monitor the occurrence, rate, and extent of matrix dipolar relaxation occurring during the excited-state lifetime of the chromophore (4, 20, 21).

Prior to absorption of a photon, a solvent shell surrounds the chromophore with an equilibrium configuration optimized to stabilize the ground-state electronic distribution of the chromophore (Fig. 4). Absorption of a photon promotes an electron into a higher molecular orbital, generating a new electronic distribution in the chromophore; in many molecules the resulting excited state is significantly more polar than the ground state. Since electronic excitation is essentially instantaneous, even on the time scale of fast molecular motion (10^{-15} s compared to 10^{-12} s), the newly formed excited singlet state is initially surrounded by a solvent shell that is not in energetic equilibrium with the chromophore. Molecular motion in the solvent shell results in dipolar relaxation that stabilizes and thus lowers the energy of the singlet excited state; the extent of this stabilization is dependent upon the relative rate of dipolar relaxation compared to the rate of de-excitation of the singlet state. If relaxation is rapid on the fluorescence time scale, the fluorescence emission will display the effects of this dipolar relaxation. In phosphorescent molecules, a second excited state (the triplet state), with a novel electronic distribution, is generated on the fluorescence time scale by intersystem crossing. This triplet state, because of its long lifetime, can be stabilized by dipolar relaxation that occurs on the phosphorescence time scale of perhaps 10^{-4} s or longer. The phosphorescence emission energy can thus provide a sensitive indicator of matrix dipolar relaxation on time scales as long as seconds.

Dipolar relaxation around the phosphorescent triplet state can be monitored either directly as a function of time following pulsed excitation (4, 20, 22, 23) or indirectly by evaluating the difference in emission energy upon excitation at the peak of the absorption and at the far red edge of the absorption band (the

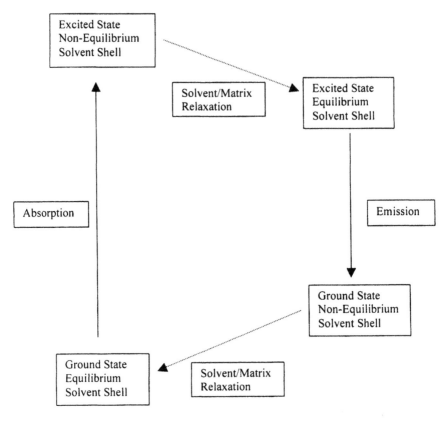

Figure 4   Energy-level diagram for understanding the effect of dipolar relaxation on the energy (wavelength) of luminescence.

so-called red-edge excitation effect) (20, 24, 25). In the former case, measurement of the emission as a function of time following excitation provides a direct measure of the rate of solvent (matrix) relaxation around the excited state. In general, the energy of the emission decays exponentially as a function of time, with a time constant equal to the dipolar (or dielectric) relaxation time (4):

$$v(t) = (v_0 - v_\infty)\exp\left(\frac{-t}{\tau_R}\right) + v_\infty \tag{12}$$

[where v_0 is the center of gravity (4) of the emission spectrum in energy units at time zero, v_∞ is the center of gravity at long times following excitation, and τ_R is the dipolar relaxation time]. In the case of red-edge excitation, the magnitude of the energy difference for the different excitation energies provides a static indication of the presence of dipolar relaxation on the lifetime of the phosphorescent probe; the energy difference is large in immobile matrixes and decreases as the molecular mobility increases.

2. Emission Intensity

The phosphorescence emission intensity is sensitive [Eq. (4)] to the product of the quantum yield for formation of the triplet state (Q_T) and the quantum efficiency for emission from the triplet state (q_P). In the absence of a heavy atom such as iodine or thallium in the local environment that may modulate k_{isc}, the rate of intersystem crossing is largely fixed by the molecular and electronic structure of the molecule (18); it and Q_T are therefore relatively insensitive to changes in the local environment. The intensity of phosphorescence is thus largely modulated by changes in the quantum efficiency for phosphorescence [Eq. (3)]:

$$q_P = \frac{k_p}{k_P + k'_{nr} + k_q[O_2]}$$

Since k_p is dependent upon the molecular structure of the probe and is insensitive to environmental factors, two photophysical processes modulate the magnitude of q_p: the rate of collisional quenching by oxygen ($k_q[O_2]$) and the rate of nonradiative decay to the ground state (k'_{nr}). Each of these perturbations will be discussed in turn.

Oxygen efficiently quenches the triplet state by a collisional quenching mechanism in which the rate constant for collisional quenching, k_q, reflects the bimolecular rate constant for molecular collisions (a modified Smoluchowski relation) (26):

$$k_q = \frac{4\pi\rho N_A D}{10^3} \tag{13}$$

(where ρ is the sum of the reaction radii of the colliding groups, D is the sum of the diffusion coefficients of these groups, and N_A is Avogadro's number). The collisional quenching constant will have units of M^{-1} s^{-1} when ρ has units of m and D has units of m^2 s^{-1}. The diffusion coefficient is related to temperature, molecular size and shape, and solution viscosity (27). For a sphere, D is a simple ratio:

$$D = \frac{k_B T}{f} \tag{14}$$

The parameter f, the translational friction coefficient, is directly proportional to solution viscosity (η, in units of Pa s for D in units of m^2 s^{-1}) and to the dimensions of the molecule. This f term, equal to $6\pi\eta r$ for a sphere of radius r, can also be calculated exactly for regular objects such as prolate and oblate ellipsoids, rigid rods, and flexible wormlike chains (27). It must be emphasized, however, that these terms apply to translational motion within fluid solutions of low viscosity; their generalized application to highly viscous amorphous solids remains to be established. In general, however, $D \propto T/\eta$.

The rate of collisional quenching is thus directly related to the oxygen (quencher) concentration and to the ratio T/η ($k_q \propto D \propto T/\eta$). Changes in the chemical and physical structure of the matrix surrounding the probe can thus modulate the phosphorescence intensity by modulating the oxygen concentration (the solubility of oxygen in the matrix) and the viscosity of the local environment.

In the absence of oxygen, the phosphorescence intensity is related to the ratio of two terms [rewriting Eq. (3) for the case where $k_q[O_2] \ll k_p + k'_{nr}$]:

$$q_P = \frac{k_p}{k_p + k'_{nr}}$$

The phosphorescence intensity in the absence of oxygen thus provides a sensitive indicator of the nonradiative decay rate. For phosphorescent molecules, k'_{nr} is determined by the extent and frequency of molecular collisions between the probe molecule and the surrounding matrix and by the chemical structure of this matrix. In general, k'_{nr} increases (and thus the quantum yield for phosphorescence decreases) as the frequency of collisions between the probe and the surrounding increase. This sensitivity is such that most luminescent molecules exhibit phosphorescence only under conditions in which the matrix surrounding the probe is a rigid solid. Although the standard matrix for phosphorescence measurements is an organic glass at 77 K, other matrices that provide a rigid environment at or near room temperature have also been used (synthetic polymers, transparent inorganic and sugar glasses, polymeric films, filter paper, silica) (14, 15). Only a few molecules exhibit appreciable phosphorescence intensity in deoxygenated liquid solution at room temperature. The best-known examples have aromatic ring systems containing heavy atoms such as bromine (eosin) or iodine (erythrosin); the structures of these and other phosphorescence molecules are shown in Figure 5.

3. Emission Polarization

Electronic transitions between states involve changes in the electron density distribution within a molecule. This change in electron density can be approximated (to high precision) as a dipole connecting the centers of electron density in the two states. Each electronic transition is thus associated with a unique dipole direction in the molecule. Only photons whose oscillating electric fields are parallel to the absorption dipole are absorbed; equivalently, luminescent photons are polarized parallel to the direction of the emission dipole.

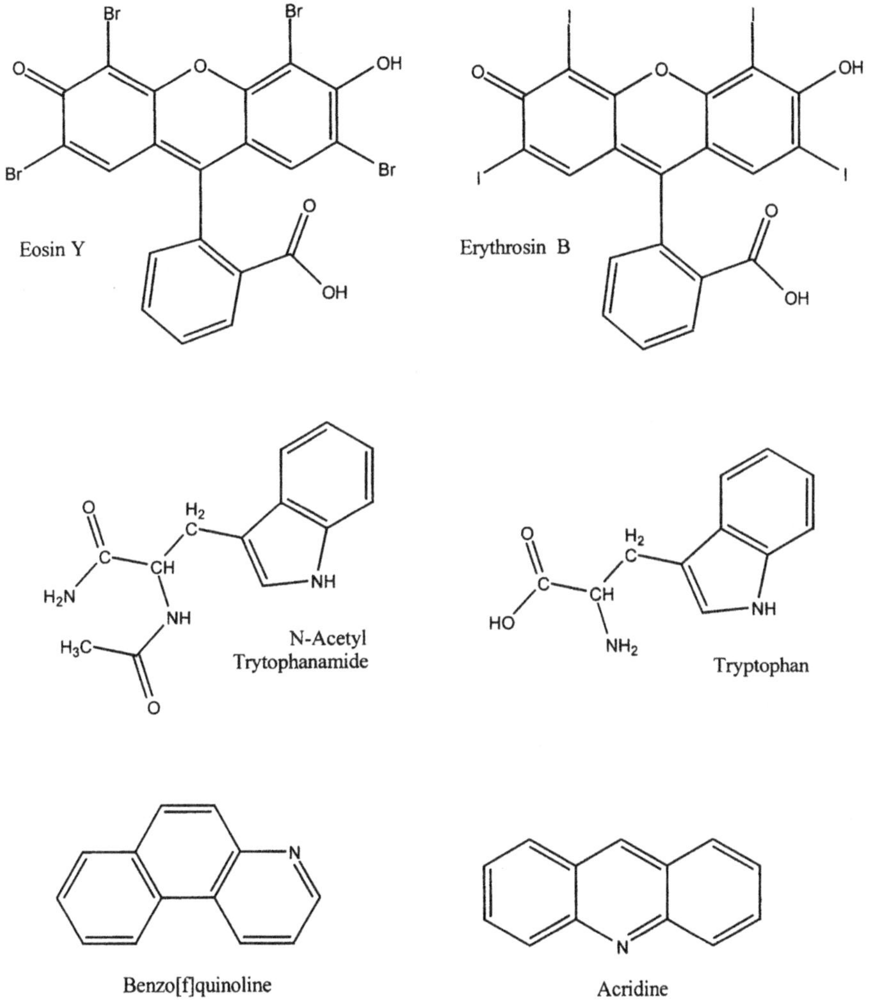

Figure 5 Chemical structures of selected phosphorescent molecules discussed in this review.

Illumination of an ensemble of molecules with polarized light will preferentially excite (photoselect) those molecules whose transition dipoles happen to be parallel to the direction of polarization (usually oriented vertically in the laboratory frame). The absorption intensity is proportional to the square of the dot product between the electric vector of the light **E** and the transition dipole μ_A:

$$I \propto (\mathbf{E} \bullet \mu_A)^2 \propto \cos^2\theta$$

(where θ is the angle between $\mathbf{E}$ and μ_A). The polarized light thus photoselects a distribution of excited chromophores that is preferentially oriented parallel to the vertical (z) axis. The excited chromophores are distributed anisotropically around the z-axis in a $\cos^2$ distribution (which has the shape of an atomic p orbital). This anisotropic distribution changes, becomes more isotropic, due to rotational motion of the molecule. Since these motions are random, the net result of stochastic thermal motion, the result is to make the excited-state distribution more isotropic by shifting the orientation of dipoles into the x-y plane. Measurement of the excited-state dipole-orientation distribution can thus provide a direct measure of chromophore molecular motion.

The orientation distribution of excited-state molecules can be monitored by measuring the intensity of emission polarized either parallel ($I_{\parallel}$) or perpendicular ($I_{\perp}$) to the excitation polarization direction. These intensities are used to calculate the emission anisotropy (4, 28):

$$r = \frac{I_{\parallel} - I_{\perp}}{I_{\parallel} + 2I_{\perp}}$$

The denominator in the anisotropy is proportional to the total intensity of emitted light (proportional to the unpolarized emission intensity); this normalization to the total intensity makes the emission anisotropy solely dependent on rotational motion(s) that occur during the excited-state lifetime.

The anisotropy has a theoretical maximum value (r_o) of 0.4 for luminescence originating from the state into which absorption occurred ($S_1 \rightarrow S_0$ fluorescence following $S_1 \leftarrow S_0$ absorption); this is the maximum value of the anisotropy for colinear absorption (μ_A) and emission (μ_E) dipoles. When the absorbing state and the emitting state are not identical, however, these dipoles are not colinear and the maximum anisotropy is less than 0.4. Since phosphorescence emission ($T_1 \rightarrow S_0$) always occurs from a different state than absorption ($S_1 \leftarrow S_0$), the r_o for phosphorescence is always less than 0.4. For many triplet chromophores, $r_o < 0$ because the transition dipole for phosphorescence (μ_P) is nearly perpendicular to μ_A ($r_o = -0.2$ for perpendicular transition dipoles).

The emission intensity can be measured using either continuous or pulsed excitation. In the former case the steady-state anisotropy (r_{ss}) monitors the average rotational motion of the probe that occurs during the excited-state lifetime; in the latter case the time-resolved anisotropy [$r(t)$] directly monitors the rotational motions of the probe as a function of time. An examination of the emission anisotropy for the simple case of isotropic rotational motion of a spherical molecule containing a chromophore with a single lifetime is illustrative. In this case the time-resolved anisotropy decays from its initial value with a simple exponential dependence:

$$r(t) = r_o \exp\left(\frac{-t}{\phi}\right) \qquad (15)$$

where the time dependence is characterized by the correlation time (ϕ) for rotational motion of a sphere (described more fully later). In the time-resolved experiment, the ability of the chromophore to report on the rotational motion (measure ϕ accurately) depends upon the rate of decay of the total luminescence intensity. If the lifetime (τ) is comparable to or longer than the correlation time (ϕ), then it is possible to detect changes in $r(t)$ during the emission (if the lifetime is very much longer than ϕ, however, there will be insufficient signal due to the low probability of emission during rotational motion); if the lifetime is shorter than the correlation time, then luminescence may decay before there are appreciable changes in $r(t)$. This dependence upon lifetime is best illustrated by the form of the steady-state anisotropy for the simple case under discussion:

$$r_{ss} = \frac{r_o}{1 + \tau/\phi} \qquad (16)$$

The value of r_{ss} is thus dependent upon the ratio of the excited-state lifetime to the rotational correlation time (τ/ϕ). When this ratio is small ($\tau \ll \phi$), the anisotropy is equal to its maximum value, r_o; when this ratio is large ($\tau \gg \phi$), the anisotropy is zero. [This general conclusion holds no matter how complex the form of $r(t)$; (29).] Measurement of the emission anisotropy thus provides a way to directly monitor the rotational motion of the chromophore occurring during the excited-state lifetime of the probe. For fluorescence probes this reflects motions that occur on time scales of about 10^{-9}–10^{-7} s, while for phosphorescence probes this reflects motions that occur on time scales from about 10^{-6} to 10 s.

The rotational correlation time is dependent upon the size and shape of the rotating species and also upon the viscosity of the matrix in which the chromophore is embedded. For a spherical particle in solution, it has a simple form (27, 28):

$$\phi = \frac{V\eta}{k_B T} \qquad (17)$$

where V is the solvated molecular volume and η is the microviscosity (the effective viscosity that the molecule experiences, often not equivalent to the macroscopic viscosity for polymer solutions, for example; see Ref. 26). For an aromatic molecule in liquid solution at 298 K, where $V \approx 10^{-27}$ m^3 and $\eta = 10^{-3}$ Pa s, $\phi \approx 0.25 \times 10^{-9}$ s; for the same molecule in a solid matrix, where $\eta \approx 10^6$ Pa s, $\phi \approx 0.25$ s. This illustrates the value of phosphorescence as a technique to monitor rotational motion in amorphous solids: only the long-lived triplet state can monitor the slow rotational motions characteristic of the highly viscous solid state.

III. PHOSPHORESCENCE AS A PROBE OF THE AMORPHOUS STATE

A. Synthetic Polymers

Luminescence techniques have been widely used to probe the structure, dynamics, and photochemical reactivity of amorphous synthetic polymers (26, 30). Although most studies have employed fluorescence probes, numerous studies demonstrate the utility of phosphorescence intensity and polarization as a probe of polymer structure and dynamics. The first systematic use of phosphorescence intensity to probe structural transitions in polymers dates from the early 1970s (31, 32). Somersall et al. (32) used the fluorescence and phosphorescence intensity of naphthalene and ketone chromophores incorporated into various polymers to monitor dynamics as a function of temperature over the range from about 90 K to room temperature. Since no attempt was made in this study to degas the samples, probe phosphorescence was susceptible to oxygen quenching. They found that although the fluorescence intensity decreased only threefold, the phosphorescence intensity (I_p) decreased over 1000-fold. Arrhenius plots of ln I_p versus $1/T$ for emission from either naphthalene or ketone groups were linear over much of the temperature range and displayed breaks at temperatures corresponding to known structural transitions in the polymers. Depending upon the specific probe and the specific polymer [polystyrene, poly(methyl methacrylate), poly(methyl acrylate)], these breaks corresponded to the onset temperatures for rotational motion of the various sidechain groups of the polymers (benzene ring, methyl group, ester group) but not to the glass transition temperature, since probe intensity was effectively quenched in the glassy state in the presence of oxygen; the probes thus reported on the various glassy-state structural transitions known to take place in these polymers.

The same lab (26, 33) conducted studies of naphthalene emission in poly(methyl methacrylate) as a function of both temperature and oxygen concentration. At low [O_2], the Arrhenius plot was flat (zero slope) up to ~240 K, indicating that the rate of nonradiative decay was constant; a break in the curve due to a transition at ~240 K introduced a novel nonradiative decay pathway for the chromophore (probably corresponding to a β transition characterizing the onset of sidechain—CO—O—CH_3 rotation in PMMA; Table 1). At higher P_{O2} (and thus higher [O_2] in the polymer matrix), the break point in the Arrhenius plot occurred at low temperature, which was interpreted as being due to the onset of oxygen diffusion into the polymer. These lower-temperature break points may correspond to structural transitions that modulate either the solubility or the diffusion rate of oxygen in PMMA.

The sensitivity of the triplet state to quenching by collision with oxygen was recognized in 1962 as a potential technique to measure oxygen diffusion in polymers (34). The diffusion equations have been solved for an experimental setup that involves exposing the polymer sample to an oxygen atmosphere at time zero and monitoring the decrease in phosphorescence intensity as a function of time $[I_p(t)]$ following introduction of oxygen (26, 34). The decay of the phosphorescence intensity after oxygen exposure follows this equation for thin films:

$$I_p(t) = B \exp\left\{-\left(\frac{\pi^2 D}{d^2}\right)t\right\} \tag{18}$$

where B is an instrumental parameter, D is the self-diffusion coefficient for O_2 plus chromophore, essentially equal to the translational diffusion coefficient for oxygen, and d is the thickness of the sample. The slope of a plot of $\ln I_p(t)$ thus provides a direct measure of the oxygen translational diffusion coefficient (D). Arrhenius plots of $\ln D$ as a function of $1/T$ were linear (34), providing a measure of the activation energy for the diffusion process.

Phosphorescence emission polarization has been used only sparingly to detect rotational motion in solid polymers. The emission anisotropy provides a way to directly monitor changes in the polymer matrix that affect the rate of probe rotational motion during the excited-triplet-state lifetime. The advantage of phosphorescence for such measurements is the long phosphorescence lifetime (typically $>10^{-3}$ s), which provides a low-frequency ($<10^3$ Hz) probe of the polymer state. [Fluorescence, on the other hand, with lifetimes of $\leq 10^{-8}$ s, provides a high-frequency ($\geq 10^8$ Hz) probe.]

Soutar and colleagues were the first to make use of phosphorescence polarization to characterize the solid-state transitions in a series of polymers (35–38). Their procedure involved measurement of the emission polarization p [$= (I_\parallel - I_\perp)/(I_\parallel + I_\perp)$] and lifetimes of various probes as a function of temperature. Their results demonstrated that, depending upon the manner in which the probe was attached to or dispersed throughout the polymer, the probe is sensitive to changes in motion occurring at the glass transition as well as to solid-solid transitions occurring in the glass below T_g. Using a naphthalene label attached to the polymer backbone at two locations so that its motion reflected backbone reorientation only, they found that in poly(methyl acrylate) the probe was sensitive to changes in motion at the glass transition (35) and immobile at lower temperatures. In poly(methyl methacrylate), however, the probe detected dynamical transitions occurring below T_g (37).

B. Sugars

Phosphorescence has been used only sparingly to characterize the physical state of amorphous sugars. Some studies have used glassy sugars as a matrix for room-temperature phosphorescence investigations of specific chromophores (39–42) or as a matrix for optical thermometry based on measurements of the delayed fluorescence intensity (43). These studies used amorphous glucose, trehalose, glucose/trehalose mixtures, or sucrose as a rigid matrix to minimize quenching of the triplet state at or near room temperature. Other reports have examined how the phosphorescence characteristics of chromophores embedded within the sugar matrices respond to variations in temperature (42, 44, 45).

Shah and Ludescher (44) used the probes erythrosin B and N-acetyl-trypto-phan amide (NATA) to monitor matrix dynamics in amorphous freeze-dried sucrose. In this study, the probes were dispersed in a concentrated aqueous sugar solution at a mole ratio of 1 probe per 1000 sugars. Fractured glasses of these samples were prepared by freeze-drying the sugar solution followed by further drying over P_2O_5; the final glasses had T_g values of 60°C (determined as the midpoint of the thermal transition measured by differential scanning calorimetry). The phosphorescence intensity decay of both probes was measured (in the absence of oxygen) over a range of temperatures below and above T_g (from about 10–80°C). Both probes exhibited complex excited-state decay behavior at all temperatures, with NATA requiring three and erythrosin two exponentials for adequate fits to the intensity decay functions at all temperatures. Plots of ln τ versus T for NATA and plots of τ versus T for erythrosin were approximately linear at temperatures below and above T_g, with all curves showing break points at about 50°C (Fig. 6). Decreases in the phosphorescence lifetimes at temperatures below T_g clearly indicated that some as-yet-unidentified collisional quenching mechanism(s) operate in the amorphous glassy sucrose and that these mechanism(s) are activated by temperature. The number and range of the probe lifetimes at low temperature also indicated that the glassy state of amorphous sucrose is dynamically heterogeneous, containing at least three distinct dynamic environments, in which the molecular mobility differs by nearly two orders of magnitude. The authors concluded that phosphorescence probes report on the dynamic changes underlying the glass transition in amorphous sugars.

Shah (46) also conducted novel measurements of dipolar relaxation in amorphous sucrose using red-edge excitation of erythrosin B embedded in freeze-dried sucrose powders (Fig. 7a). Such measurements provide an indication of the

---

Figure 6 Effect of temperature on the phosphorescence lifetimes of N-acetyl-tryptophan amide (a) and erythrosin B (b) dispersed in amorphous, freeze-dried sucrose powder. (From Ref. 44.)

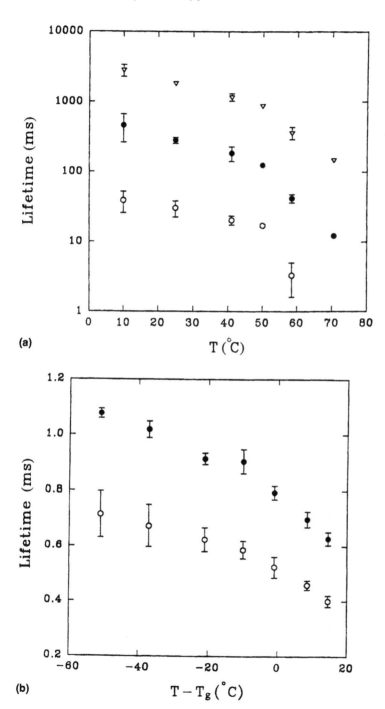

(a)

(b)

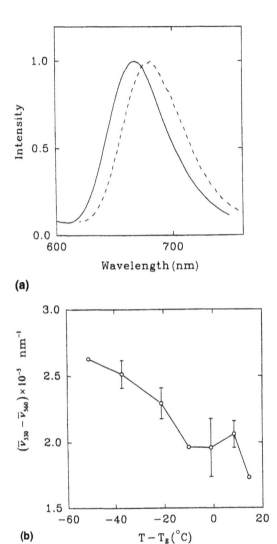

(a)

(b)

Figure 7 Red-edge excitation of erythrosin B dispersed in amorphous, freeze-dried sucrose powder. (a) Phosphorescence emission spectra with excitation at 530 nm (solid line), the absorption peak, and 560 nm (dashed line), the far red edge of the absorption. (b) Effect of temperature on the difference in emission energy with excitation at 530 and 560 nm. (From Ref. 46)

occurrence of dipolar relaxation around the excited triplet state but, unfortunately, not the rate of this relaxation. The decrease in the energy difference with increasing temperature below T_g (Fig. 7b) indicates that temperature activates as-yet-unidentified mode(s) of molecular mobility in the glassy sucrose. Direct measurements of the time evolution of the erythrosin emission spectrum as a function of time following excitation could thus be used to monitor the rate constant for dipolar relaxation in amorphous sugars.

McCaul and Ludescher (39) also characterized the phosphorescence emission from tryptophan amino acid and 4-*F*-tryptophan, 5-*F*-tryptophan, 6-*F*-tryptophan, and 5-*Br*-tryptophan amino acid embedded in fractured sucrose glasses at 20°C. Comparison of the integrated emission intensity from the probes indicated that fluorination at the 5 and 6 positions in the ring had little effect on the phosphorescence quantum yield; these analogs had quantum yields relative to tryptophan of 0.68 and 0.91, respectively. The 4-*F* and 5-*Br* analogs, on the other hand, were effectively quenched in the sucrose glass with relative quantum yields of 0.039 and 0.022, respectively. The time-resolved emission intensity decays from these probes were complex at 20°C; the decays of tryptophan and the fluorinated derivatives were well fit using three lifetimes, while the decay of the brominated derivative required two lifetimes (Table 2). The tryptophan lifetimes were 70, 419, and 2433 ms, with fractional amplitudes of 0.29, 0.34, and 0.36, respectively; the lifetimes of the fluorinated derivatives were as widely diverse and also equally distributed. The large range of the lifetimes for each probe provided further evidence that the fractured sucrose glass is dynamically heterogeneous, containing at least three environments that differ significantly in terms of their ability to quench the tryptophan triplet state by molecular collision. A detailed comparison of the lifetimes and amplitudes for tryptophan and its fluorinated analogs provides evidence that the lifetime complexity reflects environmental heterogeneity rather than probe peculiarities. Not only were the lifetimes distributed in a similar manner for all probes, but also the ratio of the lifetime of the fluorinated analog to

Table 2 Phosphorescence Lifetimes at 20°C of Tryptophan and Some Halogenated Analogs Dispersed in Amorphous Sucrose

Probe	Lifetime (ms)		
Tryptophan	70	419	2433
4-*F*-Tryptophan	2.97	20.0	119
5-*F*-Tryptophan	49	299	1754
6-*F*-Tryptophan	38	218	1205
5-*Br*-Tryptophan	0.20	1.35	—

Source: Ref. 39.

the lifetime of tryptophan was the same for each probe (Table 2). The rate constants for nonradiative decay were calculated from the probe lifetimes (Table 3); these nonradiative decay rates clearly indicate the range of molecular mobility found in the glassy sucrose matrix. These data indicate that the freeze-dried sucrose glass is a dynamically heterogeneous state, with regions of the glass differing in molecular mobility by 50-fold or more.

Fister and Harris (45) characterized the triplet-state photophysics of the aromatic dye acridine yellow embedded in clear glassy mixtures of trehalose and glucose (prepared by melting a 50:50 mixture at 115°C under vacuum); unfortunately, no effort was made to determine the glass transition temperature of this sample. Delayed emission spectra and lifetimes for delayed fluorescence were collected over the temperature interval from −78 to 21°C. The dye exhibited two lifetime components at all temperatures; careful analysis indicated that the data were well fit by discrete lifetimes rather than Gaussian distributions of lifetimes. Arrhenius plots of $\ln(k_{obs} - k_t)$ versus $1/T$ for the delayed fluorescence lifetime data (where $k_{obs} = 1/\tau$ and $k_t = k_p + k'_{nr}$) were found to be linear, suggesting that no dynamical transition occurred over the temperature range examined.

Wang and Hurtubise (42) monitored the total intensity and phosphorescence lifetimes of the probes 2-amino-1-methyl-6 phenylimidazole[4,5-b]pyridine and benzo[f]quinoline embedded in glucose glasses prepared by evaporation of aqueous glucose. Measurements of total phosphorescence intensity and lifetime over the temperature interval from 100 to 300 K provided clear evidence of a transition between 200 and 240 K. Analysis of the temperature dependence

Table 3 Collisional Quenching Rates at 20°C of Tryptophan and Some Halogenated Analogs Dispersed in Amorphous Sucrose

Probe	Collisional quenching rate (s^{-1})		
	Fast	Intermediate	Slow
Tryptophan	14	2.2	0.26
4-F-Tryptophan	336	50	8.1
	$(24)^a$	(23)	(31)
5-F-Tryptophan	20	3.2	0.42
	(1.45)	(1.45)	(1.6)
6-F-Tryptophan	26	4.4	0.68
	(1.9)	(2.0)	(2.6)

[a] The ratio of the quenching rate for the halogenated analog to that of tryptophan.
Source: Ref. 39.

of the ratio of intensity divided by lifetime (I_p/τ) suggested that the transition involved a change in the physical state of the glucose glass that decreased the triplet-state quantum yield. An Arrhenius analysis of the function $\ln(1/\tau_p - 1/\tau_p^o)$ indicated a transition for each probe around 220 K (Fig. 8); the linear data above and below the transition were used to estimate activation energies for the triplet-state deactivation processes. The small pre-exponential factors and activation energies for both probes at low temperature were interpreted as resulting from a very rigid interaction between the probes and the matrix; the larger values above the transition were thought to reflect much weaker interactions with the matrix. The authors could not unambiguously assign a physical origin to the transition itself. Although the sample contained residual moisture (estimated by gravimetry and confirmed by IR), the amount (about 1%) was insufficient to lower the glass transition temperature of glucose significantly below its anhydrous value of ~35°C (16). Possible physical origins of the transition mentioned by the authors included an onset temperature for water diffusion and a rotational relaxation of functional groups in glucose; this latter interpretation, compatible with the large activation energies measured above the transition, would imply that the phosphorescence of this probe is sensitive to a β transition in glassy glucose.

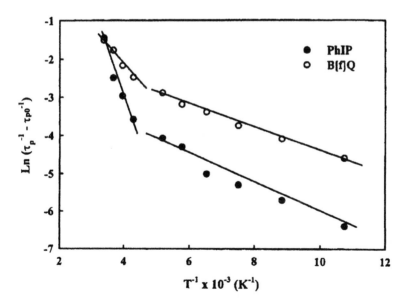

Figure 8 Arrhenius plot of the emission lifetimes of the phosphorescence probes 2-amino-1-methyl-6 phenylimidazole[4,5-b]pyridine and benzo[f]quinoline dispersed in amorphous glucose. (From Ref. 42.)

Phosphorescence probes are thus clearly sensitive to the glass transition and possibly to other lower-temperature solid-state transitions in sugars. This sensitivity has barely been exploited, however, and remains a fertile field for future study. Although there are no reports as yet on the use of triplet-state probes to investigate solid-state transitions in polymeric carbohydrates, the widespread applicability of phosphorescence to monitor solid-state transitions in polymers strongly suggests that much progress remains to be made in this area as well.

C. Proteins

Phosphorescence from proteins embedded in low-temperature glasses was first reported in 1952 (47) and subsequently verified by other studies (48). Spectroscopic studies of proteins in aqueous solution, however, soon demonstrated (49, 50) that the protein matrix could provide sufficient rigidity for tryptophan phosphorescence at room temperature. Saviotti and Galley (50) showed that, in the absence of oxygen, one of the tryptophans in liver alcohol dehydrogenase and in alkaline phosphatase emitted long-lived emission. Although such emission was considered unusual at the time, Vanderkooi and colleagues (51) later demonstrated that such emission is actually common, occurring in most of the proteins they studied. Since then, numerous studies of tryptophan phosphorescence emission have been reported in the literature; several excellent reviews exist (52–54).

Tryptophan phosphorescence from proteins is enhanced in amorphous (lyophilized) dry powders (55–57) and in amorphous dry protein films (unpublished data of Simon and Ludescher). The enhancement of phosphorescence is thought to be a direct result of removal of water and not to the formation of protein–protein contacts during drying (57). Although most tryptophan phosphorescence decays from proteins in solution display nearly single exponential decay kinetics (54), the intensity decays in the solid state are strikingly nonexponential, requiring as many as three lifetimes for an adequate fit. Such complexity probably reflects heterogeneity in the local molecular dynamics of specific protein molecules in the amorphous solid and could be due to variability in local protein packing or to random variations in the protein structure induced by dehydration stresses.

The role of hydration in modulating protein mobility and thus tryptophan phosphorescence intensity was demonstrated by a detailed study of hydration in hen egg white lysozyme powder monitored by phosphorescence (57). This protein, which contains six tryptophan residues, exhibits negligible phosphorescence emission in aqueous solution but relatively intense emission in dry powders. Measurements of total phosphorescence intensity (Fig. 9a) and of emission lifetimes (Fig. 9b) as a function of hydration level indicate that water binding initiates dynamic processes that quench the triplet state. Although hydration (h) to 0.1 g H_2O/g protein had little effect on the phosphorescence emission, subsequent hy-

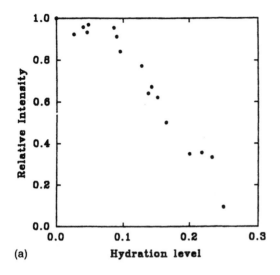

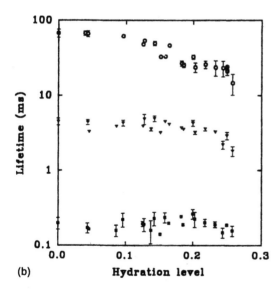

Figure 9 Effect of hydration on the (a) phosphorescence emission intensity and (b) lifetimes of the six intrinsic tryptophans in hen egg white lysozyme powders. (From Ref. 57.)

dration over the range from 0.1 to 0.3 caused a nearly linear decrease in intensity and lifetime. The authors interpreted this effect as due to an increase in the internal mobility of the protein resulting from an increased ability to exchange internal hydrogen bonds for those with surface water molecules. Such behavior is expected for tryptophan phosphorescence, since studies of indole (58) and tryptophan (59) phosphorescence in solution indicate that the probe lifetime varies linearly with viscosity over several orders of magnitude (the lifetime of tryptophan varies over 10^3-fold during a change in viscosity of over 10^4-fold).

The quenching of protein phosphorescence by water binding can also be viewed as the effect of water on the temperature of some internal dynamic transition that modulates protein mobility. Within this perspective, at $h \leq 0.1$ the experimental temperature (T_{exp}) is lower than the onset temperature for the transition (T_{trans}) and the phosphorescence intensity is high. At $h > 0.1$, however, $T_{exp} \geq T_{trans}$ and a novel quenching mechanism begins to operate, resulting in an increase in protein internal mobility and a decrease in phosphorescence intensity. Although there is a temptation to refer to this as a glass transition, there is no direct evidence that this softening transition corresponds to a change in the flow properties of the dry protein powder. A likely assignment could involve some as-yet-unidentified solid-solid transition within the glassy protein comparable to that seen in polymers. Since the entire issue of glassy states and glass transitions in proteins remains problematic, much additional work is necessary to illuminate this potentially important issue.

The use of tryptophan phosphorescence to study solid-state transitions in proteins shows great promise. The probe is nonperturbing (because intrinsic), exhibits a dynamic range of over three orders of magnitude (lifetimes in proteins range from less than 1 ms to 3 s), and has a long lifetime that has been shown in model studies to scale with solvent (matrix) viscosity (58, 59). Much additional research, however, is required to answer a number of outstanding questions. Precisely how does a change in viscosity modulate the tryptophan lifetime? Do molecular collisions per se quench the triplet state? Or must the collisions involve specific chemical groups? And, perhaps most importantly, do all molecular collisions or only specific sorts of molecular collisions at specific sites quench the triplet state? Gonnelli and Strambini (60), for example, have demonstrated that only the sidechains of histidine, tyrosine, tryptophan, cysteine, and cystine are efficient collisional quenchers of tryptophan phosphorescence in aqueous solution.

IV. PERSPECTIVES FOR FUTURE RESEARCH

A. Novel Triplet Probes

The ability of fluorescence spectroscopy and microscopy to solve a plethora of biological and biochemical problems has fueled a cottage industry in the development and application of novel fluorescence probes ranging from calcium indicators to green fluorescent protein (12); much less effort has been expended in seeking novel phosphorescent probes. There is a need for additional phosphorescence probes and for more detailed characterization of the possible quenching mechanisms for known probes. Detailed molecular knowledge about the mechanisms of triplet-state quenching in specific probes such as tryptophan, for example, could greatly enhance the utility of phosphorescence measurements to provide molecular information about dynamic transitions in proteins. In addition, the many naturally occurring chromophores of organisms, and thus of food (61), have great potential as biophysical probes of molecular structure and dynamics in food systems. Their use as spectroscopic probes, however, is largely unexplored.

B. Measurement of Oxygen Diffusion

The exquisite sensitivity of phosphorescence emission to oxygen quenching, and its established use to monitor oxygen diffusion through synthetic polymers (67), suggests that the technique could be used to monitor oxygen diffusion into and throughout a food matrix. Preliminary studies from the author's lab, for example, indicate that the phosphorescence intensity of erythrosin-5-isothiocyanate covalently attached to amino groups in porcine gelatin provides a sensitive indicator of how hydration modulates oxygen diffusion in gelatin films. Dry thin films made from labeled gelatin have high phosphorescence intensity when the film is in contact with room air, while the intensity is effectively quenched in wet films under the same conditions (unpublished data of Simon and Ludescher). Since the erythrosin probe has appreciable phosphorescence intensity in aqueous solution, the effect appears to be mediated by oxygen diffusion through wet but not dry gelatin.

C. Phosphorescence Imaging

Fluorescence imaging and microscopy have made tremendous advances in the last decade, further enhancing an already powerful tool for studying bioprocesses in situ and in vivo. Although some uses of fluorescence imaging have been made in food systems (62), the real potential of the technique to study food structure and properties remains largely unexplored. Phosphorescence imaging, on the other hand, is largely an unexplored technique in any field. Zotikov and Polyakov (63) reported phosphorescence spectra and decays from solid ribo- and deoxyribonucleosides collected through a microscope. In a later report (64), the authors were able to distinguish cell types (yeast, bacteria, protozoa, human fibroblasts) by means of phosphorescence spectra. The technique has also demonstrated its effectiveness for localizing tumors and evaluating their oxygenation state (65). Recent advances have improved the sensitivity of the technique (66).

Phosphorescence imaging has great potential for monitoring molecular mobility and dynamical transitions within the heterogeneous matrix of foods. As this review has illustrated, triplet-state probes have the potential to monitor the physical state and changes in the state of specific components in foods. The ability to make measurements of phosphorescence intensity, lifetime decays, emission spectra, or even polarization within the microscope field opens up many possible applications to study food heterogeneity, its origins in molecular structure, and its effects on macroscopic properties. Since phosphorescence probes have high quantum yields in glassy states and low quantum yields in rubbery states, it is in principle possible to image the glassy and rubbery regions of a food as a function of changes in temperature, hydration, or other environmental parameters. Phosphorescence imaging may also prove useful in monitoring the three-dimensional diffusion of oxygen into and throughout a food matrix.

REFERENCES

1. L. H. Sperling. Introduction to Physical Polymer Science. 2nd ed. New York: Wiley Interscience, 1992.
2. R. Zallen. The Physics of Amorphous Solids. New York: Wiley, 1983.
3. N. G. McCrumm, B. E. Read, G. Williams. Anelastic and Dielectric Effects in Polymeric Solids. New York: Dover, 1991.
4. J. Lakowicz. Principles of Fluorescence Spectroscopy. 2nd ed. New York: Plenum Press, 1999.
5. R. S. Becker. Theory and Interpretation of Fluorescence and Phosphorescence. New York: Wiley Interscience, 1969.
6. S. G. Schulman. Fluorescence and Phosphorescence Spectroscopy: Physicochemical Principles and Practice. Oxford: Pergamon Press, 1977.
7. C. A. Parker. Photoluminescence of Solutions. Amsterdam: Elsevier, 1968.
8. L. Brand, M. L. Johnson, eds. Fluorescence Spectroscopy. Methods in Enzymology, Vol. 278. San Diego, CA: Academic Press, 1997.
9. J. Slavik. Fluorescent Probes in Cellular and Molecular Biology. Boca Raton, FL: CRC Press, 1994.
10. E. L. Wehry. Effects of molecular structure on fluorescence and phosphorescence. In: G. G. Guilbault, ed. Practical Fluorescence. 2nd ed. New York: Marcel Dekker, 1990, pp 75–126.
11. E. L. Wehry. Effects of molecular environment on fluorescence and phosphorescence. In: G. G. Guilbault, ed. Practical Fluorescence. 2nd ed. New York: Marcel Dekker, 1990, pp 127–184.
12. R. P. Haugland. Handbook of Fluorescent Probes and Research Chemicals. 6th ed. Eugene, OR: Molecular Probes, 1996, http://www.probes.com.
13. S. P. McGlynn, T. Azumi, K. Kinoshita. Molecular Spectroscopy of the Triplet State. Englewood Cliffs, NJ: Prentice-Hall, 1969.

14. T. Vo-Dinh. Room Temperature Phosphorimetry for Chemical Analysis. New York: Wiley, 1984.

15. R. J. Hurtubise. Phosphorimetry: Theory, Instrumentation, and Applications. New York: VCH, 1990.

16. Y. H. Roos. Phase Transitions in Foods. San Diego, CA: Academic Press, 1995.

17. J. B. Birks. Photophysics of Aromatic Molecules. New York: Wiley Interscience, 1970.

18. N. Turro. Modern Molecular Photochemistry. Menlo Park, CA: Benjamin, 1978.

19. G. Strasburg, R. D. Ludescher. Theory and applications of fluorescence spectroscopy in food research. Trends Food Sci. Tech. 6:69–75, 1995.

20. A. P. Demchenko. Fluorescence Analysis of Protein Dynamics. Essays Biochem. 22:120–157, 1986.

21. A. P. Demchenko. Site-selective excitation: a new dimension in protein and membrane spectroscopy. Trends Biochemical Sci. 13:374–377, 1988.

22. L. Brand, J. R. Gohlke. Nanosecond time-resolved fluorescence spectra of a protein–dye complex. J Biol Chem 246:2317–2324, 1971.

23. R. P. DeToma, J. H. Easter, L. Brand. Dynamic interactions of fluorescence probes with the solvent environment. J. Am. Chem. Soc. 98:5001–5007, 1976.

24. J. R. Lakowicz, S. Keating-Nakamoto. Red-edge excitation of fluorescence and dynamic properties of proteins and membranes. Biochemistry 23:3013–3021, 1984.

25. A. P. Demchenko. Red-edge excitation fluorescence spectroscopy of single-tryptophan proteins. Eur. Biophys. J. 16:121–129, 1988.

26. J. Guillet. Polymer Photophysics and Photochemistry. Cambridge University Press, 1985.

27. C. R. Cantor, P. R. Schimmel. Biophysical Chemistry. Part II: Techniques for the Study of Biological Structure and Function. San Francisco: W. H. Freeman, 1980.

28. R. F. Steiner. Fluorescence anisotropy: theory and applications. In: J. R. Lakowicz, ed. Topics in Fluorescence Spectroscopy, Vol. 2: Principles. New York: Plenum Press, 1991, pp 1–52.

29. R. D. Ludescher, W. H. Ludescher. Steady-state optical polarization anisotropy of rodlike molecules undergoing torsional twisting motions. Photochem. Photobiol. 58: 881–883, 1993.

30. L. Zlatkevich. Luminescence Techniques in Solid-State Polymer Research. New York: Marcel Dekker, 1989.

31. I. Boustead. Temperature quenching of phosphorescence in polyethylene. European Polym. J. 6: 731–741, 1970.

32. A. C. Somersall, E. Dan, J. Guillet. Photochemistry of ketone polymers. XI. Phosphorescence as a probe of subgroup motion in polymers at low temperature. Macromolecules 7:233–244, 1974.

33. J. Guillet. Mass diffusion in solid polymers. In: M. A. Winnik, ed. Photophysical and Photochemical Tools in Polymer Science. Dordrecht: Reidel, 1986, pp 467–494.

34. G. Oster, N. Geacintov, A. V. Khan. Luminescence in plastics. Nature 196:1089–1090, 1962.

35. H. Rutherford, I. Soutar. Phosphorescence studies of relaxation effects in bulk polymers. I. Depolarization measurements of segmental relaxation in poly(methyl acrylate). J. Polym. Sci. Polym. Phys. Ed. 15:2213–2225, 1977.

36. H. Rutherford, I. Soutar. Phosphorescence depolarization measurements of segmental relaxation processes in poly(methyl acrylate) and poly(methyl methacrylate). J. Polym. Sci. Polym. Lett. Ed. 16:131–136, 1978.

37. H. Rutherford, I. Soutar. Phosphorescence studies of relaxation effects in bulk polymers. I. Emission epolarization study of relaxation mechanisms in poly(methyl methacrylate). J. Polym. Sci. Polym. Phys. Ed. 18:1021–1034, 1980.

38. J. Toynbee, I. Soutar. Luminescence studies of molecular motion in poly(n-butyl acrylate). In: CE Hoyle, J. M. Torkelson, eds. Photophysics of Polymers. Washington, DC:ACS Symposium Series 358, 1987, pp 123–134.

39. C. P. McCaul, R.D. Ludescher. Room-temperature phosphorescence from tryptophan and halogenated tryptophan analogs in amorphous sucrose. Photochem. Photobiol. 70:166–171, 1999.

40. Y. Chu, R. J. Hurtubise. Luminescence properties and analytical figures of merits of benzo(a)pyrene guanosine adduct adsorbed on α-, β-, and γ-cyclodextrin/NaCl, and trehelose/NaCl solid matrices. Anal. Lett. 26:1195–1209, 1993.

41. J. Wang, R. J. Hurtubise. Solid-matrix luminescence from trace organic compounds in glasses prepared from sugars. Appl. Spec. 50:53–58, 1996.

42. J. Wang, R. J. Hurtubise. Phosphorescence properties of 2-amino-1-methyl-6-phenylimidazo[4,5-b]pyridine and benzo[f]quinoline in glucose glasses via temperature variation and spectral characterization. Anal. Chem. 69:1946–1951, 1997.

43. J. Fister, D. Rank, J. M. Harris. Delayed fluorescence optical thermometry. Anal. Chem. 67:4269–4275, 1995.

44. N. K. Shah, R. D. Ludescher. Phosphorescence probes of the glassy state in amorphous sucrose. Biotech. Prog. 11:540–544, 1995.

45. J. C. Fister, J. M. Harris. Time- and wavelength-resolved delayed-fluorescence emission from acridine yellow in an inhomogeneous saccharide glass. Anal. Chem. 68:639–646, 1996.

46. N. K. Shah. Spectroscopic probes of protein hydration and glass transitions. PhD dissertation, Rutgers University, New Brunswick, NJ, 1994.

47. P. Debye, J. O. Edwards. A note on the phosphorescence of proteins. Science 116:143–144, 1952.

48. S. V. Konev. Fluorescence and Phosphorescence of Proteins and Nucleic Acids. New York: Plenum Press, 1967.

49. J. W. Hastings, Q. H. Gibson. The role of oxygen in the photoexcited luminescence of bacterial luciferase. J. Biol. Chem. 242:720–726, 1967.

50. M. L. Saviotti, W. C. Galley. Room-temperature phosphorescence and the dynamic aspects of protein structure. Proc. Natl. Acad. Sci. USA 71:4154–4158, 1974.

51. J. M. Vanderkooi, D. B. Calhoun, S. W. Englander. On the prevalence of room-temperature phosphorescence in proteins. Science 236:568–569, 1987.

52. S. Papp, J. M. Vanderkooi. Tryptophan phosphorescence at room temperature as a tool to study protein structure and dynamics. Photchem Photobiol 49:775–784, 1989.

53. N. E. Geacintov, H. C. Brenner. The triplet state as a probe of the dynamics and structure of biological molecules. Photochem. Photobiol. 50:841–858, 1989.

54. J. A. Schauerte, D. G. Steel, A. Gafni. Time-resolved room-temperature tryptophan phosphorescence in proteins. In: L. Brand, M. L. Johnson, eds. Fluorescence Spec-

troscopy. Methods in Enzymology, Volume 278. San Diego, CA: Academic Press, 1997, pp 49–71.

55. G. B. Strambini, E. Gabellieri. Intrinsic phosphorescence from proteins in the solid state. Photochem. Photobiol. 39:725–729, 1984.
56. G. J. Smith, W. H. Melhuish. Relaxation and quenching of the excited states of tryptophan in keratin. J. Photochem Photobiol. B.-Biology 17:63–68, 1993.
57. N. K. Shah, R. D. Ludescher. Influence of hydration on the internal dynamics of hen egg white lysozyme in the dry state. Photochem Photobiol 58:169–174, 1993.
58. G. B. Strambini, M. Gonnelli. The indole nucleus triplet-state lifetime and its dependence on solvent microviscosity. Chem. Phys. Lett. 115:196–200, 1985.
59. G. B. Strambini, M. Gonnelli. Tryptophan phosphorescence in fluid solution. J. Am. Chem. Soc. 117:7646–7651, 1995.
60. M. Gonnelli, G. B. Strambini. Phosphorescence lifetime of tryptophan in proteins. Biochemistry 34:13847–13857, 1995.
61. O. S. Wolfbeis. The fluorescence of organic natural products. In: S. G. Schulman, ed. Molecular Luminescence Spectroscopy, Methods and Applications: Part 1. New York: Wiley, 1985, pp 168–370.
62. L. Munck, ed. Fluorescence Analysis in Foods. England: Longman Scientific and Technical, 1989.
63. A. A. Zotikov, Y. S. Polyakov. Study of nucleoside phosphorescence with a phosphorescent microscope (in Russian). Biofizika 20(4):557–600, 1975.
64. A. A. Zotikov, Y. S. Polyakov. The use of the phosphorescence microscope for the study of the phosphorescence of various cells. Microscopica Acta 79:414–418, 1977.
65. D. F. Wilson, G. J. Cerniglia. Localization of tumors and evaluation of their state of oxygenation by phosphorescence imaging. Cancer Res. 52:3988–3993, 1992.
66. J. Hennink, R. de Haas, N. P. Verwoerd, H. J. Tanke. Evaluation of a time-resolved fluorescence microscope using phosphorescent Pt-porphine model system. Cytometry 24:312–320, 1996.
67. X. Lu, I. Manners, M.A. Winnik. Oxygen diffusion in polymer films for luminescence barometry applications. In: B. Valeur, J.-C. Brochon, ed. New Trends in Fluorescence Spectroscopy. Berlin: Springer, 2001.

15

Starch Properties and Functionalities

Lilia S. Collado
University of the Philippines Los Baños, College, Laguna, Philippines

Harold Corke
The University of Hong Kong, Hong Kong, China

I. INTRODUCTION

Cereal grains store energy in the form of starch. The proportion of starch in the grain is generally between 60 and 75% by weight (Hoseney 1986). It makes up about 90% of milled-rice dry weight (Juliano 1985) and 72% of the maize kernel dry weight (Boyer and Shannon 1987) and is the primary product obtained from wet milling of maize. In maize, most starch occurs in the endosperm, but significant amounts are also found in the embryo, bran, and tip cap (Watson 1984; Boyer and Shannon 1987). Flour or meal functionality in foods thus depends significantly on this major component. Aside from its nutritive value, the significance of starch is appreciated more for the physical properties it imparts to our foods.

Mutations that alter the levels of starch, the amylose and amylopectin ratio, and starch structure in the maize endosperm have been identified (Shannon and Garwood 1984; Wasserman et al. 1995). The availability of these mutants has also enabled study of the correlation between the molecular structure and functional properties of starches. These mutations lead to starches with new properties and functionalities and greatly expand the range of possible food and industrial applications. With the increasing power of genetic engineering techniques and a closer integration with physical property testing, there is now considerable global interest in developing highly targeted starch structure variants with specific functionality. We view the genetic engineering of starch as an important growth area, potentially providing natural materials to supplement or replace chemically modified starches in some applications.

Food starches are used in a wide range of applications as thickeners, stabilizers, carriers of low-molecular-weight substances, bulking agents for controlling consistency, and texture enhancers. The physicochemical changes that starch undergoes when it is heated in the presence of water are responsible for the unique character of many foods. The viscosity and mouthfeel of gravies and puddings and the texture of gum drops and pie fillings are classical examples. The different starch and starch derivatives in commercial use are listed in Table 1 (Lillford and Morrison 1997).

The diversity of starch applications supports ongoing research toward a

Table 1 Starch and Starch Derivatives in Commercial Use

Native starches	Corn/maize
	Wheat
	Sorghum/milo
	Rice
	Potato
	Tapioca/cassava
	Sago
	Arrowroot
	Pea
Genetically modified cultivars	Waxy maize
	Amylomaize
	Waxy sorghum
	Waxy rice
Chemically modified esters, ethers, cross-linked, cationic, oxidized	Monophosphates
	Acetates
	Hydroxypropyl
	Adipates
	Diphosphates
Hydrolyzed starches—acids, enzymes	Corn syrups
	Maltodextrins
	Glucose
Dextrins	Yellow gums
	White gums
	British gums

Source: Lillford and Morrison 1997.

better understanding of starch functionality in food systems. The purpose of this chapter is to review the chemical nature and physical nature of starch as these relate to starch functional properties and the end uses of starch. It also attempts to describe the molecular nature of starch from different botanical sources and how their amylose and amylopectin fractions relate to physical properties of starch gels. We also try to present an overview of the growing interest in the fine structure of amylose and amylopectin as they relate to starch functionality. Another aspect presented is a consideration of the advances in microscopy that led to a better understanding of the structure of the granule and its implications on resistance to enzymatic and nonenzymatic hydrolysis.

II. CHEMICAL AND PHYSICAL NATURE

A. Molecular Starch

Next to cellulose, starch is the most abundant carbohydrate in plants. Condensation of glucose units by enzymatic processes in the plant results in a linear chain of about 500–2000 glucose units known as amylose. It is a starch polysaccharide that consists primarily of long-chained α-1,4-linked glucose molecules. It may contain a few α-1,6 branch points. By a second enzymatic mechanism, some of the long linear chains break down into short lengths of ~25 glucose units, and these fragments recombine into multiple branched or treelike structures. This branched fraction, known as *amylopectin*, coexists with amylose in most common starches. It is a starch polysaccharide that consists of relatively short-chain α-1,4-linked glucose molecules interconnected by many α-1,6 branch points (Pomeranz 1991). Figure 1 presents the structure of α-1,4 and α-1,6 glucan linkages in starch and a diagrammatic representation of possible structures of amylose and amylopectin (Shewmaker and Stalker 1992).

The starch industry now demands the development of hybrids that produce native starches that behave like chemically modified starches. The maize wet-milling industry produces a number of modified-starch products important in the food industry. Genetic variability in starch structure and functional properties has led to the use of specialty starches from waxy and high-amylose genotypes (Shannon and Garwood 1984). These are essentially genetic manipulations of the relative amounts of the two polymers, amylose and amylopectin, in maize starch. The introduction of starches containing double mutant combinations with properties similar to chemically modified starches has resulted in several patents (Katz 1991). However, the mutants or double mutants may behave differently in hybrid combinations, due to differences in their expression in different inbred backgrounds (Li and Corke 1999) and in the nature of the heterotic expression. Although the amylose/amylopectin ratio is an underlying factor controlling the

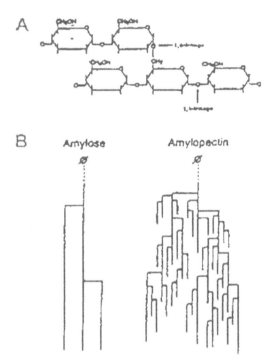

Figure 1   (A) Structure of α-1,4 and α-1,6 glucan linkages in starch. (B) Diagrammatic representation of possible structures of amylose and amylopectin. (From Shewmaker and Stalker 1992.)

rheological properties of starch pastes and gels, another genetic factor affecting rheological behavior is the fine structure of the amylose and amylopectin molecules themselves.

Starches isolated from different botanical sources are known to have different functional properties. Normal maize starch produces an opaque, short paste and a firm gel. Waxy maize and potato starches, on the other hand, produce clear and long pastes that do not set to a gel. The physical and chemical properties of common starches are presented in Table 2 (Pomeranz 1991). Traditionally this has been attributed to amylose content and the presence of phosphate derivatives. However, amylose molecular size and amylopectin branch chain lengths also differ, as reported in various studies (e.g., Hizukuri 1985; Takeda and Hizukuri 1987). Reconstitution studies conducted by Jane and Chen (1992), involving mixtures of varying molecular weights of amyloses and amylopectins fractioned from different botanical sources from potato and normal and high-amylose maize starches, showed that functional properties such as viscosity, gel strength, and

Table 2 Physical and Chemical Properties of Common Starches

Starch	Granular size (μm) Range	Granular size (μm) Average	Amylose (%)	Swelling Power (%)	Solubility at 95°C (%)	Gelatinization Range	Source	Taste	General Description of granules
Barley	2–35[a]	20	22	—	—	56–62	Cereal	Low	Round, elliptical, lenticular
Maize									
Regular	5–25[b]	15	26	24	25	62–80	Cereal	Low	Round, polygonal
Waxy	5–25	15	~1	64	23	63–74	Cereal	Low	Round, oval indentations
High amylose	—	15	Up to 80	6	12	85–87	Cereal	Low	Round
Potato	15–100	33	22	1000	82	56–69	Tuber	Slight	Egglike, oyster indentations
Rice	3–8[c]	5	17	19	—	61–80	Cereal	Low	Polygonal clusters
Rye	2–35[d]	—	23	—	—	57–70	Cereal	Low	Elliptical, lenticular
Sago	20–60	25	27	97	—	60–74	Pith	Low	Egglike, some truncate forms
Sorghum	5–25[e]	15	26	22	22	68–78	Cereal	Low	Round, polygonal
Tapioca (cassava)	5–35	20	17	71	48	52–64	Root	Fruity	Round–oval, truncated on side
Wheat	2–35[f]	15	25	21	41	53–72	Cereal	Low	Round, elliptical, lenticular
Oats	2–10	—	27	—	—	56–62	Cereal	Low	Polygonal, compound

[a] Large starch granules above 5; small granules below 5. Small granules gelatinize at 75–80°C.
[b] Mainly 10–15.
[c] Some clusters 9–30; some as large as 40.
[d] 2–8 up to 35.
[e] Mainly 10–12.
[f] 2–5, 6–15, and above 15.
Source: Pomeranz 1991.

clarity are significantly affected. Synergistic effects on paste viscosities were observed when amylose and amylopectins were mixed. Figure 2 presents gel permeation profiles of amyloses and amylopectin from the study. The long-chain amylopectin and the intermediate-molecular-size amylose produce the greatest synergistic effect on viscosity. It is suggested that the degree and size of branches

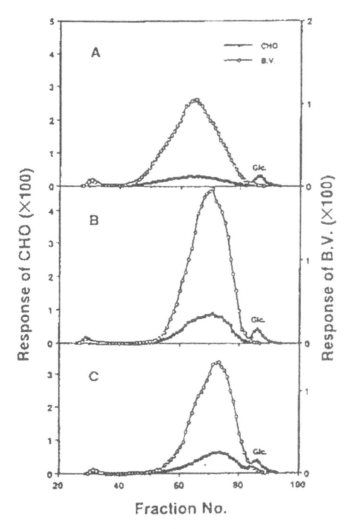

Figure 2   Sepharose CL-2B gel permeation profiles of amylases. A. Potato amylose; B. normal maize amylose; C. high amylose VII maize amylose. Glucose (Glc) was used as marker. CHO = total carbohydrate, B.V. = blue value. (From Jane and Chen 1992.)

as well as their relative pattern might be more useful predictors of starch behavior (Bradbury and Bello 1993).

B. Amylose

Amylose is essentially a linear molecule containing (α-1,4-linked glucose units with a small number of branches (Greenwood 1964; Takeda and Hizukuri 1987). The sidechains on those molecules that are branched are few and so long that they act similarly to unbranched molecules. With the linear chemical structure, amylose has the ability to change its conformation. With many hydroxyl groups, there is a high hydrogen bonding capability, with strong internal forces that permit these changes. A helical conformation is common for amylose, and a double helix form when different helices pack together. An open channel in the center of a helix permits complexing with other molecular species, such as iodine. The linear fraction displays an intense blue color with iodine, while the pure branched fraction shows only a weak violet-red color. This color difference is attributed to helical conformation of the linear molecules, with iodine in the center of the helix. The particular electronic resonance of this system gives rise to a blue color (Pomeranz 1991). The amylose content of starch is estimated by one of many variations of the classical reaction between amylose and iodine to form a stable complex, which is measured either spectrophotometrically or by potentiometric titration. The reaction is not an absolute indicator of amylose content because of interference from amylopectin and intermediate materials, e.g., long-chain amylopectins have been reported in some maize mutant genotypes.

Although the molecular composition of starch can be described approximately as a mixture of branched molecules of amylopectin (average molecular weight of at least 10^8) and amylose (average molecular weight between 10^5 and 10^6), the molecular structures of these components are quite complex. For example it has been shown that about one-half of the amylose molecules in maize starch are branched (Takeda et al. 1988). In addition, some amylopectin, such as in amylose extender—waxy (*ae wx*) maize starch, contains long chains that bind iodine in a manner similar to amylose and contribute to apparent amylose content. It is suggested that the degree and size of branches, as well as their relative distribution pattern, might be more useful predictors of starch behavior.

Methods based on the iodine reaction remain a convenient means for estimating amylose content, giving accurate and reproducible results. But such methods are generally time consuming and therefore not suited for use in quality control applications. Modification of the iodine method has been done using low-temperature gelatinization in $CaCl_2$ to shorten the time required as compared to using conventional thermal dissolution in dimethylsulfoxide (Knutson and Grove 1994). A good correlation between blue-value measurements and amylose content determined by size exclusion chromatography has been demonstrated. High-performance size exclusion chromatography (HPSEC) can be used to separate starch and debranched starch. This technique can directly monitor the effect of debranching on the molecular size distribution of starch and determine the high-molecular-weight linear amylose content of starch. In this way, the contribution of long-chain amylopectin fractions to apparent amylose content can also be estimated. It was shown that the amylose fractions in dent and amylomaize V maize starches were mainly linear because their elution times were unchanged after debranching. Long linear chains of amylopectin apparently contribute to the amylose values determined colorimetrically (Bradbury and Bello 1993).

1. Amylose–Lipid Interactions

Substantial work has been done on the formation of amylose–lipid complexes and their effects on functionality. According to Morrison (1988), lipids in starch can be classified as either true starch lipids or as starch-surface lipids. True starch lipids are lipids inside the native starch granule and are exclusively free fatty acids and monoacyl lipids (lysophospholipids). Although some investigators (Meredith et al. 1978; Galliard and Bowler 1987) assumed that starch lipids are complexed with amylose in the native starch granules, there are indications that amylose and lipids coexist independently and form complexes only during gelatinization (e.g., Kugimiya et al. 1980; Morrison et al. 1993b). Surface starch lipids are lipids acquired from the surrounding matrix. It is unclear to what extent these have penetrated the starch granule or how they are bound (Morrison 1988). These lipids can be extracted at room temperature by different solvents. The true starch lipids, however, are contained in the dense structure of the granule and not easily extracted. Although lipids are minor constituents of cereal starches (up to 1.0%), they have a remarkable influence on the gelatinization and retrogradation of starch (e.g., Lorenz 1976; Hibi et al. 1990).

The effects of polar monoacyl lipids, added as well as native, on the gelatinization behavior of starch is well recognized and exploited in many starch-containing food products. The function of monoacyl lipids is usually attributed to the formation of the helical inclusion complex between amylose and the hydrocarbon chain of the lipid. This complex-forming ability differs among different lipids and is reported to be highest for saturated monoglycerides, whereas it is very low for two hydrocarbon chains like lecithin and zero for triglycerides (Eliasson et al. 1988).

Lipids or surfactants act as texture modifiers when added to starch-containing foods. Saturated monoglycerides and sodium stearoyl lactylate are added to baked goods because of their ability to retard firming and retrogradation of starch. The complex-forming ability of amylose with monoglycerides and related surface-active monacyl lipids has been exploited in breadmaking to retard staling (Krog et al. 1989), in the manufacture of instant mashed potato to prevent stickiness (Hoover and Hadziyev 1981), and in extruded starch-containing products to control texture (Launay and Lisch 1983). It was also shown that formation of complexes prevents leaching of amylose during gelatinization, inhibits the swelling of starch granules heated in water, and reduces the water-binding capacity of starch (Eliasson 1985).

2. Resistant Starch (RS)

Amylose has also been implicated in the formation of resistant starch. *Resistant starch* (RS) refers to some starches and products of starch degradation that are not absorbed in the small intestine of healthy individuals. Resistant starch was originally considered to consist of four subcategories: (a) type I, which is based on the physical inaccessibility of starch granules trapped in the food matrix; (b) type II, which refers to native starch granules and is related to structure and conformation; (c) type III, reflecting the formation of retrograded starch material during processing; and (d) type IV, referring to some chemically modified starches that resist enzymatic hydrolysis (Englyst and MacFarlane 1987). There is considerable interest in the nutritional value of RS in foods, since a relatively slow rate of starch hydrolysis in the gastrointestinal tract of humans is associated with low glycemic responses and may confer some of the physiological effects of dietary fiber (Englyst and MacFarlane 1986).

Relevant to RS type II, X-ray diffraction crystallography has identified three possible arrangements of the crystalline structure of the granule. In each case, the starch α-glucan chains exist as left-handed, parallel-stranded double helices. In the so-called A pattern, the center of the hexagonal array is occupied by an additional helix. This pattern is found in wheat and maize. Starches of this kind are digestible when measured in vitro. When the array is occupied by water, the B pattern is observed. Potato, banana, and high-amylose maize starches have the B pattern. Starch granules from these species are resistant to digestion. The C patterns, commonly found in legumes, are considered to be a combination of the A and B arrangements and are also resistant to enzymatic digestion. The observation that amylose content and the yield of RS are positively correlated focused interest on various food applications of high-amylose starches (Sievert and Pomeranz 1989). High-amylose varieties of cereals, like amylomaize and barley, are potential sources for RS production.

For RS type III, retrogradation is favored by the presence of linear polymer chains. High-amylose content, the presence of amylopectin treated with a debranching enzyme, or acid-hydrolyzed amylopectin can increase inter- and intrahelical hydrogen bonding in retrograded starch and the formation of resistant starch (e.g., Berry 1986; Czuchajowska et al. 1991). Eerlingen et al. (1993a, b) and Eerlingen et al. (1994a, b, c) focused on several aspects of RS formation: the impact of amylopectin retrogradation, the influence of amylose chain length, the impact of incubation time and temperature of autoclaved starch, the effect of lipids, and the formation of bread enriched with RS.

Both added and endogenous lipids, although present in low quantities, have an influence on RS (type III) formation in starch. Endogenous and added lipids can form enzyme-digestible inclusion complexes with amylose. Less amylose is then available to form enzyme-resistant double helices. Thus the yield of auto-claved starch was increased by defatting starch and decreased by the addition of exogenous lipids. Minor quantities of lipids may cause significant changes in RS yield by occupying segments of the amylose chain (amylose–lipid complexation) and by causing steric hindrance for the formation of double helices in the uncom-plexed segments (Eerlingen et al. 1994b). The presence of complexing lipids affects the reassociation behavior of amylose on the retrogradation of starch and thus affects the formation of amylose–lipid complexes (Sarko and Wu 1978). Slade and Levine (1987) reported that the crystallization of amylose–lipid com-plexes is favored over amylose retrogradation.

C. Amylopectin

Amylopectin is a branched molecule with (α-1,4)-linked glucose units in linear chains and (α-1,6)-linked branched points (Greenwood 1964; Lineback 1984). Reported molecular weight values for amylopectin run into the millions (Hoseney 1986). Amylopectin is less prone to gelation, retrogradation, and syneresis be-cause of the branched structure. There is no convenient method for directly de-termining amylopectin. Amylopectin is commonly estimated by subtraction of amylose. The amylopectin model of Manners (1985) shows how branched mole-cules can be involved in the formation of alternating crystalline and amorphous regions of the granule. The packing together of clustered branches would provide crystallinity, while the branch points are considered to be in the amorphous re-gions. The crystalline regions are more resistant to enzymatic and chemical action and to penetration by water than are amorphous regions in starch granules.

The prominence given to amylose in starch studies is understandable, since retrogradation and its consequences are relatively easy to identify and follow; however, studies had revealed crystallinity in the branched amylopectin fraction. Crystallinity data indicated that native waxy starch (100% amylopectin) had 40% crystallinity, while high-amylose starch has about 15% (Zobel 1992). Imberty et al. (1991) presented a geometric arrangement where limited disruption of the original double helix would occur, further indicating that a branch point could facilitate crystallite formation.

Amylose can be leached from starches, leaving the amylopectin as well as granule crystallinity largely intact. Such findings have led to the conclusion that amylose is mainly in the amorphous phase and that amylopectin is the main component of the crystalline fraction. Since the major fraction in most starches is amylopectin, the relationships of its fine structure to utilization are increasingly investigated. Studies on amylopectin are dependent on the development of enzymatic and instrumental methodologies. It should be stressed that due to the heterogeneous nature of starch, there is a need to obtain detailed rather than average property measurements. As profile development continues and databases are enlarged, one prospective result is the development of predictive methods for amylopectin structure and starch properties. For example, the waxy rice molecular distribution profile differs markedly from that of potato (Hizukuri 1986) (Fig. 3). Waxy rice has unusual, and unexplained, freeze thaw stability (Schoch 1967). An interesting consideration is whether or not the chain profile provides a clue to this stability (Zobel 1992). As more information concerning structure/function

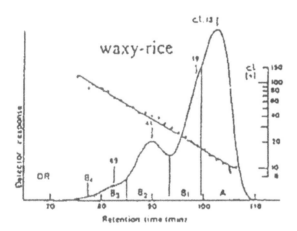

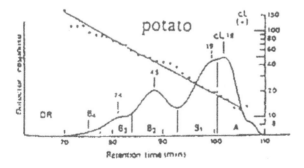

Figure 3 Molecular distribution profiles for waxy rice and potato amylopectins by gel permeation HPLC. (From Hizukuri 1986; Zobel 1992.)

relationships becomes available, researchers will look forward to modification of starch fine structure itself (Wasserman et al. 1995).

D. Intermediate Material

A third component that exists in some starches, called *intermediate material* (IM), differs with starch type and maturity. Lansky et al. (1949) found an IM characterized by a lower molecular weight than that of amylose and a slightly branched structure. These molecules showed responses intermediate between amylose and amylopectin (AP) to iodine-binding procedures, and observed mainly in amylomaize or wrinkled pea starches (Baba and Arai 1984; Colonna and Mercier 1984). Intermediate material also has been identified in oat starch, with iodine affinity values larger than that of AP and molecular weight less than that of AM (Paton 1979). Variation in the amount of AM, AP, and IM and in their structure and properties can result in starch granules with very different physicochemical and functional properties, which may affect their utilization in food products or industrial applications (Kobayashi et al. 1986; Yuan et al. 1993). Kasemsuwan et al. (1995) described intermediate material as a third population of molecules observed in SEC profiles relative to intermediate elution volumes between those of amylopectin and amylose, particularly in the case of genetically modified maize starches. Because of the different functionalities of these components, detailed characterization of the structure of starch is important.

E. Minor Components of Starch Granules

1. Lipids

The common cereal starches (maize, wheat, sorghum, rice) contain a high percentage of lipid (0.8–0.9%), compared with potato and tapioca (0.1%). Table 3 (Swinkels 1985) shows the composition of lipids in maize and wheat starch. The presence of high amounts of lipid in maize and wheat starch has unfavorable effects. The lipids repress the swelling and solubilization of maize and wheat starch granules. The lipids also increase the pasting temperatures and reduce the water-binding ability of these starches. The presence or formation of insoluble amylose–lipid complexes causes turbidity and precipitation in starch pastes and starch solutions. The amylose–lipid inclusion compounds make starch paste and films opaque or cloudy. This does not contribute to the thickening power or binding force of the gelatinized starch. The oxidation of unsaturated lipids may cause the formation of undesirable flavors in pregelatinized maize and wheat starch products. Potato starch and tapioca starch contain only a low amount of lipids (about 0.1%) and do not show the unfavorable effects mentioned.

Table 3 Minor Components of Starch Granules

Starch components	Potato starch	Maize starch	Wheat starch	Tapioca starch	Waxy maize starch
Moisture at 65% RH and 20°C	19	13	13	13	13
Lipids (%, dry basis)	0.05	0.70	0.80	0.10	0.15
Proteins (%, dry basis)	0.06	0.35	0.40	0.10	0.25
Ash (%, dry basis)	0.40	0.10	0.20	0.20	0.10
Phosphorus (%, dry basis)	0.08	0.02	0.06	0.01	0.01
Taste and odor	Low	High relative	High relative	Low	Medium

Source: Swinkels 1985.

2. Nitrogen Substances

The nitrogenous substances include proteins, peptides, amides, amino acids, nucleic acids, and enzymes that may be present in the starch granule. Potato and tapioca starch contain only small amount of nitrogenous substances (0.1%) compared with maize (0.35%), wheat (0.4%), and waxy maize (0.25%). The relatively high content of nitrogenous substances in these cereal starches may have undesirable effects, such as the formation of mealy flavor and odor in the pregelatinized starches, the tendency of cooked starch to foam, and color formation in the starch hydrolyzate (Swinkels 1985).

3. Ash

Ash is the residue of the starch product after complete combustion at a specified temperature. Native potato starch gives a relatively high ash residue because of the presence of phosphate groups in salt form. The ash residue of native potato starch contains mainly phosphate salts of potassium, sodium, calcium, and magnesium. The ash residue of the cereal starches corresponds partly with the amount of phospholipids in the starch granules.

4. Phosphorus

The phosphorus in cereal starches occurs mainly as phospholipids. Tapioca contains only very low amounts of phosphorus compounds. Potato starch is the only commercial starch that contains an appreciable amount of covalently bonded phosphate monoester group. Phosphate groups in potato starch amylopectin range from one phosphate group per 200 to one per 400 glucose units. This corresponds to a degree of substitution (DS) in potato starch of about 0.003. The counterions for the phosphoric ester groups in potato starch are mainly potassium, sodium, calcium, and magnesium. The counterion distribution in potato starch depends upon the composition of the process water in the manufacture of potato starch products. These counterions play an important role in the potato starch gelatinization process. The negatively charged phosphate groups impart some polyelectrolyte character to potato starch. Although the ionic charge is not high, in aqueous solutions the repulsion of like charges helps to untangle the individual polymer molecules and extends the sphere of influence. The bonded phosphate groups contribute to the properties of potato starch, such as low pasting temperature, rapid hydration and swelling, and high water-binding ability and a high viscosity (Swinkels 1985).

III. THE STARCH GRANULE

The basic physical structural unit of starch is the granule, which has a distinctive microscopic appearance for each botanical source. The common starches are readily identifiable by using a polarizing light microscope to determine their size, shape, and form and the position of the hilum (botanical center of the granule). Jane et al. (1994) have demonstrated that starch granules from different biological sources (roots and tubers, grains, maize, peas and beans, fruits, and nuts) could present a wide variety of fascinating morphologies. Scanning electron micrographs of grain starches from their work are presented in Figure 4.

There is also an observed set of common characteristics and those specific to each biological source, indicating substantial genetic control. Among possible genetically controlled factors are the particular types and amounts of synthetic enzymes that function in the biosynthesis of the starch molecule. Biosynthesis occurs in the amyloplast organelle, whose membranous structure and physical characteristics could impart a particular shape and morphology to the individual starch granules. This in turn may have an effect on the arrangement and association of the amylose and amylopectin molecules in the granule and on the morphology of the granule (Jane et al. 1994).

The modes of biosynthesis of starch and the nature of starch granular organization are still subjects of considerable research, and the outcome of the biosynthetic process is different in diverse plant species as well as in different genotypes of the same species. Because starch is synthesized in plastids, those structures must possess all the enzymes necessary for granule formation. The starch grows by apposition (deposition of material on the outside). The new layer deposited on the outside of the granules varies in thickness, depending upon the amount of carbohydrate available at the time (Hoseney 1986). Starch biosynthesis is a complex process that is mediated by several groups of biosynthetic enzymes. It is the balance of biosynthetic enzyme activities that defines the ultimate structure

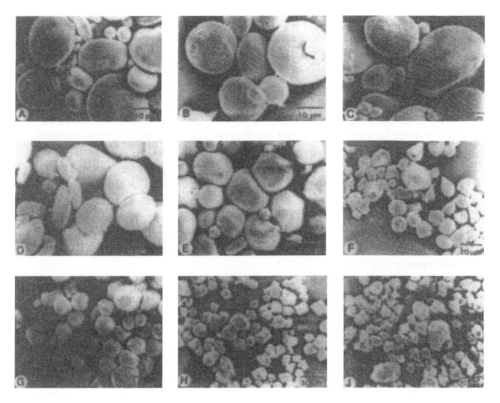

Figure 4 Scanning electron micrographs of grain starches, all at 1500×. A. wheat; B. triticale; C. rye; D. barley; E. sorghum; F. oat; G. cattail millet; H. rice; J. waxy rice. (From Jane et al. 1994.)

of a particular starch. Mutations in a specific biosynthetic enzyme activity can skew the structural balance to yield a distinctive starch biopolymer structure. Functional properties of the starches are then related not only to their structure as polymers but also to the packing of polymers within the granules (Banks and Greenwood 1973).

Molecular arrangement within a starch granule is to some extent radial, as evidenced by fibrillar fracture surfaces that can extend radially along growth rings, when starch granules are fractured (French 1984). Starch granules are birefringent, indicating a high degree of internal order. *Birefringence*, the ability to refract light in two directions, is evidenced by distinctive patterns under a polarizing microscope. The loss of birefringence indicates disruption of the molecular arrangement in the crystalline areas and is used as a major criterion for gelatinization. X-ray diffraction studies show that the molecular arrangement in a starch granule is such that there are crystalline (micellar) regions imbedded in an amorphous matrix. The molecular nature of the crystallinity within the starch granule is not fully understood but certainly involves amylopectin in both nonwaxy and waxy starches.

There is no good understanding of the state of amylose in a normal starch granule. More is understood about the ordered nature of amylopectin in a starch granule (Thompson et al. 2000). However, the linear amylose molecules are believed to be interspersed between amylopectin, rather than located in bundles (Jane et al. 1992). Findings by Morrison et al. (1993a, b) indicated that in cereal starches, there are two amorphous forms of amylose: lipid-free amylose and lipid-complexed amylose. For high-amylose maize starches (HAMS), X-ray diffraction patterns provide evidence that single-helical amylose–lipid complexes may be partially crystalline. And HAMS are even less crystalline than normal maize starch, perhaps due to a lower proportion of amylopectin (Thompson et al. 2000). Blanshard (1987) suggested that limited cocrystallinization between amylose and amylopectin may occur.

By combining old and new results provided over the years by a range of microscopic techniques, Gallant et al. (1997) gathered some of the pieces of the puzzle concerning starch granule internal structure and organization (Fig. 5). Considerable evidence now exists from SEM, TEM, and enzyme degradation studies and more recently from AFM, which indicate that the crystalline and amorphous lamellae of amylopectin are organized into larger, more or less spherical structures, termed *blocklets*. This was an idea first hinted at by Nägeli in 1858 and Bädenhuizen in 1936. The blocklets range in diameter from around 20 to 500 nm, depending on the botanical source and location in the granule.

The amorphous fraction controls the variation in granule volume due to its ability to absorb and release the free water of a raw starch granule. Figure 5 shows the existence of radial channels within starch granules, believed to be composed predominantly of semicrystalline or amorphous material and through which amylose can exit the granule structure. This also reinforces the real existence of a granule radial structure and clearly defined locales (possibly between the more crystalline blocklets) that are more easily degraded by enzymes. These have implications for determining the resistance of starch to digestion (Gallant et al. 1997).

These support earlier work on the nature of the outer surface and its relationship to the chemical and enzymatic reactivity of granules. Native granules also exhibit resistance to enzyme-catalyzed digestion, and each of the granules is attacked in a characteristic pattern (Leach and Schoch 1961). The pattern of digestion that develops when native maize and wheat starch granules are treated with amylases indicates that some areas of their surface are more susceptible to

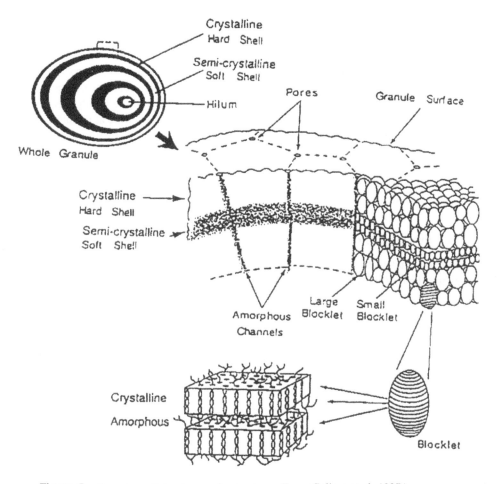

Figure 5 Overview of starch granule structure. (From Gallant et al. 1997.)

attack than are others (e.g., Evers and McDermott 1970; Evers et al. 1971; Fuwa et al. 1977). Fannon et al. (1992) showed that pores are found along the equatorial groove of large granules of wheat, rye, and barley starches but not on other starches (rice, oat, tapioca, arrowroot, canna). They proposed that the pores affect the pattern of attack by amylases and by at least some chemical reagents. Kanenaga et al. 1990 also reported that maize starch (which has pores) was more susceptible to enzymatic digestion than potato starch (on which no pores were found) and which Leach and Schoch (1961) found was digested in a different pattern.

These pores are characteristic of particular species of starch and not produced by drying (Fannon et al. 1992).

In later work (Fannon et al. 1993), it was shown that pores previously observed on the external surface of sorghum starch granules open to serpentine channels that penetrate into the granule interior. This provides evidence on observations that the enzymatic digestion of maize starch begins at the hilum (Leach and Schoch 1961). It is likely that at least some channels penetrate at the hilum. This suggests that hidden surfaces of channels for those starches with pores on the external surfaces must be considered in order to evaluate susceptibility to enzyme attack. A possible function is the regulation of the rate of starch granule conversion into D-glucose during germination (Gallant et al. 1997; Fannon et al. 1993).

A. Swelling and Solubility of Granular Starch

Starch is not water soluble, because granules are too large to form a solution. Starch has a relatively high density, about 1.45–1.64 g/cm^3, depending on its source, its prior treatment, and the measurement method (French 1984), and therefore settles out of suspension. Starch granules absorb some water in suspension at room temperature, but the amount of swelling is limited in intact granules. When heated in water suspension to progressively higher temperatures, very little happens, until a critical temperature is reached. At this point the granule begins to swell rapidly, losing the polarization crosses, a process termed *gelatinization*. Granules in a population from the same sample do not all gelatinize at the same time, but rather over a range of temperature (Pomeranz 1991).

The current methodology for the determination of swelling and solubility patterns of starch was developed by Leach et al. (1959). It was postulated that bonding forces within the granule would influence the manner of swelling. The swelling, when plotted against the temperature of pasting, would give a curve representing progressive relaxation of the bonding forces within the granule, permitting comparison of the relative bonding strengths in various starches from the temperature (energy level) necessary to cause relaxation. The method involved making a suspension of starch in a known volume of water and gently stirring to keep it in suspension while incubating it at the desired temperature for 30 min, centrifuging it at 2200 rpm for 15 min, and obtaining the weight of the gel, which is expressed as sediment paste per gram (dry basis) of the starch. Solubility of the starch is obtained by drying the supernatant and is expressed as percentage soluble. This test has always been extensively used in the characterization of starches. Modifications involved the used of a programmed shaking of the starch suspension instead of stirring (e.g., Crosbie 1991) and determination of solubles by colorimetric methods as total carbohydrate (e.g., Jacobs et al. 1996) or as amylose leached (e.g., Hoover and Vasanthan 1994).

B. Starch Ghosts

The surface of the granule remains as a remnant after cooking under light-to-moderate shear. The remnant, or *ghost* as it is commonly called, is important to paste and gel structure and properties. The ghost can be structurally strengthened by cross-linking, which strongly affects the functional properties of the starch by imparting an increased stability of the granules to heat, shear, and acid. Unmodified maize starch granules also produce pronounced ghosts when cooked, indicating a naturally stronger shell. Different cultivars of maize also produce different granule ghosts, indicating a genetic control of ghost structure. Ghosts from dull waxy maize starch, while appearing similar to those of waxy maize starch, were finer in structure and did not collapse, appearing more like a cross-linked starch. Ghosts of high-amylose starches were thick-walled, substantial, extended hollow spheres, similar to those of heavily succinylated cross-linked normal-amylose maize starch. The ghosts, like the pastes, showed structural differences that correlated strongly with the rheological behavior of each starch type (Fannon and BeMiller 1992). Using simple no-shear cooking in an autoclave, iodine staining, and light microscopy, 17 genetically modified maize starches were identified and distinguished through their ghost microstructures (Obanni and BeMiller 1995).

IV. STARCH FUNCTIONALITY

Functional properties denote characteristics that govern the behavior of a food component during processing, storage, and preparation. Functionality in a broad sense is any property of the food component, other than its nutritional value, that affects its utilization (Matil 1971; Pomeranz 1991). Starches are used as thickening agents (sauces, cream soups, pie fillings), colloidal stabilizers (salad dressings), moisture retainers (cake toppings), gel-forming agents (gum confections), binders (wafers, ice cream cones), and coating and glazing agents (nut meats, candies). The functionality of starch in various foods is dependent on its physicochemical properties.

A. Enzymatic Hydrolysis

The various kinds of enzymes that catalyze hydrolysis of amylose and amylopectin molecules are obtained from a variety of sources, such as fungi, bacteria, cereals, and other plants, and in the case of α-amylase even from animal sources. Their efficiency, specificity, and optimum conditions of activity depend on the source. A major application of enzymatic hydrolysis of starch is in the production of starch syrups. Since a starch molecule has only a reducing end group (carbon-

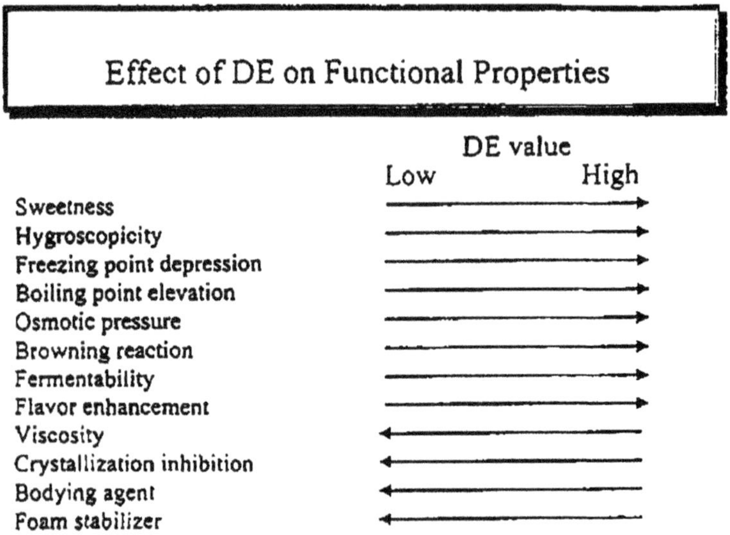

Effect of DE on Functional Properties

                                      DE value
                              Low   High

Sweetness
Hygroscopicity
Freezing point depression
Boiling point elevation
Osmotic pressure
Browning reaction
Fermentability
Flavor enhancement
Viscosity
Crystallization inhibition
Bodying agent
Foam stabilizer

Figure 6 Functional properties of starch ingredients as related to their dextrose equivalent (DE) values. (From Corn Refiners Association; Austin and Pierpoint 1998.)

1), the reducing power is used to indicate a change in molecular weight due to hydrolysis, because enzymatic hydrolysis exposes more reducing ends of the starch molecule with continued cleavage (Penfield and Campbell 1990).

The products of the partial hydrolysis of maize starch, called *hydrolysates*, are classified by their dextrose equivalent (DE). By industry definition, the DE of a maize syrup is the total amount of reducing sugars expressed as dextrose on a dry basis. Since starch contains only one reducing group per molecule, it is essentially zero DE. Complete hydrolysis, on the other hand, yields a product that is 100 DE. Partial hydrolysis produces a range of products that provide the functional properties illustrated in Figure 6. The arrow indicates the direction in which the property increases (Austin and Pierpoint 1998).

B. Starch Modification

Chemical and physical modifications provide functionalities not readily available or controllable from commercially available native starches. There is a wide range of treatments that can be applied to raw starch to change their use. Properties sought from modified starches include enhanced rheological characteristics, textural qualities, optical properties, and enhanced stability of the system. The types of modified starches and their properties are summarized in Figure 7 (Wurzburg 1986; Lillford and Morrison 1997).

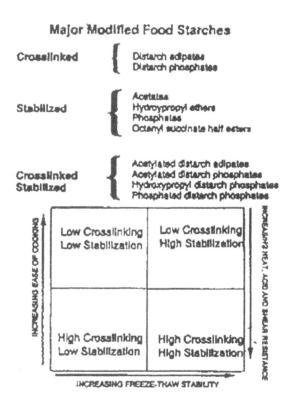

Figure 7 Types and properties of modified starches. (From Wurzburg 1986; Lillford and Morrison 1997.)

Chemical modification of starch is made possible because of the many hydroxyl groups available for reaction with the chemicals used. Chemical treatment involves holding a suspension of starch in a dilute solution of the modifying reagent at a temperature that is too low to permit appreciable granule swelling but high enough to cause some disruption of internal order and permit entry of some water and reagent. The reaction between the reagent and the starch molecules take place just within the granule and on the surface during treatment (Rogol 1986). The treated starch is dried for use as a food or industrial ingredient.

C. Physical Modification

For a wide range of starch applications (foods, plastic, paper, textile), properties of native starches do not meet industrial needs. However, native starch granules can be modified to obtain desired properties. Mostly chemical modification (such as cross-linking and/or acetylation) is used, but there is a growing interest in physical modification (heat moisture, shear, or radiation) of starch, especially for food applications. In fact such physically modified starches are considered to be natural materials with high safety. Two hydrothermal treatments that modify the physicochemical properties of starch, without destroying granule structure, are annealing (ANN) and heat-moisture treatment (HMT). Both treatments involve storage at a certain moisture level and a specific temperature, during a certain period of time.

In most publications, treatments in "excess" water (>60% w/w) or at "intermediate water content" (40–55% w/w) are referred to as annealing, whereas the term "heat-moisture treatment" is used when "low moisture levels" (<35% w/w) are applied. Both physical modifications occur at temperatures above the T_g but below the gelatinization temperature (determined at the specific moisture conditions used during the treatment) of the starch granule. Thus, hydro-thermal modifications can take place only when starch polymers in the amorphous are in the mobile rubbery state. A treatment time of only a few minutes often suffices to result in detectable changes in the physicochemical properties of the starch. The effects of annealing and heat-moisture treatment on starch are most readily differentiated using wide-angle X-ray scattering (WAXS) and differential scanning calorimetry (DSC). While annealing hardly affects WAXS patterns, heat-moisture treatment changes B-type patterns into A- (or C-) type patterns. Furthermore, annealing drastically narrows DSC gelatinization endotherms, while heat-moisture treatment causes a broadening of the endotherms (Jacobs and Delcour 1998). Depending on the botanical source of the starch, hydrothermal treatments showed marked impact on the gelatinization temperature, pasting properties, swelling and solubility, and rheological properties of the starch gel and susceptibility to acid and enzymatic hydrolysis (e.g., Sair 1967; Kulp and Lorenz 1981; Abraham 1993; Stute 1992; Hoover and Vasanthan 1994; Eerlingen et al. 1997).

Explanations offered for the observed effects of annealing and heat-mois-ture treatment on starch properties involve changes with respect to crystallite growth and increased order in the amorphous fractions, which has been attributed to increased interactions between amylose chains or between amylose and amylo-pectin (e.g., Knutson 1990; Seow and Teo 1993, Hoover and Vasanthan 1994; Kawabata et al. 1994), extra formation of amylose–lipid complexes (e.g., Lorenz and Kulp 1984; Kawabata et al. 1994), transformation of amorphous amylose into a helix (Lorenz and Kulp 1982), and alterations in the orientation of the crystallites within the amorphous matrix (Stute 1992).

D. Gelatinization and Pasting

Gelatinization of starch is responsible for its change properties during the prepara-tion and processing of food. After a survey conducted among starch scientists and technologists (Atwell et al. 1988), the following definition was proposed for *gelatinization*:

> Starch gelatinization is the collapse (disruption) of the starch granule manifested in irreversible changes in properties such as granular swelling, native crystallite melting, loss of birefringence, and starch solubilization. The point of initial gelatinization and the range over which it occurs is governed by starch concentration, method of observation, granular type, and heterogeneity within the granule population under observation.

Similarly, *pasting* had this proposed definition:

> Pasting is the phenomenon following gelatinization in the dissolution of starch. It involves granular swelling, exudation of molecular components from the granule, and, eventually, total disruption of the granules.

These definitions clearly do not support interchangeable use of the two terms. Although the terms *gelatinization* and *pasting* have often been applied to all changes that occur when starch is heated in water, gelatinization includes the early changes, according to the proposed definitions, and pasting includes later changes.

When the starch granule is heated up to gelatinization temperature in excess water, heat transfer and moisture transfer phenomena occur. The granule swells to several times its initial size as a result of loss of the crystalline order and the absorption of water inside the granular structure. The swelling behavior of cereal starches is primarily a property of their amylopectin content, and amylose acts as both a diluent and an inhibitor of swelling, specially in the presence of lipid (Tester and Morrison 1990a). Maximal swelling is also related to the molecular weight and shape of the amylopectin (Tester and Morrison 1990b).

Most studies on gelatinization and pasting have involved use of the Brabender Viscoamylograph, where a starch suspension (at a concentration depending on the swelling properties and thickness of the starch paste) is pasted over a programmed heating-and-cooling cycle as a chart of the viscosity changes is automatically recorded. Other viscometers were developed for specific purposes; however, despite fundamental problems due to measurement geometry and methodological factors, after some 60 years of use, the amylograph is the recognized industry standard for studying paste characteristics (Dengate and Meredith 1984). It has been used extensively by starch manufacturers and food processors alike. Pasting parameters have been used by many researchers for the characterization of starches from different botanical sources (species and cultivars) (e.g., Leelavathi et al. 1987; Bhattacharya and Sowbhagya 1979) and with different chemical additives (e.g., Kim and Seib 1993) and physical modifications (e.g., Abraham 1993; Stute 1992; Jacobs et al. 1996). However, the test requires a long time (typically about 90 min) and a large sample size, thereby limiting its use in many applications, such as in breeding programs, where small amounts

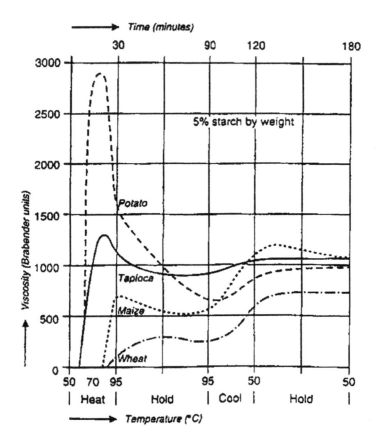

Figure 8   Pasting properties of native commercial starches. (From Swinkels 1985.)

of experimental samples are normally available. Pasting profiles of native commercial starches are presented in Figure 8 (Swinkels 1985) and those of sweet potato starches from four genotypes in Figure 9 (Collado et al. 1999).

1. Gelatinization Temperature

There are many methods employed for the determination of gelatinization temperature, such as the birefringence end-point method, the viscosity method using the amylograph (Lund 1984), X-ray diffraction (e.g., Ghiasi et al. 1982), the amylose/iodine blue value method (e.g., Lund 1984), and the differential scanning calorimetry (DSC) method (e.g., Wootton and Bamunuarachi 1979). The DSC method considers gelatinization as analogous to the melting of a crystal, and its use of a small sample (10–20 mg at 10–20% total solids) minimizes

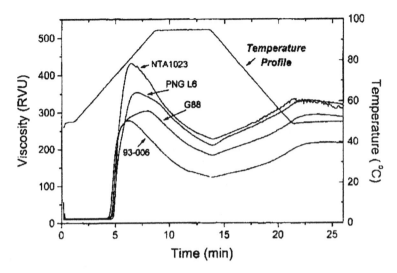

Figure 9 Representative Rapid Visco Analyzer (RVA) pasting profiles of four geno-types of sweet potato starch at 11% starch concentration. (From Collado et al. 1999.)

the thermal lag within the system. Since the pans containing the samples are hermetically sealed, water will not be lost from the system under normal operating temperatures. The purpose of the differential thermal system is to record the difference between an enthalpy change that occurs in a sample and one that occurs in a reference material when both are heated (Lund 1984).

2. Gelation

A starch gel is a solid–liquid system having a solid, continuous network in which the liquid phase is entrapped and has a characteristic structural shape resistant to flow. When starch granules swell, the amylose inside the granules leaches out. The leached amylose forms a three-dimensional network (Eliasson 1985; Tester and Morrison 1990a), and the swollen granules are embedded in this continuous matrix of amylose and amylopectin (Ring 1985). Extensive hydrogen bonding may occur during cooling. Simply stated, starch paste can be described as a two-phase system composed of a dispersed phase of swollen granules and a continuous phase of leached amylose. If the amylose phase is continuous, aggregation upon cooling will result in a strong gel due to the formation of hydrogen bonds from free amylose. The consistency of such a gel is generally firm; it appears white or cloudy and may shrink or lose water on storage (Hermansson and Svegmark 1996).

The initial stages of gelation of starch are dominated by solubilized amylose (Miles et al. 1985). The swollen gelatinized granules (containing mainly amylopectin) are embedded in and reinforce the interpenetrating amylose of the gel matrix. Svegmark and Hermansson (1993) found that the inherent amylose of potato starch did not contribute to gel formation and suggested that the starch granules caused the rheological behavior of the hot paste. The swollen starch granules formed a close-packed gel structure that possessed high shear resistance (Svegmark and Hermansson 1991). Evans and Haisman (1979) indicated that the material outside the swollen granule (e.g., amylose) had little effect on the rheology of the starch suspension.

The overall viscosity of starch pastes is governed primarily by a combination of the volume fraction of the dispersed phase and the concentration and composition of the continuous phase. At a high concentration (~10% solids), deformability of swollen particles seems to play a prevailing role (Doublier et al. 1987). It is proposed that in the dilute regime (when the volume of the swollen granules is less than the total volume), the viscosity is governed by the volume fraction of the swollen granules, whereas as in the concentrated regime it is governed by particle rigidity (Steeneken 1989).

Electron micrographs confirm the current model for the starch paste or gel, i.e., showing to varying extents a continuous phase of dispersed molecules surrounding a discontinuous phase of granule ghosts. Factors that affect the functional properties of the gels and pastes are, (a) the size of the openings in the continuous phase (matrix), (b) volume fractions of the continuous phase and the granule remnants, (c) rigidity of the continuous phase, (d) shape and rigidity of the granular ghost, and (e) interaction between the matrix and the granular ghost. Micrographs confirm that these parameters can be evaluated by this method and varied by chemical modification or genetic manipulation of starch. This may form the basis for a rapid method for screening new starches for desirable qualities (Obanni and BeMiller 1995).

E. Retrogradation

Under low energy input, as in freezing and chilling, further hydrogen bonding may occur, resulting in further tightening of structure with loss of water-holding capacity, known as *retrogradation*.

> Starch retrogradation is a process which occurs when the molecules comprising gelatinized starch begin to reassociate in an ordered structure. In its initial phases, two or more starch chains may form a simple juncture point that may then develop into more extensively ordered regions. Ultimately, under favorable conditions, a crystalline order appears (Atwell et al. 1988).

Amylose has been thought to impart stiffness to food systems, especially after retrogradation. Amylose was reported to be the main factor in short-term (several hours) development of the starch gel structure, while amylopectin was correlated with long-term (several days or weeks) development of the starch gel structure (Miles et al. 1983; Orford et al. 1987). Biliaderis and Zawistowski (1990) studied the time-dependent changes in network properties of aqueous starch gels. The storage modulus (G') time profiles revealed a two-phase gelation process: (a) an initial rapid rise due to amylose, and (b) a phase of slower G' development from amylopectin recrystallization (Biliaderis and Tonogal 1991). The G' of freshly prepared rice starch gels showed a linear relationship with the amylose content of the starch (IRRI 1991).

As shown earlier, retrograded amylose is classified as a resistant starch. It is believed that amylopectin does not substantially associate or retrograde upon standing because the outer branches are sufficiently long enough and therefore waxy starch pastes are nongelling (Pomeranz 1991) or form very weak gels. However, a role for amylopectin in retrogradation had been reported. The short amylopectin sidechains undergo a shift from coil to a helix transition (Winter and Kwak 1987). The WAXS technique has shown a slow development of crystallinity of the B form over time, which is closely related to the development of endothermic transition observed by DSC (Miles et al. 1985; Orford et al. 1987). These changes were due to a slow association of the double helices and were studied for potato and wheat starch by transmission electron microscopy, DSC, and rheology (Keetels et al. 1996). Ring et al. (1987) demonstrated that amylopectin staling within the gelatinized granule is the cause of increased firmness of the starch gel during storage. Varietal differences in rate and extent of amylopectin staling may help explain the variation in texture during storage of heat-processed rice products (Perez et al. 1993).

Shi and Seib (1992) reported that a decrease in the mole fraction of amylopectin chains between DP14 and 24 decreased the retrogradation tendency of waxy-type starches. Yuan and Thompson (1998) proposed that a greater portion of fraction chains between DP20-30 in *du wx* starch could explain the more rapid increase in G' and H during its storage. Yuan et al. (1993) reported that *du wx* starches from maize inbred lines had a greater tendency to retrograde than the *wx* starch at 10 and 30% concentrations. The rapid gelling behavior of *du wx* may be due the fact that commercial *du wx* has longer exterior chains (DP $>$ 12) than the *wx* starches and has ~3% high-molecular-weight (DP ~ 320) material (Yuan and Thompson 1998).

V. SUMMARY

Basic knowledge of the chemical nature and physical nature of starch is essential to understand the functional properties of starches in food applications. By characterization of starches we are able to identify suitable uses for it; however, this is not easily done unless a quantitative means of evaluation of quality for specific use is defined. Only then can modification be made to achieve required functionality for the specific use. When native starch cannot meet the quality required, the industry resorts to chemical or physical (or a combination) modification. Because there is growing concern over chemical modification as being "unnatural," other directions are increasingly being pursued, such as tapping nontraditional sources such as leguminous starches and the genetic modification of existing commercial varieties. With a diversity of starch uses in the food industry, there is an enormous challenge for breeders, farmers, and food manufacturers to meet consumer demands for less processed and more natural convenience foods.

REFERENCES

Abraham T.E. 1993. Stabilization of paste viscosity of cassava by heat-moisture treatment. Starch 4:131–135.

Atwell, W.A., Hood, L.F., Lineback, D.R., Varriano-Marston, E., and Zobel, H.F. 1988. The terminology and methodology associated with basic starch phenomena. Cereal Foods World 33:306–311.

Austin C.L., and Pierpoint, D.J. 1998. The role of starch-derived ingredients in beverage applications. Cereals Food World 43:748–752.

Baba, T., and Arai, Y. 1984. Structural characterization of amylopectin and intermediate material in amylomaize starch granules. Agricultural Biological Chem. 48:1763–1775.

Banks, W., and Greenwood, C.T. 1975. Fractionation of the starch granule, and the fine structures of its components. Pages 5–66. In: Starch and Its Components. Banks, W., and Greenwood, C.T., eds. Edinburgh University Press, Edinburgh, UK, pp 5–66.

Berry, C. 1986. Resistant starch: formation and measurement of starch that survives exhaustive digestion with amylolytic enzymes during the determination of dietary fiber. J. Cereal Sci. 4:301–314.

Bhattacharya, K.R., and Sowbhagya, C.M. 1979. Pasting behavior of rice: a new method of viscography. J. Food Sci. 44:797–780, 784.

Biliaderis, C.G., and Tonogal, L. 1991. Influence of lipids on the thermal and mechanical properties of concentrated starch gels. J. Agric. Food Chem. 39:833–840.

Biliaderis, C.G., and Zawistowski, J. 1990. Viscoelastic behavior of aging starch gels. Effects of concentration, temperature and starch hydrolysis on network properties. Cereal Chem. 67:240–246.

Biliaderis, C.G., Page, C.M., Maurice, T.J., and Juliano, B.O. 1986. Thermal characterization of rice starches: a polymeric approach to phase transitions of granular starch. J. Agric. Food Chem. 34:6–14.

Blanshard, J.M.V. 1987. Starch granule and function: a physicochemical approach. In: Starch: Properties and Potential. Galliard, T., ed. Wiley, New York, pp 16–54.

Boyer, C.D., and Shannon, J.C. 1987. Carbohydrates of the kernel. In: Maize Chemistry and Technology. Watson, S.A., and Ramstad, P.E., eds. Am. Assoc. Cereal Chem., St. Paul, M.N., pp 253–272.

Bradbury, A.G.W., and Bello, A.B. 1993. Determination of molecular size distribution of starch and debranched starch by a single procedure using high performance size-exclusion chromatography. Cereal Chem. 70:543–547.

Collado, L.S., Mabesa, R.C., and Corke, H. 1999. Genetic variation in the physical properties of sweet potato starch. J. Agric. Food Chem. 47:4195–4201.

Colonna, P., and Mercier, C. 1984. Macromolecular structure of wrinkled and smooth pea starch components. Carbohydrate Res. 126:233–247.

Crosbie, G.B. 1991. The relationship between starch swelling properties, paste viscosity and boiled noodle quality in wheat flours. J. Cereal Sci. 13:145–150.

Czuchajowska, Z., Sievert, D., and Pomeranz, Y. 1991. Enzyme resistant starch. IV. Effects of complexing lipids. Cereal Chem. 68:537–542.

Dengate, H.N., and Meredith, P. 1984. Wheat starch pasting measured with a "minipaster." Starch/Starke 36:200–206.

Doublier, J.L., Llamas, G., and Le Meur, M. 1987. A rheological investigation of cereal starch pastes and gels. Effects of pasting procedures. Carbohydrate Polymers 7: 251–275.

Eerlingen, R.C., Crombez, M., and Delcour, J.A. 1993a. Enzyme resistant starch. I. Incubation time and temperature of autoclaved starch quantitatively and qualitatively influence resistant starch formation. Cereal Chem. 70:339–344.

Eerlingen, R.C., Deceuninck, M., and Delcour, J.A. 1993b. Enzyme resistant starch. II. Influence of amylose chain length on resistant starch formation. Cereal Chem. 70: 345–350.

Eerlingen, R.C., Van Haesendock, I.P., De Paepe, G., and Delcour, J.A. 1994a. Enzyme resistant starch. III. The quality of straight dough breads containing varying levels of enzyme resistant starch. Cereal Chem. 71:165–170.

Eerlingen, R.C., Cillen, G., and Delcour, J.A. 1994b. Enzyme resistant starch. IV. Effect of endogenous lipid and added sodium dodecyl sulfate on formation of resistant starch. Cereal Chem. 71:170–177.

Eerlingen, R.C., Jacobs, H., and Delcour, J.A. 1994c. Enzyme resistant starch. V. Effect of retrogradation of waxy maize on enzyme susceptibility. Cereal Chem. 71:351–355.

Eerlingen R.C., Jacobs, H., Block K., and Delcour, J.A. 1997. Effects of hydrothermal treatment on the rheological properties of potato starch. Carbohydrate Res. 297: 347–356.

Eliasson, A.C. 1985. Starch gelatinization in the presence of emulsifiers: a morphological study of wheat starch. Starch/Starke 37:411–415.

Eliasson, A.C., Finstad, H., and Ljunger, G. 1988. A study of starch–lipid interactions for some native and modified maize starches. Starch/Starke 40:95–100.

Englyst, H.N., and MacFarlane, G.T. 1986. Breakdown of resistant starch and readily digestible starch by human gut bacteria. J. Sci. Food Agric. 37:699–706.

Evans, I.D., and Haisman, D.R. 1979. Rheology of gelatinized starch suspensions. J. Texture Studies 10:347–370.

Evers, A.D., and McDermott, E.E. 1970. Scanning electron microscopy of wheat starch.

II. Structures of granules modified by alpha-amylosis—preliminary report. Starch 22:23–26.

Evers, A.D., Gough, B.M., and Pybus, J.N. 1971. Scanning electron microscopy of wheat starch IV. Digestion of large granules by glucoamylase of fungal (*Aspergillus niger*) origin. Starch 23:16–18.

Fannon, J.E., and BeMiller J.N. 1992. Structure of cornstarch paste and granule remnants revealed by low-temperature scanning electron microscopy after cryopreparation. Cereal Chem. 69:456–460.

Fannon, J.E., Hauber, R.J., and BeMiller, J.N. 1992. Surface pores of starch granules. Cereal Chem. 69:384–288.

Fannon, J.E., Shull, J.M., and BeMiller J.N. 1993. Interior channels of starch granules. Cereal Chem. 70:611–613.

French, D. 1984. Organization of starch granules. In: Starch, Chemistry and Technology. 2nd ed. R.L. Whistler, J.N. BeMiller and E.F. Paschall, eds. Academic Press, Orlando, FL, pp 183–247.

Fuwa, H., Nakajima, M., Hamada, A., and Glover, D.V. 1977. Comparative susceptibility to amylases of starches from different plant species and several single endosperm mutants and their double mutant combinations with opaque-2 inbred Oh43 maize. Cereal Chem. 54:230–237.

Gallant, D.J., Bouchet, B., and Baldwin P.M. 1997. Microscopy of starch: evidence of a new level of granule organization. Carbohydrate Polymers 32:177–191.

Galliard, T., and Bowler, P. 1987. Morphology and composition of starch. In: Starch: Properties and Potential. T. Galliard, ed. Wiley, New York, pp 55–78.

Ghiasi, K., Varriano-Martson, E., and Hoseney, R.C. 1982. Gelatinization of wheat starch II. Starch–surfactant interaction. Cereal Chem. 59(2):86–88.

Greenwood, C.T. 1964. Structure, properties and amylolytic degradation of starch. Food Technol. 18:138–141.

Hermansson, A.M., and Svegmark, K. 1996. Developments in the understanding of starch functionality. Trends Food Sci. Tech. 7:345–353.

Hibi, Y., Kitamura, S., and Kuge, T. 1990. The effect of lipids on the retrogradation of cooked rice. Cereal Chem. 67:7–10.

Hizukuri, S. 1985. Relationship between the distribution of the chain length of amylopectin and the crystalline structure of starch granules. Carbohydrate Res. 141:295–306.

Hizukuri, S. 1986. Polymodal distribution of chain lengths of amylopectin and its significance. Carbohydrate Res. 147:342–347.

Hoover, R., and Hadziyev, D. 1981. The effect of monoglycerides on amylose complexing during a potato granule process. Starch/Starke 33:346–355.

Hoover, R., and Vasanthan, T. 1994. The effect of heat moisture treatment on the structure and physicochemical properties of cereal, legume and tuber starches. Carbohydrate Res. 252:33–53.

Hoseney, R.C. 1986. Principles of Cereal Science and Technology. American Association of Cereal Chemists, St Paul, MN.

Imberty, A., Buleon, A., Tran, V., and Perez, S. 1991. Recent advances in knowledge of starch structure. Starch/Starke 43:375–384.

IRRI. 1991. Annual Report of 1991. Program Report for 1990. International Rice Research Institute, Manila, Philippines, pp 210–212.

Jacob, H., and Delcour, J.A. 1998. Hydrothermal modifications of granular starch, with retention of the granular structure. A review. J. Agric. Food Chem. 46:2896–2905.

Jacobs, H., Eerlingen, R.C., and Delcour, J.A. 1996. Factors affecting the visco-amylograph and Rapid Visco-Analyzer evaluation of the impact of annealing on starch pasting properties. Starch/Starke 48:266–270.

Jane, J.L., and Chen, J.F. 1992. Effect of amylose molecular size and amylopectin branch chain length on paste properties of starch. Cereal Chem. 69:60–65.

Jane, J.L., Xu, A., Radoosavljevic, M., and Seib, P.A. 1992. Location of amylose in normal starch granules. I. Susceptibility of amylose and amylopectin in cross-linking reagents. Cereal Chem. 69:405–409.

Jane, J., Kasemsuwan, T., Leas S., Zobel, H., and Robyt, J.F. 1994. Anthology of starch granule morphology by scanning electron microscopy. Starch/Starke 46:121–129.

Juliano, B.O. 1985. Criteria and tests for rice grain qualities. In: Rice: Chemistry and Technology. B.O. Juliano, ed. Am. Assoc. Cereal Chem.: St. Paul, MN, pp 443–524.

Kanenaga, K., Harada A., and Harada, T. 1990. Actions of various amylases on starch granules from different plant origins, heated at 60°C in aqueous suspensions. Chem. Express 5:465–468.

Kasemsuwan, T., Jane, J.L., Schnable, P., Stinard, P., and Robertson, D. 1995. Characterization of the dominant mutant amylose-extender (Ael-5180) maize starch. Cereal Chem. 72:457–462.

Katz, F.R. 1991. Natural and modified starches. In: Biotechnology and Food Ingredients. I. Goldberg and R. Williams, eds. Van Nostrand Reinhold, New York, pp 315–327.

Kawabata A., Takase, N., Miyoshi, E., Sawayama, S., Kimura, T., and Kudo, K. 1994. Microscopic observation and X-ray diffractometry of heat/moisture treated starch. Starch 46:463–469.

Keetels, C.J.A.M., Oostergetel, G.T., and van Vliet, T. 1996. Recrystallization of amylopectin in concentrated starch gels. Carbohydrate Polymers 30:61–64.

Kim, W.S., and Seib, P. 1993. Apparent restriction of starch swelling in cooked noodles by lipids in some commercial wheat flours. Cereal Chem. 70:367–372.

Knutson, C.A. 1990. Annealing of maize starches at elevated temperatures. Cereal Chem. 67:376–384.

Knutson, C.A., and Grove, M.J. 1994. Rapid method for estimation of amylose in maize starches. Cereal Chem. 71:469–471.

Kobayashi, S., Schwartz, S.J., and Lineback, D.R. 1986. Comparison of the structure of amylopectins from different wheat varieties. Cereal Chem. 63:71–74.

Krog, N., Olesen, S.K., Toernaes, H., and Joensson, T. 1989. Retrogradation of the starch fraction in wheat bread. Cereal Foods World 34:281–285.

Kugimiya, M., Donovan, J.W., and Wong, R.Y. 1980. Phase transition of amylose-lipid complexes in starches: a calorimetric study. Starch/Starke 32:265–270.

Kulp, K., and Lorenz, K. 1981. Heat-moisture treatment of starches. I. Physicochemical properties. Cereal Chem. 58:46–48.

Landers, P.S., Gbur, E.E., and Sharp, R.N. 1991. Comparison of two models to predict amylose concentration in rice flours as determined by spectrophotometric assay. Cereal Chem. 68:545–548.

Lansky, S., Kooi, M., and Schoch, T.J. 1949. Properties of the fractions and linear subfractions from various starches. J. Am. Chem Soc. 71:4066–4075.

Launay, B., and Lisch, J.M. 1983. Twin-screw extrusion cooking of starches. Flow behavior of starch paste, expansion, and mechanical properties of extrudates. J. Food Eng. 2:259–280.

Leach, H.W., and Schoch, T.J. 1961. Structure of the starch granule. II. Action of various amylases on granular starches. Cereal Chem. 38:34–46.

Leach, H.W., McCowen, L.D., and Schoch, T.J. 1959. Structure of the starch granule. I Swelling and solubility patterns of various starches. Cereal Chem. 36:534–544.

Leelavathi, K., Indrani, D., and Sidhu, J.S. 1987. Amylograph pasting behavior of cereal and tubers Starch Starke 39:378–381.

Li, J.S., and Corke, H. 1999. Physicochemical properties of maize starches expressing dull and sugary-2 mutants in different genetic backgrounds. J. Agric. Food Chem. 47:4939–4943.

Lillford, P.J., and Morrison, A. 1997. Structure/function relationship of starches in food. In: Starch Structure and Functionality. Royal Society of Chemistry, Bookcraft (Bath) Ltd., Cambridge, UK, pp 1–8.

Lineback, D.R. 1984. The starch granule organization and properties. Bakers Digest 58: 16–21.

Lorenz, K. 1976. Physicochemical properties of lipid-free cereal starches. J. Food Sci. 41: 1357–1359.

Lorenz, K., and Kulp, K. 1981. Heat-moisture treatment of starches. II. Functional properties and baking potential. Cereal Chem. 56:49–52.

Lorenz, K., and Kulp, K. 1982. Cereal and root starch modification by heat-moisture treatment. I Physicochemical properties. Starch/Starke 34:50–54.

Lorenz, K., and Kulp, K. 1984. Steeping of barley starch. Effects on physicochemical properties and functional characteristics. Starch 36:116–121.

Lund, D. 1984. Influence of time, temperature, moisture, ingredients and processing conditions on starch gelatinization. CRC Crit. Rev. Food Sci. Nutrition 20:249–273.

Manners, D.J. 1985. Some aspects of the structure of starch. Cereal Foods World 30:461–467.

Matil, K.F. 1971. The functional requirements of protein for foods. J. Am. Oil Chem. Soc. 48:477–480.

Meredith, P., Dengate, H.N., and Morrison, W.R. 1978. The lipids of various sizes of wheat starch granules Starch/Starke 30:119–125.

Miles, M.J., Morris, V.J., Orford, P.D., and Ring, S.G. 1985. The roles of amylose and amylopectin in gelation and retrogradation of starch. Carbohydrate Research 135:271–281.

Morrison, W.R. 1988. Lipids in cereal starches: a review. J. Cereal Sci. 8:1–18.

Morrison, W.R., Law R.V., and Snape, C.E. 1993a. Evidence for inclusion complexes of lipids with V-amylose in maize, rice, and oat starches. J. Cereal Sci. 18:107–109.

Morrison, W.R., Tester, R.F., Snape, C.E., Law, R., and Gidley, M.J. 1993b. Swelling and gelatinization of cereal starches IV. Some effects of lipid-complexed amylose and free amylose in waxy and normal barley starches. Cereal Chem. 70:385–391.

Nagao, S., Ishibasi, S., Imai, S. Sato, T., Kanbe, T., Kaneko, Y., and Otsubo, H. 1977. Quality characteristics of soft wheat from the United States, Australia, France and Japan. Cereal Chem. 54:198–204.

Nashita, K.D., and Bean, M.M. 1982. Grinding methods: their impact on flour properties. Cereal Chem. 59:46–49.

Obanni, M., and BeMiller, J.N. 1995. Identification of starch from various maize endosperm mutants via ghost structures. Cereal Chem. 72:436–442.

Orford, P.D., Ring, S.G., Carroll, M.J., Miles, M.J., and Morris, V.J. 1987. J. Sci. Food Agric. 39:169–177.

Paton, D. 1979. On starch: some recent developments. Starch/Starke 31:184–187.

Penfield, M.P., and Campbell, A.M. 1990. Starch. In: Experimental Food Science. 3rd ed. Academic Press, San Diego, C.A., pp 358–381.

Perez, C.M., Villareal, C.P., Juliano, B.O., and Biliaderis, C.G. 1993. Amylopectin-staling of cooked non-waxy milled rices and starch gels. Cereal Chem. 70:567–571.

Pomeranz, Y. 1991. Carbohydrates: starch. In: Functional Properties of Food Components. 2nd ed. Academic Press, San Diego, CA, pp 24–77.

Ring, S.G. 1985. Observations on the crystallization of amylopectin from aqueous solution. Int J. Boil. Macromol. 7:253–254.

Ring, S.G., Colonna, P., I'Anson, K.J., Kalichevsky, M.T., Miles, M.J., Morris, V.J., and Orford, P.D. 1987. The gelation and crystallization of amylopectin. Carbohydrates Res. 162:277–293.

Rogol, S. 1986. Starch modifications: a view of the future. Cereal Foods World 31:869–874.

Sair, L. 1967. Heat-moisture treatment of starch. Cereal Chem. 44:8–26.

Sarko, A., and Wu, H.C. 1978. The crystal structures of A-, B- and C-polymorphs of amylose and starch. Starch/Starke 30:73.

Schoch, T.J. 1967. Properties and uses of rice starch. In: Starch: Chemistry and Technology. R.L. Whistler and E.F. Paschall, eds. Academic Press, New York.

Seow, C.C., and Teo, C.H. 1993. Annealing of granular rice starches. Interpretation of the effect on phase transitions associated with gelatinization. Starch/Starke 45:345–351.

Shannon, J.C., and Garwood, D.L. 1984. Genetics and physiology of starch development. In: Starch: Chemistry and Technology. R.L. Whistler, J.N. BeMiller, and E.F. Paschall, eds. Academic Press: New York, pp 25–86.

Shewmaker, C.K., and Stalker, D.M. 1992. Modifying starch biosynthesis with transgenes in potatoes. Plant Physiol 100:1083–1086.

Shi, Y.C., and Seib, P.A. 1992. The structure of four waxy starches related to gelatinization and retrogradation. Carbohydrate Res. 227:131–145.

Sievert, D., and Pomeranz, Y. 1989. Enzyme resistant starch I. Characterization and evaluation of enzymatic, thermoanalytical and microscopic methods. Cereal Chem. 66:342–347.

Slade, L., and Levine, H. 1987. Recent advances in starch retrogradation. In: Industrial Polysaccharides. The Impact of Biotechnology and Advanced Methodologies. S.S. Suvala, V. Crescenzi, and I.C. Dea, eds. Gordon and Breach Science, New York, pp 387–430.

Steeneken, P.A.M. 1989. Rheological properties of aqueous suspensions of swollen starch granules. Carbohydrate Polymers 11:23–42.

Stute, R. 1992. Hydrothermal modification of starches: the difference between annealing and heat-moisture treatment. Starch/Starke 44:205–214.

Svegmark, K., and Hermansson, A.M. 1991. Distribution of amylose and amylopectin in potato starch pastes: effects of heating and shearing. Food Structure 10:117–129.

Svegmark, K., and Hermansson, A.M. 1993. Microstructure and rheological properties of potato starch granules and amylose: a comparison of observed and predicted. Food Structure 12:181–193.

Swinkels, J.J.M. 1985. Composition and properties of commercial native starches. Starch/ Starke 40:51–54.

Takahashi, S., and Seib, P.A. 1988. Paste and gel properties of prime corn and wheat starches with and without native lipids. Cereal Chem. 65:474–483.

Takeda, Y., and Hizukuri, S. 1987. Structure of branched molecules of amyloses of various origins and molar fractions of branched and unbranched molecules. Carbohydrate Res. 165:139–145.

Takeda, Y., Shitaozono, T., and Hizukuri, S. 1988. Molecular structure of cornstarch. Starch/Starke 40:51–54.

Tester, R.F., and Morrison W.R. 1990a. Swelling and gelatinization of cereal starches. I. Effect of amylopectin, amylose and lipids Cereal Chem. 67:551–557.

Tester, R.F., and Morrison, W.R. 1990b. Swelling and gelatinization of cereal starches. II. Waxy rice starches. Cereal Chem. 67:558–563.

Thompson. D.B., Klucinec, J.D., and Boltz, K.W. 2000. Influence of amylopectin and lipid on ordering of amylose in high-amylose maize starch. In: Advanced Food Technology, Food of the 21st Century—Food and Resource, Technology, Environment (II). The 4th International Conference of Food Science and Technology. Wuxi, China, pp 353–359.

Wasserman, B.P., Harn, C., Mu-Forster, C., and Huang, R. 1995. Biotechnology: progress toward genetically modified starches. Cereal Foods World 40:801–817.

Watson, S.A. 1984. Corn and sorghum starches: production. In: Starch: Chemistry and Technology. R.L. Whistler, J.N. BeMiller, and E.F. Paschall, eds. Academic Press: Orlando, FL, pp 417–468.

Winter, W.T., and Kwak, Y.T. 1987. Rapid-scanning Raman spectroscopy: a novel approach to starch retrogradation. Food Hydrocolloids 1:461–463.

Wootton, M., and Bamunuarachi, A. 1979. Application of differential scanning calorimetry to starch gelatinization I. Commercial and native starches. 31(6):201–204.

Wurzburg, O.B. 1986. Forty years of industrial research. Cereal Foods World 31:897–899, 901–903.

Yuan, R.C., and Thompson, D.B. 1998. Rheological and thermal properties of aged starch pastes from three waxy maize genotypes. Cereal Chem. 75:117–123.

Yuan, R.C. Thompson, D.B., and Boyer, C.D. 1993. Fine structure of amylopectin in relation to gelatinization and retrogadation behavior of maize starch from wx-containing genotypes in two inbred lines. Cereal Chem. 70:81–89.

Zobel, H.F. 1992. Starch granule structure. In: Development in Carbohydrate Chemistry. R.A. Alexander and H.F. Zobel, eds. AACC, St Paul, MN, pp 1–36.

Index

Printed and bound by CPI Group (UK) Ltd, Croydon, CR0 4YY

23/10/2024

01778227-0006